GATEWAYS INTO ELECTRONICS

GATEWAYS INTO ELECTRONICS

PETER CARROLL DUNN

A Wiley-Interscience Publication

JOHN WILEY & SONS, INC.

New York • Chichester • Weinheim • Brisbane • Singapore • Toronto

This book is printed on acid-free paper. ∞

Copyright ©2000 by John Wiley & Sons, Inc. All rights reserved.

Published simultaneously in Canada.

For ordering and customer service, call 1-800-CALL-WILEY

Library of Congress Cataloging in Publication Data:

Dunn, Peter Carroll.
    Gateways into electronics / Peter Carroll Dunn.
        p. cm.
    Includes bibliographical references and index.
    ISBN 0-471-25448-7 (cloth)
    1. Electronics. I. Title.
  TK7816.D86   1999
  621.381--dc21   98-45207
                                             CIP

Printed in the United States of America

10 9 8 7 6 5 4 3 2 1

CONTENTS

PREFACE

Gateways into Electronics is addressed to students in engineering and in the experimental sciences—let us say juniors—who already possess a fair degree of sophistication in mathematics and physics and are therefore positioned to move quickly toward a quantitative understanding of electronics in its many facets. *Gateways* takes motivation for granted, but in return it makes efficient use of invested effort by carrying out a compact plan based on the following premises:

- That calculations are essential in developing insight.
- That in a short time it is possible to remove the mystery from electronics and develop the capacity to design simple instruments with elegance and assurance.
- That command of a few basic methods is sufficient to analyze the great majority of electronic circuits. The approach that follows from this premise remains vivid in the mind of the reader and promotes decisiveness in the presence of unknown circuits even if encounters are only occasional.
- That the reader should be offered the equipment to tackle the literature rather than a review of the state of the art, which is invariably well provided by industrial sources: electronics evolves quickly, and it is pointless to dwell on current examples unless principles of lasting value are involved. Concepts and points of view are therefore given precedence over a material completeness that can be found elsewhere, notably in *The Art of Electronics* by Horowitz and Hill (1989).

- That the argument that modern electronics comes in modules that simply need to be interconnected is misleading: modules are not fine-tuned to particular needs and require intelligent adaptations. This premise explains the stress on analog electronics, where subtle effects are most likely to be encountered.

- That computer packages have no place in a book that stresses the basics. On the contrary, the intent is to get the reader to judge with confidence whether a computer is making sense.

The chapters in *Gateways* can be divided into a central group (1, 3, and then 5 through 10) and a complementary group (2, 4, 11, 12, and 13). The central chapters deal with linear systems, circuit theory, operational amplifiers, semiconductor devices, and digital circuits, and require little beyond Maxwell's equations in integral form and the bare essentials of statistical mechanics. These chapters say the usual simple things, with the difference that comments are offered about where simple things fail, and consistently use a tight set of tools to the point that it becomes second nature. Integral transforms are not essential, but the notation is chosen so that no reinterpretation is necessary once they are acquired. The complementary chapters deal with integral transforms, electromagnetism, transmission lines, modulation, and noise, and are meant to underpin the central chapters and complete the basic equipment required to explore the literature. These chapters do not have extensive prerequisites—vector analysis and a clear idea of what convergence means are sufficient—but they are denser than the central chapters and can be omitted in a first reading.

The premises on which *Gateways* is based are reflected not in the selection of the material, which is standard except for the fact that fairly advanced topics are addressed, but rather in how the material is organized and treated and, particularly in the central chapters, in how inadequate or unnecessarily complex schemes are either reformulated or discarded. Three representative instances are offered in more detail:

- Circuit calculations rely almost exclusively on node equations derived from Kirchhoff's laws, on superposition, and on Thévenin's theorem, a method that always works. Nonetheless, a few general results that are both productive and easy to memorize are introduced and used when convenient, among them as a seldom quoted jewel, Blackman's theorem for its value in calculating impedances in feedback circuits.

- In the discussion of the dynamics of feedback circuits, two related ideas are made prominent: that the loop gain determines bandwidth and stability, and that the ideal loop gain is an integrator. On this basis, transistor amplifiers and servomechanisms are analyzed at a level found only in dedicated works.

- The presentation of digital electronics first emphasizes that gates have delays. Edge-triggered flip-flops are then constructed, and it is shown that if a few simple timing rules are followed, state machines—and therefore computers—are possible. Despite such a minimal program, metastability is discussed and the principal categories of arithmetic and logic circuits are covered.

The exercises in *Gateways*, with the natural exception of those that involve proofs of statements made in the text, are designed to teach something new. Some are tutorials that extend the material in the text, whereas others show the applicability of a known concept in a new setting. The guiding idea is to extract as much information as possible from simple examples and then point to generalizations. Results are clearly stated and thus provide immediate feedback as well as future reference.

In a book that ranges over many disciplines there are two choices about what to do with the notation: create a profusion of global symbols, or make symbols local. The latter approach has been favored in the interests of transparency and close agreement with the literature.

Engineers in the semiconductor and instrumentation industries have long led the way in showing how electronics can be explained in simple terms. *Gateways* has its idiosyncrasies, but it takes the lucid style used in data sheets and application notes as a model, and I must make this debt clear.

Finally, I wish to express my gratitude to my siblings Diana Gail, Donald Phillip and Patricia Lee for many kind favors, to Peter Lewis, Guido Macchi, Fernando Mariño and Bruce Thomas for encouragement in the early stages, to Guillermo Loiudice and Oscar de Sanctis for reading parts of the manuscript, and to Mariel Zapata, *prima inter pares*, for unfailing support.

PETER CARROLL DUNN

Pago de los Arroyos, 1999

1

LINEAR SYSTEMS

1-1 INTRODUCTION

The language of linear systems pervades electronics, but it is present in many other domains: it can describe the dynamics of a ship's rudder as well as it can describe an audio amplifier. For this reason we will use a context-free approach to linear systems, and in view of our current state of knowledge, we will choose examples from domains other than electronics.

In this chapter we will treat linear systems at a rather intuitive level in order to get to essential ideas without delay. Nonetheless, we will make our statements as precise and as general as possible, and we will also point out where the more formal treatment presented in Chapter 2 is required.

In the spirit just outlined, let us establish some essential terminology before proceeding to an example. As shown in block diagram form in Fig. 1-1, a linear system is characterized mathematically by a linear operator $\mathcal{O}$ that generates a well-defined *output* or *response* $x(t)$ when presented with an *input* or *excitation* $y(t)$.[†] We will write

$$x = \mathcal{O}[y]$$

For the moment we will think of the excitation and of the response as ordinary functions of the time, although we will eventually see that the full power and simplicity of the theory of linear systems can be achieved only in the context of a larger class of mathematical objects, the *generalized functions* discussed in Chapter 2.

[†]We recall that a *function* maps a number into a number, a *functional* maps a function into a number, and an *operator* maps a function into a function.

1

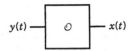

Figure 1-1. Block diagram of a linear system.

By *linear* we mean that if c_1 and c_2 are arbitrary constants, the response to the *linear combination* of excitations

$$c_1 y_1 + c_2 y_2$$

is the linear combination—with the same constants—of the responses:

$$\mathcal{O}[c_1 y_1 + c_2 y_2] = c_1 \mathcal{O}[y_1] + c_2 \mathcal{O}[y_2] = c_1 x_1 + c_2 x_2$$

A linear combination is also called a *superposition*, and we will consider that the fact that a system is linear is also expressed by saying that it obeys the *principle of superposition*.

We will require linear systems to be *time invariant, causal,* and *continuous*. By *time invariant* we mean that the parameters that define a system are constant in time or, equivalently, that inputs translated by a time τ result in outputs translated by the same time τ,

$$\mathcal{O}[y(t-\tau)] = x(t-\tau)$$

By *causal* we mean that a nonzero output must be preceded by a nonzero input,

$$y(t<0) = 0 \quad \Rightarrow \quad x(t<0) = 0$$

By *continuous* we mean roughly that small variations in the input lead to small variations in the output or, more precisely, that certain limiting procedures we will use are justified. This is a delicate question, and it is brought up at this point for the sake of completeness even though it can only be resolved in the wider context of generalized functions. We will accept without proof that the systems we will consider are continuous, but as promised above, we will make sure to point out where continuity needs to be invoked; we should therefore be able to form a fair picture of what it means and of why it is required.

To get a first impression of the nature of the problems we will be dealing with, let us consider the one-dimensional mechanical system shown in Fig. 1-2a. It consists of a mass M placed on a frictionless surface, a damper of constant D attached to the mass at one end and to a fixed point at the other, and a spring of constant K that is relaxed when $x = y$ and is attached to the mass at one end and driven by an external agent at the other. Let us assume that we want to know how the position x of the mass responds to the position y of the driven end of the spring, and that our objective is to adjust

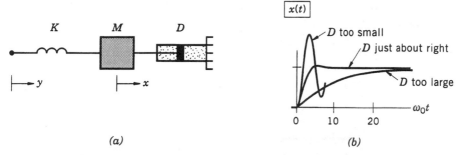

Figure 1-2. (*a*) One-dimensional linear mechanical system. (*b*) Response of x to a step in y as a function of the damping constant D.

the value of D so that x settles smoothly and quickly to a new value when y is given a sudden increment or *step*.

We know something about this system from elementary physics, and just in case memory is vague, we will refresh it with the simple experiment proposed in Exercise 1-1. We thus know, for example, that the mass will oscillate at an (angular) frequency near the *natural* or *resonant* frequency $\omega_0 = \sqrt{K/M}$ if the damping is light, and that it will respond sluggishly if the damping is heavy; we also know that if y is held steady the power into the system is zero, so that the energy loss in the damper will eventually make the mass come to rest at $x = y$. We should therefore expect, as shown in Fig. 1-2*b*, that there is an intermediate value of the damping that optimizes the response according to our requirements.

There are good reasons, other than familiarity, for choosing this mechanical system as an example: it is analogous to electronic systems that we will encounter later, and with few exceptions, as complicated a system as we will need to consider. If we can describe how it works we will have implicitly learned a good deal of electronics. Let us start toward this objective by calculating how the excitation and the response are related. Newton's second law demands

$$Mx'' = K(y-x) - Dx'$$

or after minor algebra,

$$\frac{M}{K}x'' + \frac{D}{K}x' + x = y \tag{1-1}$$

This is a linear differential equation with constant coefficients, so we are dealing with a system that is inherently linear and time invariant. Responses that are apparently noncausal—free oscillations, for example—are mathematically possible but can be rejected on physical grounds for present purposes because they imply excitations in a remote past; with this proviso, the system is also causal.

As things now stand, the description of the operator $\mathcal{O}$ for this system is the rather cumbersome "given $y(t)$, solve the differential equation for $x(t)$", and we would certainly like to have something more convenient, preferably in the form of a multiplicative operator H such that

$$X = HY \tag{1-2}$$

where X and Y are mathematical objects that are easily related to $x(t)$ and $y(t)$. In Section 1-3 we will find a partial answer to this question in terms of the response to sinusoidal excitations, and in Chapter 2 we will find a fully satisfactory answer in terms of Laplace transforms.

Most linear systems are composites of the basic system presented in Fig. 1-1, and we will look at an example to see what novelties complexity introduces. Let us consider the motor-speed controller shown in Fig. 1-3. The purpose of this system is to maintain the output speed ω_M at a constant reference value ω_R in the face of variations in the load torque τ_L. With a license on forward references, we note that this is a *feedback* system in which the difference between the reference and output speeds, $\Delta\omega = \omega_R - \omega_M$, is amplified and converted into a drive torque τ_D that is applied to the motor in order to minimize $\Delta\omega$. It will be a while before we can analyze this system in detail, but we present it notwithstanding because it exhibits in a natural manner the novelties we are looking for, *sums* of signals or, more generally, systems with multiple inputs, and *cascades* of systems.

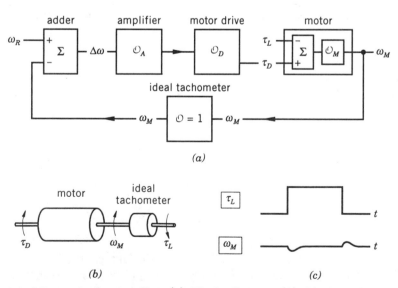

Figure 1-3. Motor-speed controller. (*a*) Block diagram. (*b*) Physical details. (*c*) Response of the speed ω_M to variations in the load torque τ_L.

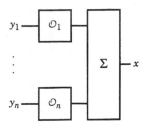

Figure 1-4. Linear system with multiple inputs.

The idea of linearity is extended straightforwardly to a system with multiple inputs by demanding that the system respond linearly to each input when all others are nulled, and that its response be the sum of the responses to each input. It follows that a linear system with inputs $y_1 \cdots y_n$ can be represented as shown in Fig. 1-4, and that its response is then of the form

$$x = \mathcal{O}_1[y_1] + \cdots + \mathcal{O}_n[y_n]$$

Two systems in this form are evident in the motor-speed controller: one is the (algebraic) adder that gives the difference between the reference and output speeds; the other is the motor, which responds to the difference between its drive and load torques in a manner that we will leave unspecified except for mentioning that it involves a linear differential equation. The motor-speed controller itself, which responds to the reference input and to the load torque, is also an example of a system with multiple inputs, but we must go a bit further and discover multiplicative operators before we can put it in the form shown in Fig. 1-4. This will be done in Exercise 1-6.

The cascade of two linear systems, illustrated in Fig. 1-5, is again a linear system, as is easily verified. Its output is given by

$$x = \mathcal{O}_1[y] = \mathcal{O}_1[\mathcal{O}_2[z]]$$

and we are faced with the problem that it is not always obvious how to obtain an operator $\mathcal{O}$ that relates x directly to z, that is, such that

$$x = \mathcal{O}[z]$$

Once more, it is clear that multiplicative operators would simplify our task; in the notation of (1-2), the overall operator H would be given by the product $H_1 H_2$:

$$X = H_1 Y = H_1 H_2 Z = HZ$$

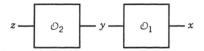

Figure 1-5. Cascade of two linear systems.

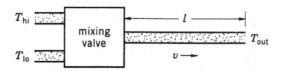

Figure 1-6. System that exhibits a delay.

Most of the systems we will consider are governed by differential equations, but there are a few exceptions, the most important of which is a *delay* by a fixed time τ, defined by

$$x(t) = y(t-\tau)$$

We will see in Chapter 11 that coaxial cables and other systems in which wave propagation plays a role are characterized primarily by such a response. A more easily visualized system that involves a delay is shown in Fig. 1-6. Flowing liquids at temperatures T_{hi} and T_{lo} are run through a mixing valve to obtain a liquid at an intermediate temperature T_{out}. To allow for thorough mixing, the sensor for T_{out} is located a distance l downstream from the valve, so that a change in valve setting is reflected at the sensor with a delay l/v, where v is the velocity of the mixture.

1-2   LINEAR DIFFERENTIAL EQUATIONS WITH CONSTANT COEFFICIENTS

In a large class of linear systems the response $x(t)$ to an excitation $y(t)$ is governed by an nth-order linear differential equation of the form

$$a_n x^{(n)} + \cdots + a_1 x' + a_0 x = b_m y^{(m)} + \cdots + b_1 y' + b_0 y \qquad (1\text{-}3)$$

where $a_0 \cdots a_n$ and $b_0 \cdots b_m$ are real constants, and $a_n \neq 0$ and $b_m \neq 0$. We will identify the most frequently encountered of these systems by name as follows.

Integrator: $\tau x' = y$

Phase lag or *first-order lag*: $\tau x' + x = y$

Phase lead: $\tau x' + x = \tau y'$

Second-order lag: $\tau_0^2 x'' + 2\zeta \tau_0 x' + x = y$

Resonator: $\tau_0^2 x'' + 2\zeta \tau_0 x' + x = \tau_0 y'$

The differential equations for a second-order lag and a resonator have been written in the notation of the appendix to this chapter, where the properties of these second-order systems are summarized. In a phase lag, τ is the *response time*, whereas in a phase lead it is the *differentiation time*. The constants in second-order systems also have names: τ_0 is the *characteristic time*, and ζ is the *damping factor*.

We will see presently that the general solution of differential equation (1-3) involves n adjustable constants; in the case of excitations that are zero for negative times, we will determine these constants by requiring that the solution satisfy n *initial conditions*, which we will take to be the limiting values of x and its derivatives through order $n-1$ at $t = 0+$, that is, as t goes to zero through positive values.

Let us first examine the case of zero excitation. We are then left with a *homogeneous* differential equation,

$$a_n x^{(n)} + \cdots + a_1 x' + a_0 x = 0$$

Using differential-operator notation, we can write this equation in factored form as

$$a_n \left(\frac{d}{dt} - s_n \right) \cdots \left(\frac{d}{dt} - s_1 \right) x(t) = 0 \qquad (1\text{-}4)$$

where $s_1 \cdots s_n$ are the n roots of the *characteristic polynomial*

$$a_n s^n + \cdots + a_1 s + a_0 \qquad (1\text{-}5)$$

Since the order of the factors in (1-4) is irrelevant, any one of them can be applied first to $x(t)$ and be required to null it; the resulting solutions are of the form

$$x_i(t) = \exp(s_i t)$$

If the roots are all distinct, the *homogeneous solution* $x_H(t)$ is a linear combination of these partial solutions:

$$x_H(t) = c_1 \exp(s_1 t) + c_2 \exp(s_2 t) + \cdots + c_n \exp(s_n t)$$

where $c_1 \cdots c_n$ are arbitrary constants. As shown in Exercise 1-8, k distinct solutions are obtained in the case of a root s_0 of multiplicity k by substituting for the k exponentials a term of the form

$$\left(c_0 + c_1 t + \cdots + c_{k-1} t^{k-1} \right) \exp(s_0 t) \qquad (1\text{-}6)$$

Real roots yield real solutions of the form

$$\exp(-t/\tau)$$

(we note that we implicitly favor *negative* real roots), whereas complex roots yield complex-conjugate pairs of solutions

$$\{\exp(-t/\tau)\exp(\pm i\lambda t)\}$$

Pairs of real solutions can be obtained by linear combinations:

$$\{\exp(-t/\tau)\exp(\pm i\lambda t)\} \rightarrow \{\exp(-t/\tau)\sin(\lambda t), \exp(-t/\tau)\cos(\lambda t)\}$$

The essential points are brought out in three brief examples:

1. $\tau^2 x'' + \tau x' = 0;$ $s_{1,2} = 0, -1/\tau;$ $x_H(t) = c_1 + c_2\exp(-t/\tau)$

2. $\tau^2 x'' + 2\tau x' + x = 0;$ $s_{1,2} = -1/\tau;$ $x_H(t) = c_1\exp(-t/\tau) + c_2 t\exp(-t/\tau)$

3. $x'' + \lambda^2 x = 0;$ $s_{1,2} = \pm i\lambda;$ $x_H(t) = c_1\exp(i\lambda t) + c_2\exp(-i\lambda t)$

The last example also has a pair of real solutions $\{\sin(\lambda t), \cos(\lambda t)\}$.

A system is *stable* if all the roots of its characteristic polynomial have a negative real part; it is otherwise *unstable* or *divergent*. If the real part of any root is negative but small in magnitude compared with the imaginary part, the system is *oscillatory*; the system is otherwise *damped*, and depending on the relative magnitudes of the real and imaginary parts of the roots, it can then be *underdamped* or *overdamped*.

A *particular solution* $x_P(t)$ of differential equation (1-3) is a solution that does not necessarily satisfy the initial conditions. The difference between two particular solutions satisfies the associated homogeneous equation, as is easily verified; the general solution can therefore be obtained by adding any particular solution and the homogeneous solution:

$$x(t) = x_P(t) + x_H(t)$$

The *n* constants in the homogeneous solution can then be chosen to satisfy the initial conditions. Thus, for example, the differential equation

$$\tau x' + x = \lambda t$$

with the initial condition

$$x(0+) = 0$$

has the homogeneous solution

$$x_H(t) = c\exp(-t/\tau)$$

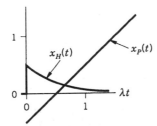 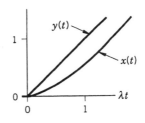

Figure 1-7. Example of how a solution $x(t)$ with the correct initial conditions [$x(0+) = 0$ in this case] is obtained by adding a particular solution $x_P(t)$ and a well-chosen instance of the homogeneous solution $x_H(t)$.

It also has a particular solution that clearly does not satisfy the initial condition,

$$x_P(t) = \lambda(t-\tau)$$

The solution that does satisfy the initial condition, shown in Fig. 1-7 for $\lambda\tau = \frac{1}{2}$, is obtained by adding $x_P(t)$ and $x_H(t)$ and making $c = \lambda\tau$:

$$x(t) = \lambda(t-\tau) + \lambda\tau \exp(-t/\tau) \tag{1-7}$$

   Linear systems can be solved for arbitrary excitations, but it turns out that such a general treatment is normally not necessary and that the response to restricted classes of excitations is sufficient to characterize a system. In particular, the frequency and step responses described in following sections are the ones to which we will pay the most attention. These responses also happen to be the ones that are most studied experimentally: step and sine-wave generators are as ubiquitous as oscilloscopes in an electronics laboratory.

1-3   FREQUENCY RESPONSE

A point we have not made up to now is that we are dealing with operators that accept complex inputs but give real outputs for real inputs. If y_1 and y_2 are real, and x_1 and x_2 are their (real) responses, we have

$$\mathcal{O}[y_1 + iy_2] = \mathcal{O}[y_1] + i\mathcal{O}[y_2] = x_1 + ix_2$$

In consequence, we can solve a system for a complex excitation and then take real or imaginary parts to obtain real responses to real excitations. We will make use of this possibility to calculate the response of linear systems to sinusoidal excitations.

Using the abbreviation

$$s = i\omega$$

we will write complex sinusoids in the form

$$x(t) = X(s) \exp(st) = X(\omega) \exp[i\psi(\omega)] \exp(st)$$

where $X(s)$ is the *complex amplitude* of $x(t)$, and $X(\omega)$ and $\psi(\omega)$ are the magnitude and phase of $X(s)$. We will indicate the correspondence between $x(t)$ and $X(s)$ by writing

$$x(t) \leftrightarrow X(s)$$

Complex amplitudes have interesting—and useful—properties; so much so, that in time we will think only in terms of complex amplitudes. In the first place, the complex amplitude of a linear combination of complex sinusoids (of the same frequency) is equal to the linear combination of the correspond-ing complex amplitudes,

$$c_1 x_1(t) + c_2 x_2(t) \leftrightarrow c_1 X_1(s) + c_2 X_2(s) \tag{1-8}$$

where c_1 and c_2 can be complex. In the second place, differentiation corresponds to multiplication by s,

$$x'(t) \leftrightarrow sX(s) \tag{1-9}$$

Finally, a delay by a time τ corresponds to multiplication by $\exp(-\tau s)$,

$$x(t-\tau) \leftrightarrow \exp(-\tau s)X(s) \tag{1-10}$$

With these preliminaries in mind, let us find the response of a linear system governed by differential equation (1-3) to a complex sinusoidal excitation

$$y(t) = Y(s) \exp(st)$$

We will propose a particular solution that is also a complex sinusoid,

$$x(t) = X(s) \exp(st)$$

In stable systems the homogeneous solution eventually decays to zero, and all solutions converge to this *steady-state* solution. Using (1-8) and (1-9) in (1-3), we obtain

$$X(s) = H(s)Y(s) \tag{1-11}$$

where $H(s)$ is the (complex) *frequency response* or *transfer function*, given by

$$H(s) = \frac{b_m s^m + \cdots + b_1 s + b_0}{a_n s^n + \cdots + a_1 s + a_0} \tag{1-12}$$

If we want the response to real excitations we can take the real or imaginary parts of $X(s)\exp(st)$ and $Y(s)\exp(st)$. Thus for

$$y(t) = Y(\omega)\cos[\omega t + \psi(\omega)]$$

which is the real part of $Y(s)\exp(st)$ and has *real amplitude* $Y(\omega)$ and phase $\psi(\omega)$, the response is the real part of $X(s)\exp(st)$,

$$x(t) = H(\omega)Y(\omega)\cos[\omega t + \psi(\omega) + \varphi(\omega)]$$

where $H(\omega)$ and $\varphi(\omega)$ are the magnitude and phase of $H(s)$. In words, $H(\omega)$ is the ratio of the (real) amplitudes $X(\omega) = H(\omega)Y(\omega)$ and $Y(\omega)$ of the response $x(t)$ and the excitation $y(t)$, and $\varphi(\omega)$ is the difference of the phases of $x(t)$ and $y(t)$.

For example, the frequency response of a phase lag

$$\tau x' + x = y$$

is

$$H(s) = \frac{1}{1+\tau s}$$

The magnitude and phase of $H(s)$ are

$$H(\omega) = \frac{1}{\sqrt{1+\omega^2\tau^2}} \quad \text{and} \quad \varphi(\omega) = -\tan^{-1}\omega\tau$$

and for $y(t) = Y(\omega)\cos\omega t$, the steady-state response is

$$x(t) = \frac{Y(\omega)}{\sqrt{1+\omega^2\tau^2}}\cos(\omega t - \tan^{-1}\omega\tau)$$

Any system that eventually responds sinusoidally to a sinusoidal input can be assigned a frequency response. In particular, it follows from (1-10) that the frequency response of a delay

$$x(t) = y(t-\tau)$$

is

$$H(s) = \exp(-\tau s)$$

It is clear from (1-11) that the frequency response $H(s)$ is a multiplicative operator that relates the complex amplitudes of $x(t)$ and $y(t)$. This means that for sinusoidal signals we have solved the problem posed in Section 1-1. We will soon see that we have gained more than is immediately apparent and that the frequency response is valuable for several other reasons. We will mention some of these reasons, even though they might not be fully appreciated at this time, because they suggest where we are headed:

1. The frequency response is sufficient to characterize a system. In fact, we want to get to the point where a mere look at a system's frequency response gives us a clear image of how the system responds to other excitations.

2. A sinusoidal excitation results in a sinusoidal response everywhere. The complex amplitudes of *all* variables are connected by frequency responses, and calculations in composite systems are thereby greatly simplified.
3. The frequency response can often be represented as a sum of responses of simpler systems such as leads and lags, whose properties we will come to know very well; responses to other excitations can then be obtained by superposing the responses of these simpler systems.
4. In some cases other responses are hard to obtain, whereas the frequency response can be calculated without excessive effort and provides sufficient information. We will experience this later in the case of the skin effect in conductors and in the treatment of periodically loaded delay lines.

In view of these reasons, we will tend to identify systems by their frequency response, and in time we will virtually forget their description in terms of differential equations and other linear operators.

1-4 STEP RESPONSE

The *step response* of a linear system, denoted by $r(t)$, is the response to the *unit step function* $\theta(t)$ defined by

$$\theta(t<0) = 0 \quad \text{and} \quad \theta(t \geq 0) = 1$$

We want $r(t)$ to be dimensionless, so we will assume that the system input and output variables are expressed in the same units. Since we are dealing with causal systems, we must also have

$$r(t<0) = 0$$

Let us first point out that for a delay

$$x(t) = y(t-\tau)$$

we have trivially that

$$r(t) = \theta(t-\tau)$$

Let us now see what happens with systems governed by differential equations. We will first consider a phase lag

$$\tau x' + x = y$$

Its step response $r(t)$ is the solution of

$$\tau r' + r = \theta$$

For $t>0$, $r(t)$ satisfies the differential equation

$$\tau r' + r = 1$$

A particular solution is $r_P(t) = 1$; the general solution is then

$$r(t) = 1 + c\exp(-t/\tau)$$

We need an initial condition. Since $r(t<0) = 0$, we have

$$r(t) = \int_0^t r'(u)\,du$$

If we assume that $r'(t)$ is bounded for $t>0$, it follows that

$$r(0+) = 0$$

and that the step response, shown in Fig. 1-8, is

$$r(t) = \theta(t)[1 - \exp(-t/\tau)] \tag{1-13}$$

We can now check that $r'(t)$ is in fact bounded for $t>0$:

$$|\tau r'(t>0)| = \exp(-t/\tau) < 1$$

Figure 1-8 also shows *Bode plots* of the frequency response: a log-log plot of $H(\omega)$ and a log-lin plot of $\varphi(\omega)$. They are placed next to the step response to start the process of association that will allow us in the future to

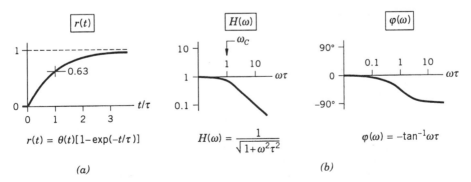

$$r(t) = \theta(t)[1 - \exp(-t/\tau)]$$

$$H(\omega) = \frac{1}{\sqrt{1+\omega^2\tau^2}}$$

$$\varphi(\omega) = -\tan^{-1}\omega\tau$$

(a)   (b)

Figure 1-8. (a) Step and (b) frequency responses of a phase lag.

infer other responses from the frequency response.[†] For now let us simply observe that the (upper) *cutoff frequency* ω_C, defined by

$$H(\omega_C) = \frac{1}{\sqrt{2}}$$

is equal to the reciprocal of the response time τ,

$$\omega_C \tau = 1$$

and that it is no coincidence that $H(0) = r(\infty) = 1$, as it is no coincidence that $H(\infty) = r(0) = 0$.

Let us next consider a phase lead

$$\tau x' + x = \tau y'$$

Its step response is given formally by

$$\tau r' + r = \tau \theta'$$

and we are faced with the problem of interpreting the derivative of $\theta(t)$. To skirt this problem we will appeal to continuity and assume that we can obtain a valid solution by means of a limiting procedure. We will thus approximate $\theta(t)$ with the function $\theta_T(t)$ shown in Fig. 1-9:

$$\theta_T(t) = \int_0^t \delta_T(u) \, du \qquad (1\text{-}14)$$

where $\delta_T(t)$, also shown in Fig. 1-9, is given by

$$\delta_T(t) = \frac{\theta(t) - \theta(t-T)}{T} \qquad (1\text{-}15)$$

[†]A dimensionless nonnegative quantity η that can be interpreted as a power ratio is often expressed in (power) decibels (dB) as

$$\eta \, (\text{dB}) = 10 \log \eta$$

Thus 10^0 corresponds to 0 dB, 10^1 to 10 dB, and so on. For consistency, a dimensionless nonnegative quantity ξ, whose *square* can be interpreted as a power ratio, is expressed in decibels as

$$\xi \, (\text{dB}) = 20 \log \xi$$

In this case 10^0 corresponds to 0 dB, 10^1 (whose square is 10^2) to 20 dB, and so on. $H(\omega)$ fits the latter description and is often expressed in decibels in the literature. Correspondingly, the cutoff frequency ω_C defined below, at which $H(\omega) = 1/\sqrt{2}$, is also known as the *half-power* or (minus) 3 dB point.

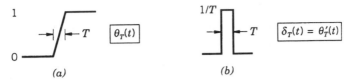

Figure 1-9. (a) The ramped step $\theta_T(t)$ and (b) its derivative $\delta_T(t)$, used to obtain the step response of a phase lead.

The differential equation for the response $r_T(t)$ to $\theta_T(t)$ is

$$\tau r'_T + r_T = \tau \theta'_T = \tau \delta_T$$

This we can handle. For $0 < t < T$ we have

$$\tau r'_T + r_T = \tau/T$$

Assuming that $r_T(t)$ remains bounded, for sufficiently small T we obtain

$$r'_T(t) \simeq 1/T \quad \text{and} \quad r_T(t) \simeq t/T$$

Evaluating $r_T(t)$ at $t = T$ and taking the limit as $T \to 0$, we get

$$r(0+) = 1$$

With this initial condition we then have to solve the homogeneous equation

$$\tau r' + r = 0$$

The result, shown in Fig. 1-10, is

$$r(t) = \theta(t) \exp(-t/\tau) \tag{1-16}$$

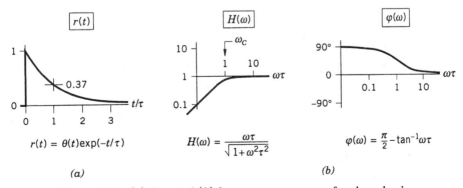

$$r(t) = \theta(t)\exp(-t/\tau)$$

$$H(\omega) = \frac{\omega\tau}{\sqrt{1+\omega^2\tau^2}}$$

$$\varphi(\omega) = \frac{\pi}{2} - \tan^{-1}\omega\tau$$

(a)   (b)

Figure 1-10. (a) Step and (b) frequency responses of a phase lead.

We observe that for $t > 0$, $r(t)$ is τ times the derivative of the step response (1-13) of a phase lag with the same time constant,

$$[r(t)]_{\text{lead}} = \tau [r'(t)]_{\text{lag}}$$

Since $\delta_T(t)$ is a difference of step functions, we could also have drawn this conclusion by observing that

$$[r_T(t)]_{\text{lead}} = \frac{\tau}{T}[r(t) - r(t-T)]_{\text{lag}}$$

These observations suggest that the derivative of $\theta(t)$ as the limit of $\delta_T(t)$ for $T \rightarrow 0$ might make sense in the proper context, and that the procedure we used to obtain $r(t)$ can be justified. That this is in fact so will be seen when we come to generalized functions in Chapter 2.

As in the case of a phase lag, Fig. 1-10 also shows Bode plots of $H(\omega)$ and $\varphi(\omega)$. This time we will limit ourselves to observing that the (lower) cutoff frequency ω_C, defined by

$$H(\omega_C) = \frac{1}{\sqrt{2}}$$

is equal to the reciprocal of the differentiation time τ,

$$\omega_C \tau = 1$$

and that it is no coincidence that $H(0) = r(\infty) = 0$ or that $H(\infty) = r(0) = 1$.

Using the methods we have developed so far, we can obtain the step responses of a second-order lag

$$\tau_0^2 x'' + 2\zeta\tau_0 x' + x = y$$

and of a resonator

$$\tau_0^2 x'' + 2\zeta\tau_0 x' + x = \tau_0 y'$$

The solutions are proposed as Exercise 1-3 and tabulated in the appendix.

Let us finally deal with the general case of differential equation (1-3). We will begin by expressing the denominator of the frequency response $H(s)$ in (1-12)—it is the characteristic polynomial (1-5)—as a product of powers of factors with *distinct* roots:

$$a_n s^n + \cdots + a_1 s + a_0 = a_n (s+s_1)^{k_1} (s+s_2)^{k_2} \cdots (s+s_l)^{k_l}$$

where the exponents $k_1 \cdots k_l$ must of course satisfy

$$k_1 + k_2 + \cdots + k_l = n$$

For $m \leq n-1$ we can then propose an expansion of $H(s)$ in *partial fractions*,

$$H(s) = \frac{b_m s^m + \cdots + b_1 s + b_0}{a_n s^n + \cdots + a_1 s + a_0} = \frac{1}{a_n} \sum_{j=1}^{l} \left[\frac{c_{j,k_j}}{(s+s_j)^{k_j}} + \cdots + \frac{c_{j,2}}{(s+s_j)^2} + \frac{c_{j,1}}{s+s_j} \right]$$

where $c_{j,k_j} \cdots c_{j,1}$ are constants. There are elegant methods for determining these constants, but we will obtain them straightforwardly by solving the nth-order system of linear equations that results from carrying out the sum on the right-hand side and equating coefficients of powers of s in the numerators on both sides.

We will accept for now—and prove in Chapter 2—that the step response of a system with frequency response $H(s)$ is obtained by adding the step responses of the simpler systems in the expansion of $H(s)$. An example of the correctness of this procedure is given in Exercise 1-7. To complete the procedure we will need the step response of a system with frequency response

$$H(s) = \frac{s_0^k}{(s+s_0)^k} \tag{1-17}$$

In Exercise 1-8 it is shown that

$$r(t) = \theta(t) - \sum_{j=0}^{k-1} \theta(t) \frac{(s_0 t)^j}{j!} \exp(-s_0 t) \tag{1-18}$$

The roots of the characteristic polynomial will almost always be distinct, and we will then only get terms with $k = 1$, which correspond to (perhaps complex) first-order lags.

It is often more expedient to factor the characteristic polynomial into linear terms with real roots and quadratic terms of the form $\omega_0^2 + \omega_1 s + s^2$ with a pair of complex-conjugate roots. If a given pair is distinct from all other pairs, the corresponding terms in the expansion in partial fractions can be replaced by a single term of the form

$$\frac{c_1 s + c_2}{\omega_0^2 + \omega_1 s + s^2}$$

Let us show, for example, that

$$\frac{\tau s}{(1+\tau s)(1+\tau s+\tau^2 s^2)} = \frac{1+\tau s}{1+\tau s+\tau^2 s^2} - \frac{1}{1+\tau s}$$

and that we can thus obtain the step response of a system with this frequency response by superposing the step responses of a second-order lag, a resonator, and a phase lag. According to our rules, we propose

$$\frac{\tau s}{(1+\tau s)(1+\tau s+\tau^2 s^2)} = \frac{c_1 s+c_2}{1+\tau s+\tau^2 s^2} + \frac{c_3}{1+\tau s}$$

Carrying out the operations on the right-hand side and equating powers of s in the numerators, we get

$$c_1 \qquad + \tau c_3 = 0$$

$$c_1 + \tau c_2 + \tau c_3 = \tau$$

$$c_2 + \quad c_3 = 0$$

and thus $c_1 = \tau$, $c_2 = 1$, and $c_3 = -1$.

With a much more practical aim, Exercise 1-12 shows that reasonably accurate sketches of the step responses of second-order and third-order systems can be made with just a few calculations and without partial-fraction expansions.

1-5 CORRESPONDENCES BETWEEN STEP AND FREQUENCY RESPONSES

We have developed a fair amount of machinery, and it is time to return to the mechanical system of Fig. 1-2 and look at it from a more informed perspective. If we rewrite differential equation (1-1) in terms of its characteristic time and its damping factor we obtain a normalized second-order lag,

$$\tau_0^2 x'' + 2\zeta\tau_0 x' + x = y$$

where

$$\tau_0 = \sqrt{M/K} \quad \text{and} \quad 2\zeta = D/\sqrt{MK}$$

Since M and K are assumed fixed, we can restate our problem by saying that we want to choose the damping factor ζ that optimizes the system step response.

Figure 1-11 shows the step response for several values of ζ. We confirm what we already knew: for a step in y, x eventually settles to a constant value $x = y$, and the responses range from smooth but slow ($\zeta \gg 1$) through smooth and fast ($\zeta \approx 1$) to fast but oscillatory and again slow to settle ($\zeta \ll 1$). Figure 1-11 also shows the magnitude of the frequency response for the same values of ζ. In all cases we have

$$H(\omega \ll \omega_0) \simeq 1 \quad \text{and} \quad H(\omega \gg \omega_0) \simeq 1/\omega^2\tau_0^2$$

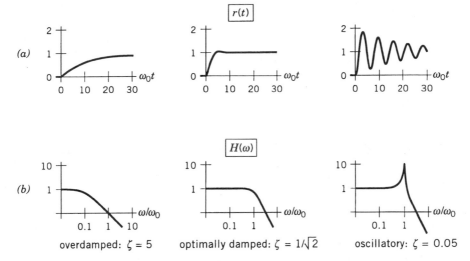

Figure 1-11. (*a*) Step and (*b*) frequency responses of a second-order lag for several values of the damping factor ζ.

What happens in between depends on the value of ζ. For small ζ, $H(\omega)$ has a sharp maximum or *resonance* at a frequency near ω_0, whereas for large ζ it falls off at frequencies well below ω_0; for ζ near 1, $H(\omega)$ stays close to 1 for frequencies up to ω_0 and then falls off to zero.

We observe interesting correspondences: if the frequency response is resonant, the step response is oscillatory, whereas for nonresonant frequency responses there is a parallel with phase lags in that the cutoff frequency determines the speed of the step response. These correspondences are explored further in Exercises 1-10 and 1-11, in which the response time of damped systems and the decay time of resonant systems are quantitatively related to characteristic widths of the frequency responses. There are other correspondences that, as in the case of phase lags, we will point out now and prove in Chapter 2: $H(s)$ goes like $1/s^2$ for $s \to \infty$ because $r(t)$ goes like t^2 for $t \to 0$, and $H(0) = 1$ because $r(\infty) = 1$.

Correspondences such as we observe in second-order systems occur in all linear systems; they might be somewhat more complicated to describe, but they are of the same nature, and no really new phenomena arise. The fact that we can often represent a frequency response as a sum of simpler responses already tells us that this must be so. It also tells us, as promised earlier, that a simple mechanical system can provide much insight into linear systems.

Differing requirements on systems determine the response to which we pay the most attention: the flatness of the frequency response might be paramount in the case of an audio amplifier, but the details of the step response are what we want if speed or a short settling time to a final value are important, as they are in the case of industrial robots.

1-6 FILTERS AND THE FREQUENCY CONTENT OF WAVEFORMS

We now want to make plausible the idea that any reasonable waveform $x(t)$ has a *frequency content*, meaning that it can be represented as a sum or integral of sine waves of the form

$$X(i\omega)\exp(i\omega t)$$

where $X(i\omega)$ is a complex amplitude and the frequency can be both positive and negative. Once this idea becomes acceptable, we will be able to view linear systems as *filters* that selectively modify the frequency content of waveforms. Our treatment will be highly qualitative and so designed that Chapter 2 can be bypassed in a first reading; for completeness, however, we note that $X(i\omega)$ will turn out to be the *Fourier transform* of $x(t)$.

Let us look again at the frequency response of a phase lag in Fig. 1-8. Sine waves with frequencies well below the cutoff frequency ω_C are essentially unmodified, whereas sine waves with frequencies well above ω_C are strongly attenuated; a phase lag is thus a *low-pass* filter, and as is clear from Fig. 1-11, so is a damped second-order lag. On the other hand, a look at Fig. 1-10 tells us that a phase lead is a *high-pass* filter, and we will see below that a resonator is a *bandpass* filter. If we could now get some idea of what the frequency content of an arbitrary waveform looks like, we might be able to offer rough guesses as to how the waveform is affected by filters.

Let us begin by considering the periodic waveform

$$x(t) = 1 + \cos(\omega_0 t) = \exp(0) + \tfrac{1}{2}\exp(i\omega_0 t) + \tfrac{1}{2}\exp(-i\omega_0 t)$$

[We can imagine $x(t)$ as a crude representation of a slowly varying signal that is masked by high-frequency noise.] In this simple case the frequency content is in plain view: it has discrete components at $\omega = 0$ and at $\omega = \pm\omega_0$. If $x(t)$ is applied to a phase lag with cutoff frequency $\omega_C \ll \omega_0$, the components at $\pm\omega_0$ are removed and the output is essentially equal to 1. On the other hand, if $x(t)$ is applied to a phase lead with the same cutoff frequency, the constant term is *blocked* and only the components at $\omega = \pm\omega_0$ survive. These effects are shown in Fig. 1-12.

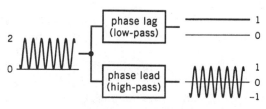

Figure 1-12. Steady-state responses of a phase lag and a phase lead with cutoff frequency $\omega_C \ll \omega_0$ to the periodic function $1 + \cos(\omega_0 t)$.

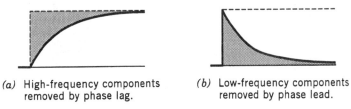

(a) High-frequency components
removed by phase lag.

(b) Low-frequency components
removed by phase lead.

Figure 1-13. Step responses of a phase lag and a phase lead with the same cutoff frequency. The shaded areas correspond to high-frequency components.

Let us next consider what happens when a unit step function $\theta(t)$ is applied to the same filters. As shown in Fig. 1-13, the low-pass filter replaces the abrupt *leading edge* of $\theta(t)$ with a smoother version, whereas the high-pass filter leaves the leading edge untouched but blocks the constant part that extends to infinity. We might thus imagine that nonperiodic waveforms also have a frequency content, and perhaps unsurprisingly, that components at high frequencies correspond to fast changes in a waveform, whereas components at low frequencies correspond to slow changes.

Let us explore this idea a little further. The square pulses shown in Fig. 1-14 are run through phase lags with response time τ. The pulse of width much larger than τ is only smoothed somewhat, whereas the pulse of width much less than τ is greatly reduced in amplitude. We are led to conclude that wide pulses have relatively more low-frequency content, whereas narrow pulses have relatively more high-frequency content.

We will now attempt to be somewhat more quantitative. To do so, we will apply a unit step function $\theta(t)$ to a resonator with frequency response

$$H(i\omega) = \frac{2\zeta\tau_0 i\omega}{1+2\zeta\tau_0 i\omega+\tau_0^2 i^2\omega^2} = \frac{2\zeta i\omega_0 \omega}{\omega_0^2-\omega^2+2\zeta i\omega_0 \omega}$$

If we assume that $\zeta \ll 1$, $H(i\omega)$ is sharply peaked around $\omega = \pm\omega_0$, as shown in Fig. 1-15a, and we have $\omega_0^2-\omega^2 \simeq 2\omega_0(\omega_0\mp\omega)$ near $\omega = \pm\omega_0$. We can

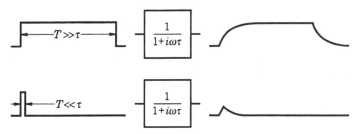

Figure 1-14. Responses of a phase lag with response time τ to square pulses of widths much larger and much smaller than τ.

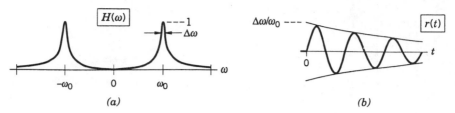

Figure 1-15. (*a*) Frequency and (*b*) step responses of a lightly damped resonator.

then write

$$H(i\omega) \simeq \frac{1}{1+2i\dfrac{\omega+\omega_0}{\Delta\omega}} + \frac{1}{1+2i\dfrac{\omega-\omega_0}{\Delta\omega}}$$

where

$$\Delta\omega = 2\zeta\omega_0$$

is the bandwidth, equal to the difference between the frequencies on either side of $\pm\omega_0$ at which $H(\omega) = 1/\sqrt{2}$. If we hold $\Delta\omega$ constant and vary ω_0, the shape of $H(i\omega)$ remains constant, and it is therefore reasonable to assume that the step response $r(t)$, shown in Fig. 1-15*b*, is proportional to the frequency content of $\theta(t)$. According to the appendix, for $\zeta \ll 1$ we have

$$r(t) \simeq \frac{\Delta\omega}{\omega_0} \exp\left(-\frac{\Delta\omega}{2}t\right) \sin(\omega_0 t)$$

The amplitude of the step response is thus inversely proportional to ω_0. Since ω_0 is arbitrary, we conclude that the frequency content $\Theta(i\omega)$ of $\theta(t)$ is a continuous *density* defined for all frequencies except zero, and that its magnitude is proportional to $1/\omega$.

If we now accept that for our purposes $\Theta(i\omega)$ is given by

$$\Theta(i\omega) = \frac{1}{i\omega}$$

and that $\Theta(i\omega)$ can be treated as if it were a complex amplitude, we can also accept that the frequency content of $\theta(t-T)$ is

$$\frac{1}{i\omega}\exp(-i\omega T)$$

We are then in a position to calculate the frequency content of the square pulse shown in Fig. 1-16*a*,

$$x_T(t) = \theta(t) - \theta(t-T)$$

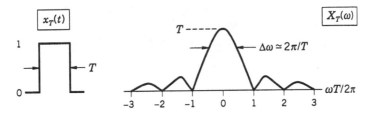

Figure 1-16. (*a*) Square pulse of unit height and width T, (*b*) magnitude of its frequency content.

We thus have

$$X_T(i\omega) = \frac{1}{i\omega}[1 - \exp(-i\omega T)] = T\frac{\sin(\omega T/2)}{(\omega T/2)}\exp(-i\omega T/2) \quad (1\text{-}19)$$

Looking at the magnitude of X_T shown in Fig. 1-16*b*, we observe that a narrow (and thus fast) pulse does in fact have more content at high frequencies and less at low frequencies than does a wide (and thus slow) pulse. We also observe—this observation applies quite generally—that the frequency content of a pulse of width T becomes small outside an interval given approximately by

$$\Delta\omega \simeq \frac{2\pi}{T}$$

To summarize, we have tried to justify rough correspondences, subliminally present in any discussion about electronics in the laboratory or in the literature, between the behavior in time of a waveform and the shape of its frequency content. Briefly stated, these correspondences are as follows:

abrupt/narrow ↔ high-frequency components

smooth/wide ↔ low-frequency components

EXERCISES

1-1. A MECHANICAL EXPERIMENT. The simple mechanical system of Fig. 1-2 is analogous to a rigid pendulum that is driven horizontally at its point of suspension. A 12-inch ruler held at one end is a handy pendulum that has a natural frequency of about 1 second and is relatively undamped in air if held lightly enough. Take y and x to be the positions of the top and bottom of the ruler. Observe that $x \simeq y$ for very slow back-and-forth motions and that $x \approx 0$ for quick motions. Also observe the large resonance for a period around 1 second. Apply a step in y and observe the oscillatory response. Now repeat everything with the tip of the ruler dipped in a bucket of water.

1-2 FREQUENCY RESPONSE THE HARD WAY. The use of complex amplitudes to represent sine waves is not mandatory, but it is certainly an efficient way of doing things. To see why, solve the differential equation

$$\tau x' + x = \cos \omega t$$

by proposing a solution

$$x = X(\omega) \cos(\omega t + \varphi)$$

Equate terms in $\sin \omega t$ and $\cos \omega t$ to obtain

$$\sin \varphi + \omega \tau \cos \varphi = 0 \quad \text{and} \quad X(\omega)(\cos \varphi - \omega \tau \sin \varphi) = 1$$

and from these,

$$X(\omega) = \frac{1}{\sqrt{1 + \omega^2 \tau^2}} \quad \text{and} \quad \varphi = -\tan^{-1}\omega\tau$$

Now imagine that you are dealing with a fourth-order differential equation and be thankful to Euler.

1-3 SECOND-ORDER DIFFERENTIAL EQUATIONS. Verify all entries for second-order lags and resonators in the appendix. Also verify that the step response of a resonator is τ_0 times the derivative of the step response of a second-order lag with the same parameters.

1-4 INITIAL CONDITIONS FROM PHYSICAL CONSIDERATIONS. In Section 1-4 the initial conditions for step responses were obtained from purely mathematical considerations. It is often the case, however, that a look at the physics of a system tells you what they are.

Suppose for example that you want initial conditions for the step response of the mechanical system shown in Fig. 1-17, in which $y(t)$ is the excitation and $x(t)$ is the response. To keep a finite force on the mass and therefore be able to see what is going on, you are inevitably led to replace $\theta(t)$ with the ramped approximation $\theta_T(t)$ given in (1-14). The input velocity for $0 < t < T$ is then

$$y'(t) = 1/T$$

If you assume that the inertia of the mass keeps x' and x bounded no matter

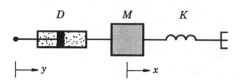

Figure 1-17. Mechanical system for Exercise 1-4. The excitation is y and the response is x.

how small T, the force on the mass in the limit $T \to 0$ is $Dy' = D/T$, and the momentum of the mass at $t = T$ is

$$Mx'(T) = (D/T)T = D$$

You can therefore conclude that the initial condition on r' is

$$r'(0+) = D/M$$

This result is compatible with the assumption that x' remains bounded; you can further conclude that

$$r(0+) = 0$$

This system is a resonator, as you should verify:

$$(M/K)x'' + (D/K)x' + x = (D/K)y'$$

Now you can check your conclusions by using the reasoning of Section 1-4.

1-5 AUTOMOBILE SUSPENSION

(a) Write down the differential equation for the crude model of an automobile suspension shown in Fig. 1-18, in which $y(t)$ is the excitation and $x(t)$ is the response. The frequency response should be

$$H(s) = \frac{1 + (D/K)s}{1 + (D/K)s + (M/K)s^2} = \frac{1 + 2\zeta\tau_0 s}{1 + 2\zeta\tau_0 s + \tau_0^2 s^2}$$

where

$$\tau_0 = \sqrt{M/K} \quad \text{and} \quad 2\zeta = D/\sqrt{MK}$$

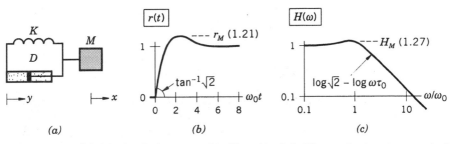

(a)   (b)   (c)

Figure 1-18. (a) Mechanical system for Exercise 1-5. The excitation is y and the response is x. (b) Step response. (c) Frequency response.

(b) From physical or mathematical considerations deduce that the initial conditions for the step response $r(t)$ are

$$r(0+) = 0 \quad \text{and} \quad r'(0+) = D/M$$

(c) For $\zeta < 1$, use the step responses given in the appendix to obtain

$$r(t>0) = 1 - \exp(-\omega_1 t)\left(\cos \omega_2 t - \frac{\omega_1}{\omega_2}\sin \omega_2 t\right)$$

where

$$\omega_0 = 1/\tau_0 \qquad \omega_1 = \zeta\omega_0 \qquad \omega_2 = \omega_0\sqrt{1-\zeta^2}$$

Verify the initial conditions.

(d) Take $\zeta = 1/\sqrt{2}$. Show that $r(t)$ has a first maximum r_M at $\omega_0 t_M = \pi/\sqrt{2} = 2.22$ and a first minimum r_m at $\omega_0 t_m = 3\pi/\sqrt{2} = 6.66$, and that

$$r_M = 1 + \exp(-\pi/2) = 1.21 \quad \text{and} \quad r_m = 1 - \exp(-3\pi/2) = 0.99$$

Show also that the magnitude of the frequency response has a maximum H_M at

$$\omega_M \tau_0 = \sqrt{(\sqrt{5}-1)/2} = 0.79$$

and that

$$H_M = \sqrt{(\sqrt{5}+1)/2} = 1.27$$

Verify that the graphs in Fig. 1-18 are correct.

1-6 FREQUENCY RESPONSE OF A MOTOR-SPEED CONTROLLER. With reference to Fig. 1-3, let $\tau_L(s)$ and $\omega_M(s)$ be the complex amplitudes of the load torque and the output speed, and let $H_A(s)$, $H_D(s)$, and $H_M(s)$ be the frequency responses of the amplifier, the motor drive, and the motor. Take the input reference speed ω_R to be zero. Show that

$$\omega_M(s) = \frac{-H_M(s)}{1+H_A(s)H_D(s)H_M(s)}\tau_L(s) \equiv H(s)\tau_L(s)$$

In principle, if you make $H_A(s) \to \infty$, you get $H(s) \to 0$, and the output speed becomes independent of the load torque. This is what every engineer would love to see happen, but there is a problem: you are dealing with a feedback system, and it can become unstable if $H_A(s)$ is made "too large". This is precisely the subject of Chapter 9.

1-7 AN EXPANSION IN PARTIAL FRACTIONS GIVES THE RIGHT STEP RESPONSE. Consider a system H with frequency response

$$H(s) = \frac{s_1 s_2}{(s+s_1)(s+s_2)} = \frac{s_1 s_2}{(s_2-s_1)(s+s_1)} + \frac{s_1 s_2}{(s_1-s_2)(s+s_2)}$$

$$= H_1(s) + H_2(s)$$

Show that the differential equation corresponding to $H(s)$ can be obtained from the differential equations corresponding to $H_1(s)$ and $H_2(s)$. *Suggestion:* Use differential operators as in (1-4). Now show that the step response $r(t)$ of H is equal to the sum of the step responses $r_1(t)$ and $r_2(t)$ of H_1 and H_2. In particular, observe that the initial conditions for $r(t)$ are satisfied automatically, that is, that

$$r_1(0+) + r_2(0+) = 0 \quad \text{and} \quad r_1'(0+) + r_2'(0+) = 0$$

1-8 REPEATED ROOTS. The frequency response with k repeated roots given in (1-17),

$$H(s) = \frac{s_0^k}{(s+s_0)^k}$$

corresponds to the differential equation

$$\left(\frac{d}{dt} + s_0\right)^k x(t) = s_0^k y(t)$$

(a) Show that

$$\left(\frac{d}{dt} + s_0\right)^k \left[t^{k-1} \exp(-s_0 t)\right] = (k-1)\left(\frac{d}{dt} + s_0\right)^{k-1} \left[t^{k-2} \exp(-s_0 t)\right]$$

$$\vdots$$

$$= (k-1)! \left(\frac{d}{dt} + s_0\right) \exp(-s_0 t)$$

$$= 0$$

and that the solution of the associated homogeneous equation is the one given in (1-6),

$$x_H(t) = \left(c_0 + c_1 t + \cdots + c_{k-1} t^{k-1}\right) \exp(-s_0 t)$$

(b) For $t > 0$ the step response $r(t)$ is the solution of

$$\left(\frac{d}{dt} + s_0\right)^k r(t) = s_0^k$$

such that $r(t)$ and its derivatives through order $k-1$ are zero at $t = 0+$. Verify that $r(t)$ is given correctly by (1-18),

$$r(t) = \theta(t) - \sum_{j=0}^{k-1} \theta(t) \frac{(s_0 t)^j}{j!} \exp(-s_0 t)$$

Do this by observing that $x_P(t) = 1$ is a particular solution and that

$$r'(t) = \theta(t) s_0 \frac{(s_0 t)^{k-1}}{(k-1)!} \exp(-s_0 t)$$

1-9   STYLIZED BODE PLOTS. The representation of a frequency response by means of Bode plots—a log-log plot of magnitude versus frequency and a log-lin plot of phase versus frequency—is convenient when dealing with cascaded systems: the overall plots are obtained by graphical addition, since magnitudes multiply and phases add. In this exercise you will become familiar with *stylized* Bode plots, easily memorized graphs that consist only of straight lines and thus allow quick representation and straightforward graphical addition.

Take the case of a phase lag

$$H(s) = \frac{1}{1 + \tau s}$$

The magnitude and phase of $H(s)$ are

$$H(\omega) = \frac{1}{\sqrt{1 + \omega^2 \tau^2}} \quad \text{and} \quad \varphi(\omega) = -\tan^{-1} \omega \tau$$

Look at the asymptotes. For $\omega \tau \ll 1$ you have

$$H(\omega) \simeq 1 \quad \text{and} \quad \varphi(\omega) \approx 0$$

whereas for $\omega \tau \gg 1$ you have

$$\log[H(\omega)] \simeq -\log \omega \tau \quad \text{and} \quad \varphi(\omega) \simeq -\frac{\pi}{2}$$

The stylized Bode plots shown in Fig. 1-19 take these asymptotes into account and maximize ease of construction while trying to give a reasonable account of the intermediate region. The magnitude plot is a horizontal line for $\omega \tau \leq 1$ and a line with slope -1 (-20 dB per decade) for $\omega \tau \geq 1$; the phase plot is a horizontal line at $\varphi = 0$ for $\omega \tau \leq 0.1$, a horizontal line at $\varphi = -90°$ for $\omega \tau \geq 10$, and a line with a slope of $-45°$ per decade in the

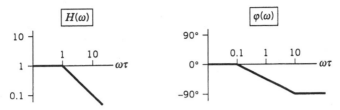

Figure 1-19. Stylized Bode plots of a phase lag (Exercise 1-9).

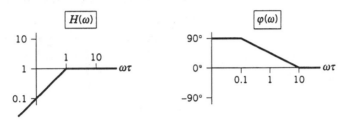

Figure 1-20. Stylized Bode plots of a phase lead (Exercise 1-9).

intermediate region. The stylized plots are reasonably accurate: the maximum error in the magnitude plot occurs at $\omega\tau = 1$, where the exact value is $1/\sqrt{2}$ rather than 1, whereas the exact phases at $\omega\tau = 0.1$ and $\omega\tau = 10$ are $-6°$ and $-84°$ rather than $0°$ and $-90°$.

As another example, Fig. 1-20 shows stylized Bode plots of a phase lead

$$H(s) = \frac{\tau s}{1 + \tau s}$$

The magnitude and phase of $H(s)$ are

$$H(\omega) = \frac{\omega\tau}{\sqrt{1 + \omega^2\tau^2}} \quad \text{and} \quad \varphi(\omega) = \frac{\pi}{2} - \tan^{-1}\omega\tau$$

For $\omega\tau \gg 1$ you have the asymptotes

$$H(\omega) \simeq 1 \quad \text{and} \quad \varphi(\omega) \approx 0$$

whereas for $\omega\tau \ll 1$ you have

$$\log[H(\omega)] \simeq \log(\omega\tau) \quad \text{and} \quad \varphi(\omega) \simeq \pi/2$$

You can now check that the stylized plots in Fig. 1-21 represent the cascade

$$H(s) = \frac{10\tau s}{1 + 10\tau s} \frac{10}{1 + \tau s}$$

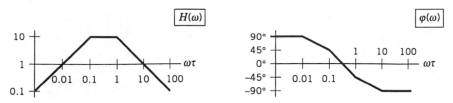

Figure 1-21. Stylized Bode plots of a lead–lag cascade (Exercise 1-9).

1-10   BANDWIDTH AND RISE TIME IN DAMPED SYSTEMS. In systems like the one shown in Fig. 1-22, in which $H(\omega) = 1$ at $\omega = 0$ and drops off to zero at high frequencies, the *bandwidth* or *cutoff frequency* ω_C is defined as the frequency at which the magnitude of the frequency response is $1/\sqrt{2}$ times its value at $\omega = 0$:

$$H(\omega_C) = \frac{1}{\sqrt{2}}$$

The step response of such systems is of the form

$$r(t) = \theta(t)\left[1 - x_H(t)\right]$$

where $x_H(t)$ eventually decays to zero. For these systems, the *rise time* t_R is defined as the time it takes the step response to go from 10% to 90% of its final value. It is a much used parameter in electronics because it is usually well defined and therefore easy to measure. In fact, most oscilloscopes have special marks on the screen to facilitate the measurement.

In a phase lag with frequency response

$$H(s) = \frac{1}{1 + \tau s}$$

the cutoff frequency is $\omega_C = 1/\tau$ and the step response is

$$r(t) = \theta(t)\left[1 - \exp(-t/\tau)\right]$$

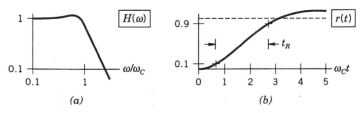

Figure 1-22. Representative (*a*) frequency and (*b*) step responses of stable systems whose step response settles to a constant value at large times (Exercise 1-10).

If $t_{0.1}$ and $t_{0.9}$ are the times at which $r(t)$ is equal to 0.1 and 0.9, you can verify that

$$t_{0.1} = 0.105\tau \quad \text{and} \quad t_{0.9} = 2.303\tau$$

so that the rise time t_R is related to the bandwidth ω_C by

$$\omega_C t_R \simeq 2.2$$

This relation holds within 5% for second-order lags with $\zeta > \frac{1}{2}$ or, more generally, if the resonance in the frequency response is not more than a few percent above the value at $\omega = 0$. Use the data in the appendix to verify that $\omega_C t_R = 2.09$ in a second-order lag with $\zeta = \frac{1}{2}$. You will see further examples in Chapter 9.

1-11 BANDWIDTH AND DECAY TIME IN RESONANT SYSTEMS. Consider a resonator

$$H(s) = \frac{\tau_0 s}{1 + 2\zeta\tau_0 s + \tau_0^2 s^2}$$

for $\zeta < 1$. Its step response, given in the appendix and shown in Fig. 1-23, is

$$r(t) = \frac{\omega_0}{\omega_2} \exp(-\omega_1 t) \sin(\omega_2 t)$$

where

$$\omega_0 = \frac{1}{\tau_0} \qquad \omega_1 = \zeta\omega_0 \qquad \omega_2 = \omega_0\sqrt{1 - \zeta^2}$$

The amplitude of the oscillation decays by a factor $e \simeq 2.72$ in a time

$$\tau_D = \frac{1}{\omega_1} = \frac{1}{\zeta\omega_0}$$

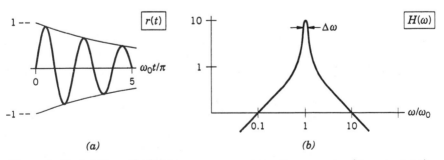

Figure 1-23. (a) Step and (b) frequency responses of a resonator (Exercise 1-11).

The frequencies at which $H(\omega)$, also shown in Fig. 1-23, is $1/\sqrt{2}$ times its value at resonance are the *half-power* frequencies. Show that the resonance occurs at $\omega = \omega_0$ and that the half-power frequencies are given by

$$\frac{\omega_\pm}{\omega_0} = \sqrt{1+\zeta^2}\pm\zeta$$

The bandwidth $\Delta\omega$, defined as the full width of the resonance at the half-power points, is given by

$$\frac{\Delta\omega}{\omega_0} \equiv \frac{\omega_+-\omega_-}{\omega_0} = 2\zeta$$

It follows that the bandwidth and the decay time are related by

$$\tau_D\,\Delta\omega = 2$$

More generally, this relation holds within 10% for the decay of an oscillation due to a resonance whose maximum is at least a factor of 2 above any nearby background response. Show that this is true for a second-order lag if $\zeta < 0.3$.

$H(\omega)$ has a maximum of value $Q = 1/2\zeta$ at $\omega = \omega_0$. Q is often used in place of ζ to characterize a resonance. For $Q \gg 1$ and frequencies near ω_0, $(1-\omega^2\tau_0^2)$ is approximately $2(1-\omega\tau_0)$. Show that the frequency response can be rewritten in terms of Q as

$$H(s) \simeq \frac{Q}{1+2iQ\dfrac{\omega-\omega_0}{\omega_0}}$$

and that the half-power frequencies and the bandwidth are given approximately by

$$\frac{\omega_\pm}{\omega_0} \simeq 1\pm\frac{1}{2Q} \quad\text{and}\quad \frac{\Delta\omega}{\omega_0} \equiv \frac{\omega_+-\omega_-}{\omega_0} \simeq \frac{1}{Q}$$

The decay time τ_D is $2Q/\omega_0$, equivalent to Q/π cycles. You thus have a simple rule for measuring Q: Count the number of cycles it takes for the amplitude of the oscillations to decay by a factor $e \simeq 2.72$, and multiply by π.

1-12 QUICK VISUALIZATION OF HIGHER-ORDER STEP RESPONSES. It is convenient to have a quick way of visualizing the step response that corresponds to a given frequency response. This is of course no problem in the case of integrators or of phase leads or lags, but it is less obvious if you are dealing with responses of second and higher order.

In the case of underdamped second-order lags, you know the maximum in the step response; this, plus the fact that $r'(0) = 0$, is enough to sketch $r(t)$.

Except at small times of order τ_2^2/τ_1 or less, an overdamped second-order lag with $\zeta \geq 1$ can be approximated by the cascade of a first-order lag and a delay:

$$\frac{1}{1+\tau_1 s+\tau_2^2 s^2} \rightarrow \frac{\exp\left(-\dfrac{\tau_2^2}{\tau_1} s\right)}{1+\left(\tau_1 - \dfrac{\tau_2^2}{\tau_1}\right) s}$$

Figure 1-24 shows that the approximation is quite good even in the marginal case $\tau_2 = \tau_1/2$ ($\zeta = 1$). Notice that $\tau_1 = 2\zeta\tau_2$, so that the bandwidth, as given in the appendix, is very well approximated:

$$\tau_1 - \frac{\tau_2^2}{\tau_1} = \tau_2\left(2\zeta - \frac{1}{2\zeta}\right) \simeq \frac{1}{\omega_C}$$

A bit of freehand drawing to make the waveform start off with zero slope at $t = 0$ results in a very good likeness of the real waveform.

More generally, for waveforms that do not have an excessive overshoot, a small lag can be replaced by a delay if its time constant τ is at least twice as small as the characteristic times of the rest of the frequency response. That is, one can make the replacement

$$\frac{1}{1+\tau s} \rightarrow \exp(-\tau s)$$

The responses $r(t)$ and $r_D(t)$ in Fig. 1-25 correspond to

$$H(s) = \frac{1}{(1+\tau s/2)(1+\tau s+\tau^2 s^2)} \quad \text{and} \quad H_D(s) = \frac{\exp(-\tau s/2)}{1+\tau s+\tau^2 s^2}$$

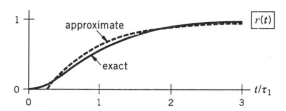

Figure 1-24. Except at small times, the step response of an overdamped second-order lag is well approximated by the delayed step response of a first-order lag (Exercise 1-12). The marginal case $\tau_2 = \tau_1/2$ is illustrated.

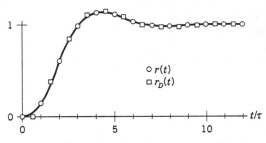

Figure 1-25. Step response $r(t)$ of a third-order system and its close simulation by the delayed step response $r_D(t)$ of a second-order system (Exercise 1-12).

The differences are almost imperceptible, particularly if the graph of $r_D(t)$ is retouched so as to make it start off with zero slope at $t = 0$. Show that

$$\frac{1}{(1+\tau s/2)(1+\tau s+\tau^2 s^2)} = \left(\frac{2}{3}\right)\frac{1-\tau s}{1+\tau s+\tau^2 s^2} + \left(\frac{1}{3}\right)\frac{1}{1+\tau s/2}$$

and that

$$r(t) = \theta(t)\left[1 - \frac{1}{3}\exp(-\omega_3 t) - \exp(-\omega_1 t)\left(\frac{2}{3}\cos\omega_2 t + \frac{2}{\sqrt{3}}\sin\omega_2 t\right)\right]$$

where

$$\omega_1 = \frac{1}{2\tau} \qquad \omega_2 = \frac{\sqrt{3}}{2\tau} \qquad \omega_3 = \frac{2}{\tau}$$

Also show that with $u = t - \tau/2$,

$$r_D(t) = \theta(u)\left[1 - \exp(-\omega_1 u)\left(\cos\omega_2 u + \frac{1}{\sqrt{3}}\sin\omega_2 u\right)\right]$$

Plot enough points of $r(t)$ and $r_D(t)$ to verify that Fig. 1-25 is correct.

1-13   OTHER RESPONSES OF LEADS AND LAGS. For $\zeta > 1$ the frequency response of a second-order lag is

$$H(s) = \frac{1}{(1+\tau_1 s)(1+\tau_2 s)}$$

This response can be considered as the response of two cascaded phase lags, as shown in Fig. 1-26. The step response of the first lag is

$$\theta(t)\left[1 - \exp(-t/\tau_1)\right]$$

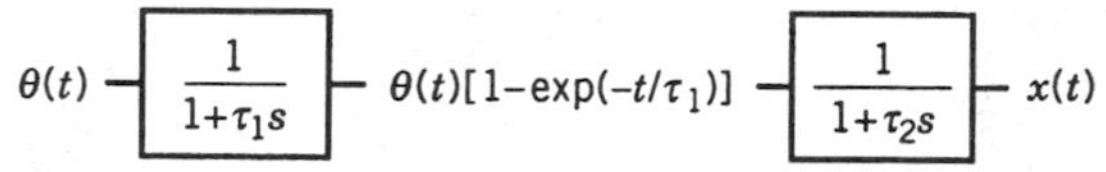

$$\theta(t) \longrightarrow \boxed{\dfrac{1}{1+\tau_1 s}} \longrightarrow \theta(t)[1-\exp(-t/\tau_1)] \longrightarrow \boxed{\dfrac{1}{1+\tau_2 s}} \longrightarrow x(t)$$

Figure 1-26. The step response of an overdamped second-order lag can be used to obtain the response of a phase lag to excitations of the form $\theta(t)[1-\exp(-t/\tau)]$ (Exercise 1-13).

The step response of a second-order lag for $\zeta > 1$ can therefore be used to obtain the response of a phase lag to excitations of the form $\theta(t)[1-\exp(-t/\tau)]$.

Now convince yourself that the step response of a resonator for $\zeta > 1$ can be used to obtain the response of a phase lead to excitations of the form $\theta(t)[1-\exp(-t/\tau)]$ and of a phase lag to excitations of the form $\theta(t)\exp(-t/\tau)$.

1-14 THE FREQUENCY CONTENT OF A SHARP PULSE IS CONSTANT. Consider the square pulse of unit area given in (1-15),

$$\delta_T(t) = \frac{\theta(t) - \theta(t-T)}{T}$$

According to (1-19), its frequency content $\Delta_T(i\omega)$ is

$$\Delta_T(i\omega) = \frac{1-\exp(-i\omega T)}{i\omega T} = \frac{\sin(\omega T/2)}{(\omega T/2)}\exp(-i\omega T/2)$$

If you now let T go to zero, and remember that

$$\lim_{\eta \to 0} \frac{\sin \eta}{\eta} = 1$$

you obtain

$$\lim_{T \to 0} \Delta_T(i\omega) = 1$$

You may tentatively conclude that the frequency content of an extremely sharp pulse is constant and that the constant is 1 if the pulse has unit area. This question is addressed more precisely in Chapter 2, where you will see that $\delta_T(t)$ converges to the generalized function δ.

APPENDIX: STEP AND FREQUENCY RESPONSES OF SECOND-ORDER SYSTEMS

A1-1   Second-Order Lag

The differential equation for a second-order lag is

$$\tau_0^2 x'' + 2\zeta\tau_0 x' + x = y$$

and the corresponding frequency response is

$$H(s) = \frac{1}{1 + 2\zeta\tau_0 s + \tau_0^2 s^2}$$

The roots of the characteristic polynomial are real for $\zeta > 1$, real and equal for $\zeta = 1$, and complex conjugate for $\zeta < 1$.

The magnitude and phase of $H(s)$ are given by

$$H(\omega) = \frac{1}{\left[(1 - \omega^2\tau_0^2)^2 + 4\zeta^2\tau_0^2\omega^2\right]^{1/2}} \quad \text{and} \quad \varphi(\omega) = \tan^{-1}\frac{1 - \omega^2\tau_0^2}{2\zeta\omega\tau_0} - \frac{\pi}{2}$$

If $\zeta < 1/\sqrt{2}$, $H(\omega)$ has a resonance of magnitude $1/(2\zeta\sqrt{1-\zeta^2})$ at $\omega = \omega_0\sqrt{1-2\zeta^2}$, where $\omega_0 = 1/\tau_0$.

The cutoff frequency ω_C, defined by $H(\omega_C) = 1/\sqrt{2}$, is given by

$$\omega_C^2\tau_0^2 = \sqrt{(1-2\zeta^2)^2 + 1} + (1 - 2\zeta^2)$$

The following approximation is good to 3% for $\zeta \geq 1$:

$$\frac{1}{\omega_C\tau_0} \approx 2\zeta - \frac{1}{2\zeta}$$

The differential equation for the step response is

$$\tau_0^2 r'' + 2\zeta\tau_0 r' + r = 1$$

with initial conditions

$$r(0) = 0 \quad \text{and} \quad r'(0) = 0$$

The step response for $\zeta \geq 1$ is given by

$$r(t > 0, \zeta > 1) = 1 - \frac{\tau_1}{\tau_1 - \tau_2}\exp\left(-\frac{t}{\tau_1}\right) - \frac{\tau_2}{\tau_2 - \tau_1}\exp\left(-\frac{t}{\tau_2}\right)$$

$$r(t > 0, \zeta = 1) = 1 - \exp\left(-\frac{t}{\tau_0}\right) - \frac{t}{\tau_0}\exp\left(-\frac{t}{\tau_0}\right)$$

where

$$\tau_{1,2} = \tau_0\left(\zeta \pm \sqrt{\zeta^2-1}\right)$$

The step response for $\zeta < 1$ is given by

$$r(t>0, \zeta<1) = 1-\exp(-\omega_1 t)\left[\cos \omega_2 t + \frac{\omega_1}{\omega_2}\sin(\omega_2 t)\right]$$

where

$$\omega_0 = 1/\tau_0 \qquad \omega_1 = \zeta\omega_0 \qquad \omega_2 = \omega_0\sqrt{1-\zeta^2}$$

If $\zeta<1$, $r(t)$ is oscillatory and has a maximum r_M at $\omega_2 t = \pi$ given by

$$r_M = 1+\exp\left(-\pi\frac{\omega_1}{\omega_2}\right) = 1+\exp\left(-\pi\frac{\zeta}{\sqrt{1-\zeta^2}}\right)$$

A1-2   Resonator

The differential equation for a resonator is

$$\tau_0^2 x'' + 2\zeta\tau_0 x' + x = \tau_0 y'$$

and the corresponding frequency response is

$$H(s) = \frac{\tau_0 s}{1+2\zeta\tau_0 s+\tau_0^2 s^2}$$

The roots of the characteristic polynomial are real for $\zeta>1$, real and equal for $\zeta=1$, and complex conjugate for $\zeta<1$.

The magnitude and phase of $H(s)$ are given by

$$H(\omega) = \frac{\omega\tau_0}{\left[(1-\omega^2\tau_0^2)^2+4\zeta^2\tau_0^2\omega^2\right]^{1/2}} \quad \text{and} \quad \varphi(\omega) = \tan^{-1}\frac{1-\omega^2\tau_0^2}{2\zeta\omega\tau_0}$$

$H(\omega)$ has a maximum of value $Q = 1/2\zeta$ at the resonant frequency $\omega_0 = 1/\tau_0$. The differential equation for the step response is

$$\tau_0^2 r'' + 2\zeta\tau_0 r' + r = 0$$

with initial conditions

$$r(0) = 0 \quad \text{and} \quad \tau_0 r'(0) = 1$$

The step response for $\zeta \geq 1$ is

$$r(t>0, \zeta>1) = \frac{\tau_0}{(\tau_1 - \tau_2)}\left[\exp\left(-\frac{t}{\tau_1}\right) - \exp\left(-\frac{t}{\tau_2}\right)\right]$$

$$r(t>0, \zeta=1) = \frac{t}{\tau_0}\exp\left(-\frac{t}{\tau_0}\right)$$

where

$$\tau_{1,2} = \tau_0\left(\zeta \pm \sqrt{\zeta^2-1}\right)$$

The step response for $\zeta < 1$ is

$$r(t>0, \zeta<1) = \frac{\omega_0}{\omega_2}\exp(-\omega_1 t)\sin(\omega_2 t)$$

where

$$\omega_0 = 1/\tau_0 \qquad \omega_1 = \zeta\omega_0 \qquad \omega_2 = \omega_0\sqrt{1-\zeta^2}$$

2

GENERALIZED FUNCTIONS AND INTEGRAL TRANSFORMS

2-1 INTRODUCTION

In Chapter 1 we assumed that it was legitimate to obtain the step response of a phase lead by using a limiting procedure; we will now put this assumption on a firm footing by introducing generalized functions. Despite their imposing name, generalized functions have surprisingly simple properties, and we will see that they offer a freedom of action that ordinary functions cannot match. We will also complete our program of obtaining a multiplicative operator for arbitrary inputs to a linear system by introducing Laplace transforms, whose natural setting is precisely among generalized functions, and we will give the frequency content of waveforms a formal basis by introducing Fourier transforms. Our examples will be straightforward, but our approach will be abstract; the payoff will be a powerful set of tools, applicable far beyond the context in which they will be developed.

We will be interested primarily in concepts and mechanics of use; this is why we will prove many minor theorems and accept a few difficult theorems: simple proofs will prepare us to understand what we are accepting. The difficult theorems generally involve the integrability of functions and the legitimacy of exchanging the order of limits; they will thus be intuitively reasonable on the one hand, and verifiable in specific cases on the other.

2-2 MATHEMATICAL BACKGROUND

The results we will obtain in this chapter are quite general, and are best stated in a mathematical language of corresponding generality. To do so, we need to introduce several recurrent terms of this broader language.

A function is *smooth* if it has derivatives of all orders everywhere on the interval $(-\infty, +\infty)$. Polynomials and exponentials, for example, are smooth.

A function is *absolutely integrable* (on a given interval) if its absolute value is integrable. An example of a function that is integrable but not absolutely integrable on $(-\infty, +\infty)$ is $(1/t)\sin(\lambda t)$.

A function is *locally integrable* if it is integrable on any bounded interval $t_1 \leq t \leq t_2$. Smooth functions are clearly locally integrable. Two examples of functions that are locally integrable but not integrable on $(-\infty, +\infty)$ are the unit step function $\theta(t)$, whose integral diverges, and $\exp(i\lambda t)$, whose integral does not converge. For the sake of simple proofs, and with no tangible loss, we will assume that locally integrable functions are locally *absolutely* integrable.

For us it will suffice to think of integrability in terms of the familiar Riemann integral. The results that we will quote or obtain, however, will be stated so that they are valid if integrability is interpreted in the wider sense of Lebesgue, briefly described below. As we might expect, the Riemann integral, when it exists, coincides with the Lebesgue integral.

In Lebesgue's charming invention, a statement can be true *almost everywhere*, meaning that it is true except on a set of *measure zero*. The integers, the rational numbers, and in fact any countable set, are examples of sets of measure zero. Thus the function that is 1 on irrational numbers and has arbitrary values on rational numbers is equal to 1 almost everywhere; more importantly, functions that are integrable in the sense of Lebesgue have the same integral if they are equal almost everywhere. In view of the specific instances that we will encounter, such as the set of points of discontinuity of a function, we need only imagine that a set is of measure zero if the number of points contained in any bounded interval is finite.

With these preliminaries, we can quote a few theorems that we will need later on and in the exercises. We will start by noting that if $x(t)$ is locally integrable, its integral

$$y(t) = \int_0^t x(u)\,du$$

is continuous; in addition, $y(t)$ is differentiable almost everywhere, and $y'(t) = x(t)$ almost everywhere. If $x(t)$ is continuous, $y(t)$ is differentiable.

When we come to Fourier transforms, we will make use of the *Riemann–Lebesgue lemma*, which should be intuitively quite obvious in the case of continuous functions: If $x(t)$ is absolutely integrable on $(-\infty, +\infty)$, then

$$\lim_{\lambda \to \infty} \int_{-\infty}^{+\infty} x(t)\exp(i\lambda t)\,dt = 0$$

Finally, we note that we will exchange the order of integration of repeated integrals on the basis of *Fubini's theorem*, which states that if either

$$\int_{-\infty}^{+\infty} du \int_{-\infty}^{+\infty} dv \, |x(u,v)| \quad \text{or} \quad \int_{-\infty}^{+\infty} dv \int_{-\infty}^{+\infty} du \, |x(u,v)|$$

exists, then

$$\int_{-\infty}^{+\infty} du \int_{-\infty}^{+\infty} dv \, x(u,v) \quad \text{and} \quad \int_{-\infty}^{+\infty} dv \int_{-\infty}^{+\infty} du \, x(u,v)$$

both exist and are equal. In applications, we should remember our assumption that a locally integrable function is locally absolutely integrable.

2-3 CONVOLUTIONS OF FUNCTIONS

The *convolution* $x_1 * x_2$ of two locally integrable functions x_1 and x_2, when it exists, is given by[†]

$$x_1 * x_2(t) = \int_{-\infty}^{+\infty} x_1(u) x_2(t-u) \, du \tag{2-1}$$

We will accept that the integrand $x_1(u)x_2(t-u)$ is locally integrable, so that existence is only a question of the convergence of the integral.

To give a simple example, we will convolve $t\theta(t)$ and $\theta(t)$; as a visual aid, Fig. 2-1 shows $u\theta(u)$ and $\theta(t-u)$ at time t. We then have

$$t\theta(t) * \theta(t) = \int_{-\infty}^{+\infty} u\theta(u) \theta(t-u) \, du = \theta(t) \int_{0}^{t} u \, du = \frac{t^2}{2} \theta(t)$$

We observe here an important property of the convolution of two functions: it is smoother than either of its *factors*. Thus $\theta(t)$ and the derivative of $t\theta(t)$ are discontinuous at $t = 0$, whereas $t\theta(t) * \theta(t)$ has a continuous derivative.

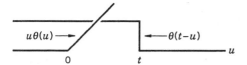

Figure 2-1. The integrands $u\theta(u)$ and $\theta(t-u)$ in the convolution of $t\theta(t)$ and $\theta(t)$ at time t.

[†] $x_1 * x_2$ should be read as a whole, and $x_1 * x_2(t)$ as the value of the function $x_1 * x_2$ at time t. When x_1 and x_2 are given explicitly, however, in the interest of simplicity we will write $x_1(t) * x_2(t)$ rather than a complicated formal expression and depend on context if necessary.

As another example, we will convolve $\exp(i\omega t)$ with a square pulse of unit area:

$$\exp(i\omega t) * \frac{\theta(t) - \theta(t-\tau)}{\tau} = \frac{1 - \exp(-i\omega\tau)}{i\omega\tau}\exp(i\omega t)$$

In this case we observe that $\exp(i\omega t)$ is almost unperturbed by the convolution if $\omega\tau \ll 1$, whereas it is multiplied by a small amplitude if $\omega\tau \gg 1$. This behavior is quite general, and in Section 2-9 it should become apparent that it characterizes low-pass filters: convolution with narrow pulses of unit area is like multiplication by 1 for slowly varying functions, whereas fast fluctuations are reduced by convolution with wide pulses of unit area.

Setting $v = t - u$ in (2-1), we obtain that convolution is commutative:

$$x_1 * x_2 = x_2 * x_1$$

Associativity, however, is not in general guaranteed; that is, it is not in general true that

$$(x_1 * x_2) * x_3 = x_1 * (x_2 * x_3)$$

For example, the fact that $2t(1+t^2)^{-2}$ is an odd function implies that

$$\left[1 * \frac{2t}{(1+t^2)^2}\right] * \theta(t) = 0 * \theta(t) = 0$$

But on the other hand, we have

$$1 * \left[\frac{2t}{(1+t^2)^2} * \theta(t)\right] = 1 * \int_{-\infty}^{t}\frac{2u}{(1+u^2)^2}\,du = -1 * \frac{1}{1+t^2} = -\pi$$

There are two cases that interest us in which existence and associativity are guaranteed within a set of functions; in both cases, the convolution integral is over a bounded interval (see Exercise 2-1):

1. A set of locally integrable functions of which all but one are zero outside a bounded interval
2. A set of locally integrable functions that are zero for $t < \tau$, where τ is a constant

In Exercise 2-14 it is shown that $\psi * x$ is smooth if ψ is smooth and the convolution integral is over a bounded interval, and that

$$(\psi * x)^{(m)} = \psi^{(m)} * x$$

If ψ_1 and ψ_2 are smooth, the derivative of their convolution can be obtained by differentiating either factor. Thus if $m = j + k$,

$$(\psi_1 * \psi_2)^{(m)} = \psi_1^{(j)} * \psi_2^{(k)}$$

Translations in time can also be applied to either factor of a convolution:

$$x_1 * x_2(t - \tau) = x_1(t - \tau) * x_2(t) = x_1(t) * x_2(t - \tau)$$

2-4 LAPLACE TRANSFORMS OF FUNCTIONS

The *Laplace transform* $\mathcal{L}$ is a linear operator defined on the set of locally integrable functions $x(t)$ that are zero for $t < 0$ and such that for some real σ_C (it can actually be $-\infty$), the integral

$$\int_0^\infty |x(t)| \exp(-\sigma t)\, dt$$

converges if $\sigma > \sigma_C$ and diverges if $\sigma < \sigma_C$. The Laplace transform $\mathcal{L}(x)$ of such a function is the complex function $X(s)$ defined for $\sigma > \sigma_C$ by

$$X(s) = \mathcal{L}(x) = \int_0^\infty x(t) \exp(-st)\, dt$$

where s is now the complex variable

$$s = \sigma + i\omega$$

As shown in Exercise 2-15, $X(s)$ has a derivative with the same *abscissa of convergence* σ_C; that is, it is *analytic* on the half-plane $\sigma > \sigma_C$.

We want to see as quickly as possible why Laplace transforms are useful. To do so, we need to calculate some elementary transforms and define the inverse transform. Let us first get the transforms. For $\sigma > \sigma_C = -\tau^{-1}$ we have

$$\int_0^\infty \exp(-t/\tau) \exp(-st)\, dt = \frac{1}{\tau^{-1} + s} \int_0^\infty \exp(-u)\, du = \frac{\tau}{1 + \tau s}$$

From this result we obtain the transform of $\theta(t) \exp(-t/\tau)$,

$$\mathcal{L}[\theta(t) \exp(-t/\tau)] = \frac{\tau}{1 + \tau s} \tag{2-2}$$

and for $\tau = \infty$, the transform of the unit step function $\theta(t)$,

$$\mathcal{L}(\theta) = \frac{1}{s} \qquad (2\text{-}3)$$

For $\sigma > 0$ we also have

$$\int_0^\infty t \exp(-st)\, dt = \frac{1}{s^2} \int_0^\infty u \exp(-u)\, du = \frac{1}{s^2}$$

The transform of $t\theta(t)$ is therefore

$$\mathcal{L}(t\theta) = \frac{1}{s^2} \qquad (2\text{-}4)$$

We will accept that functions that have the same Laplace transform are equal almost everywhere. If we then define the inverse Laplace transform $\mathcal{L}^{-1}$ of $X(s)$ as the function $x(t)$ whose Laplace transform is $X(s)$,

$$x(t) = \mathcal{L}^{-1}(X) \quad \Leftrightarrow \quad \mathcal{L}(x) = X(s)$$

we understand that it is one of an infinite family of such functions, among which we will not bother to distinguish. A moment's consideration will show that the inverse Laplace transform is also a linear operator.

Now we are in a position to do something useful. Let us consider a phase lag

$$\tau x' + x = y$$

For the moment let us assume that $x'(t)$ is continuous, so that $x(t)$ is also continuous and $x(0) = 0$. An integration by parts, fully allowed under these conditions, then yields

$$\mathcal{L}(x') = sX(s)$$

Transforming both sides of the differential equation, we obtain

$$(\tau s + 1)X(s) = Y(s)$$

and thus

$$X(s) = H(s)Y(s)$$

where

$$H(s) = \frac{1}{1 + \tau s}$$

This is precisely what we have been looking for: a multiplicative operator—the *transfer function* $H(s)$—that relates the excitation and the response. We note that $H(i\omega)$ is identical to the frequency response because of the way that derivatives transform; we will soon see that this identification holds much more generally.

Let us obtain the response to a ramp $y(t) = \lambda t\theta(t)$. We solved this problem in Section 1-2, and we can check in (1-7) that $x(t)$ has a continuous derivative, so we are safe. From (2-4) we have

$$Y(s) = \mathcal{L}(\lambda t\theta) = \frac{\lambda}{s^2}$$

The transform $X(s) = H(s)Y(s)$ of the response is therefore

$$X(s) = \frac{\lambda}{s^2(1+\tau s)} = \frac{\lambda}{s^2} - \frac{\lambda\tau}{s} + \frac{\lambda\tau^2}{1+\tau s}$$

Taking inverse transforms using (2-2), (2-3), and (2-4), we arrive at the correct result:

$$x(t) = \theta(t)\left[\lambda(t-\tau) + \lambda\tau\exp(-t/\tau)\right]$$

We observe in this example a decided advantage offered by Laplace transforms: initial conditions are inherent in the solution and therefore need not be determined in a separate operation.

We would certainly like to have a transfer function for arbitrary excitations, and not only for systems governed by differential equations. To achieve this objective we will have to deal with derivatives of functions with jump discontinuities. Before proceeding, however, we will state some theorems.

First, four theorems whose only requirement for validity, other than some minor restrictions on the abscissa of convergence, is that $X(s)$ exist:

$$\mathcal{L}\left[\int_0^t x(u)\,du\right] = \frac{1}{s}X(s) \tag{2-5}$$

$$\mathcal{L}[x(t-\tau)] = \exp(-\tau s)X(s) \tag{2-6}$$

$$\mathcal{L}[\exp(s_0 t)x(t)] = X(s-s_0) \tag{2-7}$$

$$\mathcal{L}[t^n x(t)] = (-1)^n X^{(n)}(s) \tag{2-8}$$

Theorems (2-5), (2-6), and (2-7), which are quite straightforward, are proved in Exercise 2-2. Theorem (2-8) is a restatement of the analyticity of $X(s)$. In Exercise 2-4 it is shown that most of the entries in the table of Laplace transforms presented in the appendix to this chapter can be obtained using these theorems.

Often useful are the *initial-value* and *final-value* theorems, which relate the behavior of $X(\sigma)$ for $\sigma \to +\infty$ and $\sigma \to 0+$ to the behavior of $x(t)$ for $t \to 0+$ and $t \to +\infty$:

(initial value) $x(0+) = \lim_{t \to 0+} x(t) = \lim_{\sigma \to +\infty} \sigma X(\sigma)$ (2-9)

(final value) $x(+\infty) = \lim_{t \to +\infty} x(t) = \lim_{\sigma \to 0+} \sigma X(\sigma)$ (2-10)

These two theorems require, in addition to the existence of $X(s)$, that the limit of $x(t)$ at $0+$ or $+\infty$ exist or be $\pm\infty$. Cases in which this requirement is not met and the theorems fail are given in Exercise 2-7. As an example, we get correct answers by applying the initial-value and final-value theorems to the ramp $x(t) = t\theta(t)$ whose transform $1/s^2$ we obtained in (2-4):

$$\lim_{\sigma \to +\infty} \sigma \frac{1}{\sigma^2} = x(0+) = 0$$

$$\lim_{\sigma \to 0+} \sigma \frac{1}{\sigma^2} = x(+\infty) = +\infty$$

Finally, we have the *convolution* theorem, which is central in the theory of linear systems: If x_1 and x_2 have Laplace transforms $X_1(s)$ and $X_2(s)$, their convolution has a Laplace transform equal to the product of $X_1(s)$ and $X_2(s)$,

$$\mathcal{L}(x_1 * x_2) = X_1(s) X_2(s)$$

The proof of this theorem is outlined in Exercise 2-3.

2-5   GENERALIZED FUNCTIONS

Following Schwartz (1965), we will now generalize the concepts of function and derivative in order to remove restrictions on the theorem that

$$\mathcal{L}(x') = sX(s)$$

and to make sense of the limit for $T \to 0$ of functions such as $\delta_T(t)$ in (1-15).

We will first define a *test function* as a smooth function that is zero outside a bounded interval. We will then define a *generalized function* or *distribution* x_G as a sequence of smooth functions

$$x_G = \{x_n\}$$

such that the *functional* $\langle x_G, \phi \rangle$, defined for a test function ϕ by

$$\langle x_G, \phi \rangle = \lim_{n \to \infty} \int_{-\infty}^{+\infty} x_n(t) \phi(t) \, dt$$

exists and is finite for all test functions. There is nothing obvious about the rather abstract requirement that the functional $\langle x_G, \phi \rangle$ exist; mathematical experience has shown, however, that it rejects anomalies while conserving the properties that make generalized functions useful.

Strictly speaking, generalized functions do not have a *value* at any particular time; we will see, however, that intuition will win out over rigor, and that we will end up thinking of them as ordinary functions with quaint derivatives but much easier to handle in other respects. In any event, we will often write $x_G(t)$ to distinguish a generalized function from its translation by a time τ, given by

$$x_G(t-\tau) = \{x_n(t-\tau)\}$$

Let us move toward a fundamental generalized function, the *delta function* δ. We will start by introducing the function $\delta_0(t)$ given by

$$\delta_0(|t| < T) = \frac{K}{T} \exp\left[\frac{-1}{1-(t/T)^2}\right] \quad \text{and} \quad \delta_0(|t| \geq T) = 0$$

where T is arbitrary and K is chosen such that

$$\int_{-\infty}^{+\infty} \delta_0(t) \, dt = 1$$

The derivatives of $\delta_0(t)$ for $|t| < T$ are $\delta_0(t)$ times rational functions of t. This, plus the fact that

$$\lim_{t \to \infty} t^m \exp(-\lambda t) = 0$$

for $\lambda > 0$ implies that $\delta_0(t)$ and all its derivatives tend to zero for $t \to \pm T$, and that $\delta_0(t)$ is therefore a smooth function; since it is zero for $t > |T|$, $\delta_0(t)$ is also a test function. We can now spawn an infinity of test functions by convolving $\delta_0(t)$ with any integrable function that is zero outside a bounded interval.

The functions $\delta_n(t)$, defined by

$$\delta_n(t) = n\delta_0(nt)$$

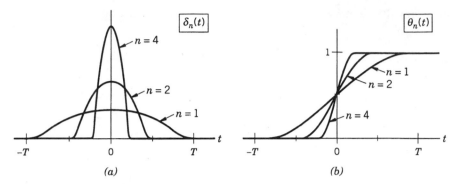

Figure 2-2. (a) The smooth functions $\delta_n(t)$ used to construct the delta function; (b) their integrals $\theta_n(t)$.

and shown for several values of n in Fig. 2-2, are zero for $|t| > T/n$ and have unit area,

$$\int_{-\infty}^{+\infty} \delta_n(t)\, dt = 1$$

As shown in Exercise 2-13, it follows that if $\psi(t)$ is smooth (it actually need only be continuous),

$$\lim_{n \to \infty} \int_{-\infty}^{+\infty} \psi(t)\,\delta_n(t)\, dt = \psi(0)$$

We will now define the delta function δ as

$$\delta = \{\delta_n\} \tag{2-11}$$

The defining functional for δ is

$$\langle \delta, \phi \rangle = \lim_{n \to \infty} \int_{-\infty}^{+\infty} \delta_n(t)\,\phi(t)\, dt = \phi(0)$$

and it is clear that it exists for all test functions ϕ.

We will define the mth derivative of a generalized function $x_G = \{x_n\}$ as

$$x_G^{(m)} = \{x_n^{(m)}\}$$

Its defining functional is

$$\langle x_G^{(m)}, \phi \rangle = \lim_{n \to \infty} \int_{-\infty}^{+\infty} x_n^{(m)}(t)\,\phi(t)\, dt$$

Integrating by parts m times yields

$$\langle x_G^{(m)}, \phi \rangle = (-1)^m \langle x_G, \phi^{(m)} \rangle$$

The defining functional for $x_G^{(m)}$ exists for all test functions ϕ because $\phi^{(m)}$ is also a test function. We conclude that generalized functions are smooth, that is, that they have derivatives of all orders. This is a first indication of the fact that generalized functions have simpler properties than ordinary functions. A generalized derivative we will often see is δ'. Its defining functional is

$$\langle \delta', \phi \rangle = -\langle \delta, \phi' \rangle = -\phi'(0)$$

Generalized functions can clearly be added, or multiplied by a constant, but they cannot be freely multiplied by one another. It does make sense, however, to define the product of a smooth function ψ and a generalized function x_G:

$$\psi x_G = \{ \psi x_n \}$$

This product obeys the familiar rule for differentiation:

$$(\psi x_G)' = \psi' x_G + \psi x_G'$$

The defining functional for ψx_G is

$$\langle \psi x_G, \phi \rangle = \lim_{n \to \infty} \int_{-\infty}^{+\infty} \psi(t) x_n(t) \phi(t) \, dt = \langle x_G, \psi \phi \rangle$$

and it exists for all test functions ϕ because $\psi\phi$ is also a test function. We note in the case of the delta function that

$$\psi \delta = \psi(0) \delta$$

We also note that for reasons rooted in physics, a generalized function of the form $a\delta(t)$, where a is a constant, is called an *impulse* of amplitude a. The delta function is thus an impulse of amplitude 1.

We now want to lift the ordinary functions with which we are familiar into the wider class of generalized functions. This is easily obtained by the following artifice: if $x(t)$ is a locally integrable function, the convolution

$$\delta_n * x(t) = \int_{-\infty}^{+\infty} x(u) \, \delta_n(t-u) \, du$$

is well defined because $\delta_n(t)$ is zero outside a bounded interval; it is also smooth (it is called a *regularization* of x) because $\delta_n(t)$ is smooth. We will then define the *regular* generalized function

$$x_G = \{ \delta_n * x \} \tag{2-12}$$

as the generalized function associated with $x(t)$. Its defining functional is

$$\langle x_G, \phi \rangle = \lim_{n \to \infty} \int_{-\infty}^{+\infty} \phi(t)\, dt \int_{-\infty}^{+\infty} x(u)\, \delta_n(t-u)\, du$$

Exchanging the order of integration, we obtain

$$\langle x_G, \phi \rangle = \lim_{n \to \infty} \int_{-\infty}^{+\infty} x(t)\, dt \int_{-\infty}^{+\infty} \phi(t+u)\, \delta_n(u)\, du$$

In Exercise 2-14 it is shown that the limit exists and that

$$\langle x_G, \phi \rangle = \int_{-\infty}^{+\infty} x(t)\, \phi(t)\, dt$$

It follows that $\langle x_G, \phi \rangle$ exists for any test function ϕ because x is locally integrable and ϕ is zero outside a bounded interval.

Let us take $x(t) = \theta(t)$. Its regularizations are the smooth functions

$$\theta_n(t) = \delta_n * \theta(t) = \int_{-\infty}^{t} \delta_n(u)\, du$$

shown for several values of n in Fig. 2-2. Since $\theta_n'(t) = \delta_n(t)$, we conclude that

$$\theta_G = \{\theta_n\}$$

is a generalized function such that

$$\theta_G' = \delta$$

We have thus finally made sense of the derivative of $\theta(t)$.

If a generalized function x_G and a locally integrable function x are such that for any $\tau > 0$,

$$\langle x_G, \phi \rangle = \int_{-\infty}^{+\infty} x(t)\, \phi(t)\, dt$$

for all test functions ϕ that differ from zero only inside closed intervals $t_1 + \tau \le t \le t_2 - \tau$, we will identify x_G with x on the open interval $t_1 < t < t_2$. We will thus say that $\delta(t) = 0$ for $t \neq 0$, that is, on the open intervals $-\infty < t < 0$ and $0 < t < \infty$, and that $\theta_G(t) = \theta(t)$ everywhere.

In view of this identification, and because all locally integrable functions are now implicitly *generalized* functions, we will not distinguish regular generalized functions with a subscript when there is no danger of confusion: θ will thus stand for θ_G. Furthermore, we will henceforth consider that the excitations and responses of linear systems are generalized functions.

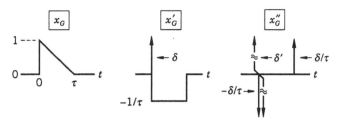

Figure 2-3. A piecewise smooth function with an isolated jump discontinuity, and its first two generalized derivatives.

The identification of a generalized function with an ordinary function does not extend to the derivatives of these functions, which exist to all orders for the generalized function and might not exist at all for the ordinary function. To see this essential point in an example more productive than $\theta(t)$, let us consider the *piecewise smooth* function with an isolated jump discontinuity shown in Fig. 2-3,

$$x_G(t) = \left(1 - \frac{t}{\tau}\right)[\theta(t) - \theta(t-\tau)]$$

The generalized derivative is equal to the ordinary derivative except at the jump, where it generates a delta function times the amplitude of the jump, whereas the ordinary derivative simply does not exist at the jump. In the same way we can obtain higher generalized derivatives, but we must then take into account the delta functions generated previously. It is of course possible to arrive at these results analytically: use of the fact that $\psi\delta = \psi(0)\delta$ yields

$$x_G'(t) = \left(1 - \frac{t}{\tau}\right)[\delta(t) - \delta(t-\tau)] - \frac{1}{\tau}[\theta(t) - \theta(t-\tau)]$$

$$= \delta(t) - \frac{1}{\tau}[\theta(t) - \theta(t-\tau)]$$

This result agrees with Fig. 2-3, as does

$$x_G''(t) = \delta'(t) - \frac{1}{\tau}[\delta(t) - \delta(t-\tau)]$$

2-6 INTEGRALS AND CONVOLUTIONS OF GENERALIZED FUNCTIONS

The integral $\int x_G$ of a generalized function x_G is any generalized function whose derivative is x_G:

$$\left(\int x_G\right)' = x_G$$

In Exercise 2-8 it is shown that if $x_n(t < \tau) = 0$, where τ is a constant, one possibility for $\int x_G$ is

$$\int x_G = \left\{ \int_{-\infty}^{t} x_n(u)\, du \right\}$$

and in Exercise 2-9 it is shown that two generalized functions that have the same derivative can differ only by a constant K:

$$x'_{G1} = x'_{G2} \quad \Rightarrow \quad x_{G1} - x_{G2} = K$$

If there are open intervals where the integral of a generalized function can be identified with a continuous function, we can take definite integrals if we choose limits in these open intervals. For $t \neq 0$ we thus have

$$\int_{-\infty}^{t} \delta(u)\, du = \theta(u)\Big|_{-\infty}^{t} = \theta(t)$$

The integral of $\delta(t)$ from $-\infty$ to $+\infty$ is 1. It follows that the amplitude a of an impulse $a\delta(t)$ is in fact the *area* of the impulse. More generally, we have the *sifting theorem* for smooth functions:

$$\int_{-\infty}^{+\infty} \psi(u)\, \delta(t-u)\, du = \psi(t) \int_{-\infty}^{+\infty} \delta(t-u)\, du = \psi(t)$$

We now have a first taste of how generalized functions behave like ordinary functions when they are in proper company. This, we correctly surmise, means they are next to smooth function and under integrals.

The convolution of two generalized functions $x_{G1} = \{x_{1n}\}$ and $x_{G2} = \{x_{2n}\}$ is defined by

$$x_{G1} * x_{G2} = \{x_{1n} * x_{2n}\}$$

Existence is guaranteed if either all x_{1n} or all x_{2n} are zero outside a fixed bounded interval, or if $x_{1n}(t < \tau) = x_{2n}(t < \tau) = 0$ for some fixed τ and for all n. It follows from the definition that generalized derivatives can be applied to either factor of a convolution:

$$(x_{G1} * x_{G2})' = x'_{G1} * x_{G2} = x_{G1} * x'_{G2}$$

Translations can also be applied to either factor:

$$x_{G1} * x_{G2}(t-\tau) = x_{G1}(t-\tau) * x_{G2}(t) = x_{G1}(t) * x_{G2}(t-\tau)$$

Finally, in Exercise 2-14 it is shown that the sifting theorem for smooth functions can be expressed as a convolution:

$$\psi * \delta = \psi$$

2-7 CONVERGENCE OF GENERALIZED FUNCTIONS

A sequence of generalized functions x_{Gn} converges to a generalized function x_G if

$$\lim_{n \to \infty} \langle x_{Gn}, \phi \rangle = \langle x_G, \phi \rangle$$

exists for all test functions ϕ. We will then write

$$\lim_{n \to \infty} x_{Gn} = x_G \quad \text{or} \quad x_{Gn} \to x_G$$

For example, it is easy to show that

$$\delta(t - n\tau) \to 0$$

because test functions are zero outside bounded intervals. More to the point, the smooth functions x_n that define a generalized function

$$x_G = \{x_n\}$$

are themselves generalized functions, so that

$$x_n \to x_G$$

The concept of convergence also applies to a family of generalized functions that depend on a parameter. Thus, for example, the functions $\theta_T(t)$ and $\delta_T(t)$, defined in (1-14) and (1-15) and shown in Fig. 1-9, converge for $T \to 0$ to the generalized functions $\theta(t)$ and $\delta(t)$ because we get the correct defining functionals for any test function ϕ:

$$\lim_{T \to 0} \langle \theta_T, \phi \rangle = \int_0^\infty \phi(t)\, dt = \langle \theta, \phi \rangle$$

and

$$\lim_{T \to 0} \langle \delta_T, \phi \rangle = \phi(0) = \langle \delta, \phi \rangle$$

The result for θ_T is obvious. The case of δ_T is taken up in Exercise 2-13, where it is also shown that

$$\frac{\theta(t) - 2\theta(t-T) + \theta(t-2T)}{T^2}$$

converges to $\delta'(t)$.

For any test function ϕ we have

$$\lim_{n \to \infty} \langle x_{Gn}^{(m)}, \phi \rangle = (-1)^m \lim_{n \to \infty} \langle x_{Gn}, \phi^{(m)} \rangle = (-1)^m \langle x_G, \phi^{(m)} \rangle = \langle x_G^{(m)}, \phi \rangle$$

It follows that if $x_{Gn} \to x_G$,

$$x_{Gn}^{(m)} \to x_G^{(m)}$$

That the order of limits and derivatives can be exchanged is one of the most valuable properties of generalized functions, and one that ordinary functions rarely have. For example, we know that

$$\lim_{T \to 0} \delta_T = \lim_{T \to 0} \frac{\theta(t) - \theta(t-T)}{T} = \delta$$

We can easily check that

$$\lim_{T \to 0} \delta_T' = \lim_{T \to 0} \frac{\delta(t) - \delta(t-T)}{T} = \delta'$$

because we obtain the correct defining functional:

$$\lim_{T \to 0} \langle \delta_T', \phi \rangle = \lim_{T \to 0} \frac{\phi(0) - \phi(T)}{T} = -\phi'(0) = \langle \delta', \phi \rangle$$

Let us now consider two sequences of generalized functions x_{G1n} and x_{G2n}, and let us assume that they converge to generalized functions x_{G1} and x_{G2}. We will accept, under conditions noted immediately below and by now quite familiar, that convolution is continuous, that is, that $x_{G1}*x_{G2}$ exists and that

$$x_{G1n}*x_{G2n} \to x_{G1}*x_{G2}$$

This vital theorem holds if either all x_{G1n} or all x_{G2n} are zero outside a fixed bounded interval, or if $x_{G1n}(t<\tau) = x_{G2n}(t<\tau) = 0$ for some fixed τ and for all n. Looking back at the definition of a convolution in Section 2-6, we might feel that existence depends somewhat circularly on continuity; this, however, is an artifact of our presentation, and it disappears if a convolution is defined in an equivalent but for us more inconvenient manner.

For any generalized function $x_G = \{x_n\}$, continuity implies that

$$\delta * x_n \to \delta * x_G$$

But from the sifting theorem for smooth functions we have

$$\delta * x_n = x_n$$

and we obtain the sifting theorem for generalized functions,

$$\delta * x_G = x_G$$

The delta function is thus the unity of an algebra in which convolution plays the role of multiplication. It follows from the sifting theorem that an mth-order derivative can be expressed as a convolution with $\delta^{(m)}$:

$$x_G^{(m)} = \delta^{(m)} * x_G$$

From the continuity of the convolution, and the sifting theorem, it also follows that derivatives of generalized functions are given by the usual definition:

$$\lim_{T \to 0} \frac{x_G(t) - x_G(t-T)}{T} = \lim_{T \to 0} x_G(t) * \frac{\delta(t) - \delta(t-T)}{T}$$

$$= x_G(t) * \delta'(t) = x_G'(t)$$

2-8 LAPLACE TRANSFORMS OF GENERALIZED FUNCTIONS

To simplify some proofs, we will define the Laplace transform for generalized functions that are equal to the mth generalized derivative of a continuous function $\xi(t)$ such that $\xi(t<0) = 0$ and $\mathcal{L}(\xi)$ exists:

$$x_G = \xi_G^{(m)}$$

We will further require representation in *explicit form*:

$$x_G = \left\{ \delta_n^{(m)} * \xi \right\}$$

It turns out that these are not really limitations because it can be shown that generalized functions have Laplace transforms if and only if they can be put in explicit form. What is more, it can be shown that *all* generalized functions are *locally* equal to the mth generalized derivative of a continuous function. In any event, before proceeding we will show that the generalized functions that we have described so far can be put in explicit form.

Let us first point out that our definition (2-11) of the delta function can be put in explicit form with $\xi(t) = t\theta(t)$: an integration by parts yields

$$\delta_n = \delta_n'' * t\theta$$

More elegantly, remembering that $\psi\delta = \psi(0)\delta$ if ψ is smooth, we have

$$\delta'' * t\theta = (t\theta)'' = (t\delta + \theta)' = \theta' = \delta$$

If x is locally integrable and has a Laplace transform, the function

$$\xi(t) = \int_0^t x(u) \, du$$

is continuous and also has a Laplace transform. If ψ is smooth, $\psi'\xi$ can be integrated by parts:

$$\int_0^t \psi'(u) \, du \int_0^u x(v) \, dv = \int_0^t x(u) \, du \int_u^t \psi'(v) \, dv$$

$$= \psi(t) \int_0^t x(u) \, du - \int_0^t \psi(u) x(u) \, du$$

or in a more recognizable form,

$$\int_0^t \psi'(u) \xi(u) \, du = \psi(t) \xi(t) - \int_0^t \psi(u) x(u) \, du$$

Applying this result to $\delta_n * x$, we obtain

$$\delta_n * x = \delta_n' * \xi$$

The regular generalized function associated with x, defined in (2-12), can thus be put in explicit form.

If x_{G1} and x_{G2} are in explicit form, we have

$$x_{G1} * x_{G2} = \{\delta_n^{(m)} * \xi_1 \cdot \delta_n^{(l)} * \xi_2\} = \{(\delta_n * \delta_n)^{(l+m)} * \xi_1 * \xi_2\} = \{\delta_n^{(l+m)} * \xi_1 * \xi_2\}$$

so that $x_{G1} * x_{G2}$ can be put in explicit form with $\xi = \xi_1 * \xi_2$. We have used the fact that according to the sifting theorem, the delta function is also given by

$$\delta = \delta * \delta = \{\delta_n * \delta_n\}$$

The Laplace transform of a generalized function $x_G = \{x_n\}$ that can be put in explicit form is defined by

$$X(s) = \mathcal{L}(x_G) = \lim_{n \to \infty} \int_{-\infty}^{+\infty} x_n(t) \exp(-st) \, dt$$

Setting $x_n = \delta_n^{(m)} * \xi$, we have

$$\mathcal{L}(x_G) = \lim_{n \to \infty} \int_{-\infty}^{+\infty} \delta_n^{(m)} * \xi(t) \exp(-st) \, dt$$

$$= \mathcal{L}(\xi) \lim_{n \to \infty} \int_{-\infty}^{+\infty} \delta_n^{(m)}(t) \exp(-st) \, dt = s^m \mathcal{L}(\xi)$$

If the abscissa of convergence of $\mathcal{L}(\xi)$ is σ_C, the abscissa of convergence of $X(s)$ is at most σ_C. Several important results follow.

1. The Laplace transform of the delta function is equal to 1:

$$\mathcal{L}(\delta) = \lim_{n \to \infty} \int_{-\infty}^{+\infty} \delta_n(t) \exp(-st) \, dt = \exp(0) = 1$$

2. The Laplace transform of a derivative is obtained by multiplication by s:

$$\mathcal{L}(x_G') = s\mathcal{L}(x_G)$$

As an example, we note the consistent result that

$$\mathcal{L}(\delta) = \mathcal{L}(\theta') = s\mathcal{L}(\theta) = s(1/s) = 1$$

3. A locally integrable function and its associated generalized function have the same Laplace transform.

4. If x_{G1} and x_{G2} have Laplace transforms, the convolution $x_{G1}*x_{G2}$ exists and has a Laplace transform equal to the product of the Laplace transforms of x_{G1} and x_{G2}:

$$\mathcal{L}(x_{G1}*x_{G2}) = \mathcal{L}(x_{G1})\mathcal{L}(x_{G2})$$

We observe that $\mathcal{L}(\delta) = 1$ corresponds neatly to the fact that δ is the unity in the algebra of convolutions:

$$\mathcal{L}(\delta*x_G) = \mathcal{L}(\delta)\mathcal{L}(x_G) = \mathcal{L}(x_G)$$

Transforming both sides of differential equation (1-3), we obtain that the transfer function of a system governed by this differential equation is

$$H(s) = \frac{b_m s^m + \cdots + b_1 s + b_0}{a_n s^n + \cdots + a_1 s + a_0}$$

As suggested in Section 2-4, $H(s)$ is formally equal to the frequency response (1-12), and again this is true because of the way that derivatives transform. It should now be clear that adding the step responses of an expansion in partial fractions, as we did in Section 1-4, is justified by the linearity of the inverse Laplace transform.

As an example, let us use Laplace transforms to obtain the step response $r(t)$ of a phase lag. We must solve

$$\tau r' + r = \theta$$

Transforming both sides yields

$$(\tau s+1)R(s) = \frac{1}{s}$$

and thus

$$R(s) = \frac{1}{s(1+\tau s)} = \frac{1}{s} - \frac{\tau}{1+\tau s}$$

Taking inverse Laplace transforms using (2-2) and (2-3), we arrive at the answer we already know, given in (1-13):

$$r(t) = \theta(t)[1-\exp(-t/\tau)]$$

Let us check this result. The generalized derivative of $\tau r(t)$ is

$$\tau r'(t) = \tau\delta(t)[1-\exp(-t/\tau)] + \theta(t)\exp(-t/\tau) = \theta(t)\exp(-t/\tau)$$

so that $r(t)+\tau r'(t)$ is in fact equal to $\theta(t)$. Similarly, we can calculate the step response of a phase lead. We have

$$\tau r'+r = \tau\theta' = \tau\delta$$

so that

$$R(s) = \frac{\tau}{1+\tau s}$$

and we recover (1-16),

$$r(t) = \theta(t)\exp(-t/\tau)$$

Again we can check the result. We have

$$\tau r'(t) = \tau\delta(t)\exp(-t/\tau) - \theta(t)\exp(-t/\tau) = \tau\delta(t) - \theta(t)\exp(-t/\tau)$$

so that $\tau r'$ and r add up to $\tau\delta$.

2-9   IMPULSE RESPONSE AND CONVOLUTION

Physically, the impulse response $h(t)$ of a linear system $\mathcal{O}$ is the response to a pulse that has unit area and a width much smaller than the response time of the system. Mathematically, we will define $h(t)$ to be the response to a delta function, that is, to an impulse of unit area. Let us consider a delay

$$x(t) = y(t-\tau)$$

Its impulse response is

$$h(t) = \delta(t-\tau)$$

On the other hand, for an input $y(t)$ with Laplace transform $Y(s)$ we have

$$X(s) = \mathcal{L}[y(t-\tau)] = \exp(-\tau s)Y(s)$$

The transfer function of a delay by τ is thus

$$H(s) = \exp(-\tau s)$$

We note that our choice of notation is self-fulfilling because

$$H(s) = \mathcal{L}(h) = \exp(-\tau s)$$

That is, the transfer function $H(s)$ is the Laplace transform of the impulse response $h(t)$. A moment's consideration shows that this is also true for systems governed by differential equation (1-3). Let us see if this idea carries any further.

Let us consider a smooth input $y(t)$ to a linear system with impulse response $h(t)$. Looking at Fig. 2-4, it seems plausible that the response $x(T)$ at a given time T is the limit of the sum of the responses $h(T-u)y(u)\Delta u$ to impulses of area $y(u)\Delta u$, where u ranges from $-\infty$ to T. We might thus expect that

$$x(T) = \int_{-\infty}^{T} h(T-u)y(u)\,du = h*y(T)$$

that is, that the output of a linear system is the convolution of the input and the impulse response. We will now argue that this theorem, also called the *convolution* theorem, holds under broad conditions for continuous linear systems that operate on generalized functions. We note that the convolution

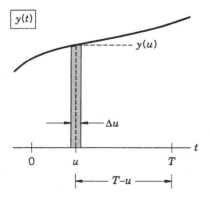

Figure 2-4. A narrow pulse of height $y(u)$ and width Δu can be replaced approximately by an impulse of area $y(u)\Delta u$.

theorem holds trivially for delays and, at least for Laplace-transformable inputs, for systems governed by differential equation (1-3). To simplify our notation, we will not distinguish generalized functions with a subscript.

Let us thus consider a (causal time-invariant) linear system $\mathcal{O}$ and a sequence of inputs y_n such that $y_n(t<0) = 0$, and let us assume that y_n converges to an input y,

$$y_n \to y$$

We will then say that $\mathcal{O}$ is *continuous* if the sequence of responses $x_n = \mathcal{O}[y_n]$ converges to a limit x,

$$x_n \to x$$

and

$$x = \mathcal{O}[y]$$

Delays are clearly continuous, and as we will see below, so are linear systems governed by differential equation (1-3).

Let us now consider a continuous linear system $\mathcal{O}$ with impulse response $h(t)$. For $\tau > 0$, the sequence of inputs

$$\delta_n^- = \frac{n}{\tau}\left[\delta(t) - \delta\left(t - \frac{\tau}{n}\right)\right]$$

converges to δ'. By superposition and time invariance, the sequence of responses to δ_n^- is

$$h_n^- = \frac{n}{\tau}\left[h(t) - h\left(t - \frac{\tau}{n}\right)\right]$$

This sequence converges to h' and, since $\mathcal{O}$ is continuous, to $\mathcal{O}[\delta']$. Repeating this procedure, we obtain

$$\mathcal{O}[\delta^{(m)}] = h^{(m)}$$

Let us further consider an input y of the form

$$y = \xi^{(m)} = \delta^{(m)} * \xi$$

where $\xi(t)$ is continuous, and zero for $t < 0$, and for $\tau > 0$ let us define

$$y_n(t) = \frac{\tau}{n} \sum_{j=1}^{n^2} \xi\left(j\frac{\tau}{n}\right)\delta^{(m)}\left(t - j\frac{\tau}{n}\right)$$

For any test function ϕ we have

$$\langle y_n, \phi \rangle = (-1)^m \frac{\tau}{n} \sum_{j=1}^{n^2} \xi\left(j\frac{\tau}{n}\right)\phi^{(m)}\left(j\frac{\tau}{n}\right)$$

The right-hand side of this equation tends to the integral of $(-1)^m \xi(t)\phi^{(m)}(t)$ because ϕ is zero outside a bounded interval, and y_n thus converges to y:

$$\langle y_n, \phi \rangle \to (-1)^m \int_{-\infty}^{+\infty} \xi(t)\phi^{(m)}(t)\, dt = (-1)^m \langle \xi, \phi^{(m)} \rangle = \langle y, \phi \rangle$$

By superposition and time invariance, the responses to y_n are

$$x_n(t) = \frac{T}{n} \sum_{j=1}^{n^2} \xi\left(j\frac{T}{n}\right) h^{(m)}\left(t - j\frac{T}{n}\right)$$

Using the fact that $h^{(m)} = h * \delta^{(m)}$, we can write x_n as a convolution:

$$x_n(t) = h(t) * \left[\frac{T}{n} \sum_{j=1}^{n^2} \xi\left(j\frac{T}{n}\right) \delta^{(m)}\left(t - j\frac{T}{n}\right) \right] = h * y_n(t)$$

The continuity of $\mathcal{O}$ implies that x_n converges to a limit x and that $x = \mathcal{O}[y]$. But on the other hand, the continuity of the convolution implies that $h * y_n$ converges to $h * y$, and we obtain the convolution theorem,

$$x = h * y$$

More generally, we will accept that this theorem holds for arbitrary inputs as long as the convolution makes sense. If $\mathcal{L}(h)$ and $\mathcal{L}(y)$ exist, we obtain

$$X(s) = \mathcal{L}(h)Y(s)$$

and we conclude, again in self-fulfilling notation, that the transfer function $H(s)$, when it exists, is the Laplace transform of the impulse response,

$$H(s) = \mathcal{L}(h)$$

The fact that $h * y_n \to x$ if $y_n \to y$ justifies the procedure we used in Section 1-4 to obtain the step response of a phase lead. Since $x' = h * y'$, we would also have been justified in obtaining it as τ times the derivative of the step response of a phase lag with the same time constant τ.

As a converse to the convolution theorem, it is clear that any causal linear system in which the response is obtained by convolving the input and the impulse response is continuous. For arbitrary inputs, not necessarily Laplace-transformable, it is easy to show that $x = h * y$ is a (causal) solution of differential equation (1-3). It follows that systems governed by this differential equation are continuous if we accept, as is the case, that causal solutions are unique.

To connect with what we already know about the frequency response, let us see what happens if $H(s)$ exists for $\sigma = 0$ and the input is sinusoidal,

$$y(t) = \exp(i\omega t)$$

Under these conditions the convolution is well defined even though $y(t)$ is not zero anywhere, and

$$x(t) = h(t) * \exp(i\omega t) = \exp(i\omega t)\left\{\int_{-\infty}^{+\infty} \exp(-i\omega u)h_n(u)\,du\right\}$$

The quantity in braces—it is actually a generalized function of $i\omega$—is independent of t. Assuming that h_n is in explicit form, for any test function $\phi(t)$ we have

$$\langle x, \phi \rangle = \langle \exp(i\omega t), \phi \rangle \lim_{n\to\infty} \int_{-\infty}^{+\infty} \exp(-i\omega u)h_n(u)\,du = \langle \exp(i\omega t), \phi \rangle H(i\omega)$$

It follows that we can identify $x(t)$ with $H(i\omega)\exp(i\omega t)$:

$$x(t) = H(i\omega)\exp(i\omega t)$$

We conclude that the frequency response is equal to $H(i\omega)$ and therefore, as we will soon see, that it is the *Fourier transform* of the impulse response. As an example, let is do the calculation explicitly for a phase lead

$$\tau x' + x = \tau y'$$

with transfer function

$$H(s) = \frac{\tau s}{1 + \tau s}$$

The impulse response can be obtained as the derivative of the step response,

$$h(t) = r'(t) = \left[\theta(t)\exp\left(-\frac{t}{\tau}\right)\right]' = \delta(t) - \frac{1}{\tau}\theta(t)\exp\left(-\frac{t}{\tau}\right)$$

This time we will take a sinusoidal input starting at $t = 0$,

$$y(t) = \theta(t)\exp(i\omega t)$$

We then have

$$x(t) = h*y(t) = \theta(t)\exp(i\omega t) - \frac{1}{\tau}\theta(t)\exp(i\omega t)\int_0^t \exp\left(-\frac{u}{\tau} - i\omega u\right)du$$

Performing the integral, we obtain

$$x(t) = \theta(t)\left[\frac{i\omega\tau}{1 + i\omega\tau}\exp(i\omega t) + \frac{1}{1 + i\omega\tau}\exp\left(-\frac{t}{\tau}\right)\right]$$

The real exponential decays to zero, so that for $t \to \infty$,

$$x(t) \to \frac{i\omega\tau}{1+i\omega\tau} \exp(i\omega t) = H(i\omega)\exp(i\omega t)$$

This agrees with what we obtained for stable systems in Section 1-3.

2-10 FOURIER TRANSFORMS

In Section 1-6 we argued that waveforms have a frequency content and that they can be represented as a superposition of sine waves of different frequencies. We will now show that Fourier transforms give us such a representation. Our treatment will be aimed mainly at Shannon's sampling theorem in Section 2-11 and at our needs in Chapters 12 and 13, but we will nonetheless obtain results of great interest. We will deal only briefly with the transforms of ordinary functions because the powerful theorems, as we have seen, apply to generalized functions.

Let us start by considering an absolutely integrable piecewise smooth function $x(t)$. We will define the (direct) *Fourier transform* $\mathcal{F}(x)$ of $x(t)$ as the function $\hat{x}(\omega)$ given by

$$\hat{x}(\omega) = \mathcal{F}(x) = \int_{-\infty}^{+\infty} x(t)\exp(-i\omega t)\, dt$$

In Exercise 2-18 it is shown that $\hat{x}(\omega)$ has an inverse $\mathcal{F}^{-1}(\hat{x})$, and that except at its points of discontinuity, $x(t)$ is given by the *inversion formula*,

$$x(t) = \mathcal{F}^{-1}(\hat{x}) = \frac{1}{2\pi}\int_{-\infty}^{+\infty} \hat{x}(\omega)\exp(i\omega t)\, d\omega \qquad (2\text{-}13)$$

The Fourier transform $\hat{x}(\omega)$ is thus precisely the frequency content of $x(t)$ or, more formally, the *amplitude spectrum* of $x(t)$. As an example, let us calculate the Fourier transform of the square pulse

$$x_T(t) = \theta(t) - \theta(t-T)$$

We have

$$\hat{x}_T(\omega) = \int_0^T \exp(-i\omega t)\, dt = T \cdot \frac{\sin(\omega T/2)}{(\omega T/2)} \exp(-i\omega T/2)$$

But this is what we obtained in (1-19) by guessing that the Fourier transform of $\theta(t)$ was equal to $1/i\omega$. As shown in Exercise 2-20, our guess was actually rather good.

Observing that for $\lambda > 0$ we have

$$\int_{-\infty}^{+\infty} \exp(-\lambda|t|) \exp(-i\omega t)\, dt = \int_{0}^{\infty} \exp(-\lambda t)\left[\exp(i\omega t) + \exp(-i\omega t)\right] dt$$

$$= \frac{1}{\lambda - i\omega} + \frac{1}{\lambda + i\omega}$$

we obtain a transform we will need in Chapter 12,

$$\mathscr{F}[\exp(-\lambda|t|)] = \frac{2\lambda}{\lambda^2 + \omega^2} \tag{2-14}$$

As a final example, we quote the result of Exercise 2-17,

$$\mathscr{F}\left[\frac{1}{\sqrt{2\pi}\sigma} \exp\left(-\frac{t^2}{2\sigma^2}\right)\right] = \exp\left(-\frac{\sigma^2 \omega^2}{2}\right)$$

In all the examples so far we observe, in agreement with our discussion in Section 1-6, that $|\hat{x}(\omega)|$ is broad if $|x(t)|$ is narrow, and vice versa.

2-10-1   Functions of Rapid Descent and Generalized Functions of Slow Growth

To get simple results for Fourier transforms—and with no loss for our purposes—we will restrict generalized functions somewhat by widening our notion of a test function. We will thus define a *function of rapid descent* as a smooth function $\rho(t)$ such that $\rho(t)$ and its derivatives of any order go to zero faster than any power of t as $|t|$ goes to infinity; for nonnegative m and n we thus require

$$\lim_{|t| \to \infty} t^m \rho^{(n)}(t) = 0$$

The test functions defined in Section 2-5 are zero outside a bounded interval and thus clearly of rapid descent. More to the point, $\exp(-\lambda^2 t^2)$ is of rapid descent. As shown in Exercise 2-19, if $\rho(t)$ is of rapid descent, then so are $t^m \rho(t)$ and $\rho^{(n)}(t)$.

We now want to show that $\hat{\rho}(\omega)$ is of rapid descent. We first observe that

$$\hat{\rho}^{(n)}(\omega) = \int_{-\infty}^{+\infty} (-it)^n \rho(t) \exp(-i\omega t)\, dt \tag{2-15}$$

because the derivatives of $\rho(t)\exp(-i\omega t)$ with respect to ω are continuous and well behaved for $|t| \to \infty$, so that differentiation under the integral sign is allowed. Integrating by parts n times, we also obtain

$$(i\omega)^n \hat{\rho}(\omega) = \int_{-\infty}^{+\infty} \rho^{(n)}(t) \exp(-i\omega t)\, dt \tag{2-16}$$

We thus have

$$(i\omega)^m \hat{\rho}^{(n)}(\omega) = \int_{-\infty}^{+\infty} [(-it)^n \rho(t)]^{(m)} \exp(-i\omega t)\, dt$$

But the function multiplying $\exp(-i\omega t)$ in the integrand is of rapid descent and therefore absolutely integrable, so that according to the Riemann–Lebesgue lemma, the integral goes to zero as $|\omega|$ goes to infinity. It is thus clear that $\hat{\rho}(\omega)$ is of rapid descent.

The inversion formula holds everywhere for functions of rapid descent. It follows that $\rho_1(t) \neq \rho_2(t)$ implies that $\hat{\rho}_1(\omega) \neq \hat{\rho}_2(\omega)$. Since the arguments we have used can be applied to the inverse Fourier transform, it also follows that there is an *isomorphism* or one-to-one correspondence between the set of functions of rapid descent and the set of direct or inverse Fourier transforms of functions of rapid descent. Put differently, any function of rapid descent is the direct Fourier transform of a unique function of rapid descent, and it is also the inverse Fourier transform of a unique function of rapid descent.

If we denote the inverse Fourier transform of a function of rapid descent $\rho(t)$ by $\bar{\rho}(\omega)$, we have

$$\bar{\rho}(\omega) = \mathcal{F}^{-1}(\rho) = \frac{1}{2\pi} \int_{-\infty}^{+\infty} \rho(t) \exp(i\omega t)\, dt$$

From (2-15) and (2-16), and from brief calculations, we obtain the following properties of the direct and inverse Fourier transforms of functions of rapid descent:

$$\mathcal{F}[\rho^{(n)}(t)] = (i\omega)^n \hat{\rho}(\omega) \qquad \mathcal{F}^{-1}[\rho^{(n)}(t)] = (-i\omega)^n \bar{\rho}(\omega)$$

$$\mathcal{F}[(-it)^n \rho(t)] = \hat{\rho}^{(n)}(\omega) \qquad \mathcal{F}^{-1}[(it)^n \rho(t)] = \bar{\rho}^{(n)}(\omega)$$

$$\mathcal{F}[\rho(t-\tau)] = \exp(-i\omega\tau)\hat{\rho}(\omega) \qquad \mathcal{F}^{-1}[\rho(t-\tau)] = \exp(i\omega\tau)\bar{\rho}(\omega)$$

$$\mathcal{F}[\rho(t)\exp(-i\lambda t)] = \hat{\rho}(\omega+\lambda) \qquad \mathcal{F}^{-1}[\rho(t)\exp(i\lambda t)] = \bar{\rho}(\omega+\lambda)$$

$$(2\text{-}17)$$

A generalized function $x(t)$ is of *slow growth* if the functional $\langle x, \rho \rangle$ exists for all functions of rapid descent $\rho(t)$. It follows that the derivatives of a generalized function of slow growth are also of slow growth because

$$\langle x', \rho \rangle = -\langle x, \rho' \rangle$$

and ρ' is of rapid descent if ρ is of rapid descent.

As a simple example, let us consider a regular generalized function $x(t)$ such that

$$|x(t)| < K(1+|t|^m)$$

Since $|x(t)\rho(t)| < K(1+|t|^m)|\rho(t)|$ and the latter is integrable, it follows that $\langle x, \rho \rangle$ exists for any function $\rho(t)$ of rapid descent and that $x(t)$ is of slow growth. In contrast, it is clear that $\exp(\lambda t)$ is not of slow growth.

2-10-2   Fourier Transforms of Generalized Functions of Slow Growth

We will define the Fourier transform $\mathcal{F}(x)$ of a generalized function of slow growth $x(t)$ by

$$\langle \mathcal{F}(x), \rho \rangle = \langle x, \mathcal{F}(\rho) \rangle$$

It follows that $\mathcal{F}(x)$ is a generalized function of a slow growth because $\mathcal{F}(\rho)$ is of rapid descent. It also follows that $x_1 \neq x_2$ implies that $\mathcal{F}(x_1) \neq \mathcal{F}(x_2)$ because the set of Fourier transforms of functions of rapid descent is identical to the set of functions of rapid descent.

The Fourier transform of an ordinary absolutely integrable function $x(t)$ is, as consistency demands, a particular case of the generalized transform; we thus have

$$\langle \mathcal{F}(x), \rho \rangle = \int_{-\infty}^{+\infty} \rho(\omega)\, d\omega \int_{-\infty}^{+\infty} x(t) \exp(-i\omega t)\, dt$$

$$= \int_{-\infty}^{+\infty} x(t)\, dt \int_{-\infty}^{+\infty} \rho(\omega) \exp(-i\omega t)\, d\omega = \langle x, \mathcal{F}(\rho) \rangle$$

From the definition let us immediately obtain two important results. To begin with, we have

$$\mathcal{F}(\delta) = 1$$

because

$$\langle 1, \rho(\omega) \rangle = \int_{-\infty}^{+\infty} \rho(\omega)\, d\omega = \hat{\rho}(0) = \langle \delta(t), \hat{\rho}(t) \rangle$$

We also have

$$\mathcal{F}(1) = 2\pi\delta$$

because

$$\langle 2\pi\delta(\omega), \rho(\omega) \rangle = 2\pi\rho(0) = \int_{-\infty}^{+\infty} \hat{\rho}(t)\, dt = \langle 1, \hat{\rho}(t) \rangle$$

To connect with more informal discussions in the literature, we note that

$$\int_{-\infty}^{+\infty} \delta(t) \exp(\pm i\omega t)\, dt = 1$$

and that in practice, we can heuristically say that

$$\frac{1}{2\pi}\int_{-\infty}^{+\infty} \exp(\pm i\omega t)\, dt = \delta(\omega)$$

We will define the inverse Fourier transform $\mathcal{F}^{-1}(x)$ of a generalized function of slow growth by

$$\langle \mathcal{F}^{-1}(x), \rho \rangle = \langle x, \mathcal{F}^{-1}(\rho) \rangle$$

The inverse Fourier transform clearly exists and is of slow growth; it is also properly named because

$$\langle \mathcal{F}^{-1}[\mathcal{F}(x)], \rho \rangle = \langle \mathcal{F}(x),\ \mathcal{F}^{-1}(\rho) \rangle = \langle x, \mathcal{F}[\mathcal{F}^{-1}(\rho)] \rangle = \langle x, \rho \rangle$$

Different generalized functions of slow growth have different inverse Fourier transforms; we conclude, as we did in the case of functions of rapid descent, that the direct and inverse Fourier transforms establish isomorphisms on the set of generalized functions of slow growth.

A fundamental property of the Fourier transform is that it is continuous on the set of generalized functions of slow growth, meaning that if $x_n \to x$, then $\hat{x}_n \to \hat{x}$. This result follows by observing that

$$\langle \hat{x}_n, \rho \rangle = \langle x_n, \hat{\rho} \rangle \to \langle x, \hat{\rho} \rangle = \langle \hat{x}, \rho \rangle$$

This means, for example, that we can freely exchange the order in which we take Fourier transforms and infinite sums of generalized functions of slow growth.

The properties of Fourier transforms of functions of rapid descent given in (2-17) extend to the Fourier transforms of generalized functions of slow growth. Using $\bar{x}(\omega)$ to denote the inverse Fourier transform of $x(t)$, we have

$$\mathcal{F}[x^{(n)}(t)] = (i\omega)^n \hat{x}(\omega) \qquad\qquad \mathcal{F}^{-1}[x^{(n)}(t)] = (-i\omega)^n \bar{x}(\omega)$$

$$\mathcal{F}[(-it)^n x(t)] = \hat{x}^{(n)}(\omega) \qquad\qquad \mathcal{F}^{-1}[(it)^n x(t)] = \bar{x}^{(n)}(\omega)$$

$$\mathcal{F}[x(t-\tau)] = \exp(-i\omega\tau)\hat{x}(\omega) \qquad \mathcal{F}^{-1}[x(t-\tau)] = \exp(i\omega\tau)\bar{x}(\omega)$$

$$\mathcal{F}[x(t)\exp(-i\lambda t)] = \hat{x}(\omega+\lambda) \qquad \mathcal{F}^{-1}[x(t)\exp(i\lambda t)] = \bar{x}(\omega+\lambda)$$

Working out one entry it will become clear how the extension is obtained:

$$\langle \mathcal{F}[x^{(n)}(t)], \rho(\omega) \rangle = \langle x^{(n)}(t), \hat{\rho}(t) \rangle = (-1)^n \langle x(t), \hat{\rho}^{(n)}(t) \rangle$$

$$= \langle x(t), (-1)^n \mathcal{F}[(-i\omega)^n \rho(\omega)] \rangle$$

$$= \langle \mathcal{F}[x(t)], (i\omega)^n \rho(\omega) \rangle$$

$$= \langle (i\omega)^n \hat{x}(\omega), \rho(\omega) \rangle$$

We can now obtain a result that will see much use in Chapters 12 and 13. We have

$$\mathcal{F}[\exp(\mp i\lambda t)] = 2\pi\delta(\omega \pm \lambda)$$

and thus

$$\mathcal{F}(\cos \lambda t) = \pi[\delta(\omega+\lambda) + \delta(\omega-\lambda)]$$

A generalized function $x(t)$ that is zero outside a bounded interval is *ipso facto* of slow growth. We will accept that its Fourier transform $\hat{x}(\omega)$ is smooth [a result that is quite easy to obtain if $x(t)$ is an ordinary piecewise smooth function], that $|\hat{x}(\omega)|$ grows no faster than some nonnegative power of $|\omega|$ for $|\omega| \to \infty$, and that

$$\hat{x}(\omega) = \langle x(t), \exp(-i\omega t)\rangle$$

If $x(t)$ is zero outside a bounded interval, it makes sense to define the products $\hat{x}\hat{y}$ and $\bar{x}\bar{y}$ because $\hat{x}$ and $\bar{x}$ are smooth, and we will also accept that

$$\mathcal{F}(x*y) = \hat{x}\hat{y}$$
$$\mathcal{F}^{-1}(x*y) = 2\pi\bar{x}\bar{y} \tag{2-18}$$

More generally, we will accept that these relations hold if both members make sense; under the same conditions, we also have

$$2\pi\mathcal{F}(xy) = \hat{x}*\hat{y}$$
$$\mathcal{F}^{-1}(xy) = \bar{x}*\bar{y} \tag{2-19}$$

2-10-3 Functions of Integrable Square

Let us now consider the important class of ordinary (complex) functions $x(t)$ of *integrable square*, that is, such that

$$\int_{-\infty}^{+\infty} |x(t)|^2 \, dt < \infty$$

We will accept that such functions are locally integrable; it is then clear that they are generalized functions of slow growth. We will also accept the *Parseval–Plancherel* theorem, which states that the direct and inverse Fourier transforms of a function of integrable square are also of integrable square (this implies yet another isomorphism), and that for any two functions $x(t)$ and $y(t)$ of integrable square we have

$$\int_{-\infty}^{+\infty} x(t)y^*(t) \, dt = \frac{1}{2\pi} \int_{-\infty}^{+\infty} \hat{x}(\omega)\hat{y}^*(\omega) \, d\omega$$

In particular, we have

$$\int_{-\infty}^{+\infty} |x(t)|^2 \, dt = \frac{1}{2\pi} \int_{-\infty}^{+\infty} |\hat{x}(\omega)|^2 \, d\omega$$

2-10-4   Alternative Expressions for the Transforms

Fourier transforms can be expressed in terms of the frequency f rather than the angular frequency $\omega = 2\pi f$; we will use this alternative when convenient. The direct transform is then given by

$$\hat{x}(f) = \int_{-\infty}^{+\infty} x(t) \exp(-i2\pi ft)\, dt$$

and the inversion formula (2-13) becomes

$$x(t) = \int_{-\infty}^{+\infty} \hat{x}(t) \exp(+i2\pi ft)\, df$$

To account for the effect of this reformulation on the direct transforms we have obtained so far, we need only keep in mind that

$$\delta(\omega) = \delta(2\pi f) = \frac{1}{2\pi}\delta(f)$$

We note two cases of interest,

$$\mathcal{F}(1) = \delta(f)$$
$$\mathcal{F}[\cos(2\pi f_0 t)] = \tfrac{1}{2}[\delta(f+f_0) + \delta(f-f_0)]$$

(2-20)

We also note that the formulas in (2-18) and (2-19) become

$$\mathcal{F}(x*y) = \hat{x}\hat{y} \qquad \mathcal{F}(xy) = \hat{x}*\hat{y}$$
$$\mathcal{F}^{-1}(x*y) = \bar{x}\bar{y} \qquad \mathcal{F}^{-1}(xy) = \bar{x}*\bar{y}$$

and that the Parseval–Plancherel theorem reads

$$\int_{-\infty}^{+\infty} x(t)y^*(t)\, dt = \int_{-\infty}^{+\infty} \hat{x}(f)\hat{y}^*(f)\, df$$

2-11   SHANNON'S SAMPLING THEOREM

In many electronic systems a signal $x(t)$ is sampled at a uniform rate, and the resulting sampled signal can be written in the form[†]

$$T \sum_n x(nT)\, \delta(t-nT)$$

[†] Unless otherwise indicated, sums in this section are from $n = -\infty$ to $n = +\infty$.

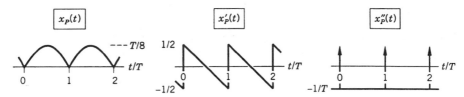

Figure 2-5. The periodic function given by $x_P(t) = (T/2)(t/T - t^2/T^2)$ in the interval $0 \le t \le T$, and its first two generalized derivatives.

We will now show that $x(t)$ can be fully recovered if $\hat{x}(\omega)$ is bandlimited, that is, if $\hat{x}(\omega)$ is zero for $|\omega| > \omega_0$, and $x(t)$ is therefore smooth. Let us start by considering the function $x_P(t)$ given by

$$x_P(t) = \frac{T}{12} - \sum_{n \neq 0} \frac{T}{(n\lambda T)^2} \exp(in\lambda t)$$

where $\lambda = 2\pi/T$. As shown in Exercise 2-21, this infinite trigonometric series converges in the ordinary sense to the continuous periodic function shown in Fig. 2-5 and given for $0 \le t < T$ by

$$x_P(t) = \frac{T}{2}\left(\frac{t}{T} - \frac{t^2}{T^2}\right)$$

Taking generalized derivatives of the two expressions for $x_P(t)$, we obtain

$$-\frac{1}{T} + \sum_n \delta(t - nT) = \frac{1}{T} \sum_{n \neq 0} \exp(in\lambda t)$$

and thus

$$\sum_n \delta(t - nT) = \frac{1}{T} \sum_n \exp(in\lambda t)$$

If we take Fourier transforms on both sides of this result, we finally obtain

$$\mathscr{F}\left[\sum_n \delta(t - nT)\right] = \lambda \sum_n \delta(\omega - n\lambda) \qquad (2\text{-}21)$$

With these preliminaries, let us consider the sampled signal corresponding to $x(t)$. Recalling that $x(t)$ is smooth, we can write

$$T \sum_n x(nT)\delta(t - nT) = \left[T \sum_n \delta(t - nT)\right] x(t)$$

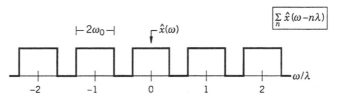

Figure 2-6. Fourier transform (assumed real for simplicity) of a sampled waveform that satisfies the Nyquist criterion for sampling.

Taking the Fourier transform of the right-hand side and using (2-19), which is allowed because we are assuming that $\hat{x}(\omega)$ is zero outside a bounded interval, and then using (2-21), we get

$$\mathcal{F}\left\{\left[T\sum_n \delta(t-nT)\right]x(t)\right\} = \left[\sum_n \delta(\omega-n\lambda)\right] * \hat{x}(\omega) = \sum_n \hat{x}(\omega-n\lambda) \quad (2\text{-}22)$$

Let us now assume that λ satisfies the *Nyquist criterion* for sampling, meaning that

$$\lambda > 2\omega_0$$

As shown in Fig. 2-6, the transforms $\hat{x}(\omega-n\lambda)$ in the last term in (2-22) do not overlap in this case, and we can therefore multiply the last two terms in (2-22) by $\theta(\omega+\omega_0)-\theta(\omega-\omega_0)$ to obtain

$$\hat{x}(\omega) = [\theta(\omega+\omega_0)-\theta(\omega-\omega_0)]\left\{\left[\sum_n \delta(\omega-n\lambda)\right] * \hat{x}(\omega)\right\} \quad (2\text{-}23)$$

We note that the product makes sense because $\theta(\omega+\omega_0)-\theta(\omega-\omega_0)$ is smooth where (2-22) is nonzero. Using (2-19) we obtain, at least formally,

$$x(t) = \mathcal{F}^{-1}[\theta(\omega+\omega_0)-\theta(\omega-\omega_0)] * \mathcal{F}^{-1}\left\{\left[\sum_n \delta(\omega-n\lambda)\right] * \hat{x}(\omega)\right\}$$

and after a brief calculation,

$$x(t) = \frac{\sin(\omega_0 t)}{\pi t} * \mathcal{F}^{-1}\left\{\left[\sum_n \delta(\omega-n\lambda)\right] * \hat{x}(\omega)\right\} \quad (2\text{-}24)$$

From the first two terms in (2-22) we then obtain

$$x(t) = \frac{\sin(\omega_0 t)}{\pi t} * \left\{\left[T\sum_n \delta(t-nT)\right]x(t)\right\}$$

$$= \frac{\sin(\omega_0 t)}{\pi t} * \left[\frac{2\pi}{\lambda}\sum_n x(nT)\delta(t-nT)\right]$$

and finally

$$x(t) = \frac{2\pi}{\lambda} \sum_n x(nT) \frac{\sin[\omega_0(t-nT)]}{\pi(t-nT)}$$

We will accept that this expression makes sense because $|x(t)|$ grows no faster than some nonnegative power of $|t|$ for $|t| \to \infty$, and that our use of (2-19) to obtain (2-24) is therefore justified. We have thus recovered $x(t)$ in terms of its sampled values. In particular, if we go to *critical* sampling, that is, if we set

$$\lambda = 2\omega_0$$

we obtain the *interpolation formula,*

$$x(t) = \sum_n x(nT) \frac{\sin(\omega_0 t - n\pi)}{\omega_0 t - n\pi}$$

As discussed in Exercise 2-23, $x(t)$ can be recovered in practice by applying a low-pass filter with bandwidth $\lambda/2$ to the function

$$\sum_n x(nT)\{\theta(t-nT) - \theta[t-(n+1)T]\}$$

Recovery can be almost perfect if $x(t)$ is heavily *oversampled*, that is, if λ is made much larger than $2\omega_0$.

If $x(t)$ is *undersampled* by making λ less than $2\omega_0$, the transforms $\hat{x}(\omega - n\lambda)$ in the last term in (2-22) overlap, and the result is *aliasing*, a phenomenon in which high-frequency components of $x(t)$ are misconstrued as components at lower frequencies. Aliasing is made evident in a simple case by sampling a sine wave of frequency ω at a frequency λ just slightly under ω; the result is as if a sine wave of the much lower frequency $\omega - \lambda$ had been sampled.

EXERCISES

2-1 ASSOCIATIVITY OF THE CONVOLUTION. Assuming that $x(t)$, $y(t)$, and $z(t)$ are locally integrable, and zero for $t < 0$, show that

$$x*(y*z)(t) = \int_{-\infty}^{+\infty} \int_{-\infty}^{+\infty} du\, dv\, x(u) z(v) y[t-(u+v)]$$

where the integral is over the domain defined by $u>0$, $v>0$, and $u+v<t$. From the commutativity of the convolution, deduce that

$$x*(y*z) = (x*y)*z$$

2-2 PROPERTIES OF THE LAPLACE TRANSFORM. Assume that $x(t<0) = 0$ and that $X(s)$ has an abscissa of convergence σ_C.

(a) As you saw in Section 2-8, integration by parts is valid for the integrand

$$\psi'(t) \int_0^t x(u) \, du$$

if ψ is smooth and x is locally integrable. Using this fact, prove that

$$\mathcal{L}\left[\int_0^t x(u) \, du \right] = \frac{1}{s} X(s)$$

If $\sigma_C \leq 0$, the abscissa of convergence is at most 0. Use the fact that

$$\int_0^t |x(u)| \exp(-\sigma u) \, du > \exp(-\sigma t) \int_0^t |x(u)| \, du$$

for $\sigma > 0$ to show that it is at most σ_C if $\sigma_C > 0$.

(b) Prove that

$$\mathcal{L}[x(t-\tau)] = \exp(-\tau s) X(s)$$

For $\tau < 0$, $x(t)$ must be zero for $t < -\tau$.

(c) Prove that

$$\mathcal{L}[\exp(s_0 t) x(t)] = X(s-s_0)$$

If σ_C is finite, the abscissa of convergence is $\sigma_C + \sigma_0$.

2-3 LAPLACE TRANSFORM OF A CONVOLUTION. With the aid of Fubini's theorem, show that

$$\mathcal{L}(x_1 * x_2) = X_1(s) X_2(s)$$

if $\sigma > \max\{\sigma_{C1}, \sigma_{C2}\}$ by following through the indicated steps:

$$\mathcal{L}(x_1 * x_2) = \int_0^\infty \exp(-st) \, dt \int_0^t x_1(u) x_2(t-u) \, du$$

$$= \int_0^\infty x_1(u) \, du \int_u^\infty \exp(-st) x_2(t-u) \, dt$$

$$= \int_0^\infty x_1(u) \exp(-su) \, du \int_0^\infty x_2(t) \exp(-st) \, dt$$

$$= X_1(s) X_2(s)$$

2-4   TABLE OF LAPLACE TRANSFORMS. Starting from

$$\mathcal{L}[\theta(t)] = \frac{1}{s}$$

and using theorems (2-7) and (2-8), obtain

$$\mathcal{L}[\theta(t)t^n] = \frac{n!}{s^{n+1}} \qquad \mathcal{L}[\theta(t)\exp(\pm i\lambda t)] = \frac{1}{s \mp i\lambda}$$

From the expressions for $\mathcal{L}[\theta(t)\exp(\pm i\lambda t)]$, obtain

$$\mathcal{L}[\theta(t)\cos(\lambda t)] = \frac{s}{s^2+\lambda^2} \qquad \mathcal{L}[\theta(t)\sin(\lambda t)] = \frac{\lambda}{s^2+\lambda^2}$$

Finally, use (2-7) to obtain

$$\mathcal{L}[\theta(t)\exp(-\omega_1 t)\cos(\omega_2 t)] = \frac{s+\omega_1}{(s+\omega_1)^2+\omega_2^2}$$

$$\mathcal{L}[\theta(t)\exp(-\omega_1 t)\sin(\omega_2 t)] = \frac{\omega_2}{(s+\omega_1)^2+\omega_2^2}$$

2-5   EXCITATION AT RESONANCE. Consider an undamped resonator with transfer function

$$H(s) = \frac{\omega_0 s}{s^2+\omega_0^2}$$

Using Laplace transforms, show that the response to $y(t) = \theta(t)\exp(i\omega_0 t)$ is

$$x(t) = \tfrac{1}{2}\theta(t)[\omega_0 t \exp(i\omega_0 t) + \sin(\omega_0 t)]$$

Obtain the same result by solving for an excitation $y(t) = \theta(t)\exp(i\lambda t)$ and applying L'Hôpital's rule for $\lambda \to \omega_0$. This is one more example of continuity in action.

2-6   AN ALTERNATIVE LAPLACE TRANSFORM. In many engineering textbooks the Laplace transform presented is the so-called $\mathcal{L}_+$ transform defined by

$$\mathcal{L}_+(x) = \int_{0+}^{\infty} x(t)\exp(-st)\,dt$$

This transform is useful for solving linear differential equations when the initial conditions are specified, but it has the drawback that impulses at $t = 0$

are ignored, so that the automatic generation of the initial conditions for step or impulse responses is lost.

For locally integrable functions it is clear that

$$\mathbf{L}_+(x) = \mathbf{L}(x)$$

For functions that are smooth for $t>0$, show that

$$\mathbf{L}_+(x') = s\mathbf{L}_+(x) - x(0+)$$

To see where trouble starts, consider the transform of $\theta'(t)$,

$$\mathbf{L}_+(\theta') = s(1/s) - 1 = 0$$

or directly from the definition,

$$\mathbf{L}_+(\theta') = \int_{0+}^{\infty} 0 \cdot \exp(-st)\, dt = 0$$

This is *not* the transform of the delta function. However, if you apply $\mathbf{L}_+$ to a resonator and feed in the initial conditions

$$r(0+) = 0 \quad \text{and} \quad \tau_0 r'(0+) = 1$$

you will obtain the correct step response:

$$R_+(s) = \frac{\tau_0}{1 + 2\zeta\tau_0 s + \tau_0^2 s^2} = R(s)$$

2-7 THE INITIAL-VALUE AND FINAL-VALUE THEOREMS.

To see where the initial-value theorem (2-9) and the final-value theorem (2-10) come from, consider

$$\sigma X(\sigma) = \sigma \int_0^{\infty} x(t) \exp(-\sigma t)\, dt = \int_0^{\infty} x\left(\frac{u}{\sigma}\right) \exp(-u)\, du$$

The integrand in the last integral tends to $x(0+)\exp(-u)$ for $\sigma \to +\infty$, and you can thus expect that if $x(0+)$ exist you can pull it out of the integral. For $\sigma \to 0+$, the integrand tends to $x(+\infty)\exp(-u)$, and similar considerations hold.

The final-value theorem demands the existence of the limit at $t = +\infty$. A simple function that does not satisfy this condition is $\theta(t)\exp(i\lambda t)$. Note, however, that the limit for $\sigma \to 0+$ does exist,

$$\lim_{\sigma \to 0+} \sigma X(\sigma) = \lim_{\sigma \to 0+} \frac{\sigma}{\sigma - i\lambda} = 0$$

Also note that the initial value $\exp(i\lambda 0+) = 1$ is correctly given by

$$\lim_{\sigma \to +\infty} \sigma X(\sigma) = 1$$

As another example, consider the square wave

$$x(t) = \theta(t) - \theta(t-\tau) + \theta(t-2\tau) - \theta(t-3\tau) + \cdots$$

Its Laplace transform is

$$X(s) = \frac{1}{s}[1 - \exp(-\tau s) + \exp(-2\tau s) - \exp(-3\tau s) + \cdots] = \frac{1}{s[1+\exp(-\tau s)]}$$

Again, the limit at $t = +\infty$ does not exist even though the limit for $\sigma \to 0+$ does exist,

$$\lim_{\sigma \to 0+} \sigma X(\sigma) = \tfrac{1}{2}$$

In contrast, verify the correct result for the initial value,

$$\lim_{\sigma \to +\infty} \sigma X(\sigma) = 1 = x(0+)$$

An example of a function that does not satisfy the initial-value theorem is $\theta(t)\sin(\tau/t)$, which has no limit at $t = 0+$, but for which

$$\lim_{\sigma \to +\infty} \sigma X(\sigma) = 0$$

To see this last point, use the substitution $t = 1/\sigma u$ to obtain

$$\lim_{\sigma \to +\infty} \sigma X(\sigma) = \lim_{\sigma \to +\infty} \int_0^\infty \frac{1}{u^2} \exp\left(-\frac{1}{u}\right) \sin(\sigma \tau u)\, du$$

It follows from the Riemann–Lebesgue lemma that the limit is zero.

2-8   INTEGRALS OF GENERALIZED FUNCTIONS EXIST. Consider a generalized function $x = \{x_n\}$ such that $x_n(t<\tau) = 0$ for some fixed τ, and define $y = \int x$ by

$$y = \left\{ \int_{-\infty}^t x_n(u)\, du \right\}$$

Exchanging the order of integration, show that for any test function ϕ,

$$\langle y, \phi \rangle = \langle x, \chi \rangle$$

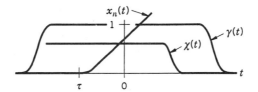

Figure 2-7. The smooth functions $x_n(t)$, $\chi(t)$, and $\gamma(t)$ defined in Exercise 2-8; the product $\chi(t)\gamma(t)$ is a test function.

where

$$\chi(t) = \int_t^\infty \phi(u)\, du$$

As indicated in Fig. 2-7, $\langle x, \chi \rangle$ is not modified if $\chi(t)$ is multiplied by a smooth function $\gamma(t)$ that is equal to 1 wherever $x_n(t)\chi(t)$ differs from zero. The product $\gamma(t)\chi(t)$ is clearly a test function, and $\langle \int x, \phi \rangle$ therefore exists for all ϕ. You can obtain $\gamma(t)$ by convolving a sufficiently wide square pulse and one of the functions $\delta_n(t)$ that define the delta function in (2-11).

2-9 INTEGRALS OF GENERALIZED FUNCTIONS DIFFER BY A CONSTANT. Consider a generalized function x such that $x' = 0$, that is, such that

$$\langle x', \phi \rangle = 0$$

for all test functions ϕ. Choose a test function ϕ_0 such that

$$\int_{-\infty}^{+\infty} \phi_0(u)\, du = 1$$

Any test function ϕ can then be rewritten as

$$\phi = \phi_0 \int_{-\infty}^{+\infty} \phi(u)\, du + \chi$$

where

$$\chi = \phi - \phi_0 \int_{-\infty}^{+\infty} \phi(u)\, du$$

It follows that

$$\int_{-\infty}^{+\infty} \chi(u)\, du = 0$$

and that

$$\zeta(t) = \int_t^\infty \chi(u)\, du$$

is a test function because it is smooth, and zero outside a bounded interval, in fact, the same bounded interval outside of which χ is zero. Observing that

$$\langle x, \chi \rangle = \langle x', \zeta \rangle = 0$$

you obtain

$$\langle x, \phi \rangle = \langle x, \phi_0 \rangle \int_{-\infty}^{+\infty} \phi(u) \, du + \langle x, \chi \rangle = \langle x, \phi_0 \rangle \int_{-\infty}^{+\infty} \phi(u) \, du$$

It should now be clear that x can only be an arbitrary constant.

2-10   THE SCHWARZ INEQUALITY FOR FUNCTIONS OF INTE-GRABLE SQUARE. Starting from

$$0 \le \int_{-\infty}^{+\infty} \left[\frac{1}{\sqrt{\eta}} |x(t)| - \sqrt{\eta} |y(t)| \right]^2 dt$$

where $\eta > 0$, you clearly have

$$2 \int_{-\infty}^{+\infty} |x(t)y(t)| \, dt \le \frac{1}{\eta} \int_{-\infty}^{+\infty} |x(t)|^2 \, dt + \eta \int_{-\infty}^{+\infty} |y(t)|^2 \, dt$$

Now choose η such that

$$\eta^2 \int_{-\infty}^{+\infty} |y(t)|^2 \, dt = \int_{-\infty}^{+\infty} |x(t)|^2 \, dt$$

You then obtain the Schwarz inequality,

$$\int_{-\infty}^{+\infty} |x(t)y(t)| \, dt \le \left[\int_{-\infty}^{+\infty} |x(t)|^2 \, dt \int_{-\infty}^{+\infty} |y(t)|^2 \, dt \right]^{1/2}$$

It follows that you also have

$$\left| \int_{-\infty}^{+\infty} x(t)y(t) \, dt \right|^2 \le \int_{-\infty}^{+\infty} |x(t)|^2 \, dt \int_{-\infty}^{+\infty} |y(t)|^2 \, dt$$

This last inequality is used to obtain optimal filters in Chapter 13.

2-11   CONTINUITY: IMPULSE RESPONSE OF A PHASE LAG. Consider a phase lag with step response

$$r(t) = \theta(t) \left[1 - \exp\left(-\frac{t}{\tau} \right) \right]$$

Show that for $t > T$ the response to

$$\delta_T(t) = \frac{\theta(t) - \theta(t-T)}{T}$$

is

$$h_T(t) = \frac{1}{\tau} \exp\left(-\frac{t}{\tau}\right) \frac{\exp(T/\tau) - 1}{(T/t)}$$

and that the impulse response is given correctly by

$$h(t) = \lim_{T \to 0} h_T(t) = \frac{1}{\tau} \theta(t) \exp\left(-\frac{t}{\tau}\right) = r'(t)$$

2-12 CONTINUITY: IMPULSE RESPONSE OF A PHASE LEAD. The aim of this exercise is to use a limiting process to obtain the impulse response of a phase lead

$$\tau x' + x = \tau y'$$

Replace $\delta(t)$ with the approximation $\Delta_T(t)$ shown in Fig. 2-8a,

$$\Delta_T(t) = \int_0^t \Delta'_T(u) \, du$$

where $\Delta'_T(t)$, shown in Fig. 2.8b, is given by

$$\Delta'_T(t) = \frac{\theta(t) - 2\theta(t-T) + \theta(t-2T)}{T^2}$$

If $h_T(t)$ is the response to $\Delta_T(t)$, show that

$$\int_0^{2T} h_T(t) \, dt = \frac{[1 - \exp(-T/\tau)]^2}{(T/\tau)^2}$$

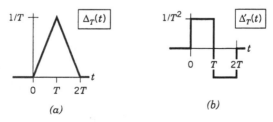

(a)   *(b)*

Figure 2-8. (*a*) The function $\Delta_T(t)$ used to approximate the delta function; (*b*) its derivative $\Delta'_T(t)$. (Exercise 2-12.)

and for $t > 2T$, that

$$h_T(t) = -\frac{1}{\tau}\exp\left(-\frac{t}{\tau}\right)\frac{[1-\exp(T/\tau)]^2}{(T/\tau)^2}$$

Conclude that the impulse response of a phase lead is

$$h(t) \equiv \delta(t) - \frac{1}{\tau}\theta(t)\exp\left(-\frac{t}{\tau}\right)$$

and that in the sense of generalized functions,

$$h(t) = r'(t)$$

where $r(t)$ is the step response given in (1-16),

$$r(t) = \theta(t)\exp\left(-\frac{t}{\tau}\right)$$

Remember that if $\psi(t)$ is a smooth function,

$$\psi(t)\delta(t) = \psi(0)\delta(t)$$

2-13   DEFINING FUNCTIONAL FOR THE DELTA FUNCTION. In this exercise you will show that if $\psi(t)$ is continuous, then

$$\lim_{n \to \infty} \int_{-\infty}^{+\infty} \psi(t)\delta_n(t)\,dt = \psi(0)$$

First observe that

$$\int_{-\infty}^{+\infty} \psi(t)\delta_n(t)\,dt = \psi(0) + \int_{-T/n}^{+T/n} [\psi(t) - \psi(0)]\delta_n(t)\,dt$$

But $\psi(t)$ is continuous, so that for $\varepsilon > 0$ and n large enough, $|\psi(t) - \psi(0)| < \varepsilon$ for $|t| < T/n$. With similar arguments, show that the functions δ_T defined in (1-16) converge to δ,

$$\lim_{T \to 0} \delta_T = \delta$$

and, by considering its integral, that

$$\lim_{T \to 0} \frac{\theta(t) - 2\theta(t-T) + \theta(t-2T)}{T^2} = \delta'$$

2-14   EXCHANGING THE ORDER OF LIMITS AND INTEGRALS

First some terminology:

A function $x(t)$ is *uniformly continuous* on a set if given $\varepsilon > 0$, one can find $T > 0$ such that for *every* t and $t + u$ in that set, $0 < |u| < T$ implies

$$|x(t+u) - x(t)| < \varepsilon$$

A function that is continuous on a closed and bounded interval $t_1 \le t \le t_2$ is also uniformly continuous. For example, $x(t) = \tau/t$ is not uniformly continuous on $0 < t \le \tau$, but it is uniformly continuous on $t_0 \le t \le \tau$ if $0 < t_0 < \tau$.

A sequence of functions $x_n(t)$ converges *pointwise* at t to a function $x(t)$ if

$$\lim_{n \to \infty} x_n(t) = x(t)$$

For example, $(t/\tau)^n$ converges pointwise to zero for $0 \le t < \tau$, but converges to 1 for $t = \tau$.

A sequence of functions $x_n(t)$ converges *uniformly* to a function $x(t)$ on a set if given $\varepsilon > 0$, it is possible to find N such that for *every* t in that set

$$n > N \quad \Rightarrow \quad |x_n(t) - x(t)| < \varepsilon$$

For example, if $x(t)$ is continuous, $x(t + \tau/n)$ converges uniformly to $x(t)$ on any closed and bounded interval because $x(t)$ is uniformly continuous on any closed and bounded interval that also contains $t + \tau/n$. In contrast, $(t/\tau)^n$ does not converge uniformly to zero on the interval $0 \le t/\tau < 1$.

Now a theorem:

Assume that x is absolutely integrable, that ψ is smooth, that ψ_n are smooth and uniformly bounded, and that $\psi_n \to \psi$ uniformly on bounded intervals. Then

$$\int_{-\infty}^{+\infty} x\psi_n \to \int_{-\infty}^{+\infty} x\psi$$

Given $\varepsilon > 0$, choose τ so that

$$\left| \int_{+\tau}^{+\infty} x(\psi_n - \psi) \right| + \left| \int_{-\infty}^{-\tau} x(\psi_n - \psi) \right| < \frac{\varepsilon}{2}$$

This is possible because ψ is also bounded. Now choose N such that

$$n > N \quad \Rightarrow \quad |\psi - \psi_n| \left| \int_{-\tau}^{+\tau} |x| \right| < \frac{\varepsilon}{2}$$

in the interval $-\tau \le t \le \tau$. Then, for $n > N$,

$$\left| \int_{-\infty}^{+\infty} x(\psi_n - \psi) \right| \le \left| \int_{+\tau}^{+\infty} x(\psi_n - \psi) \right| + \left| \int_{-\infty}^{-\tau} x(\psi_n - \psi) \right| + \left| \int_{-\tau}^{+\tau} x(\psi_n - \psi) \right| < \varepsilon$$

With this equipment you can obtain the following results.

(a) If ψ is smooth and x is locally integrable, $\psi * x$ is smooth if the integral that defines it is over a bounded interval. From the mean-value theorem obtain

$$(\psi * x)' = \lim_{\tau \to 0} \int_{-\infty}^{+\infty} \frac{\psi(t-u+\tau) - \psi(t-u)}{\tau} x(u)\, du$$

$$= \lim_{\tau \to 0} \int_{-\infty}^{+\infty} \psi'(t-u+\varepsilon\tau) x(u)\, du$$

where $0 < \varepsilon < 1$. The integral is over a (closed) bounded interval, so that $\psi'(t-u+\varepsilon\tau)$ converges uniformly to $\psi'(t-u)$, and

$$(\psi * x)' = \psi' * x$$

(b) If x is locally integrable and $x_G = \{x * \delta_n\}$,

$$\langle x_G, \phi \rangle = \int_{-\infty}^{+\infty} x(t) \phi(t)\, dt$$

First exchange the order of integration and obtain

$$\int_{-\infty}^{+\infty} \phi(t)\, dt \int_{-\infty}^{+\infty} x(u) \delta_n(t-u)\, du = \int_{-\infty}^{+\infty} x(u)\, du \int_{-\infty}^{+\infty} \phi(t+u) \delta_n(t)\, dt$$

But the smooth functions

$$\int_{-\infty}^{+\infty} \phi(t+u) \delta_n(t)\, dt$$

are uniformly bounded and converge uniformly to $\phi(u)$ everywhere.

(c) If ψ is smooth, $\psi * \delta = \psi$. This is case (b) again:

$$\int_{-\infty}^{+\infty} \phi(t)\, dt \int_{-\infty}^{+\infty} \psi(u) \delta_n(t-u)\, du = \int_{-\infty}^{+\infty} \psi(u)\, du \int_{-\infty}^{+\infty} \phi(t+u) \delta_n(t)\, dt$$

2-15 THE LAPLACE TRANSFORM IS ANALYTIC. Assume that $X(s)$ has an abscissa of convergence σ_C. For any $\sigma > \sigma_C$, define

$$\varepsilon = \frac{\sigma - \sigma_C}{2}$$

For sufficiently large t, you have

$$t|x(t)|\exp(-\sigma t) < |x(t)|\exp[-(\sigma-\varepsilon)t]$$

But $\sigma-\varepsilon>\sigma_C$; this implies that

$$D(s) = -\int_0^\infty tx(t)\exp(-st)\,dt$$

exists for $\sigma>\sigma_C$. The differential quotient of $X(s)$ is

$$\frac{\Delta X}{\Delta s} = \int_0^\infty x(t)\frac{\exp[-(s+\Delta s)t]-\exp(-st)}{\Delta s}\,dt$$

$$= -\int_0^\infty tx(t)\exp(-st)\frac{1-\exp(-t\Delta s)}{t\Delta s}\,dt$$

Using the same trick as above, you can obtain

$$\left|\frac{\Delta X}{\Delta s}-D(s)\right|$$

$$\leq \int_0^\infty t|x(t)|\exp[-(\sigma-\varepsilon)t]\left[\left|\frac{1-\exp(-t\Delta s)}{t\Delta s}-1\right|\exp(-\varepsilon t)\right]dt$$

For $|\Delta s|<\varepsilon$, the quantity in large brackets is bounded for $t\geq 0$ and converges uniformly to zero on any bounded interval. You can now follow the steps that led to the theorem in Exercise 2-14 to show that

$$X'(s) = D(s) = -\int_0^\infty tx(t)\exp(-st)\,dt$$

It is clear that the abscissa of convergence of $X(s)$ and $X'(s)$ are equal. It follows that if $X(s)$ exists and has an abscissa of convergence σ_C, it is analytic on the half-plane $\sigma>\sigma_C$.

2-16 POINTWISE CONVERGENCE AND GENERALIZED CONVERGENCE. The *pointwise* convergence of a sequence of ordinary functions to a limit function does not imply that the generalizations of these functions converge to the generalization of the limit function. The sequence δ'_n, for example, converges pointwise to zero in the sense of functions, but it converges to δ' in the sense of generalized functions. Observe in this case that generalized convergence makes more physical sense in that it conserves information that is lost in pointwise convergence. On the other hand, the family $\sin\lambda t$ fails to converge in the sense of functions for $\lambda\to\infty$, but it converges to zero in the sense of generalized functions: for any test function ϕ you have

$$\lim_{\lambda \to \infty} \int_{-\infty}^{+\infty} \phi(t) \sin \lambda t \, dt = \lim_{\lambda \to \infty} \frac{1}{\lambda} \int_{-\infty}^{+\infty} \phi'(t) \cos \lambda t \, dt = 0$$

because $\phi'(t) \cos \lambda t$ is bounded on a finite interval (this is a particular case of the Riemann–Lebesgue lemma). Again, this result makes physical sense because any system stops responding at high enough frequencies.

2-17   CHARACTERISTIC FUNCTIONS. The Fourier transform of a probability density distribution is known as the *characteristic function* of the distribution. For a normal distribution with mean μ and standard deviation σ you have

$$p(\zeta) = \frac{1}{\sqrt{2\pi}\,\sigma} \exp\left[-\frac{(\zeta-\mu)^2}{2\sigma^2} \right]$$

and

$$\int_{-\infty}^{+\infty} p(\zeta) \, d\zeta = 1$$

The characteristic function is given by

$$\hat{p}(\nu) = \frac{1}{\sqrt{2\pi}\,\sigma} \int_{-\infty}^{+\infty} \exp\left[-\frac{(\zeta-\mu)^2}{2\sigma^2} \right] \exp(-i\nu\zeta) \, d\zeta$$

With the transformation $\zeta - \mu \to \zeta$ you get

$$\hat{p}(\nu) = \frac{1}{\sqrt{2\pi}\,\sigma} \exp(-i\mu\nu) \int_{-\infty}^{+\infty} \exp\left(-\frac{\zeta^2}{2\sigma^2} - i\nu\zeta \right) d\zeta$$

By completing the square in the argument of the exponential in the integral, you get

$$\hat{p}(\nu) = \exp(-i\mu\nu) \exp\left(-\frac{\sigma^2\nu^2}{2} \right) \frac{1}{\sqrt{2\pi}\,\sigma} \int_{-\infty}^{+\infty} \exp\left[-\frac{(\zeta+i\nu\sigma^2)^2}{2\sigma^2} \right] d\zeta$$

Since the exponential function has no poles, the path of integration in the complex plane can be shifted down to the real axis, and you finally get

$$\hat{p}(\nu) = \exp(-i\mu\nu) \exp\left(-\frac{\sigma^2\nu^2}{2} \right)$$

2-18   THE INVERSION FORMULA. Assume that $x(t)$ is piecewise smooth and absolutely integrable. Now consider the function $x_+(t)$ given by

$$x_+(t) = \int_0^\infty x(t+u)\frac{\sin \lambda u}{\pi u}\,du$$

For any $T>0$ you have

$$\lim_{\lambda \to \infty} x_+(t) = \lim_{\lambda \to \infty} \int_0^T x(t+u)\frac{\sin \lambda u}{\pi u}\,du + \lim_{\lambda \to \infty} \int_T^\infty x(t+u)\frac{\sin \lambda u}{\pi u}\,du$$

But the second integral on the right-hand side is zero by the Riemann–Lebesgue lemma. You can then write

$$\lim_{\lambda \to \infty} x_+(t) = \lim_{\lambda \to \infty} \int_0^T \frac{x(t+u)-x(t+)}{\pi u}\sin \lambda u\,du + \lim_{\lambda \to \infty} x(t+)\int_0^T \frac{\sin \lambda u}{\pi u}\,du$$

Since $x(t)$ is piecewise smooth, $(1/u)[x(t+u)-x(t+)]$ tends to the right-hand derivative of $x(t)$, so that the first term on the right-hand side is zero by the Riemann–Lebesgue lemma. Using the fact that

$$\int_{-\infty}^{+\infty} \frac{\sin \eta}{\eta}\,d\eta = \pi$$

you then obtain

$$\lim_{\lambda \to \infty} x_+(t) = \lim_{\lambda \to \infty} x(t+)\int_0^{\lambda T} \frac{\sin \eta}{\pi \eta}\,d\eta = \frac{1}{2}x(t+)$$

With analogous arguments on

$$x_-(t) = \int_0^\infty x(t-u)\frac{\sin \lambda u}{\pi u}\,du$$

you finally obtain

$$\lim_{\lambda \to \infty} \int_{-\infty}^{+\infty} x(t+u)\frac{\sin \lambda u}{\pi u}\,du = \frac{1}{2}[x(t-) + x(t+)]$$

Note incidentally that you have also obtained a valuable representation of the delta function,

$$\delta(t) = \lim_{\lambda \to \infty} \frac{\sin \lambda t}{\pi t}$$

Now consider

$$x_\lambda(t) = \frac{1}{2\pi} \int_{-\lambda}^{+\lambda} \exp(i\omega t) \, d\omega \int_{-\infty}^{+\infty} x(u) \exp(-i\omega u) \, du$$

According to Fubini's theorem, the order of integration can be exchanged because exp($i\omega t$) is absolutely integrable on the finite interval $(-\lambda, +\lambda)$, and you get

$$x_\lambda(t) = \frac{1}{2\pi} \int_{-\infty}^{+\infty} x(u) \, du \int_{-\lambda}^{+\lambda} \exp[i\omega(t-u)] \, d\omega = \int_{-\infty}^{+\infty} x(u) \frac{\sin[\lambda(t-u)]}{\pi(t-u)} \, du$$

You then have

$$\lim_{\lambda \to \infty} x_\lambda(t) = \lim_{\lambda \to \infty} \int_{-\infty}^{+\infty} x(t+u) \frac{\sin \lambda u}{\pi u} \, du = \frac{1}{2}[x(t-) + x(t+)]$$

You have thus established that

$$\lim_{\lambda \to \infty} \frac{1}{2\pi} \int_{-\lambda}^{+\lambda} \hat{x}(\omega) \exp(i\omega t) \, d\omega = \frac{1}{2}[x(t-) + x(t+)]$$

and where $x(t)$ is smooth, that

$$x(t) = \lim_{\lambda \to \infty} \frac{1}{2\pi} \int_{-\lambda}^{+\lambda} \hat{x}(\omega) \exp(i\omega t) \, d\omega$$

This is not quite the inversion formula (2-13) because the limits on the integral are symmetric, whereas where $x(t)$ is smooth you should actually require

$$x(t) = \lim_{\nu \to \infty} \lim_{\lambda \to \infty} \frac{1}{2\pi} \int_{-\nu}^{+\lambda} \hat{x}(\omega) \exp(i\omega t) \, d\omega$$

$$= \lim_{\lambda \to \infty} \lim_{\nu \to \infty} \frac{1}{2\pi} \int_{-\nu}^{+\lambda} \hat{x}(\omega) \exp(i\omega t) \, d\omega$$

If you wish to complete the proof, first obtain

$$\int_{0}^{+\lambda} \hat{x}(\omega) \exp(i\omega t) \, d\omega = \int_{-\infty}^{+\infty} x(t+u) \frac{\sin \lambda u}{u} \, du - i \int_{-\infty}^{+\infty} x(t+u) \frac{1-\cos \lambda u}{u} \, du$$

$$\int_{-\nu}^{0} \hat{x}(\omega) \exp(i\omega t) \, d\omega = \int_{-\infty}^{+\infty} x(t+u) \frac{\sin \nu u}{u} \, du + i \int_{-\infty}^{+\infty} x(t+u) \frac{1-\cos \nu u}{u} \, du$$

Now you need to show that the second term on the right-hand side of these expressions tends to a finite limit. Use the Riemann–Lebesgue lemma to obtain

$$\lim_{\lambda \to \infty} \int_{+T}^{+\infty} x(t+u) \frac{1 - \cos \lambda u}{u} \, du = \int_{+T}^{+\infty} \frac{x(t+u)}{u} \, du$$

$$\lim_{\lambda \to \infty} \int_{-\infty}^{-T} x(t+u) \frac{1 - \cos \lambda u}{u} \, du = \int_{-\infty}^{-T} \frac{x(t+u)}{u} \, du$$

and assuming that $x(t)$ is smooth,

$$\lim_{\lambda \to \infty} \int_{-T}^{+T} x(t+u) \frac{1 - \cos \lambda u}{u} \, du = \int_{-T}^{+T} \frac{x(t+u) - x(t)}{u} \, du$$

You thus have a finite limit, which you can write succinctly as

$$\lim_{\lambda \to \infty} \int_{-\infty}^{+\infty} x(t+u) \frac{1 - \cos \lambda u}{u} \, du = \text{p.v.} \int_{-\infty}^{+\infty} \frac{x(t+u)}{u} \, du$$

where for any function $y(t)$,

$$\text{p.v.} \int_{-\infty}^{+\infty} y(t) \, dt = \lim_{T \to 0} \left[\int_{-\infty}^{-T} y(t) \, dt + \int_{+T}^{+\infty} y(t) \, dt \right]$$

is the *Cauchy principal value* of the integral (for another case involving a principal value, see Exercise 2-20).

The limits on the integral in the inversion formula thus need not be symmetric where $x(t)$ is smooth, and you obtain

$$x(t) = \frac{1}{2\pi} \int_{-\infty}^{+\infty} \hat{x}(\omega) \exp(i\omega t) \, d\omega$$

2-19 FUNCTIONS OF RAPID DESCENT. Let $\rho(t)$ be a function of rapid descent. Starting from

$$[t\rho(t)]' = \rho(t) + t\rho'(t)$$

show that

$$[t\rho(t)]^{(n)} = n\rho^{(n-1)}(t) + t\rho^{(n)}(t)$$

For any m you then have

$$t^m [t\rho(t)]^{(n)} = nt^m \rho^{(n-1)}(t) + t^{m+1} \rho^{(n)}(t)$$

It follows that $t\rho(t)$ is of rapid descent and that $t^m\rho(t)$ is also of rapid descent. Now show that $\rho^{(m)}(t)$ is of rapid descent. It follows that

$$[t^m\rho(t)]^{(n)}$$

is of rapid decent.

2-20 FOURIER TRANSFORM OF THE UNIT STEP FUNCTION. The *Cauchy principal value* of an integral, denoted by

$$\text{p.v.} \int_{-\infty}^{+\infty} x(t)\, dt$$

is defined as

$$\lim_{T \to 0} \left[\int_{-\infty}^{-T} x(t)\, dt + \int_{+T}^{+\infty} x(t)\, dt \right]$$

Show that the defining functional for the generalized derivative of $\ln|\omega\tau|$ is given by

$$\langle (\ln|\omega\tau|)', \phi \rangle = -\langle \ln|\omega\tau|, \phi' \rangle = \text{p.v.} \int_{-\infty}^{+\infty} \frac{1}{\omega} \phi(\omega)\, d\omega$$

It is customary to denote the generalized function $(\ln|\omega\tau|)'$ by

$$\text{p.v.} \frac{1}{\omega}$$

Show directly that the Fourier transform of the unit step function $\theta(t)$ is

$$\mathcal{F}(\theta) = \pi\delta(\omega) + \text{p.v.} \frac{1}{i\omega}$$

Use the fact that any function of rapid descent can be represented as the inverse Fourier transform of a function of rapid descent. You thus have to show that

$$\langle \mathcal{F}(\theta), \mathcal{F}^{-1}(\rho) \rangle = \langle \theta, \rho \rangle = \int_0^\infty \rho(t)\, dt$$

Clearly you have

$$\left\langle \pi\delta(\omega), \frac{1}{2\pi} \int_{-\infty}^{+\infty} \rho(t) \exp(i\omega t)\, dt \right\rangle = \frac{1}{2} \int_{-\infty}^{+\infty} \rho(t)\, dt$$

Now show that

$$\left\langle \text{p.v.} \frac{1}{i\omega}, \frac{1}{2\pi} \int_{-\infty}^{+\infty} \rho(t) \exp(i\omega t)\, dt \right\rangle = \frac{1}{\pi} \int_{-\infty}^{+\infty} \rho(t)\, dt \int_0^\infty \frac{1}{\omega} \sin(\omega t)\, d\omega$$

Complete the proof using the fact that

$$\int_0^\infty \frac{1}{\omega} \sin(\omega t)\, d\omega = \frac{\pi}{2} \operatorname{sgn}(t)$$

2-21 A CONVENIENT FOURIER SERIES. Consider the continuous periodic function $x(t)$ shown in Fig. 2-5 and given for $0 \le t < T$ by

$$x(t) = \frac{T}{2}\left(\frac{t}{T} - \frac{t^2}{T^2} \right)$$

Propose an expansion of $x(t)$ in a Fourier series,

$$x(t) = \sum_{n=-\infty}^{+\infty} a_n \exp\left(2\pi i n \frac{t}{T} \right)$$

If you multiply both sides by $\exp(-2\pi i m t/T)$ and integrate from 0 to T, only the term $n = m$ survives, and you get

$$a_n = \frac{1}{T} \int_0^T x(t) \exp\left(-2\pi i n \frac{t}{T} \right) dt$$

Do the integrals and obtain

$$a_0 = \frac{T}{12}$$

and for $n \ne 0$,

$$a_n = -\frac{T}{(2\pi n)^2}$$

You thus get

$$\frac{T}{2}\left(\frac{t}{T} - \frac{t^2}{T^2} \right) = \frac{T}{12} - \sum_{n \ne 0} \frac{T}{(2\pi n)^2} \exp\left(2\pi i n \frac{t}{T} \right)$$

In the language of Exercise 2-14, the series

$$x_F(t) = \sum_{n \ne 0} \frac{T}{(2\pi n)^2} \exp\left(2\pi i n \frac{t}{T} \right) \exp\left(-2\pi i m \frac{t}{T} \right)$$

converges uniformly in the interval $0 \le t \le T$ because the series formed by the absolute values of the terms is proportional to

$$\sum_{n \ne 0} \frac{1}{n^2}$$

and is thus convergent. It follows that $x_F(t)$ can be integrated term by term as you did above.

2-22 THE CONTINUITY OF THE FOURIER TRANSFORM IS USE-FUL. Start from the fact that for $\varepsilon > 0$,

$$\lim_{\varepsilon \to 0} \theta(t) \exp(-\varepsilon t) = \theta(t)$$

Now consider

$$\frac{1}{\omega - i\varepsilon} = \mathbf{F}\left[i\theta(t) \exp(-\varepsilon t)\right]$$

The continuity of the Fourier transform on generalized functions of slow growth implies that

$$\lim_{\varepsilon \to 0} \frac{1}{\omega - i\varepsilon} = \mathbf{F}\left[i\theta(t)\right]$$

If you now accept the result of Exercise 2-20, you can obtain the *Plemelj relations*, much used in theoretical physics:

$$\lim_{\varepsilon \to 0} \frac{1}{\omega \pm i\varepsilon} = \text{p.v.} \frac{1}{\omega} \mp i\pi\delta(\omega)$$

Similarly, you can show that

$$\lim_{\varepsilon \to 0} \frac{\varepsilon}{\omega^2 + \varepsilon^2} = \pi\delta(\omega)$$

2-23 RECOVERY OF SAMPLED SIGNALS. Consider a bandlimited signal $x(t)$ such that $\hat{x}(\omega)$ is zero for $|\omega| > \omega_0$, and assume that $x(t)$ is oversampled at a rate $1/T = \lambda/2\pi > \omega_0/\pi$, so that the transforms $\hat{x}(\omega - n\lambda)$ in the last term in (2-22), shown schematically in Fig. 2-6, do not overlap. Now consider a causal filter with an impulse response $h(t)$ such that $\hat{h}(\omega)$ is essentially 1 for $|\omega| < \omega_0$ and essentially 0 for $|\omega| > \lambda/2$. If you multiply the last two terms in (2-22) by $\hat{h}(\omega)$ rather than $\theta(\omega + \omega_0) - \theta(\omega - \omega_0)$, you get

$$\hat{x}(\omega) \simeq \hat{h}(\omega) \left\{ \left[\sum_n \delta(\omega - n\lambda) \right] * \hat{x}(\omega) \right\}$$

Using (2-19) and the first equality in (2-22), you can then obtain

$$x(t) \simeq h(t) * \left[T \sum_n x(nT) \delta(t - nT) \right]$$

and finally,

$$x(t) \simeq T \sum_n x(nT) h(t - nT)$$

But this is precisely what you get if you run the sampled signal

$$x_T(t) = T\sum_n x(nT)\,\delta(t-nT)$$

through a filter with impulse response $h(t)$. Since $h(t)$ is zero for $t<0$, the reconstructed signal at any time is obtained using only samples from the past.

Sampled signals are often obtained in a *sample-and-hold* circuit in the form

$$\sum_n x(nT)\{\theta(t-nT) - \theta[t-(n+1)T]\}$$

This is equivalent to running x_T through a filter with impulse response

$$h_T(t) = \frac{\theta(t) - \theta(t-T)}{T}$$

and frequency response

$$\hat{h}_T(\omega) = \frac{\sin(\omega T/2)}{\omega T/2}\,\exp(-i\omega T/2)$$

If $x(t)$ is heavily oversampled, that is, if $\lambda \gg 2\omega_0$, $\hat{h}_T(\omega)$ is essentially 1 where $\hat{x}(\omega)$ is nonzero and thus does not affect the reconstructed signal.

APPENDIX: TABLE OF LAPLACE TRANSFORMS

$x(t)$	$X(s)$	$x(t)$	$X(s)$
$\int_0^t x(u)\,du$	$\dfrac{1}{s}X(s)$	$\theta(t)\exp(-\dfrac{t}{\tau})$	$\dfrac{1}{s+1/\tau}$
$x(t-\tau)$	$\exp(-\tau s)X(s)$	$\theta(t)\exp(i\lambda t)$	$\dfrac{1}{s-i\lambda}$
$\exp(s_0 t)x(t)$	$X(s-s_0)$	$\theta(t)\cos(\lambda t)$	$\dfrac{s}{s^2+\lambda^2}$
$t^n x(t)$	$(-1)^n X^{(n)}(s)$	$\theta(t)\sin(\lambda t)$	$\dfrac{\lambda}{s^2+\lambda^2}$
$x_1 * x_2$	$X_1(s)X_2(s)$	$\theta(t)\exp(-\omega_1 t)\cos(\omega_2 t)$	$\dfrac{s+\omega_1}{(s+\omega_1)^2+\omega_2^2}$
$\theta(t)$	$\dfrac{1}{s}$	$\theta(t)\exp(-\omega_1 t)\sin(\omega_2 t)$	$\dfrac{\omega_2}{(s+\omega_1)^2+\omega_2^2}$
$\theta(t)t^n$	$\dfrac{n!}{s^{n+1}}$	$\dfrac{\theta(t)}{\sqrt{\pi t}}$	$\dfrac{1}{\sqrt{s}}$

3

CIRCUIT THEORY

3-1 CIRCUIT ELEMENTS

High-frequency electromagnetic structures such as waveguides must be described in terms of quantities that vary in space as well as in time; a complete description might thus involve specification of the electric and magnetic fields everywhere. There exist structures, however, called *lumped elements*, that are described quite accurately at low frequencies in terms of integrals over the spatial variables; these, as we will now see, are usually voltages [line integrals of the electric field, expressed in volts (V)] and currents [flow of charge per unit time through a given surface, expressed in coulombs (C) per second, or amperes (A)]. We will first examine the low-frequency behavior of isolated lumped elements and then proceed to *networks* or *circuits*, structures that are obtained by interconnecting lumped elements and that can also be described in terms of voltages and currents.

Let us start by recalling three of the four fundamental laws of electromagnetism, valid as shown in vacuum and in the materials we will consider for now (the fourth law states that there are no magnetic charges, but we will not need it). In integral form these laws are as follows.

1. Gauss's law for a closed surface,

$$\int \mathbf{E} \cdot d\mathbf{a} = \frac{q}{\varepsilon_0}$$

where $\mathbf{E}$ is the electric field, q is the charge enclosed in the volume bounded by the surface, and ε_0 is the vacuum permittivity, given in terms of the speed of light $c = 3 \cdot 10^8$ m/s by

$$\frac{1}{\varepsilon_0 c^2} = \mu_0 \equiv 4\pi \cdot 10^{-7}$$

2. Faraday's law for a closed path,

$$\oint \mathbf{E} \cdot d\mathbf{l} = -\frac{d\Phi}{dt}$$

where Φ is the flux of the magnetic field **B** through a surface bounded by the closed path,

$$\Phi = \int \mathbf{B} \cdot d\mathbf{a}$$

If $d\Phi/dt = 0$ we are in the electrostatic case, and Faraday's law reduces to

$$\oint \mathbf{E} \cdot d\mathbf{l} = 0$$

In the electrostatic case, the *voltage* of a point B with respect to a point A (or voltage drop from B to A), given by

$$V_{BA} = -\int_A^B \mathbf{E} \cdot d\mathbf{l}$$

is well defined because the line integral of the electric field is independent of the path chosen between A and B.

3. The Ampère–Maxwell law for a closed path,

$$c^2 \oint \mathbf{B} \cdot d\mathbf{l} = \frac{i}{\varepsilon_0} + \frac{d\Psi}{dt}$$

where i and Ψ are the current and the flux of the electric field through a surface bounded by the closed path. In the cases we will study in this chapter the electric flux term does not contribute, and the Ampère–Maxwell law reduces to Ampère's circuital law,

$$\oint \mathbf{B} \cdot d\mathbf{l} = \mu_0 i$$

In Chapter 4 we will show that these laws imply conservation of charge or *continuity*, which we will write as

$$i = -\frac{dq}{dt}$$

where i is the current leaving a closed surface and q is the charge in the volume bounded by the surface.

For later use in semiconductors and transmission lines, we note that everything we have said so far is valid in a homogeneous isotropic medium with dielectric constant κ and permittivity $\varepsilon = \kappa \varepsilon_0$ if we make the substitutions

$$\varepsilon_0 \to \varepsilon \quad \text{and} \quad c^2 \to c^2/\kappa$$

and ignoring the charges bound in the dielectric as well as their currents, we take q and i to be the *free charge* and the *free current*.

Before proceeding to lumped elements, we should remind ourselves of the notion of a *perfect conductor*, best exemplified in practice by metals such as copper and gold. We will imagine a perfect conductor as containing a sea of free charge carriers that rearrange themselves instantaneously in the presence of an external electric field and generate surface charges that keep the internal electric field at zero. The voltage between any two points in a perfect conductor is therefore zero, or in other words, perfect conductors are equipotentials. We will assume that the wires that interconnect the lumped elements in a circuit are perfect conductors.

Let us start with the lumped element that is perhaps easiest to approach, the circular parallel-plate *capacitor* shown in Fig. 3-1. We will assume that the plates are perfect conductors of area A and that they carry charges $\pm q$. If the plate separation d is much smaller than the plate diameter, we can ignore edge effects and assume that the charges are uniformly distributed on the inner surfaces of the plates and that the electric field E is uniform and normal to the plates in the interior of the capacitor and zero outside.

Applying Gauss's law to the volume indicated by dashed lines in Fig. 3-1, we obtain

$$AE = \frac{q}{\varepsilon_0}$$

The voltage between the plates is

$$v = Ed$$

Figure 3-1. Cross section of a circular parallel-plate capacitor. The charge on the top plate is obtained by applying Gauss's law to the volume indicated in dashed lines.

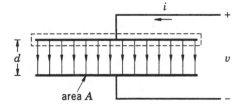

We can now obtain the electrostatic equation for the capacitor,

$$q = Cv \qquad (3\text{-}1)$$

where C [in farads (F)] is the *capacitance*, given by

$$C = \varepsilon_0 \frac{A}{d}$$

For reference, we note that

$$\varepsilon_0 = 8.85 \text{ pF/m}$$

If we assume—we will justify this assumption in Chapter 4—that slow variations in v and q do not affect (3-1), we can take time derivatives and obtain

$$i = Cv' \qquad (3\text{-}2)$$

Generalizing, we will accept as a capacitor of capacitance C any two-terminal device—of arbitrary shape and perhaps containing dielectric materials—that obeys (3-2) at sufficiently low frequencies.

Let us now turn to *inductors*, of which we will choose one of the easiest to analyze, the perfectly conducting N-turn solenoid shown in Fig. 3-2. If the solenoid is such that its length l is much larger than its diameter, we can assume that the magnetic field B generated by a current i is axial and uniform in its interior and zero outside.

Applying Ampère's law to the closed path shown in dashed lines in Fig. 3-2, we obtain

$$lB = N\mu_0 i$$

The flux through a surface bounded by a closed path that starts at the positive terminal, follows the conductor to the negative terminal, and returns to the positive terminal through a path external to the solenoid, is

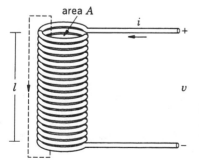

Figure 3-2. An N-turn solenoidal inductor. The axial magnetic field is obtained by applying Ampère's law to the closed path in dashed lines. Outside the solenoid the magnetic field is negligible, and the voltage v between the terminals is therefore well defined.

$$\Phi = NAB$$

where A is the cross-sectional area of the solenoid. We can now express the flux as

$$\Phi = Li \qquad (3\text{-}3)$$

where L [in henrys (H)] is the *inductance* of the solenoid,

$$L = \mu_0 \frac{A}{l} N^2 \qquad (3\text{-}4)$$

We note that although the vacuum permeability μ_0 is dimensionless, we have

$$\mu_0 = 4\pi \cdot 10^{-7} \text{ H/m} = 1.257 \ \mu\text{H/m}$$

If we apply Faraday's law to the path for which we defined the flux, we get no contribution from the part that follows the conductor because the electric field there is zero, and since the external part of the path is in a region where the magnetic field is zero, the voltage between the terminals is well defined and given by[†]

$$v = -\int_{-}^{+} \mathbf{E} \cdot d\mathbf{l} = -\oint \mathbf{E} \cdot d\mathbf{l} = \Phi'$$

Substituting $\Phi = Li$ from (3-3), we obtain

$$v = Li' \qquad (3\text{-}5)$$

In arriving at this result we have assumed—this assumption too we will justify in Chapter 4—that (3-3) is not affected by slow variations in Φ and i, and that taking time derivatives is therefore legitimate. As in the case of capacitors, we will accept as an inductor of inductance L any two-terminal device—again, of arbitrary shape and perhaps containing magnetic materials —that obeys (3-5) at sufficiently low frequencies.

Finally we come to *resistors*, which are actually harder to describe at an elementary level because they involve properties of materials. Briefly stated, the charge carriers in a resistive material have limited mobility because they are continually colliding with the atoms that make up the material, and in a long wire we find that the current is proportional to the voltage between the ends,

$$v = Ri \qquad (3\text{-}6)$$

[†]The path external to the solenoid must be essentially a direct run between the terminals. Thus a path that spirals around the solenoid on its way from the negative to the positive terminal is inadmissible because it effectively increases or decreases the number of turns by 1 and changes Φ by $\pm AB$. In less idealized circumstances the magnetic field is not quite zero in the vicinity of the terminals, and Φ is therefore slightly path dependent. The voltage between the terminals is then as well defined as the external paths are restricted.

where R [in ohms (Ω)] is the *resistance*, given in terms of the *conductivity* σ [in reciprocal ohms or siemens (S), per meter], the length l, and the cross-sectional area A of the wire by

$$R = \frac{1}{\sigma}\frac{l}{A}$$

Once again, we will describe as a resistor of resistance R any two-terminal device that obeys (3-6) at sufficiently low frequencies. In Chapter 4 we will return to this subject and develop a simple model that accounts for the conductivity of materials. To close a minor loop, we note that a perfect conductor has infinite conductivity and that a perfectly conducting wire has zero resistance.

Capacitors, inductors and resistors are governed by (extremely simple) linear differential equations, so that relations between voltages and currents for sinusoidal variations can be described in terms of complex amplitudes and frequency responses. Figure 3-3 shows symbols for the lumped elements we have discussed as well as the relevant differential equations and their corresponding frequency responses. We note the convention that v (whatever its actual sign) is the voltage of the positive terminal with respect to the negative terminal, and that i (again, whatever its actual sign) is the current entering the positive terminal. We also note that we use the same letters to denote time-dependent variables and complex amplitudes. We will see that context is sufficient to avoid confusion.

We will complete our initial list of circuit elements with the *independent* voltage and current sources represented in Fig. 3-4. At this time we will consider a voltage source to be a two-terminal device that maintains a stated (perhaps time-dependent) voltage between its terminals no matter what current it must source or sink, and we will not worry about how this is achieved; there are enough everyday objects that do this to a high degree of perfection, household electrical outlets and automobile batteries being good

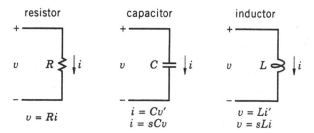

Figure 3-3. Symbols and sign conventions for lumped elements. Also shown are the differential equations for these elements and the corresponding frequency responses.

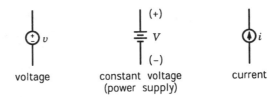

voltage   constant voltage   current
               (power supply)

Figure 3-4. Symbols and sign conventions for independent sources.

examples, that our imagination should not be strained. Considerably less familiar is a current source, a two-terminal device that delivers a stated (also perhaps time-dependent) current no matter what voltage it is subjected to. We will accept the existence of current sources for now, but we will soon be in a position to exhibit devices that are almost ideal.

Voltage sources are subject to the same constraints as inductors: the voltage between the terminals is well defined if the path of integration is restricted to regions where there are no significant time-varying magnetic fields such as one might find inside automobile alternators and other electric machines.

3-2   THE BASICS: KIRCHHOFF'S LAWS AND SUPERPOSITION

Circuits consist of interconnected elements, and we need rules that will allow us to solve for the voltages and the currents. Let us start by establishing the terminology we will use: a *branch* is a circuit element and a *node* is the interconnection of two or more branches. Defining a branch as a circuit element might seem superfluous at this point, but it will make sense later when a branch can be any two-terminal circuit.

Given the demands we have made on ideal circuit elements, we have ensured that $\oint \mathbf{E} \cdot d\mathbf{l} = 0$; but on the other hand,

$$\oint \mathbf{E} \cdot d\mathbf{l} = \text{sum of the voltage drops in a closed path}$$

We thus obtain *Kirchhoff's voltage law*: The sum of the voltage drops in a closed path is zero or, equivalently, the voltage between two given nodes is the sum of the voltage drops along any path that joins the nodes in question; these two points of view are exhibited in Fig. 3-5.

It is worth pointing out that although the paths we use to define the voltage do not in general go *physically* through a circuit element, there is no harm in thinking as if they did—it is almost unavoidable—once we have committed a circuit to paper.

We will now make the assumption that there are no charges of any importance on the terminals of circuit elements or on wires that interconnect

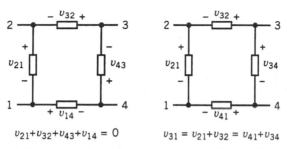

$$v_{21}+v_{32}+v_{43}+v_{14} = 0 \qquad v_{31} = v_{21}+v_{32} = v_{41}+v_{34}$$

Figure 3-5. Two equivalent ways of expressing Kirchhoff's voltage law.

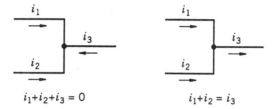

$$i_1+i_2+i_3 = 0 \qquad i_1+i_2 = i_3$$

Figure 3-6. Two equivalent ways of expressing Kirchhoff's current law.

elements. There is therefore, at least at low enough frequencies, no charge buildup at a node. From conservation of charge we then obtain Kirchhoff's current law: The sum of the currents entering a node is zero or, equivalently, the sum of the currents entering a node is equal to the sum of the currents leaving the node. These two points of view are exhibited in Fig. 3-6.

The final basic rule we need is the *principle of superposition*, which follows from the linearity of the laws of electromagnetism: The effect of several sources is equal to the sum of the effects of each source when all others are nulled (a nulled voltage source is a short circuit, whereas a nulled current source is an open circuit). In the case of voltage sources it is well worth stressing that *null* means *make zero* and by no means *remove*, either physically or on paper.

Let us use these rules to solve some elementary circuits and get an idea of what they are used for. We will start with an extremely important circuit, the resistive voltage divider shown in Fig. 3-7. The circuit has been drawn in two forms, one that shows all interconnections explicitly, and another, which we will favor heavily, that only gives voltages with respect to a *common* (point) or *ground* indicated by a horizontal or vertical dash.

One way of solving the divider is to observe that the currents in R_1 and R_2 must be equal by Kirchhoff's current law, that the voltage drops in R_1 and R_2 must add up to the input voltage v_I by Kirchhoff's voltage law,

$$v_I = iR_1 + iR_2$$

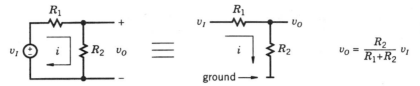

Figure 3-7. Basic resistive divider.

and that the output voltage v_O is therefore given by

$$v_O = iR_2 = \frac{R_2}{R_1 + R_2} v_I \tag{3-7}$$

However, in order to introduce the *node equations* that we will emphasize, we will use Kirchhoff's current law at node v_O but write the currents directly in terms of the voltage differences and the resistances in the branches:

$$\frac{v_I - v_O}{R_1} = \frac{v_O}{R_2}$$

From this equation we easily recover v_O as given in (3-7).

The Wheatstone bridge shown in Fig. 3-8 is formed by two voltage dividers fed by a constant voltage V. We will use it to measure the relative variation α of the resistance in the lower left leg. (Measurements that involve detecting *variations* in a resistance as a function of an external variable are extremely common: *thermistors*, for example, are widely used to measure temperatures near room temperature.) For small α we have from (3-7) that

$$\frac{v}{V} = \frac{(1+\alpha)R}{R+(1+\alpha)R} - \frac{R}{R+R} = \frac{1}{2}\left(\frac{1+\alpha}{1+\alpha/2} - 1\right) \approx \frac{\alpha}{4}$$

The point here is that the bridge configuration bucks out the base value of the variable resistor, and we get a direct measurement of the variation.

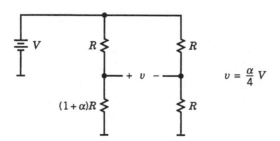

Figure 3-8. Wheatstone bridge.

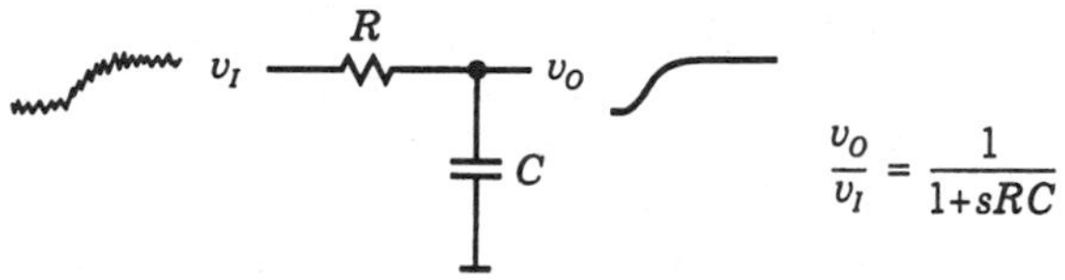

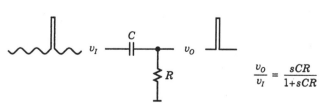

$$\frac{v_O}{v_I} = \frac{1}{1+sRC}$$

Figure 3-9. RC voltage divider. It is a phase lag, and therefore a low-pass filter.

The RC circuit of Fig. 3-9 is also a voltage divider, but in this case the output leg has been replaced by a capacitor, and the division is therefore frequency dependent. Summing currents at the output node (node v_O), we obtain

$$\frac{v_I - v_O}{R} = sCv_O$$

or, solving for v_O,

$$\frac{v_O}{v_I} = \frac{1}{1+sRC} \tag{3-8}$$

This is interesting, because we have obtained a phase lag or low-pass filter which we can use, among many other things, to remove high-frequency noise from the input without significantly affecting low-frequency signals.

Our final example of a divider is the CR circuit of Fig. 3-10. Once again, if we sum currents at the output node, we obtain

$$(v_I - v_O)sC = \frac{v_O}{R}$$

or, solving for v_O,

$$\frac{v_O}{v_I} = \frac{sCR}{1+sCR} \tag{3-9}$$

$$\frac{v_O}{v_I} = \frac{sCR}{1+sCR}$$

Figure 3-10. CR voltage divider. It is a phase lead, and therefore a high-pass filter.

This time we obtain a phase lead or high-pass filter. This circuit can be used to remove a constant or *direct-current*[†] (dc) voltage that is superposed on a signal of interest; in this case, we will say that C is a *blocking* capacitor. More generally, as shown in Fig. 3-10, a CR circuit can be used to remove low-frequency noise (such as power-line hum) from the input.

3-3 IMPEDANCES AND ADMITTANCES

By combining lumped elements we can form composite two-terminal circuits whose voltage and current, like those of lumped elements, are related by a frequency response. In the circuit shown in Fig. 3-11, for instance, we have

$$i = \frac{v}{R} + sCv = \frac{v}{R}(1+sRC)$$

or

$$\frac{v}{i} = \frac{R}{1+sRC}$$

We can now broaden our definition of a branch to include any two-terminal combination of lumped elements, shown symbolically in Fig. 3-12, and we will define the *impedance Z* (in ohms) of such a circuit as the ratio of the complex amplitudes of the voltage and the current:

$$Z = \frac{v}{i}$$

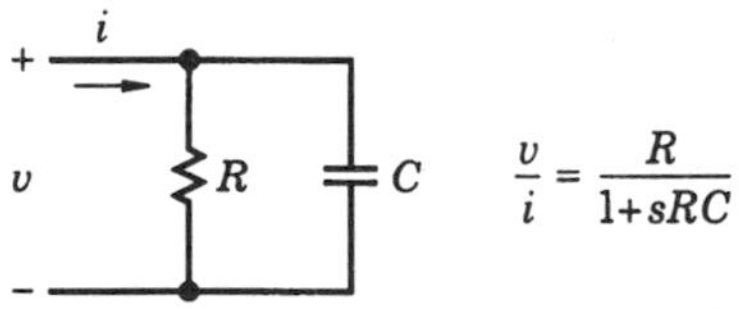

Figure 3-11. Parallel combination of a capacitor and a resistor. Its voltage and current, like those of basic lumped elements, are related by a frequency response.

$\dfrac{v}{i} = \dfrac{R}{1+sRC}$

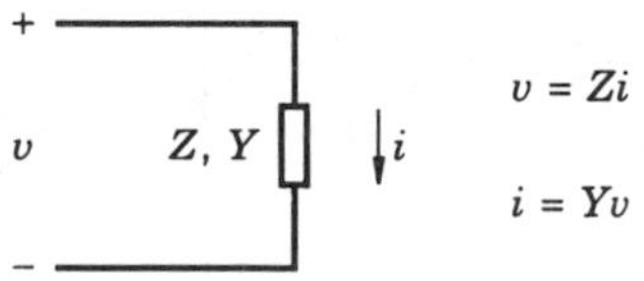

Figure 3-12. Symbol and sign conventions for an arbitrary two-terminal combination of lumped elements, loosely called an impedance or an admittance.

$v = Zi$

$i = Yv$

[†]The counterpart of *direct current* is *alternating current* (ac), originally meaning *sinuoidal* but now loosely used to signify *variable*.

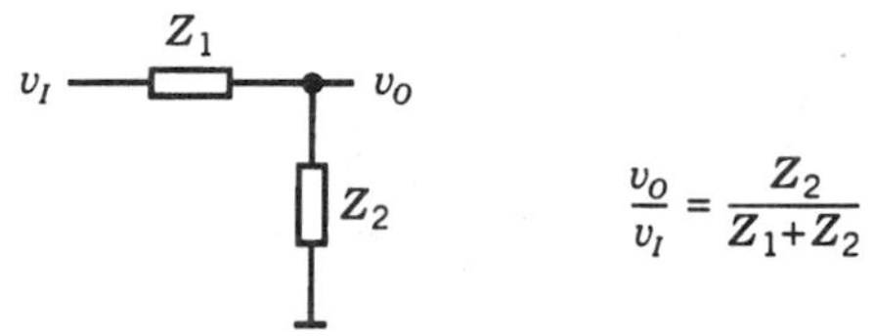

$$\frac{v_O}{v_I} = \frac{Z_2}{Z_1 + Z_2}$$

Figure 3-13. Divider with arbitrary impedances in the input and output legs.

We will also define the *admittance Y* (in siemens) as the reciprocal of the impedance:

$$Y = \frac{1}{Z} = \frac{i}{v}$$

The impedances of a resistor, an inductor, and a capacitor are R, sL, and $1/sC$, and the corresponding admittances are $1/R$, $1/sL$, and sC. The real part of an impedance Z is the *resistance R*, and the imaginary part is the *reactance X*; the real part of an admittance Y is the *conductance G*, and the imaginary part is the *susceptance B*. Thus the conductance of a resistor R is $G = 1/R$, and the reactance of an inductor L is $X = \omega L$.

We can now solve the generalized voltage divider of Fig. 3-13. Following the steps that led to (3-7), we obtain

$$\frac{v_O}{v_I} = \frac{Z_2}{Z_1 + Z_2}$$

From this formula we immediately obtain the transfer functions of RC and CR circuits given in (3-8) and (3-9).

Two admittances connected in *parallel*, as shown in Fig. 3-14a, have the same voltage, so that

$$i = (Y_1 + Y_2)v$$

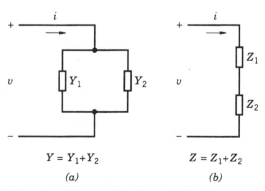

$$Y = Y_1 + Y_2$$

(a)

$$Z = Z_1 + Z_2$$

(b)

Figure 3-14. Parallel and series combinations. (*a*) Admittances in parallel add. (*b*) Impedances in series add.

and the admittance Y of the combination is given by

$$Y = Y_1 + Y_2$$

In contrast, two impedances connected in *series*, as shown in Fig. 3-14b, have the same current, so that

$$v = (Z_1 + Z_2)i$$

and the impedance Z of the combination is given by

$$Z = Z_1 + Z_2$$

We conclude that admittances in parallel add, as do impedances in series. In particular, the capacitances of capacitors in parallel add, as do the inductances of inductors in series and the resistances of resistors in series.

We will denote the parallel combination of two impedances Z_1 and Z_2 by $Z_1\|Z_2$. We then have

$$Z_1\|Z_2 = \frac{Z_1 Z_2}{Z_1 + Z_2}$$

To simplify notation, we will consider that a parallel combination takes precedence over a sum, so that rather than $Z_0 + (Z_1\|Z_2)$ we can write $Z_0 + Z_1\|Z_2$.

3-4 DEPENDENT SOURCES: THE TRANSFORMER

Let us consider the solenoidal *transformer* shown in Fig. 3-15. The fluxes in the two *windings* of the transformer are

$$\Phi_1 = -L_1 i_1 + M i_2$$
$$\Phi_2 = -M i_1 + L_2 i_2$$

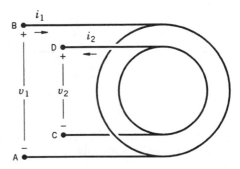

Figure 3-15. Transformer consisting of two coaxial solenoids: collapsed view along the axis.

where M is *the* mutual inductance. The fact that the mutual inductances between the windings are equal and that

$$M^2 = k^2 L_1 L_2$$

where

$$k^2 \leq 1$$

is the subject of Exercise 4-10.

Following the reasoning we used to calculate the voltage of an inductor in Section 3-1, we obtain

$$v_1 = v_{BA} = -\int_A^B \mathbf{E} \cdot d\mathbf{l} = \int_B^A \mathbf{E} \cdot d\mathbf{l} = \oint \mathbf{E} \cdot d\mathbf{l} = -\Phi_1' = L_1 i_1' - M i_2'$$

$$v_2 = v_{DC} = -\int_C^D \mathbf{E} \cdot d\mathbf{l} = \int_D^C \mathbf{E} \cdot d\mathbf{l} = \oint \mathbf{E} \cdot d\mathbf{l} = -\Phi_2' = M i_1' - L_2 i_2'$$

In terms of complex amplitudes and frequency responses, we then have

$$v_1 = s L_1 i_1 - s M i_2$$
$$v_2 = s M i_1 - s L_2 i_2$$

Solving for i_1 and v_2 in terms of i_2 and v_1, we get

$$i_1 = \frac{1}{sL_1} v_1 + \frac{M}{L_1} i_2$$

$$v_2 = \frac{M}{L_1} v_1 + s\left(\frac{M^2}{L_1} - L_2\right) i_2$$

Substituting $M = k\sqrt{L_1 L_2}$, we finally obtain

$$i_1 = \frac{1}{sL_1} v_1 + k\sqrt{\frac{L_2}{L_1}} i_2$$

$$v_2 = k\sqrt{\frac{L_2}{L_1}} v_1 - s(1 - k^2) L_2 i_2$$

(3-10)

These two equations can be represented by the circuit shown in Fig. 3-16. We note that the sources in this circuit are *dependent*, meaning that they are

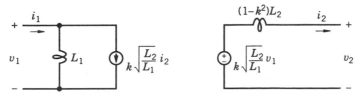

Figure 3-16. Equivalent circuit of the transformer of Fig. 3-15.

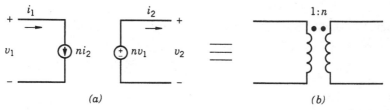

Figure 3-17. Ideal transformer. (*a*) Structure. (*b*) Standard symbol. The dots indicate terminals with the same polarity.

Figure 3-18. An admittance can be represented as a dependent source governed by its
own voltage drop: in both circuits $i = Yv$.

proportional to a voltage or to a current; we will return to this interesting development, but let us first abstract a new lumped element, the *ideal transformer*, shown in Fig. 3-17*a* as a pair of dependent sources but commonly represented in the literature by the symbol in Fig. 3-17*b*. The *turns ratio* $n = k\sqrt{L_2/L_1}$ is the ratio of the output (or *secondary*) voltage v_2 to the input (or *primary*) voltage v_1 or, equivalently, it is the ratio of the primary current i_1 to the secondary current i_2. In transformers in which the windings have the same cross section and k is close to 1, n is in fact approximately equal to the ratio of the number of turns in the secondary winding to the number of turns in the primary winding.

Dependent sources in a circuit are governed by variables in the circuit itself; as we will see in Section 3-5, they are therefore, like impedances and unlike independent sources, part of the structure of the circuit. This point is brought out quite directly by the fact, illustrated in Fig. 3-18, that an admittance can be represented as a dependent current source governed by its own voltage drop.

Dependent sources will arise naturally in the description of transistors; as a preview, Exercise 3-14 shows how they can be used to represent vacuum triodes, which are functionally quite similar to transistors.

3-5   STRUCTURAL CONSIDERATIONS

Most of the circuits we will encounter can be solved quite simply by sticking to the basics: summing currents at nodes and, where convenient, invoking superposition. There are, however, a few general results well worth the small investment that we will now make. To obtain these results, we must look at the structure of the *node equations* that we get when we sum currents at

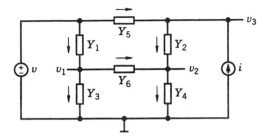

Figure 3-19. Simple circuit used to infer the general form of node equations.

nodes. Let us start from the simple circuit of Fig. 3-19 and infer the general behavior. Our procedure will be to solve for the voltages at all nodes. Once we have these voltages we can, if we wish, solve for the currents in the branches. We will take the voltage at one node, the junction of Y_3 and Y_4, to be zero (ground) and measure all other voltages with respect to this node.

Summing currents at nodes 1, 2, and 3 yields

$$(v-v_1)Y_1 = v_1Y_3 + (v_1-v_2)Y_6$$
$$(v_1-v_2)Y_6 + (v_3-v_2)Y_2 = v_2Y_4$$
$$i + (v-v_3)Y_5 = (v_3-v_2)Y_2$$

Rearranging terms, we get

$$
\begin{array}{llll}
v_1(Y_1+Y_3+Y_6) & -v_2Y_6 & 0 & = vY_1 \\
-v_1Y_6 & +v_2(Y_2+Y_4+Y_6) & -v_3Y_2 & = 0 \\
0 & -v_2Y_2 & +v_3(Y_2+Y_5) & = vY_5+i
\end{array}
$$

From this example it is not hard to infer that the node equations for an arbitrary circuit without dependent sources are of the form

$$v_1Y_{11} + v_2Y_{12} + \cdots + v_nY_{1n} = i_1$$
$$v_1Y_{21} + v_2Y_{22} + \cdots + v_nY_{2n} = i_2$$
$$\vdots \qquad\qquad\qquad\qquad\qquad (3\text{-}11)$$
$$v_1Y_{n1} + v_2Y_{n2} + \cdots + v_nY_{nn} = i_n$$

We observe the following points:

1. The *admittance matrix* formed by the coefficients Y_{jk} of the unknown node voltages depends only on admittances and expresses the structure of the circuit.

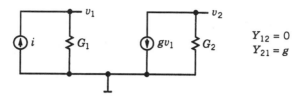

Figure 3-20. Reciprocity does not necessarily hold in circuits with dependent sources.

2. Independent sources occur only in the independent terms $i_1 \cdots i_n$.

3. The admittance matrix is symmetric,

$$Y_{jk} = Y_{kj}$$

We will not exploit this symmetry (called *reciprocity*) to its full extent, although we will find it useful as a check on our calculations.

Let us now consider the circuit with a dependent source shown in Fig. 3-20. Summing currents at nodes 1 and 2, we obtain

$$G_1 v_1 + 0 \quad = i$$

$$g v_1 + G_2 v_2 = 0$$

We conclude from this example, in which $Y_{12} = 0$ and $Y_{21} = g$, that the symmetry of the admittance matrix can be broken by dependent sources. We also conclude that dependent sources are structural, meaning that they contribute only to the admittance matrix.

An important point we want to stress is that the admittance matrix remains unaltered if, as illustrated in Fig. 3-21, an *independent* voltage source is inserted in series with a branch or an *independent* current source is inserted in parallel with a branch. The series voltage source of Fig. 3-21*a* modifies the sum of the currents at node 1 only in the term involving the branch between nodes 1 and 2:

$$\cdots + [v_2 - (v_1 + v)] Y + \cdots = \cdots$$

Figure 3-21. The admittance matrix is unaltered (*a*) if an independent voltage source is inserted in series with a branch or (*b*) if an independent current source is inserted in parallel with a branch.

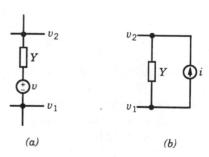

(a)   (b)

The term in v becomes part of the independent term i_1 in (3-11),

$$\cdots + (v_2 - v_1)Y + \cdots = \cdots + vY$$

and the admittance matrix is therefore unaltered. A similar argument holds for node 2. The parallel current source of Fig. 3-21b contributes directly to the independent terms i_1 and i_2 in (3-11), and thus does not alter the admittance matrix.

3-6 THÉVENIN'S THEOREM

Let us consider the circuit shown in Fig. 3-22a, and let us assume that it contains any combination of sources, dependent sources, and impedances, and that its only connection to the outside world is by means of two terminals. Let us take the positive terminal as node 1 and the negative terminal as ground. If we assume that we know i, we obtain

$$vY_{11} + v_2 Y_{12} + \cdots + v_n Y_{1n} = i_1 - i$$
$$vY_{21} + v_2 Y_{22} + \cdots + v_n Y_{2n} = i_2$$
$$\vdots$$
$$vY_{n1} + v_2 Y_{n2} + \cdots + v_n Y_{nn} = i_n$$

where $i_1 \cdots i_n$ do not depend on i. If we solve for v, we obtain

$$v = v_T - i z_T \tag{3-12}$$

where z_T depends only on the admittances Y_{jk}, that is, on circuit structure defined by impedances and dependent sources, and v_T is a linear combination of independent sources. This relation, represented by the equivalent circuit of Fig. 3-22b, is Thévenin's theorem: A linear two-terminal circuit is equivalent to a voltage source v_T in series with an impedance z_T. It follows

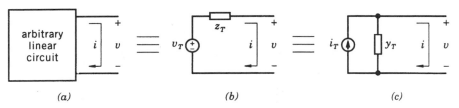

Figure 3-22. Thévenin's theorem and Norton's theorem. (a) Arbitrary two-terminal linear circuit. (b) Thévenin representation. (c) Norton representation.

necessarily that v_T is the open-circuit voltage (obtained by making $i = 0$) and that z_T is the impedance seen at the terminals when $v_T = 0$, that is, when all internal sources are nulled.

Going further, the short-circuit current i_T (obtained by making $v = 0$) is given by

$$i_T z_T = v_T$$

It follows that the Thévenin impedance z_T can be obtained by taking the ratio of the open-circuit voltage v_T and the short-circuit current i_T.

On the other hand, recalling that $y_T = 1/z_T$, we can rewrite (3-12) as

$$i = i_T - v y_T$$

and represent this relation with the equivalent circuit of Fig. 3-22c. We then obtain an alternative form of Thévenin's theorem, usually called Norton's theorem: A linear two-terminal circuit is equivalent to a current source i_T in parallel with an admittance y_T, and it follows necessarily that i_T is the short-circuit current (obtained by making $v = 0$) and that y_T is the admittance seen at the terminals when $i_T = 0$, that is, when all internal sources are nulled.

As a simple example, let us obtain the Thévenin equivalents of the resistive divider shown in Fig. 3-23. Summing currents at the positive output terminal yields

$$\frac{v_I - v}{R_1} = \frac{v}{R_2} + i$$

After minor algebra, we get

$$v = \frac{R_2}{R_1 + R_2} v_I - \frac{R_1 R_2}{R_1 + R_2} i$$

From this expression it follows that

$$v_T = \frac{R_2}{R_1 + R_2} v_I \quad \text{and} \quad z_T = \frac{R_1 R_2}{R_1 + R_2} = R_1 \| R_2$$

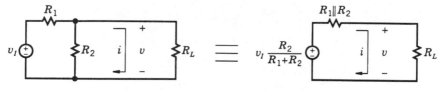

Figure 3-23. Thévenin equivalents of a resistive divider driven by a voltage source.

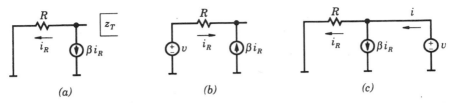

Figure 3-24. Alternative procedures for obtaining the output impedance of a circuit with dependent sources. (*a*) Original circuit. (*b*) An independent voltage source is inserted in series with branch *R*. (*c*) An independent source is applied to the output.

On the other hand, these are clearly the values we obtain for the open-circuit voltage and for the impedance looking back into the divider when $v_I = 0$.

The Thévenin impedance of a two-terminal circuit without dependent sources can often be calculated by taking series and parallel combinations of branches. In circuits with dependent sources this method is usually inapplicable. There are various ways of proceeding. If there are internal sources, we can calculate the open-circuit voltage and the short-circuit current and take their ratio. In the absence of internal sources, we can insert a voltage source in series with a branch or a current source in parallel with a branch and do the same calculation. In particular, the branch can be the two-terminal circuit itself, and this procedure is then equivalent to the obviously correct procedure of driving the circuit with a voltage source and calculating the current or driving it with a current source and calculating the voltage.

As an example, let us calculate the output impedance of the circuit shown in Fig. 3-24a. On the one hand, we can insert the voltage source v shown in Fig. 3-24b and calculate the short-circuit current i_T and the open-circuit voltage v_T:

$$i_T = \frac{v}{R} + \frac{\beta v}{R} = (\beta+1)\frac{v}{R}$$
$$v_T = v$$

Here we have used the fact that $i_R = v/R$ when the output is short circuited and that $i_R + \beta i_R = 0$ when the output is open circuited. It follows that

$$z_T = \frac{v_T}{i_T} = \frac{R}{\beta+1}$$

On the other hand, if we apply a voltage source v to the output, as shown in Fig. 3-24c, the current i delivered by v is

$$i = \frac{v}{R} + \beta\frac{v}{R} = (\beta+1)\frac{v}{R}$$

and we immediately obtain z_T as given above.

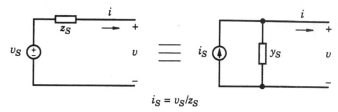

$$i_S = v_S/z_S$$

Figure 3-25. Source transformations that are valid even when v_S and i_S are dependent sources.

3-7 SOURCE TRANSFORMATIONS

The transformation shown in Fig. 3-25 is *always* possible, even when v_S or i_S depends on v or on i, or on variables elsewhere in the circuit; it simply offers two ways of saying the same thing, because

$$v_S - iz_S = v$$

and

$$i_S - vy_S = i$$

are equivalent expressions if we assume that

$$i_S = v_S y_S = \frac{v_S}{z_S}$$

3-8 TWO-PORTS

Figure 3-26 shows two circuit configurations that occur so frequently that they merit separate consideration. These configurations represent circuits that contain no independent sources and are connected to the external world

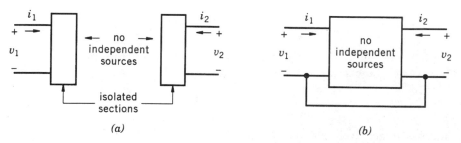

Figure 3-26. Two-ports. (*a*) Version with isolated two-terminal sections. (*b*) Three-terminal version.

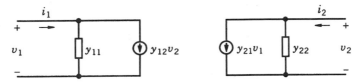

Figure 3-27. The y-parameter model of a two-port.

by means of two pairs of terminals or *ports*. The configuration of Fig. 3-26*a* consists of two two-terminal sections that are isolated from each other, although either one can contain dependent sources governed by voltages or currents in the other. In the configuration of Fig. 3-26*b*, two of the terminals are short circuited. In both cases, we will see that knowledge of any pair of the variables v_1, i_1, v_2, and i_2 is sufficient to determine the other pair.

If we assume that v_1 and v_2 are known, we can apply Norton's theorem to both ports and obtain the equivalent circuit shown in Fig. 3-27, usually referred to as the *admittance-parameter* or *y-parameter model*, in which

$$i_1 = y_{11}v_1 + y_{12}v_2$$
$$i_2 = y_{21}v_1 + y_{22}v_2$$

Thinking in terms of signal flow from left to right, y_{11} and y_{21} are the (*short-circuit*) *input admittance* and *forward transadmittance*, and it is clear that they can be obtained by calculating i_1 and i_2 when v_2 is zero, that is, when the output is short circuited. In the same vein, y_{22} and y_{12} are the (*short-circuit*) *output admittance* and *reverse transadmittance*, obtained by calculating i_2 and i_1 when v_1 is zero, that is, when the input is short circuited.

As an example, let us consider the pi network of Fig. 3-28. When $v_2 = 0$, we have

$$i_1 = (G_1 + G_3)v_1 \quad \text{and} \quad i_2 = -G_3v_1$$

and when $v_1 = 0$, we have

$$i_2 = (G_2 + G_3)v_2 \quad \text{and} \quad i_1 = -G_3v_2$$

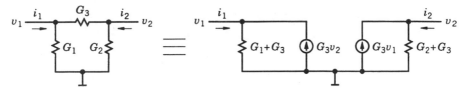

Figure 3-28. The y-parameter model of a pi network.

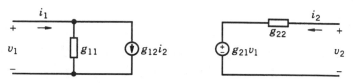

Figure 3-29. The g-parameter model of a two-port.

The y-parameters are thus

$$y_{11} = G_1 + G_3$$
$$y_{22} = G_2 + G_3$$
$$y_{21} = y_{12} = -G_3$$

We note that $y_{21} = y_{12}$ in this case because there are no dependent sources.

Other equivalent circuits or models can be obtained by applying either Thévenin's or Norton's theorem to the input and to the output. Figure 3-29 shows the *g-parameter* model obtained by assuming that v_1 and i_2 are known and applying Norton's theorem to the input and Thévenin's theorem to the output. In this case the equations read[†]

$$i_1 = g_{11}v_1 + g_{12}i_2$$
$$v_2 = g_{21}v_1 + g_{22}i_2$$

Thinking again in terms of signal flow from left to right, g_{11} and g_{21} are the (*open-circuit*) *input admittance* and *forward voltage gain*, obtained when i_2 is zero, that is, when the output is open circuited, and g_{22} and g_{12} are the (*short-circuit*) *output impedance* and *reverse current gain*, obtained when v_1 is zero, that is, when the input is short circuited.

If we now look at the equations in (3-10) and at Fig. 3-16, it is clear that except for a sign reversal in the output current, we have used the g-parameter model to represent the transformer of Fig. 3-15. Further models are considered in Exercise 3-21.

3-9 POWER AND ENERGY

If we remove a charge Δq from the negative plate of a capacitor and slowly transport it to the positive plate, as shown in Fig. 3-30a, the work $v\Delta q$ we do appears as an increase Δu in the stored energy u of the capacitor,

$$\Delta u = v\Delta q$$

[†]We normally use the letter g to denote a conductance. Its use for a different purpose is forced in this case by the literature.

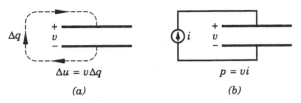

Figure 3-30. (*a*) Moving a charge Δq from the negative to the positive plate of a capacitor at voltage v increases the stored energy u of the capacitor by $\Delta u = v\Delta q$. (*b*) The power p into a capacitor is given by the vi product.

This leads us to surmise that the power p delivered to a capacitor that is at voltage v and is being charged by a current i, as shown in Fig. 3-30*b*, is given by

$$p = u' = vq' = vi$$

In Chapter 4 we will see that this relation, while not strictly true, does hold at sufficiently low frequencies.

From conservation of energy it follows that the product vi is also the power delivered by the current source, and furthermore, that the power delivered to any two-terminal network, even if it contains nonlinear elements, is given by

$$p = vi$$

It also follows that the power into an n-terminal network is the sum of the products of the voltages at $n-1$ of the terminals relative to the nth, taken as reference, times the currents into these terminals. Thus, for the network shown in Fig. 3-31, we have

$$p = v_1 i_1 + v_2 i_2$$

Let us look at some particular cases. The power dissipated in a resistor is

$$p = vi = Ri^2 = Gv^2$$

The power into a capacitor is the time derivative of the stored energy,

$$u' = p = vi = vCv'$$

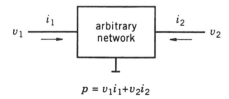

$$p = v_1 i_1 + v_2 i_2$$

Figure 3-31. The power into a three-terminal network is given by the sum of the vi products at two of the terminals when the third terminal is grounded.

Integrating with respect to time, we obtain

$$u = \tfrac{1}{2}Cv^2$$

The power into an inductor is also the time derivative of the stored energy,

$$u' = p = vi = Li'i$$

and the stored energy is

$$u = \tfrac{1}{2}Li^2$$

For sinusoidal signals in linear two-terminal networks we have

$$v(t) = v_0 \cos(\omega t) \quad \text{and} \quad i(t) = i_0 \cos(\omega t + \varphi)$$

and the instantaneous power is

$$v(t)i(t) = v_0 i_0 \cos^2(\omega t) \cos(\varphi) - v_0 i_0 \cos(\omega t) \sin(\omega t) \sin(\varphi)$$

The average power is thus

$$\langle p \rangle = \langle v(t)i(t) \rangle = \tfrac{1}{2}v_0 i_0 \cos(\varphi)$$

or in terms of the complex amplitudes of the voltage and the current,

$$\langle p \rangle = \tfrac{1}{2}\text{Re}(vi^*)$$

EXERCISES

3-1 LABORATORY I. You will need a square-wave generator with a $50\,\Omega$ output impedance, a sine-wave generator with an output impedance under $50\,\Omega$, and a current source capable of delivering 1A. Do Exercises 3-9 and 3-15 before proceeding.

(**a**) Using enameled copper wire of diameter 1 mm (18AWG), wind a 50-turn solenoid on a wood or Lucite cylinder of radius 25 mm. Use (3-4) to estimate that $L \simeq 100\,\mu\text{H}$. Obtain also that the resistance r of the solenoid is $0.17\ \Omega$ (the resistance per unit length of copper wire of diameter 1 mm is 22 Ω/km). Run a 1A current through the solenoid and measure the voltage to obtain r.

(**b**) The response of the circuit of Fig. 3-32a to a step of amplitude V is

$$v_O(t) = \frac{V}{2}\left[1 - \exp\left(-\frac{t}{\tau}\right)\right]$$

where $\tau = L/R$ and $R = 100\ \Omega$. Use this circuit to measure L.

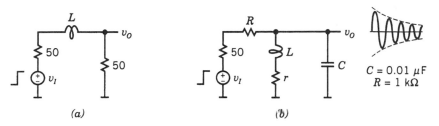

Figure 3-32. Circuits for Exercise 3-1.

(c) If $Q>5$, the response of the circuit of Fig. 3-32b to a step of amplitude V is approximately

$$v_O(t) \simeq V\frac{R_0}{R}\exp\left(-\frac{\omega_0 t}{2Q}\right)\sin(\omega_0 t)$$

where $\omega_0 = 1/\sqrt{LC}$, $R_0 = \sqrt{L/C}$, and $Q = R/R_0$. Use this circuit to measure Q and L. (Observe that $Q = n\pi$, where n is the number of cycles in which the amplitude of the oscillation decays by a factor $e = 2.72$.) According to Exercise 3-15, the equivalent parallel resistance of the solenoid is $R_0^2/r \simeq 60$ kΩ and thus much larger than R. According to Exercise 3-22, the losses in the capacitor can be represented near the resonant frequency by a parallel resistance that is at least 100 times larger than $1/\omega_0 C$ and can therefore also be ignored.

(d) Measure the resonant frequency f_0 and the width of the resonance at the half-power points of the circuit of Fig. 3-32b. From these, obtain L and Q.

3-2 LABORATORY II.

(a) Show that the frequency responses of the circuit of Fig. 3-33a are

$$\frac{v_C}{v_I} \simeq \frac{1}{1+\tau s} \quad\text{and}\quad \frac{v_O}{v_I} \simeq \frac{1}{(1+\tau s)(1+\tau s/10)}$$

(b) For v_C as well as v_O, use a sine-wave generator to measure the cutoff frequency ω_C, and a square-wave generator to measure the rise time t_R. The step responses should be as given in Fig. 3-33b. Verify in both cases that

$$t_R \omega_C \simeq 2.2$$

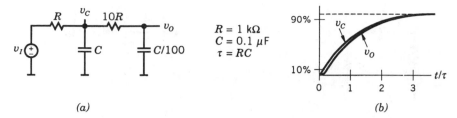

R = 1 kΩ
C = 0.1 μF
τ = RC

(a)   (b)

Figure 3-33. Circuits for Exercise 3-2.

(c) According to Exercise 1-12, the step response of v_O for $t>\tau/10$ can be obtained by making the substitution

$$\frac{1}{1+\tau s/10} \to \exp(-\tau s/10)$$

Verify that this is so by measuring the delay between the step responses of v_O and v_C at several values of t between $\tau/10$ and 2τ.

3-3 OLD-FASHIONED ELECTRONICS: POTENTIOMETERS. The circuit of Fig. 3-34a shows how a divider can be used to measure an unknown voltage v if there is available a calibrated source v_{cal}. Resistor R is a potentiometer, that is, a resistor with a tap or pick-off point. You may assume that the position of the tap (the value of α) can be accurately set. If α is adjusted so that Δv is zero, you have

$$v = \frac{1}{\alpha}v_{cal}$$

This circuit, in addition to being somewhat old-fashioned, has the drawback that it loads the source of v with a current v/R. In contrast, the improved potentiometer shown in Fig. 3-34b draws no current from the source of the unknown voltage v when properly adjusted, that is, when v_{var} and α are set so that $\Delta v_1 = \Delta v_2 = 0$, in which case $v = \alpha n v_{cal}$.

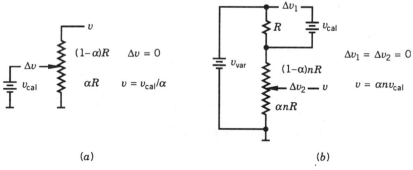

(a)   (b)

Figure 3-34. Potentiometer circuits (Exercise 3-3).

3-4 *R–2R* NETWORKS AND DIGITAL-TO-ANALOG CONVERTERS.
Consider the *R–2R* divider shown in Fig. 3-35. If you load it with a resistor
$R_L = 2R$, the output voltage is half the input voltage and the input resistance
R_I is equal to R_L. This allows you to repeatedly substitute copies of the
loaded divider in place of the load resistor, and you end up with the *R–2R*
network of Fig. 3-36, which generates a decreasing series of voltages equal to
the input voltage divided by powers of 2. In *digital-to-analog converters*, these
voltages are selectively added to generate 2^N voltages from zero to $(1 - 2^{-N})V$
in steps of size $2^{-N}V$.

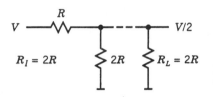

Figure 3-35. *R–2R* divider with a load resistance $R_L = 2R$ (Exercise 3-4).

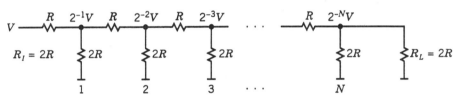

Figure 3-36. *R–2R* network (Exercise 3-4).

3-5 SPLITTER FOR SIGNALS ON COAXIAL CABLES. Coaxial cables
are used to transmit high-frequency signals with minimal distortion. If a
coaxial cable is terminated (loaded) at the receiving end in its *characteristic
resistance R_C* (usually 50 to 100 ohms) there is no reflection from that end
and the cable looks like a resistance R_C. Figure 3-37 shows a *splitter* used to
distribute signals on coaxial cables. Convince yourself that the impedance
into any input is R_C if the remaining inputs are terminated in R_C, and that
the transmitted amplitudes are half the input amplitude.

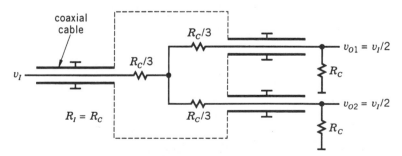

Figure 3-37. Splitter for signals on coaxial cables (Exercise 3-5).

3-6 KELVIN–VARLEY DIVIDER. The Kelvin–Varley divider shown in Fig. 3-38 can divide accurately by 10,000 or more. Convince yourself that its input resistance is equal to $10R$, whatever the positions of the switches, if the output load is infinite.

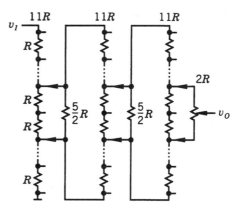

Figure 3-38. Kelvin–Varley divider (Exercise 3-6).

3-7 IMPEDANCE OF A WHEATSTONE BRIDGE. Convince yourself that the impedance of the Wheatstone bridge shown in Fig. 3-39 cannot be obtained by series and parallel combinations unless $n = 0, 1,$ or ∞.

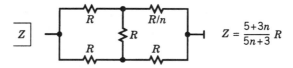

$$Z = \frac{5+3n}{5n+3} R$$

Figure 3-39. Wheatstone bridge (Exercise 3-7).

3-8 INPUT CIRCUIT OF A COMMON-EMITTER AMPLIFIER. The common-emitter amplifier is one of the basic transistor amplifiers you will meet later; as you will see, its input circuit is as given in Fig. 3-40. Show that

$$\frac{v_B}{v_T} = \frac{1+sR_MC_M}{1+s(R_MC_M+R_TC_T+R_TC_M)+s^2R_MC_MR_TC_T}$$

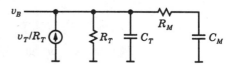

Figure 3-40. Circuit for Exercise 3-8.

3-9 FIRST-ORDER AND SECOND-ORDER CIRCUITS. Solve for v_O in the circuits shown in Fig. 3-41 and check that the normalized differential equations and frequency responses are correct as given.

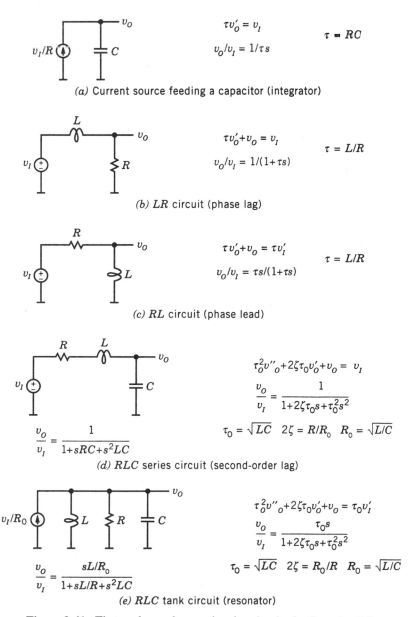

$$\tau v_0' = v_I$$
$$v_0/v_I = 1/\tau s$$
$$\tau = RC$$

(a) Current source feeding a capacitor (integrator)

$$\tau v_0' + v_0 = v_I$$
$$v_0/v_I = 1/(1+\tau s)$$
$$\tau = L/R$$

(b) LR circuit (phase lag)

$$\tau v_0' + v_0 = \tau v_I'$$
$$v_0/v_I = \tau s/(1+\tau s)$$
$$\tau = L/R$$

(c) RL circuit (phase lead)

$$\frac{v_O}{v_I} = \frac{1}{1+sRC+s^2LC}$$

$$\tau_0^2 v''_0 + 2\zeta\tau_0 v_0' + v_0 = v_I$$
$$\frac{v_0}{v_I} = \frac{1}{1+2\zeta\tau_0 s+\tau_0^2 s^2}$$
$$\tau_0 = \sqrt{LC} \quad 2\zeta = R/R_0 \quad R_0 = \sqrt{L/C}$$

(d) RLC series circuit (second-order lag)

$$\frac{v_O}{v_I} = \frac{sL/R_0}{1+sL/R+s^2LC}$$

$$\tau_0^2 v''_0 + 2\zeta\tau_0 v_0' + v_0 = \tau_0 v_I'$$
$$\frac{v_0}{v_I} = \frac{\tau_0 s}{1+2\zeta\tau_0 s+\tau_0^2 s^2}$$
$$\tau_0 = \sqrt{LC} \quad 2\zeta = R_0/R \quad R_0 = \sqrt{L/C}$$

(e) RLC tank circuit (resonator)

Figure 3-41. First-order and second-order circuits for Exercise 3-9.

3-10 EQUIVALENT CIRCUIT OF A QUARTZ CRYSTAL. A quartz crystal is a piezoelectric device that is used to make narrowband filters and stable oscillators. Its equivalent circuit is shown in Fig. 3-42; the numerical values are for a typical 4 MHz crystal. The impedance Z between the terminals, given by

$$Z = \frac{1+sRC+s^2LC}{sC_P(1+sRC_S+s^2LC_S)}$$

has a very sharp series resonance at $\omega_S = 1/\sqrt{LC}$ and an equally sharp parallel resonance at $\omega_P = 1/\sqrt{LC_S}$, which is about 0.1% higher than ω_S. Show that the impedances at the series and parallel resonances are

$$Z(i\omega_S) \simeq R \quad \text{and} \quad Z(i\omega_P) \simeq \frac{1}{R}\frac{L}{C_0}\frac{C}{C_0} \simeq 430\ \text{k}\Omega$$

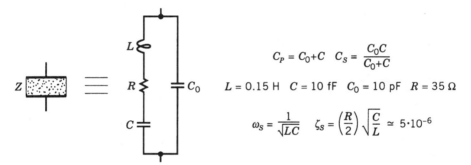

$$C_P = C_0+C \quad C_S = \frac{C_0C}{C_0+C}$$

$$L = 0.15\ \text{H} \quad C = 10\ \text{fF} \quad C_0 = 10\ \text{pF} \quad R = 35\ \Omega$$

$$\omega_S = \frac{1}{\sqrt{LC}} \quad \zeta_S = \left(\frac{R}{2}\right)\sqrt{\frac{C}{L}} \simeq 5\cdot10^{-6}$$

Figure 3-42. Equivalent circuit of a quartz crystal (Exercise 3-10).

3-11 THÉVENIN EQUIVALENTS IN CIRCUITS WITH DEPENDENT SOURCES. The output impedance z_T of the circuit of Fig. 3-43 cannot be obtained by series and parallel combinations because of the presence of the dependent source. However, z_T can be obtained by dividing the open-circuit voltage v_T by the short-circuit current i_T. Observe that $v_T = v_B$ because you cannot have nonzero currents of opposite signs in a branch. For the record, you are dealing with an *emitter follower*, a transistor circuit with a large input impedance (infinite in this case), a small output impedance, and a gain close to 1 (exactly 1 in this case).

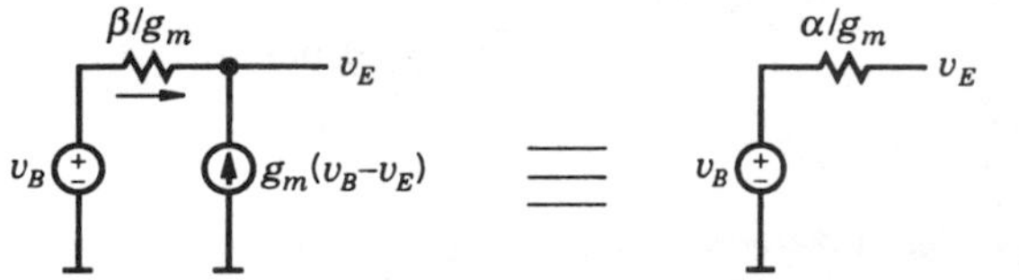

$$v_T = v_B$$
$$i_T = v_B(g_m+g_m/\beta)$$
$$z_T = v_T/i_T = \alpha/g_m$$
$$\alpha = \beta/(\beta+1)$$

Figure 3-43. Circuit for Exercise 3-11.

3-12   IMPEDANCE EQUALIZER. Show that the impedance of the equalizer of Fig. 3-44 is equal to R if the time constants RC and L/R are equal.

As shown in Fig. 3-45, this network can compensate for a load capacitance C_L, so that a coaxial cable can be terminated in its characteristic resistance R_C (see Exercise 3-5); the price is a longer output rise time. For a load capacitance $C_L = 100$ pF and a characteristic resistance $R_C = 50\Omega$, you must make $L = 250$ nH. With no compensating network you observe a reflected pulse about 2 ns wide.

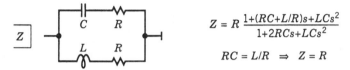

$$Z = R\,\frac{1+(RC+L/R)s+LCs^2}{1+2RCs+LCs^2}$$

$$RC = L/R \ \Rightarrow\ Z = R$$

Figure 3-44. Impedance equalizer (Exercise 3-12).

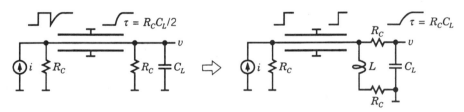

Figure 3-45. Impedance equalizer used to neutralize the load capacitance C_L at the receiving end of a coaxial cable (Exercise 3-12).

3-13   COMPENSATION OF A ×10 OSCILLOSCOPE PROBE. Oscilloscopes typically have an input impedance of 1 MΩ in parallel with 15 pF; in addition, oscilloscope probes normally include about 1 meter of coaxial cable with a capacitance of order 100 pF. The circuit of Fig. 3-46a is the first step required to compensate for these capacitances. If $R_1C_1 = R_2C_2$, you have $v_O = v_I/10$, and the circuit under test is loaded with 10 MΩ in parallel with 10 pF. The frequency response is

$$\frac{v_O}{v_I} = \frac{R_2(1+sR_1C_1)}{R_1(1+sR_2C_2)+R_2(1+sR_1C_1)}$$

Show that the step responses indicated in Fig. 3-46b are correct.

This arrangement is satisfactory for frequencies up to about 20 MHz. At higher frequencies reflections in the coaxial cable become important, and a more elaborate network is required.

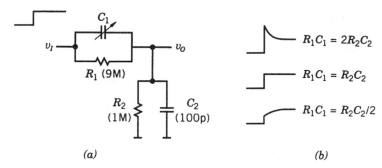

(a) *(b)*

Figure 3-46. Low-frequency compensation of a ×10 oscilloscope probe (Exercise 3-13). (*a*) Circuit. (*b*) Step responses.

3-14 DEPENDENT SOURCES: VACUUM-TRIODE AMPLIFIER. Consider the nonstandard—but nonetheless viable—parallel-plate *vacuum triode* shown in Fig. 3-47a. It consists of two conducting plates, the top one being the *anode* or *plate* and the bottom one the *cathode*, and a wire *grid* that is located close to the cathode; the ensemble is in vacuum.

Assume initially that the grid is removed and that the plate is made positive relative to the cathode; assume also that the cathode is heated by an external agent to the point that it emits a substantial amount of electrons. As shown in Fig. 3-47b, these electrons have an energy distribution dn/du. If

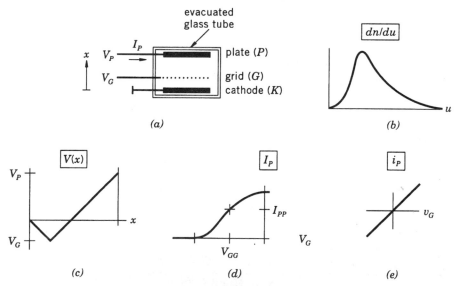

(c) *(d)* *(e)*

Figure 3-47. Vacuum triode (Exercise 3-14). (*a*) Structure. (*b*) Energy distribution of electrons emitted from the cathode. (*c*) Potential between the plates when the grid voltage is negative. (*d*) Plate current as a function of the grid voltage in the temperature-limited regime. (*e*) Small-signal dependence of the plate current on the grid voltage around a fixed grid voltage V_{GG}.

the plate voltage V_P is made sufficiently high, all the electrons emitted are collected by the plate; in this *temperature-limited regime*, the plate current I_P becomes independent of V_P.

If the grid is now inserted and made negative relative to the cathode, the potential $V(x)$ between the plates looks as shown in Fig. 3-47c. A fraction of the electrons do not have enough energy to pass beyond the grid, and as indicated in Fig. 3-47d, the plate current is reduced, although it is still independent of minor variations in the plate voltage. If the grid is made negative enough, the plate current is completely cut off. Since the grid is negative relative to the cathode it repels electrons, and the grid current is therefore zero.

While admittedly sketchy, the preceding description should be sufficient to convince you that this vacuum triode is a (nonlinear) dependent current source. The plate current in commercial triodes is not independent of the plate voltage, but the somewhat more complicated *pentodes* are in fact good grid-controlled current sources.

If the grid voltage is varied by an amount v_G around a given *bias* voltage V_{GG}, the plate current exhibits corresponding variations i_P around a value I_{PP} (the *operating point*), as shown in Fig. 3-47e. For small v_G, these variations are given by

$$i_P = g_m v_G$$

where g_m is the *transconductance*,

$$g_m = \frac{dI_P}{dV_G}$$

Figure 3-48a shows the symbol for a vacuum triode, and Fig. 3-48b shows how small variations in the variables can be represented by a linear *small-signal model* or *incremental circuit*, in this case a current source controlled by the grid voltage.

Consider now the circuit of Fig. 3-49a, in which a resistance R_P has been inserted between the plate and a fixed *plate supply voltage* V_{PP}, and assume

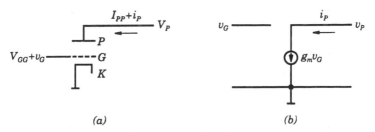

(a)   *(b)*

Figure 3-48. Vacuum triode (Exercise 3-14). (*a*) Symbol and sign conventions. (*b*) Small-signal model.

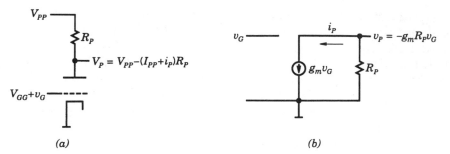

Figure 3-49. Vacuum-triode amplifier (Exercise 3-14). (*a*) Structure. (*b*) Incremental circuit.

that $V_P = V_{PP} - I_P R_P$ remains high enough that the triode is still a current source. The incremental circuit for small variations in the grid voltage is shown in Fig. 3-49*b*. The incremental plate voltage v_P is given by

$$v_P = -g_m R_P v_G$$

If $g_m R_P > 1$, the incremental plate voltage v_P is larger in magnitude than the incremental grid voltage v_G, and you have amplifier.

Real amplifiers require additional circuitry to establish the operating point, and capacitors to couple ground-referenced signals to the grid and to the plate; in addition, as mentioned, triodes are not quite what you have been shown in that they are not ideal current sources, so that their small-signal model has a nonnegligible resistor in parallel with the current source. Nonetheless, the essential ideas remain unaltered in a more complete analysis.

3-15   TANK CIRCUITS. Verify that the responses given for the circuits in Fig. 3-50 are correct, and convince yourself that for frequencies such that

$$R_S \ll \omega L$$

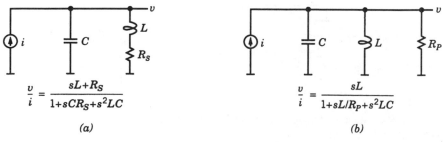

Figure 3-50. Tank circuits for Exercise 3-15.

the series resistance R_S of the inductor in a tank circuit (Fig. 3-50a) can be replaced by a parallel resistance R_P (Fig. 3-50b) given by

$$R_P R_S = R_0^2 = L/C$$

3-16 INITIAL CONDITIONS FROM PHYSICAL CONSIDERATIONS. The initial conditions for the step response of the RLC series circuit of Fig. 3-41d can be obtained by observing that $r_0(0+) = 0$ because the capacitor needs a finite time to change its voltage and that $r_0'(0+) = 0$ because the current in the inductor—and therefore in the capacitor—is initially zero.

With similar arguments, show that the initial conditions for the step response of the tank circuit of Fig. 3-41e are $r_0(0+) = 0$ and $\tau_0 r_0'(0+) = 1$.

3-17 CIRCUITS WITH INITIAL CONDITIONS IN CAPACITORS AND INDUCTORS. In the circuit of Fig. 3-51a the capacitor is initially charged to a voltage V, whereas in the circuit of Fig. 3-51b the capacitor is initially discharged and a voltage source $V\theta(t)$ has been inserted in series with the capacitor. In both cases, for $t > 0$ you have

$$v = V - \frac{1}{C}\int_0^t i(\eta)\, d\eta$$

You may thus conclude that the behavior for $t > 0$ of circuits with initially charged capacitors can be obtained from step responses, which always involve initially discharged capacitors.

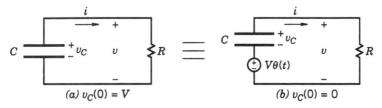

(a) $v_C(0) = V$ (b) $v_C(0) = 0$

Figure 3-51. (a) A capacitor with an initial voltage can be replaced by (b) an initially discharged capacitor in series with a stepped voltage source (Exercise 3-17).

To see a particular case, write down and solve the differential equation for the circuit of Fig. 3-51a with the initial condition $v(0+) = V$ and obtain that

$$v(t > 0) = V \exp(-t/RC)$$

This is clearly what you obtain for the equivalent CR circuit of Fig. 3-51b when you apply the voltage step $V\theta(t)$.

With similar arguments convince yourself that as shown in Fig. 3-52, an inductor with an initial current I can be replaced by an inductor with zero initial current in parallel with a stepped current source $I\theta(t)$.

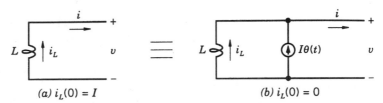

(a) $i_L(0) = I$ (b) $i_L(0) = 0$

Figure 3-52. (a) An inductor with an initial current can be replaced by (b) an inductor with zero initial current in parallel with a stepped current source (Exercise 3-17).

3-18   THE PI–TEE TRANSFORMATION. Show that the pi and tee networks of Fig. 3-53 are equivalent if

$$R_A R_2 = R_B R_1 = R_C R_3 = R^2$$

where, depending on where you start from,

$$R^2 = R_1 R_2 + R_1 R_3 + R_2 R_3 \quad \text{or} \quad R^2 = \frac{R_A R_B R_C}{R_A + R_B + R_C}$$

Suggestion: Calculate the y-parameters of both networks and equate them.

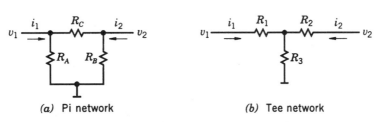

(a) Pi network   (b) Tee network

Figure 3-53. Circuits for Exercise 3-18.

3-19   CONSERVATION OF ENERGY. Show that the initial energy stored in the capacitor of Fig. 3-54a or in the inductor of Fig. 3-54b is equal to the energy dissipated by the resistor:

$$u_C(0) = \int_0^\infty p_R(t)\, dt \quad \text{and} \quad u_L(0) = \int_0^\infty p_R(t)\, dt$$

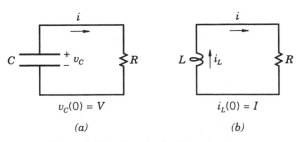

Figure 3-54. Circuits for Exercise 3-19.

3-20 TRANSFORMERS. Show that if the ideal transformer of Fig. 3-17 is loaded with an impedance Z_S in the secondary, the input impedance Z_P on the primary side is

$$Z_P = Z_S/n^2$$

Use this result to show that if you short-circuit the secondary of the transformer of Fig. 3-16, the input impedance Z_{SC} on the primary side is, as you should expect from symmetry,

$$Z_{SC} = s(1-k^2)L_1$$

With the secondary open circuited, the secondary-to-primary voltage ratio is the turns ratio n,

$$n = k\sqrt{L_2/L_1}$$

and the input impedance on the primary side is

$$Z_{OC} = sL_1$$

L_1, L_2, and k can therefore be determined by measuring Z_{SC}, Z_{OC}, and n. Convince yourself that this procedure can be extended to the case in which the primary and secondary windings have series resistances R_1 and R_2.

3-21 TWO-PORT MODELS. Apply Thévenin's theorem to both ports of the circuits of Fig. 3-26 to obtain the *impedance-parameter* (or *z-parameter*) model for two-ports shown in Fig. 3-55, in which

$$v_1 = z_{11}i_1 + z_{12}i_2$$

$$v_2 = z_{21}i_1 + z_{22}i_2$$

Thinking in terms of signal flow from left to right, z_{11} and z_{21} are the (*open-circuit*) *input impedance* and *forward transimpedance*, obtained by calculating v_1 and v_2 when i_2 is zero, that is, when the output is open circuited. In the same vein, z_{22} and z_{12} are the (*open-circuit*) *output impedance* and *reverse transimpedance*, obtained by calculating v_2 and v_1 when i_1 is zero, that is, when the input is open circuited.

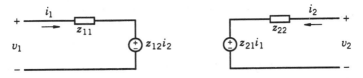

Figure 3-55. The z-parameter model (Exercise 3-21).

Show that the z-parameters of the tee network of Fig. 3-53b are given by

$$z_{11} = R_1 + R_3$$
$$z_{22} = R_2 + R_3$$
$$z_{12} = z_{21} = R_3$$

Now apply Thévenin's theorem to the input and Norton's theorem to the output of the circuits of Fig. 3-26 to obtain the *hybrid-parameter* (or *h-parameter*) model of Fig. 3-56, in which

$$v_1 = h_{11}i_1 + h_{12}v_2$$
$$i_2 = h_{21}i_1 + h_{22}v_2$$

The nomenclature should be clear by now: h_{11} and h_{21} are the (*short-circuit*) *input impedance* and *forward current gain*, whereas h_{22} and h_{12} are the (*open-circuit*) *output admittance* and *reverse voltage gain*. As you will eventually see, the h-parameter model is used in the literature to describe the low-frequency properties of transistors.

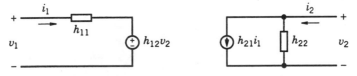

Figure 3-56. The h-parameter model (Exercise 3-21).

3-22 DIELECTRIC LOSSES. Consider a capacitor C filled with a nonconducting dielectric and subjected to a sinusoidal voltage $V_0 \sin(\omega t)$. Accept as a rough experimental fact that in analogy with hysteresis losses in magnetic materials, the energy loss in one cycle is proportional to V_0^2 but independent of ω. It follows that the power loss is proportional to ω and that it can be accounted for by placing in parallel with C a conductance G_C given by

$$G_C = \omega CD$$

where D is the *loss factor*. Again roughly, D is about 10^{-2} in the G10 glass epoxy used in printed circuits, 10^{-3} in Mylar, and 10^{-4} in exceptional materials such as polyethylene, polystyrene, and Teflon.

One consequence you can draw is that the Q of a tank circuit, defined in Exercise 1-11, is limited by dielectric losses, and that at best you have

$$Q = \frac{1}{D}$$

The memory effects that underlie dielectric losses manifest themselves in the form of *dielectric absorption* when a charged capacitor is shorted for a while: after the short is removed, the voltage builds up to a fraction (of order D) of its initial value. In addition to the hazards it implies in the handling of high-voltage capacitors, dielectric absorption limits the accuracy of circuits such as *integrating analog-to-digital converters* in which the voltage on a capacitor measures the integrated current into the capacitor.

4

FROM FIELDS TO CIRCUITS

4-1 INTRODUCTION

The concepts of an ideal capacitor or of an ideal inductor are widely used, and rightfully so, because they are fertile abstractions. In this chapter we will look into some aspects of the physics of lumped circuit elements in order to understand the context in which these abstractions are valid. In doing so, we will incidentally lay the groundwork for later discussions on transistors, transmission lines, and noise. To get to the point without hindrance, we will not consider lumped elements containing dielectric or magnetic materials; we will thus deal with metals, resistive materials with permittivity ε_0, and air (vacuum).

As we proceed, we will see that the range over which lumped elements are ideal is actually quite broad: we might have to do some rethinking beyond 100 MHz, but at 100 kHz we seldom have anything to worry about. More specifically, we will show that a lumped element can be considered ideal only for frequencies such that

$$\left(\frac{\omega l}{c} \right)^2 \ll 1$$

where l is a characteristic dimension of the element. This is not a sufficient condition, but it is certainly necessary. With this condition in mind, we will review Maxwell's equations and see what approximations are valid if it holds.

We will also see that fair representations of physical lumped elements can be obtained by adding corrections—in the form of further ideal elements—to

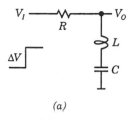
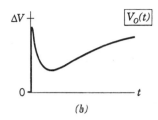

Figure 4-1. (*a*) *RC* circuit with a parasitic inductance in series with the capacitor. (*b*) The output voltage $V_O(t)$ exhibits an initial spike if the input voltage $V_I(t)$ is a sharp step.

the basic idealized elements. A physical capacitor can thus be imagined as an ideal capacitor with a small *parasitic* inductor in series, and a physical inductor as an ideal inductor with a small parasitic capacitor in parallel. With such corrections, physical circuits can be realistically transcribed to paper as equivalent circuits with ideal lumped elements at frequencies well beyond 100 MHz.

The wiring in circuits introduces *stray* capacitances and inductances that are often more important than the corrections to the lumped elements themselves. By means of examples, we will come to the conclusion that these stray elements are of the order of ε_0 or μ_0 times the length of the wires.

Not paying attention to corrections can result in unexpected or even faulty behavior. For example, the *RC* circuit of Fig. 4-1 (it is the subject of the laboratory proposed in Exercise 4-1) might be intended to produce a delay, but if the rise time of the step input is sufficiently short, the series inductance of the capacitor, either intrinsic or due to the wiring, generates a spike that might prematurely trigger a subsequent circuit.

4-2 MAXWELL'S EQUATIONS IN VACUUM

The force on a point charge q that moves with velocity **v** in an electromagnetic field is the Lorentz force **F**, given in terms of the electric and magnetic fields **E** and **B** by

$$\mathbf{F} = q(\mathbf{E} + \mathbf{v} \times \mathbf{B})$$

The fields are connected to each other and to the charge and current densities ρ and $\mathbf{J}$ by Maxwell's equations[†]:

$$\nabla \cdot \mathbf{E} = \frac{\rho}{\varepsilon_0} \qquad \text{(Gauss's law)}$$

$$\nabla \times \mathbf{E} = -\partial \mathbf{B} \qquad \text{(Faraday's law)}$$

$$\nabla \cdot \mathbf{B} = 0 \qquad \text{(Absence of magnetic charges)}$$

$$c^2 \nabla \times \mathbf{B} = \frac{\mathbf{J}}{\varepsilon_0} + \partial \mathbf{E} \qquad \text{(Ampère–Maxwell law)}$$

Maxwell's equations are linear, which implies that superpositions of solutions are also solutions.

At the surface separating two media with different properties, the fields must satisfy the following boundary conditions:

$$\mathbf{E}_\perp - \mathbf{E}'_\perp = \frac{\Sigma}{\varepsilon_0} \hat{\mathbf{n}}$$

$$\mathbf{E}_\| - \mathbf{E}'_\| = 0$$

$$\mathbf{B}_\perp - \mathbf{B}'_\perp = 0$$

$$\mathbf{B}_\| - \mathbf{B}'_\| = \frac{\mathbf{K}}{\varepsilon_0 c^2} \times \hat{\mathbf{n}}$$

As shown in Fig. 4-2, $\hat{\mathbf{n}}$ is a unit vector normal to the surface, Σ and $\mathbf{K}$ are the surface charge and current densities, $\mathbf{B}_\perp$ and $\mathbf{B}_\|$ are the normal and tangential components of $\mathbf{B}$ at the surface, and ditto for $\mathbf{E}_\perp$ and $\mathbf{E}_\|$. We note that the boundary conditions are derived from Maxwell's equations and thus do not represent new information.

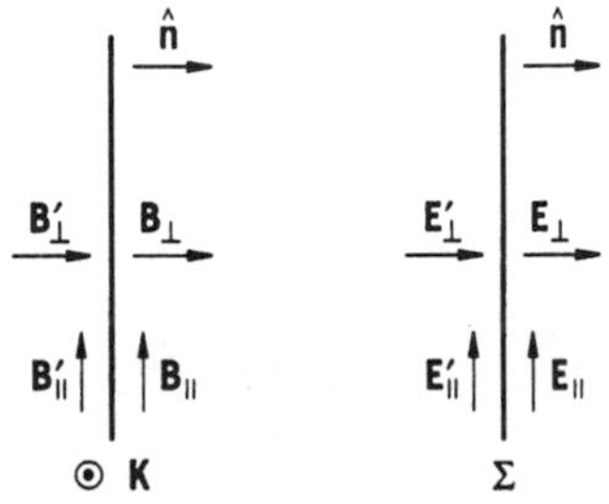

Figure 4-2. Sign conventions at a surface separating two media with different properties.

[†]We will use ∂ to indicate $\partial/\partial t$ and consider that, like ∇, it takes precedence over operations such as scalar products. If $\mathbf{r}$ is a vector, we will denote the magnitude of $\mathbf{r}$ by r and a unit vector in the direction of $\mathbf{r}$ by $\hat{\mathbf{r}}$. In this and the following section, $\mathbf{x}$ denotes a point in space, and we can imagine that it has Cartesian components (x_1, x_2, x_3). Elsewhere the Cartesian components of a vector are (x, y, z), and $\hat{\mathbf{x}}$ denotes a unit vector in the x-direction.

As a general rule, we will not develop solutions but rather state them and verify that they satisfy Maxwell's equations and the boundary conditions.

If we take the divergence of the Ampère–Maxwell law and recall that $\nabla \cdot \nabla \times \mathbf{B} = 0$, we obtain

$$\nabla \cdot \mathbf{J} + \varepsilon_0 \, \partial \nabla \cdot \mathbf{E} = 0$$

Substituting $\rho = \varepsilon_0 \nabla \cdot \mathbf{E}$ from Gauss's law, we deduce that conservation of charge is implicit in Maxwell's equations:

$$\nabla \cdot \mathbf{J} + \partial \rho = 0$$

The divergence of the magnetic field $\mathbf{B}$ is zero. We can therefore express $\mathbf{B}$ as the curl of a vector potential $\mathbf{A}$:

$$\mathbf{B} = \nabla \times \mathbf{A} \tag{4-1}$$

Using this expression in Faraday's law, we get

$$\nabla \times (\mathbf{E} + \partial \mathbf{A}) = 0$$

Since the quantity in parentheses has zero curl, it can be expressed as minus the gradient of a scalar potential ϕ, and we obtain

$$\mathbf{E} = -\nabla \phi - \partial \mathbf{A} \tag{4-2}$$

Except in unusual circumstances, such as when there are dipole layers on a surface, the boundary conditions on $\mathbf{A}$ and ϕ are that they be continuous.

We are free to *choose a gauge*, that is, to specify the divergence of $\mathbf{A}$. We will choose the *Lorentz gauge*, defined by

$$c^2 \nabla \cdot \mathbf{A} + \partial \phi = 0$$

In Exercise 4-2 it is shown that the potentials in the Lorentz gauge obey the *inhomogeneous wave equations*,

$$\nabla^2 \phi - \frac{1}{c^2} \partial^2 \phi = -\frac{\rho}{\varepsilon_0}$$

$$\nabla^2 \mathbf{A} - \frac{1}{c^2} \partial^2 \mathbf{A} = -\frac{\mathbf{J}}{\varepsilon_0 c^2} \tag{4-3}$$

It is further shown in Exercise 4-4 that these equations are satisfied by the *retarded potentials*[†]

$$\phi(\mathbf{x}, t) = \frac{1}{4\pi\varepsilon_0} \int \frac{[\rho]}{r} \, dv'$$

$$c^2 \mathbf{A}(\mathbf{x}, t) = \frac{1}{4\pi\varepsilon_0} \int \frac{[\mathbf{J}]}{r} \, dv' \tag{4-4}$$

where

$$[\rho] = \rho\left(\mathbf{x}', t - \frac{r}{c}\right)$$

$$[\mathbf{J}] = \mathbf{J}\left(\mathbf{x}', t - \frac{r}{c}\right)$$

are the sources evaluated at an earlier or *retarded* time such that a signal from the *source* point $\mathbf{x}'$ arrives at the *field* point $\mathbf{x}$ at time t, and $\mathbf{r}$ is the vector from the source point to the field point,

$$\mathbf{r} = \mathbf{x} - \mathbf{x}'$$

Retarded potentials for a simple configuration are calculated in Exercise 4-5.

The flow of energy per unit area and per unit time is given by the Poynting vector $\mathbf{S}$,

$$\mathbf{S} = \varepsilon_0 c^2 \mathbf{E} \times \mathbf{B}$$

and the field energy density u is given by

$$u = \frac{\varepsilon_0}{2} \mathbf{E} \cdot \mathbf{E} + \frac{\varepsilon_0 c^2}{2} \mathbf{B} \cdot \mathbf{B}$$

Conservation of energy is then expressed by Poynting's theorem (Exercise 4-6),

$$-\partial u = \nabla \cdot \mathbf{S} + \mathbf{E} \cdot \mathbf{J} \tag{4-5}$$

where $\mathbf{E} \cdot \mathbf{J}$ is the power dissipated per unit volume. In words, the field energy loss in a volume equals the sum of the energy emitted from the volume plus the energy dissipated in the volume.

[†]The inhomogeneous wave equations are also satisfied by the formally similar *advanced potentials*, but these are rejected on physical grounds.

4-3   LOW-FREQUENCY POTENTIALS AND FIELDS

Let us assume that we are dealing with a system in which the sources are of finite extent and vary slowly enough that they can be expanded in a Taylor series in r/c. From expressions (4-4) for the retarded potentials we then have

$$\phi(\mathbf{x},t) = \frac{1}{4\pi\varepsilon_0}\int\frac{1}{r}\left[\rho(\mathbf{x}',t)\quad\frac{r}{c}\partial\rho(\mathbf{x}',t) +\frac{r^2}{2c^2}\partial^2\rho(\mathbf{x}',t) +\cdots\right]dv'$$

$$c^2\mathbf{A}(\mathbf{x},t) = \frac{1}{4\pi\varepsilon_0}\int\frac{1}{r}\left[\mathbf{J}(\mathbf{x}',t) -\frac{r}{c}\partial\mathbf{J}(\mathbf{x}',t) +\frac{r^2}{2c^2}\partial^2\mathbf{J}(\mathbf{x}',t) +\cdots\right]dv'$$

The second term in the expansion of the scalar potential is proportional to

$$\int\partial\rho(\mathbf{x}',t)\,dv'$$

But this is the time derivative of the total charge of the system, and therefore zero. The corresponding integral in the expansion of the vector potential,

$$\int\partial\mathbf{J}(\mathbf{x}',t)\,dv'$$

is independent of $\mathbf{x}$ and therefore does not contribute to $\mathbf{B} = \nabla\times\mathbf{A}$. Let us see what it contributes to $\mathbf{E}$. Integrating by parts (easily done in Cartesian coordinates for sources of finite extent) and using conservation of charge, we get

$$\int\mathbf{J}(\mathbf{x}',t)\,dv' = -\int\mathbf{x}'\nabla\cdot\mathbf{J}(\mathbf{x}',t)\,dv' = \int\mathbf{x}'\partial\rho(\mathbf{x}',t)\,dv'$$

We note that this is the time derivative of the system dipole moment. Again using the fact that the total charge is constant, we arrive at

$$\int\partial\mathbf{J}(\mathbf{x}',t)\,dv' = -\int(\mathbf{x}-\mathbf{x}')\,\partial^2\rho(\mathbf{x}',t)\,dv' = -\int r\hat{\mathbf{r}}\,\partial^2\rho(\mathbf{x}',t)\,dv'$$

The contribution to $\mathbf{E}$ from the second term in the expansion of $\mathbf{A}$ is thus

$$-\frac{1}{4\pi\varepsilon_0 c^3}\int r\hat{\mathbf{r}}\,\partial^3\rho(\mathbf{x}',t)\,dv'$$

For sources that vary sinusoidally with frequency ω it becomes

$$\frac{i}{4\pi\varepsilon_0}\int\left(\frac{\omega r}{c}\right)^3\frac{\rho(\mathbf{x}')}{r^2}\hat{\mathbf{r}}\,dv'$$

The integrand in this contribution to $\mathbf{E}$ is of order $(\omega r/c)^3$ relative to the integrand in the contribution to $\mathbf{E}$ from the first term in the expansion of the scalar potential; this assertion follows from the observation that the latter contribution is

$$-\frac{1}{4\pi\varepsilon_0}\nabla\int\frac{\rho(\mathbf{x}')}{r}dv' = \frac{1}{4\pi\varepsilon_0}\int\frac{\rho(\mathbf{x}')}{r^2}\hat{\mathbf{r}}\,dv'$$

We conclude that there are no contributions of order $\omega r/c$ to the fields, and that use of the *quasistatic* potentials

$$\phi(\mathbf{x},t) = \frac{1}{4\pi\varepsilon_0}\int\frac{\rho(\mathbf{x}',t)}{r}dv'$$

$$c^2\mathbf{A}(\mathbf{x},t) = \frac{1}{4\pi\varepsilon_0}\int\frac{\mathbf{J}(\mathbf{x}',t)}{r}dv'$$

(4-6)

rather than the exact retarded potentials is justified as long as $(\omega r/c)^2$ is sufficiently small. That this will generally be the case for us can be seen by observing that $(\omega r/c)^2$ is about 0.01 for $r = 5$ cm and $\omega/2\pi = 100$ MHz. Even though we will not prove it here, it is worth pointing out that making $(\omega r/c)^2$ small also guarantees that radiated power is negligible. We also conclude that the fields are given in terms of the sources by

$$\mathbf{B} = \nabla\times\mathbf{A} = \frac{1}{4\pi\varepsilon_0 c^2}\int\frac{\mathbf{J}(\mathbf{x}',t)}{r^2}\times\hat{\mathbf{r}}\,dv'$$

$$\mathbf{E} = -\nabla\phi - \partial\mathbf{A} = \frac{1}{4\pi\varepsilon_0}\int\frac{\rho(\mathbf{x}',t)}{r^2}\hat{\mathbf{r}}\,dv' - \frac{1}{4\pi\varepsilon_0 c^2}\int\frac{\partial\mathbf{J}(\mathbf{x}',t)}{r}dv'$$

The expression for $\mathbf{B}$ is the quasistatic version of the Biot–Savart law, and in the expression for $\mathbf{E}$ the contribution from the charge density is given by the quasistatic version of Coulomb's law.

4-4   THE DRUDE–LORENTZ MODEL OF CONDUCTORS[†]

We will use a simple but fertile model of conductors, the *Drude–Lorentz* or *free-electron* model. In this picture, a conductor consists of a lattice of fixed charged ions, and of electrons that are free to roam except for occasional collisions with the lattice ions. The collisions restore to the electrons an

[†]The elementary theory of probability and random variables used in the following discussion is reviewed in Section 12-2.

invariant spherically symmetrical thermal momentum distribution. If the electrons acquire energy from external fields between collisions, the restoration is effective only immediately after a collision. Inside the conductor, Maxwell's equations in vacuum are valid.

The net effect of these constraints, as we will see next, is that we can often ignore the fact that the electrons have a thermal momentum distribution and consider that they are stopped dead after each collision. At any given time, an electron has a momentum

$$\mathbf{p} = \mathbf{p}_T + \mathbf{p}_E$$

where $\mathbf{p}_T$ is thermal momentum and $\mathbf{p}_E$ is momentum acquired from external fields since the last collision. The average (mean) momentum $\langle \mathbf{p} \rangle$ and the average square momentum $\langle p^2 \rangle$ are given by

$$\langle \mathbf{p} \rangle = \langle \mathbf{p}_T \rangle + \langle \mathbf{p}_E \rangle$$

and

$$\langle p^2 \rangle = \langle p_T^2 \rangle + \langle p_E^2 \rangle + 2\langle \mathbf{p}_T \cdot \mathbf{p}_E \rangle$$

Since $\langle \mathbf{p}_T \rangle = 0$ and $\langle \mathbf{p}_T \cdot \mathbf{p}_E \rangle = \langle \mathbf{p}_T \rangle \cdot \langle \mathbf{p}_E \rangle$ because $\mathbf{p}_E$ is independent of $\mathbf{p}_T$, we get

$$\langle \mathbf{p} \rangle = \langle \mathbf{p}_E \rangle$$

and

$$\langle p^2 \rangle = \langle p_T^2 \rangle + \langle p_E^2 \rangle$$

The average *excess* momentum and energy can thus be calculated as if electrons started from rest after each collision.

The time t between collisions follows a Poisson distribution with probability density

$$\frac{1}{\tau_F} \theta(t) \exp\left(-\frac{t}{\tau_F} \right)$$

where τ_F is the *mean free time*. This distribution is also valid, at any given time, for the time since the *last* collision. The moments of this distribution are

$$\langle t^n \rangle = \frac{1}{\tau_F} \int_0^\infty t^n \exp\left(-\frac{t}{\tau_F} \right) dt = n!\, \tau_F^n \tag{4-7}$$

With the assumptions above, we can calculate the current density generated in a conductor by a slowly varying electric field $\mathbf{E}$. In a time t after its last collision, an electron of charge $-e$ and mass m acquires an excess velocity

$$\mathbf{v}(t) = -\frac{e\mathbf{E}t}{m}$$

The average excess velocity or *drift velocity* $\langle \mathbf{v} \rangle$ is then

$$\langle \mathbf{v} \rangle = -\frac{e \langle t \rangle}{m} \mathbf{E} = -\frac{e \tau_F}{m} \mathbf{E}$$

If n is the number of electrons per unit volume, the current density $\mathbf{J}$ is

$$\mathbf{J} = -en \langle \mathbf{v} \rangle$$

Substituting the drift velocity in terms of the electric field, we obtain the low-frequency version of Ohm's law,

$$\mathbf{J} = \sigma \mathbf{E}$$

where σ is the *conductivity*,

$$\sigma = \frac{ne^2 \tau_F}{m} \tag{4-8}$$

This expression is valid for frequencies such that $\omega \tau_F \ll 1$. At frequencies such that $\omega \tau_F \gg 1$ we can ignore the fact that electrons suffer occasional collisions; the excess velocity of *all* electrons is then given by

$$i \omega \mathbf{v} = -\frac{e \mathbf{E}}{m}$$

and the current density is

$$\mathbf{J} = -en\mathbf{v} = \frac{ne^2}{i \omega m} \mathbf{E}$$

The high-frequency and low-frequency limits are included in the generalization of Ohm's law obtained in Exercise 4-11,

$$\mathbf{J} = \frac{\sigma \mathbf{E}}{1 + i \omega \tau_F} \tag{4-9}$$

4-5 RELAXATION TIMES IN CONDUCTORS

Following Saslow and Wilkinson (1971), let us consider a conductor in the Drude–Lorentz model, and let us write Ohm's law (4-9) as a differential equation,

$$\tau_F \, \partial \mathbf{J} + \mathbf{J} = \sigma \mathbf{E}$$

Taking the divergence of both sides of this equation yields

$$\tau_F \, \partial \nabla \cdot \mathbf{J} + \nabla \cdot \mathbf{J} = \sigma \nabla \cdot \mathbf{E}$$

If we then make use of Gauss's law and of conservation of charge,

$$\nabla \cdot \mathbf{E} = \frac{\rho}{\varepsilon_0}$$

$$\nabla \cdot \mathbf{J} = -\partial \rho$$

we obtain a differential equation for the charge density,

$$\tau_F \frac{\varepsilon_0}{\sigma} \partial^2 \rho + \frac{\varepsilon_0}{\sigma} \partial \rho + \rho = 0 \tag{4-10}$$

We note that there are no spatial derivatives in this expression; the charge thus decays to zero uniformly over the whole conductor. Let us first assume that

$$\frac{\varepsilon_0}{\sigma} \ll \tau_F$$

This is amply true in copper, in which $\sigma = 5.6 \cdot 10^7$ S/m and $\tau_F = 2.4 \cdot 10^{-14}$ s:

$$\frac{\varepsilon_0}{\sigma} = 7 \cdot 10^{-6} \tau_F$$

In this case the solutions are lightly damped oscillations ($Q \simeq 400$),

$$\rho_\pm(\mathbf{x}, t) = \rho_\pm(\mathbf{x}, 0) \exp\left(-\frac{t}{2\tau_F}\right) \exp(\pm i\omega_p t) \tag{4-11}$$

where ω_p is the *plasma frequency*, given by

$$\omega_p^2 = \frac{\sigma}{\varepsilon_0 \tau_F} = \frac{ne^2}{\varepsilon_0 m}$$

The charge relaxes with a time constant τ_R that is twice the mean free time,

$$\tau_R = 2\tau_F$$

Mean free times are of order 10^{-13} seconds; the relaxation time is thus far smaller than any time we will be dealing with. On the other hand, if the conductivity is small enough that

$$\tau_F \ll \frac{\varepsilon_0}{\sigma}$$

we can ignore the second derivative in (4-10), and we are then left with a first-order differential equation,

$$\frac{\varepsilon_0}{\sigma}\partial\rho+\rho = 0$$

In this case the charge density decays exponentially,

$$\rho(\mathbf{x},t) = \rho(\mathbf{x},0)\exp\left(-\frac{\sigma}{\varepsilon_0}t\right)$$

and the *dielectric* relaxation time τ_D is

$$\tau_D = \frac{\varepsilon_0}{\sigma}$$

This time can also be small. Let us take a material with $\sigma = 10^{-2}$ S/m, typical of what we might find in a 100 kΩ resistor. We then have

$$\tau_D \simeq 10^{-9} \text{ s}$$

which is usually much less than the time constants of circuits using resistors of a value this high.

The fields inside an ideal conductor must always be zero; this requirement implies currents and a rearrangement of surface charges when the conductor is subjected to an external field. The preceding discussion was meant to prepare us to accept that these processes can be considered to be instantaneous in good conductors such as copper. To take a simple example, let us see what happens when the infinite slab of Fig. 4-3 is subjected to an external electric field E_0 normal to its faces. In terms of E_0 and the field E_C in the conductor, the surface charge density Σ on the top face is

$$\Sigma = \varepsilon_0(E_0 - E_C)$$

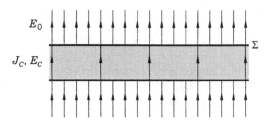

Figure 4-3. Infinite conducting slab subjected to an external electric field E_0 normal to its faces.

Ohm's law for the current density J_C in the conductor reads

$$\tau_F \, \partial J_C + J_C = \sigma E_C$$

and conservation of charge demands

$$J_C = \partial \Sigma$$

It follows that

$$\tau_F \frac{\varepsilon_0}{\sigma} \partial^2 \Sigma + \frac{\varepsilon_0}{\sigma} \partial \Sigma \mid \Sigma - \varepsilon_0 E_0$$

Comparing with (4-10), we see in this case—and we will accept it as true in general—that surface charges track the external electric field with the same short time constants that characterize the relaxation of a volume charge.

4-6 SURFACE CHARGES IN INDUCTORS

The surface charges that null the electric field inside good conductors do not contribute to the curl of the electric field except through their associated currents, whose effect is negligible at low frequencies, but they can contribute significantly to the electric field itself. In capacitors, in fact, surface charges are the sole source of the electric field at low frequencies. Less obviously, surface charges are also important in structures such as inductors in which the magnetic field is prominent at low frequencies. In Section 3-1 we saw that the voltage at the terminals of an inductor is well defined if it is taken along suitably restricted paths. We will now see that surface charges play a crucial role in establishing this voltage.

The contribution $-\partial \mathbf{A}$ from the current in an inductor to the electric field $\mathbf{E} = -\nabla \phi - \partial \mathbf{A}$ is not negligible at low frequencies as it is in capacitors, and has a nonzero curl $-\nabla \times \partial \mathbf{A} = -\partial \mathbf{B}$ that makes line integrals of $\mathbf{E}$ path dependent. Thus despite its origin as an induced effect, required only to cancel $-\partial \mathbf{A}$ inside the conductor that carries the current, the contribution $-\nabla \phi$ from the surface charges on the conductor must be overwhelming in the region around the terminals if the voltage is to be well defined. In addition, $-\nabla \phi$ necessarily alters the configuration of $\mathbf{E}$ beyond recognition because at the surface of the conductor, $\partial \mathbf{A}$ is generally along the direction of current flow, whereas $\mathbf{E}$ must be normal to the surface.

To see the effects of surface charges in a manageable geometry, let us consider the two-dimensional current distribution of Fig. 4-4a, composed of two infinite current sheets with linearly increasing surface current densities $\mathbf{K}(y = \pm a) = \mp K_0 \lambda t \hat{\mathbf{x}}$, and let us assume initially that there are no conductors present. According to Exercise 4-10, the magnetic field is

$$\mathbf{B}(|y| < a) = \mu_0 K_0 \lambda t \hat{\mathbf{z}}$$

$$\mathbf{B}(|y| > a) = 0$$

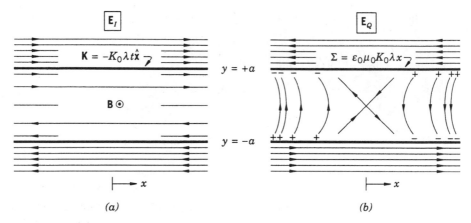

Figure 4-4. (a) Electric field $\mathbf{E}_I$ generated by two infinite current sheets with linearly increasing surface current densities $\mathbf{K}(y = \pm a) = \mp K_0 \lambda t \hat{\mathbf{x}}$. ($b$) Electric field $\mathbf{E}_Q$ generated by two infinite charge sheets surface charge densities $\Sigma(y = \pm a) = \pm \varepsilon_0 \mu_0 K_0 \lambda x$. The sum of $\mathbf{E}_I$ and $\mathbf{E}_Q$ is the electric field of the parallel-plate inductor shown in Fig. 4-5.

It is easy to verify that the vector potential

$$\mathbf{A}(|y| < a) = -\mu_0 K_0 \lambda t y \hat{\mathbf{x}}$$

$$\mathbf{A}(|y| > a) = -\mu_0 K_0 \lambda t a \hat{\mathbf{x}} \, \mathrm{sgn}(y)$$

is continuous and satisfies $\mathbf{B} = \nabla \times \mathbf{A}$. Since there are no charges present we have $\phi = 0$, and the electric field $\mathbf{E}_I$ due to the currents, also shown in Fig. 4-4a, is given by $\mathbf{E}_I = -\partial \mathbf{A}$:

$$\mathbf{E}_I(|y| < a) = \mu_0 K_0 \lambda y \hat{\mathbf{x}}$$

$$\mathbf{E}_I(|y| > a) = \mu_0 K_0 \lambda a \hat{\mathbf{x}} \, \mathrm{sgn}(y)$$

As indicated in Fig. 4-5, the electric field changes radically if the current sheets are carried by perfectly conducting plates. Assuming antisymmetry

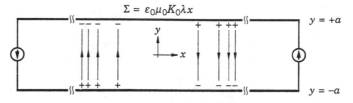

Figure 4-5. Electric field in a parallel-plate inductor driven antisymmetrically by linearly increasing current sources at $x = \pm\infty$.

around $x = 0$, Maxwell's equations and the boundary conditions are satisfied by the electric field

$$\mathbf{E}(|y| < a) = -\mu_0 K_0 \lambda x \hat{\mathbf{y}}$$

$$\mathbf{E}(|y| > a) = 0$$

and by the surface charge density

$$\Sigma(y = \pm a) = \pm \varepsilon_0 \mu_0 K_0 \lambda x$$

It is easy to verify that Σ generates a continuous potential ϕ given by

$$\phi(|y| < a) = \mu_0 K_0 \lambda xy$$

$$\phi(|y| > a) = \mu_0 K_0 \lambda ax \, \text{sgn}(y)$$

and that the electric field $\mathbf{E}_Q = -\nabla\phi$ due to Σ, shown in Fig. 4-4b, is

$$\mathbf{E}_Q(|y| < a) = -\mu_0 K_0 \lambda (y\hat{\mathbf{x}} + x\hat{\mathbf{y}})$$

$$\mathbf{E}_Q(|y| > a) = -\mu_0 K_0 \lambda a\hat{\mathbf{x}} \, \text{sgn}(y)$$

Consistency demands that

$$\mathbf{E} = -\nabla\phi - \partial\mathbf{A} = \mathbf{E}_I + \mathbf{E}_Q$$

and this is in fact the case.

For $|y| < a$ and $|x| \gg a$, $\mathbf{E}_I = -\partial\mathbf{A}$ is negligible compared with $\mathbf{E}_Q = -\nabla\phi$, and since $\nabla \times \mathbf{E}_Q = -\nabla \times \nabla\phi = 0$, the voltage between, say, opposite points on the plates is well defined as long as paths deviate little from a direct run across the plates. This is essentially the situation at the terminals of an inductor, whatever its geometry.

4-7   THE SKIN EFFECT IN CONDUCTORS

Let us consider the long cylindrical conductor of radius a shown in Fig. 4-6, and let us assume that it carries a sinusoidal current at a frequency ω such that

$$\omega \tau_F \ll 1 \quad \text{and} \quad \frac{\omega \varepsilon_0}{\sigma} \ll 1$$

In good conductors such as copper these conditions hold for frequencies as high as 10^{12} Hz, that is, well beyond the microwave region. Under these conditions, Ohm's law (4-9) reads

$$\mathbf{J} = \sigma\mathbf{E}$$

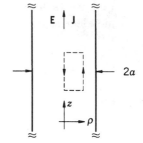

Figure 4-6. Cross section of an infinite cylindrical conductor carrying a sinusoidal current. The path in dashed lines is used to show that the current density cannot be constant.

and the Ampère–Maxwell law reduces to Ampère's law,

$$\nabla \times \mathbf{B} = \frac{1}{\varepsilon_0 c^2}\mathbf{J} + \frac{i\omega}{c^2}\mathbf{E} = \mu_0 \mathbf{J}\left(1 + \frac{i\omega\varepsilon_0}{\sigma}\right) \simeq \mu_0 \mathbf{J}$$

We want to show that except at $\omega = 0$, the current is crowded toward the surface of the conductor. To see this, let us assume for a moment that the current density $\mathbf{J}$ is constant,

$$J_z = J_0$$

The magnetic field is then

$$B_\varphi(\rho) = \frac{\mu_0 J_0}{2\pi\rho}\pi\rho^2 = \frac{\mu_0 J_0 \rho}{2}$$

If we consider the path shown in dashed lines in Fig. 4-6, it is clear that it encloses an alternating flux and that the electric fields on its axial legs must be different; more precisely, Faraday's flux law gives

$$E_z(\rho) - E_z(0) = i\omega \int_0^\rho B_\varphi(\xi)\, d\xi = i\omega\frac{\mu_0 J_0 \rho^2}{4}$$

and we therefore have

$$J_z(\rho) = J_0\left(1 + i\omega\frac{\sigma\mu_0 \rho^2}{4}\right) \tag{4-12}$$

The magnitude of the current density thus increases as the surface of the conductor is approached. We will now see that this effect is intensified at higher frequencies and that the fields and the current density are confined to a thin *skin layer* at the surface. The fields are given in terms of the current density by

$$B_\varphi(\rho) = \frac{\mu_0}{\rho}\int_0^\rho \xi J_z(\xi)\, d\xi$$

and

$$E_z(\rho) = \frac{J_z(\rho)}{\sigma}$$

From Faraday's law we also have

$$\frac{dJ_z}{d\rho} = i\omega\sigma B_\varphi$$

We can then obtain a differential equation for the current density,

$$\frac{1}{\rho}\frac{d}{d\rho}\left(\rho\frac{dJ_z}{d\rho}\right) = \frac{2i}{\delta^2}J_z(\rho) \tag{4-13}$$

where δ is the *skin depth*,

$$\delta = \sqrt{\frac{2}{\sigma\omega\mu_0}}$$

In copper, for example, we have $\sigma \approx 5.6 \cdot 10^7$ S/m, and the skin depth at 10 GHz is

$$\delta \approx 0.7 \ \mu\text{m}$$

If we assume that $\delta \ll a$ and that all the action is near the surface, ρ does not vary appreciably, and (4-13) reduces to

$$\frac{d^2J_z}{d\rho^2} = \frac{2i}{\delta^2}J_z(\rho)$$

The solution of this much simpler differential equation is

$$J_z(\rho) = J_z(a)\exp\left[(1+i)\frac{\rho-a}{\delta}\right] \tag{4-14}$$

The current density and the fields decay exponentially as a function of the distance from the surface, which is consistent with our assumption. As a result, we will frequently ignore the fields inside copper conductors, for even at 1 MHz the skin depth is less than one-tenth of a millimeter. At the other extreme of the frequency spectrum, we can see why there is a limit to the useful thickness of copper in power distribution cables: at 50 or 60 Hz the skin depth is about 1 cm, and copper further in is just wasted.

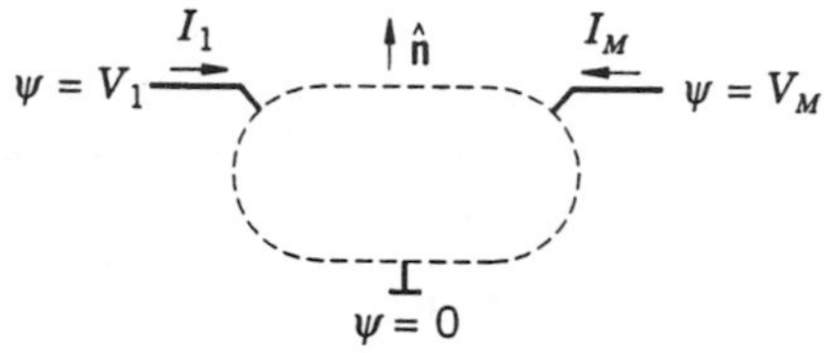

Figure 4-7. Surface enclosing a circuit, crossed only by $M+1$ wires that connect the circuit to the external world. A surface potential ψ can be defined if $\partial \mathbf{B} \cdot \hat{\mathbf{n}} = 0$ on the surface.

$$\psi = V_1 \xrightarrow{I_1} \qquad \uparrow \hat{\mathbf{n}} \qquad \xleftarrow{I_M} \psi = V_M$$
$$\psi = 0$$

4-8 POWER IN LUMPED CIRCUITS

In Chapter 3 we accepted that the power into a circuit is given by products of voltages and currents at the terminals of the circuit. We will now establish more precisely the conditions under which this is true. Let us consider a surface surrounding a circuit connected to the external world by $M+1$ thin wires that carry currents I_m, as shown in Fig. 4-7, and let us assume that the normal component of the time derivative of the magnetic field $\mathbf{B}$ is zero on the surface:

$$\partial \mathbf{B} \cdot \hat{\mathbf{n}} = 0$$

It follows that the tangential component $\mathbf{E}_\parallel$ of the electric field $\mathbf{E}$ at the surface can be expressed in terms of a potential ψ. If we assume that ψ can be continued in a sensible way into the interior of the volume surrounded by the surface, we can write

$$\mathbf{E}_\parallel = -\nabla\psi_\parallel$$

As indicated in Fig. 4-7, we will make $\psi = 0$ at one of the wires. The power P flowing *into* the volume is given by

$$P = -\varepsilon_0 c^2 \int \mathbf{E} \times \mathbf{B} \cdot d\mathbf{a}$$

Since only the tangential components of $\mathbf{E}$ and $\mathbf{B}$ contribute to the integral, we can write

$$P = \varepsilon_0 c^2 \int \nabla\psi \times \mathbf{B} \cdot d\mathbf{a}$$

Assuming now that $\mathbf{B}$ is well-behaved inside the volume, and recalling that the divergence of a curl is zero, we can use the vector identity

$$\nabla \times (\psi \mathbf{B}) = \psi \nabla \times \mathbf{B} + \nabla\psi \times \mathbf{B}$$

and the divergence theorem applied to $\nabla \times (\psi \mathbf{B})$,

$$\int \nabla \times (\psi \mathbf{B}) \cdot d\mathbf{a} = \int \nabla \cdot [\nabla \times (\psi \mathbf{B})] \, dv = 0$$

to obtain

$$P = -\varepsilon_0 c^2 \int \psi \, \nabla \times \mathbf{B} \cdot d\mathbf{a}$$

If we now use the Ampère–Maxwell law, we get

$$P = -\int \psi \mathbf{J} \cdot d\mathbf{a} - \varepsilon_0 \int \psi \, \partial \mathbf{E} \cdot d\mathbf{a}$$

This expression can be rewritten partly in terms of the currents *into* the volume and the voltages at the wires:

$$P = \sum_{m=1}^{M} V_m I_m - \varepsilon_0 \int \psi \, \partial \mathbf{E} \cdot d\mathbf{a}$$

Finally, if we also assume that the normal component of the time derivative of **E** is zero on the surface,

$$\partial \mathbf{E} \cdot \hat{\mathbf{n}} = 0$$

we obtain that the power into the circuit is given by a sum of VI products at the terminals,

$$P = \sum_{m=1}^{M} V_m I_m$$

The surface potential ψ does not necessarily coincide with the scalar potential ϕ since the vector potential **A** can contribute to ψ. To illustrate this point, let us consider the shorted coaxial cable shown in cross section in Fig. 4-8, and let us assume that the current grows linearly in time:

$$I = 2\pi a K_0 \lambda t$$

The fields given by

$$B_\varphi = -\mu_0 \frac{a}{\rho} K_0 \lambda t \quad \text{and} \quad E_\rho = \mu_0 \frac{a}{\rho} K_0 \lambda z$$

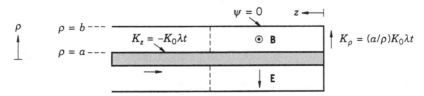

Figure 4-8. Shorted coaxial cable used to show that the surface potential ψ is not identical to the scalar potential ϕ.

satisfy Maxwell's equations and the boundary conditions on **E** and **B**. On a plane perpendicular to the axis, **E** can be obtained from a potential ψ that is well defined everywhere,

$$\psi = -\mu_0 K_0 \lambda az \ln\frac{\rho}{b}$$

We thus have

$$\mathbf{E} = -\frac{\partial \psi}{\partial \rho}\hat{\rho}$$

yet ψ differs from ϕ because

$$\mathbf{E} = -\nabla\phi - \partial\mathbf{A}$$

and **A** has a nonzero component in the ρ-direction due to the surface current density K_ρ in the shorting plate. The voltage from the center conductor to the shield is

$$V = \psi(a) - \psi(b) = \mu_0 K_0 \lambda az \ln\frac{b}{a}$$

and we can check that the power into the short-circuited end is VI:

$$P = -\varepsilon_0 c^2 \int \mathbf{E}\times\mathbf{B}\cdot d\mathbf{a} = \varepsilon_0 c^2 \int_a^b 2\pi\rho\mu_0^2 \frac{a^2}{\rho^2}K_0^2\lambda^2 zt\,d\rho = 2\pi\mu_0 a^2 K_0^2 \lambda^2 zt \ln\frac{b}{a} = VI$$

Let us consider three more examples. The first example, shown in Fig. 4-9a, merely generalizes what we saw in the case of a shorted coaxial cable: in long two-conductor transmission lines of arbitrary cross section, the components of **E** and **B** along the axis are zero, so that the power transmitted across a plane perpendicular to the axis is $P = VI$.

The second example, shown in Fig. 4-9b, is a circular parallel-plate capacitor fed by infinitely long axial wires. We will make the enclosing surface parallel to the plates on small disks just outside the plates and then let it follow field lines of the electric field due to the charges; we note that this surface includes most of the fringing field. If the disks are small enough they cover areas of the plates on which surface charges are negligible; this makes $\nabla\phi\cdot\hat{\mathbf{n}} = 0$ on the whole surface, and we have

$$\mathbf{E}\cdot\hat{\mathbf{n}} = -\partial\mathbf{A}\cdot\hat{\mathbf{n}}$$

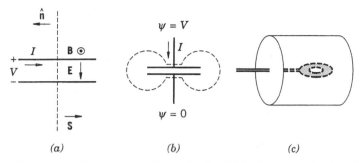

Figure 4-9. Examples of two-terminal circuits in which the power is given by the VI product. (a) Transmission line. (b) Circular parallel-plate capacitor with infinitely long axial terminals. (c) Toroidal inductor.

On the other hand, by symmetry we have $\partial \mathbf{B} \cdot \hat{\mathbf{n}} = 0$, and we can define a potential ψ on the surface. The vector potential can be written as

$$\mathbf{A} = \mu_0 I \mathbf{f}$$

where $\mathbf{f}$ is a dimensionless vector function. For sinusoidal signals we have

$$\varepsilon_0 \int \psi i \omega \mathbf{E} \cdot d\mathbf{a} = \varepsilon_0 \mu_0 \omega^2 I \int \psi \mathbf{f} \cdot d\mathbf{a} = \pm \left(\frac{\omega l}{c} \right)^2 VI$$

where l is a characteristic length and the sign does not matter, and we conclude that $P = VI$ at sufficiently low frequencies.

The third example, shown in Fig. 4-9c, is a toroidal inductor fed by a closely spaced pair of wires. The enclosing surface is a large cylinder coaxial with the wires. The magnetic field due to the toroid is zero at the surface and the magnetic field due to the wires is parallel to the surface, so that a surface potential can be defined. The electric field due to the wires is also parallel to the surface, and for sinusoidal signals the electric field $\mathbf{E}_T$ generated by the toroid is given in terms of a dimensionless vector function $\mathbf{f}$ by

$$\mathbf{E}_T = i \omega \mu_0 I \mathbf{f}$$

In terms of a characteristic length l we then have

$$-\varepsilon_0 \int \psi i \omega \mathbf{E} \cdot d\mathbf{a} = \varepsilon_0 \mu_0 \omega^2 I \int \psi \mathbf{f} \cdot d\mathbf{a} = \pm \left(\frac{\omega l}{c} \right)^2 VI$$

As in the second example, we have $P = VI$ at sufficiently low frequencies.

4-9 PARASITIC CIRCUIT ELEMENTS

Looking back at the parallel-plate capacitor in Fig. 3-1, let us consider the closed rectangular path formed by the terminal wires plus a segment crossing the plates and a segment joining the terminals. The path encloses a flux proportional to the current, so that whatever the details, the voltage at the terminals has a (first) additive correction proportional to the derivative of the current. From this observation we infer that any capacitor C has a (small) parasitic inductance L in series.

Looking now at the solenoidal inductor in Fig. 3-2, and recalling from the discussion in Section 4-6 that a voltage induces surface charges on the conductor, we infer that any inductor L has a (small) parasitic capacitance C in parallel.

For either a capacitor or an inductor the corrections are negligible if $\omega^2 LC \ll 1$. This is the justification for assuming in Section 3-1 that the static expressions (3-1) for the charge on a capacitor and (3-3) for the flux in an inductor remain valid for slow variations in time.

The dominant correction for a resistor is generally either a parasitic capacitance in parallel if the resistance is high, or a parasitic inductance in series if the resistance is low, although a combination of both is occasionally necessary.

By means of two examples we will now see that corrections to the basic lumped elements are in fact small (further examples are given in Exercises 4-15 and 4-16).

As a first example, let us consider the rather large solenoid shown in Fig. 4-10. Its inductance is approximately

$$L = \mu_0 N^2 \frac{\pi a^2}{l} \simeq 250 \ \mu\text{H}$$

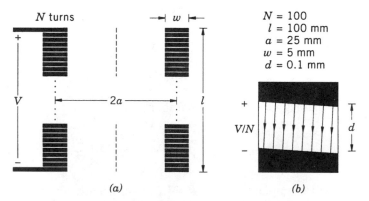

Figure 4-10. Solenoid wound with conductors of rectangular cross section. (*a*) Axial cross section. (*b*) Radial view of the electric field between adjacent turns.

To make calculations easy, we have made the cross section of the conductor rectangular. The voltage between two adjacent turns is N times smaller than the terminal voltage V, so that the surface charge densities $\Sigma_\pm$ in the capacitorlike structure formed by two adjacent turns are

$$\Sigma_\pm = \pm\varepsilon_0 \frac{V}{Nd}$$

Ignoring charges on the terminals, the charge that the external circuit has to supply is the charge on the positive side of one turn,

$$Q = 2\pi a w \Sigma_+ = \frac{2\pi a w}{Nd}\varepsilon_0 V$$

The parasitic capacitance C that appears in parallel with the inductance is thus quite small, despite the tight packing of the coil:

$$C = \frac{Q}{V} = \frac{2\pi a w}{Nd}\varepsilon_0 \simeq 0.7 \text{ pF}$$

An inductor this large might be used in audio circuits at frequencies up to 30 kHz; for this purpose it is virtually ideal, as indicated by the fact that its resonant frequency $f_0 = 1/2\pi\sqrt{LC} \simeq 12$ MHz is far above the audio range.

As a second example let us consider the somewhat artificial—but therefore calculable—resistor of Fig. 4-11 (Marcus, 1941). It consists of a resistive cylinder centered in a conducting cylindrical cup. This example is interesting because it shows us that resistors must acquire surface charges in the regions that are not in contact with conductors in order to make the normal component of the electric field zero just inside the surface in these regions and thus inhibit runaway charge accumulation. Such charges contribute to the parasitic capacitance in parallel with a resistor.[†]

The potential ϕ given by

$$\phi(a<\rho<b) = E_0 z \frac{\ln(\rho/b)}{\ln(a/b)}$$

$$\phi(\rho<a) = E_0 z$$

generates a uniform field of magnitude E_0 along the axis of the resistive cylinder. In addition, ϕ is zero at $\rho = b$ and continuous at $\rho = a$. In

---

[†]The much neglected question of fields and charges associated with resistors is discussed further in Heald (1984); Jackson (1996); Jefimenko (1989), pp. 299–305 and Plates 6–9.

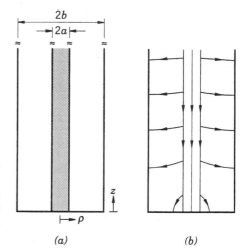

Figure 4-11. (*a*) Resistor consisting of a resistive cylinder centered in a conducting cylindrical cup; (*b*) configuration of its electric field, indicating that the resistive cylinder acquires surface charges.

(a)   *(b)*

cylindrical coordinates we have

$$-\nabla \cdot \mathbf{E} = \nabla^2 \phi = \frac{1}{\rho} \frac{\partial}{\partial \rho} \left(\rho \frac{\partial \phi}{\partial \rho} \right) + \frac{\partial^2 \phi}{\partial z^2}$$

and we can easily check that $\nabla \cdot \mathbf{E} = 0$ for $a < \rho < b$. We also clearly have $\nabla \cdot \mathbf{E} = 0$ in the interior of the resistive cylinder. We conclude that ϕ correctly describes the resistor.

The surface charge density Σ on the resistive cylinder is given by

$$\Sigma = -\varepsilon_0 \frac{\partial \phi}{\partial \rho} \bigg|_{\rho = a} = \frac{\varepsilon_0 E_0 z}{a \ln(b/a)}$$

We can now obtain the surface charge Q between 0 and z,

$$Q = \frac{\pi \varepsilon_0 E_0 z^2}{\ln(b/a)}$$

The voltage between z and 0 is $E_0 z$; the parasitic capacitance C of a resistor of length z is thus

$$C = \frac{\pi \varepsilon_0 z}{\ln(b/a)}$$

For reasonable values $z = 1$ cm, $b = 2$ mm, and $a = 1$ mm, we have

$$C \simeq 0.4 \text{ pF}$$

Such a capacitance is small, but it is not always negligible: it can represent a problem in voltage dividers with large resistors.

EXERCISES

4-1 LABORATORY: PARASITIC INDUCTANCE OF CAPACITORS. The experimental setup shown in Fig. 4-12 will allow you to set an upper limit on the parasitic inductance of capacitors. You will need:

1. A pulse generator with a $50\,\Omega$ output impedance and a rise time around 5 ns
2. An oscilloscope with a bandwidth above 150 MHz
3. A BNC T
4. A 10 ns (2 meter) length of $50\,\Omega$ RG-58 coaxial cable
5. A $50\,\Omega$ feedthrough terminator at the oscilloscope input

(a) Convince yourself that the equivalent circuit in Fig. 4-12 is correct; keep in mind that a terminated $50\,\Omega$ cable looks like a $50\,\Omega$ resistor. Short-circuit the center conductor of the BNC T to ground. You should observe a small spike due to residual inductance.

(b) Connect capacitors of several types (mica, ceramic, film) with capacitances between 1000 pF and 1 μF from the center conductor of the BNC T to ground (pressure connections are good enough) and observe the spikes. Compare with what you get when you replace the capacitors with wires of equal length. You should come to the conclusion that most of what you see is due to wiring inductance, and be able to set an upper limit of about 10 nH on the intrinsic inductance of the capacitors.

(c) Model the rising portion of the generator waveform as a ramp and show that the height of the spikes is proportional to L if L/R is much less than RC and several times less than the generator rise time t_R.

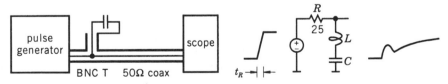

Figure 4-12. Circuit used in Exercise 4-1 to set an upper limit on the parasitic inductance of capacitors.

4-2 THE INHOMOGENEOUS WAVE EQUATIONS FOR THE SCALAR AND VECTOR POTENTIALS. From $\nabla \cdot \mathbf{B} = 0$ and Faraday's law, we have already obtained (4-1) and (4-2),

$$\mathbf{B} = \nabla \times \mathbf{A} \quad \text{and} \quad \mathbf{E} = -\nabla \phi - \partial \mathbf{A}$$

Insert these expressions in the Gauss and Ampère–Maxwell laws and use the vector identity

$$\nabla \times \nabla \times \mathbf{A} = \nabla(\nabla \cdot \mathbf{A}) - \nabla^2 \mathbf{A}$$

to obtain

$$\nabla^2 \mathbf{A} - \frac{1}{c^2}\partial^2 \mathbf{A} - \nabla\left(\nabla \cdot \mathbf{A} + \frac{1}{c^2}\partial\phi\right) = -\frac{\mathbf{J}}{\varepsilon_0 c^2}$$

$$\nabla^2 \phi + \nabla \cdot \partial \mathbf{A} = -\frac{\rho}{\varepsilon_0}$$

Finally, apply the Lorentz condition

$$c^2 \nabla \cdot \mathbf{A} + \partial\phi = 0$$

to get the inhomogeneous wave equations (4-3). If you get stuck, see Feynman (1964).

4-3 JEFIMENKO'S FORMULAS (Jefimenko, 1989; Griffiths and Heald, 1991; Jackson, 1999). If a scalar ψ and a vector $\mathbf{V}$ depend on a vector $\mathbf{x}$ via a scalar τ,

$$\psi = \psi[\tau(\mathbf{x})] \quad \text{and} \quad \mathbf{V} = \mathbf{V}[\tau(\mathbf{x})]$$

show that

$$\nabla\psi = \frac{d\psi}{d\tau}\nabla\tau \quad \text{and} \quad \nabla \times \mathbf{V} = -\frac{d\mathbf{V}}{d\tau} \times \nabla\tau$$

Apply these results and the vector identity

$$\nabla \times (\psi\mathbf{V}) = \psi\nabla \times \mathbf{V} - \mathbf{V} \times \nabla\psi$$

to the retarded potentials (4-4) and obtain Jefimenko's formulas for the fields in terms of the sources,

$$\mathbf{B} = \nabla \times \mathbf{A} = \frac{1}{4\pi\varepsilon_0 c^2}\int\left(\frac{[\mathbf{J}]}{r^2} + \frac{[\partial\mathbf{J}]}{rc}\right) \times \hat{\mathbf{r}}\, dv'$$

$$\mathbf{E} = -\nabla\phi - \partial\mathbf{A} = \frac{1}{4\pi\varepsilon_0}\int\left(\frac{[\rho]}{r^2} + \frac{[\partial\rho]}{rc}\right)\hat{\mathbf{r}}\, dv' - \frac{1}{4\pi\varepsilon_0 c^2}\int\frac{[\partial\mathbf{J}]}{r}\, dv'$$

Note that $[\partial\rho] = \partial[\rho]$ and $[\partial\mathbf{J}] = \partial[\mathbf{J}]$, and remember that $\mathbf{r} = \mathbf{x} - \mathbf{x}'$ points from the source point to the field point. Also make use of the fact that

$$\nabla r = \hat{\mathbf{r}} \quad \text{and} \quad \nabla\frac{1}{r} = -\frac{\hat{\mathbf{r}}}{r^2}$$

Expand the retarded sources in Jefimenko's formulas and show that the first-order terms drop out:

$$\frac{[\rho]}{r^2} + \frac{[\partial\rho]}{rc} = \frac{\rho(\mathbf{x}',t)}{r^2} - \frac{r^2}{2c^2}\frac{\partial^2\rho(\mathbf{x}',t)}{r^2} + \cdots$$

and similarly for [**J**].

Since it was shown in Section 4-3 that the term $-\partial\mathbf{A}$ does not contribute in first order to the electric field, you see again that retardation is a second-order effect.

4-4 GENERAL SOLUTION OF THE INHOMOGENEOUS WAVE EQUATIONS. First do Exercise 4-3 as a warm-up. Then use the vector identity

$$\nabla\cdot(\psi\mathbf{V}) = \psi\nabla\cdot\mathbf{V} + \mathbf{V}\cdot\nabla\psi$$

and the fact that

$$\nabla\cdot\frac{\hat{\mathbf{r}}}{r^2} = 4\pi\delta(\mathbf{r}) \quad \text{and} \quad \nabla\cdot\frac{\hat{\mathbf{r}}}{r} = \frac{1}{r^2}$$

to show that the retarded scalar potential given in (4-4),

$$\phi(\mathbf{x},t) = \frac{1}{4\pi\varepsilon_0}\int\frac{[\rho]}{r}\,dv'$$

satisfies the inhomogeneous wave equation given in (4-3),

$$\nabla^2\phi - \frac{1}{c^2}\partial^2\phi = -\frac{\rho}{\varepsilon_0}$$

4-5 EXAMPLE OF RETARDED POTENTIALS. Consider an infinite metallic slab parallel to the x–y plane and of thickness $2d$. Assume that it is undergoing (very lightly damped) plasma oscillations, so that the top and bottom faces have surface charge densities of amplitudes $\mp\Sigma_0$. By symmetry **J** must be along the z-direction; conservation of charge further requires

$$J_z = -i\omega\Sigma_0$$

This problem can of course be solved by elementary methods. First observe that **A** is along the z-direction and depends only on z, which implies

$$\mathbf{B} = \nabla\times\mathbf{A} = 0 \text{ everywhere}$$

On the other hand, from Gauss's law, symmetry for each charge sheet, and superposition, you have

$$E_z(|z|<d) = \frac{\Sigma_0}{\varepsilon_0} \quad \text{and} \quad E_z(|z|>d) = 0$$

By direct integration over the retarded sources, obtain

$$\phi(z>d) = -\frac{\Sigma_0 c}{\varepsilon_0 \omega} \exp\left(-\frac{i\omega z}{c}\right) \sin\left(\frac{\omega d}{c}\right)$$

$$\phi(z<-d) = -\phi(-z)$$

$$\phi(|z|<d) = -\frac{\Sigma_0 c}{\varepsilon_0 \omega} \exp\left(-\frac{i\omega d}{c}\right) \sin\left(\frac{\omega z}{c}\right)$$

Verify that ϕ is continuous at $z = \pm d$ and that it satisfies the inhomogeneous wave equation

$$\nabla^2\phi + \left(\frac{\omega}{c}\right)^2 \phi = -\frac{\rho}{\varepsilon_0}$$

Calculate A_z from the Lorentz condition

$$c^2 \nabla \cdot \mathbf{A} + i\omega\phi = 0$$

and determine the constant of integration by demanding that $\mathbf{A}$ satisfy the inhomogeneous wave equation

$$\nabla^2\mathbf{A} + \left(\frac{\omega}{c}\right)^2 \mathbf{A} = -\frac{\mathbf{J}}{\varepsilon_0 c^2}$$

You should obtain

$$A_z(z>d) = -\frac{\Sigma_0}{\varepsilon_0 \omega} \exp\left(-\frac{i\omega z}{c}\right) \sin\left(\frac{\omega d}{c}\right)$$

$$A_z(z<-d) = A_z(-z)$$

$$A_z(|z|<d) = -\frac{i\Sigma_0}{\varepsilon_0 \omega} \exp\left(-\frac{i\omega d}{c}\right) \cos\left(\frac{\omega z}{c}\right) + \frac{i\Sigma_0}{\varepsilon_0 \omega}$$

(You can also obtain A_z by integration if you are willing to use a convergence factor to eliminate an oscillating term at infinity.) Verify that $\mathbf{A}$ is continuous at $z = \pm d$ and that the electric field is given by

$$\mathbf{E} = -\nabla\phi - i\omega\mathbf{A}$$

Observe that the expansion of ϕ has a term that is of first order in $\omega d/c$ or $\omega z/c$; this can be so because the source is of infinite extent.

4-6 CONSERVATION OF ENERGY. Use Maxwell's equations (Faraday and Ampère–Maxwell) and the vector identity

$$\nabla \cdot (\mathbf{E} \times \mathbf{B}) = \mathbf{B} \cdot (\nabla \times \mathbf{E}) - \mathbf{E} \cdot (\nabla \times \mathbf{B})$$

to prove Poynting's theorem (4-5). For an interesting discussion, see Feynman (1964).

4-7 INDUCTANCE OF A LOOP. Consider the conducting loop shown in cross section in Fig. 4-13a. Assume that it carries a current I at a frequency high enough that the skin effect is well developed and all the current is on the surface.

For $\rho \simeq a$ the field in the plane of the loop is the field of an infinite wire,

$$B_z(\rho) \simeq \frac{\mu_0 I}{2\pi(a-\rho)}$$

As indicated in Fig. 4-13b, if you extrapolate all the way to the center, you get $B_z(0) = \mu_0 I / 2\pi a$ rather than the exact value $\mu_0 I / 2a$ obtained from the Biot–Savart law. Since the flux integral is weighted by ρ, you can ignore the error if $a \gg b$ and obtain

$$\Phi \simeq \frac{\mu_0 I}{2\pi} \int_0^{a-b} \frac{2\pi\rho}{a-\rho}\, d\rho = \mu_0 I\left(b-a+a\ln\frac{a}{b}\right)$$

This is an underestimate. If you compensate by taking the inductance to be

$$L = \mu_0 a \ln\frac{a}{b}$$

the result is quite accurate. For a more general discussion of this and related problems, see Jackson (1999), Section 5-17.

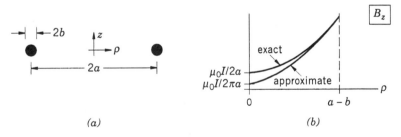

Figure 4-13. One-turn inductor (Exercise 4-7). (a) Cross section (b) Exact z-component of the magnetic field in the plane of symmetry, and the approximation obtained by extrapolating the field of an infinite wire.

4-8 POTENTIAL OF A PAIR OF LINE CHARGES. Consider a pair of line charges separated by a distance $2\rho_0$ and with densities $\pm\lambda$, as shown in Fig. 4-14. By superposition, the potential is

$$\phi = \frac{\lambda}{2\pi\varepsilon_0} \ln\frac{\rho_-}{\rho_+}$$

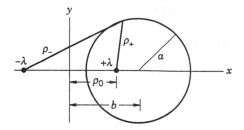

Figure 4-14. Equipotentials of a pair of line charges (Exercise 4-8).

where ρ_+ and ρ_- are the distances from the field point to the charges. The equipotentials in the right half-plane are given by

$$\frac{\rho_-}{\rho_+} = k$$

where $k > 1$ is a constant. In Cartesian coordinates you have

$$k^2 \left[(x - \rho_0)^2 + y^2 \right] = (x + \rho_0)^2 + y^2$$

Prove that the equipotentials are circles given by

$$(x - b)^2 + y^2 = a^2$$

where

$$a = \frac{2k}{k^2 - 1} \rho_0 \quad \text{and} \quad b = \frac{k^2 + 1}{k^2 - 1} \rho_0$$

Solving for k in terms of a and b, obtain

$$k = b/a + \sqrt{(b/a)^2 - 1}$$

Deduce that the capacitance per unit length of a pair of conductors of radius a separated by $2b$ between centers is

$$\mathcal{C} = \varepsilon_0 \frac{\pi}{\cosh^{-1}(b/a)}$$

and that the capacitance per unit length of a conductor of radius a whose center is located a distance b above a ground plane is twice this value. These results can be generalized: the capacitance per unit length of a pair of conductors of any shape is a few times ε_0, that is, about 50 pF/m.

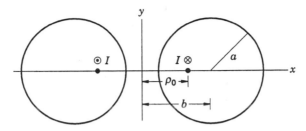

Figure 4-15. Magnetic field lines of two long wires (Exercise 4-9).

4-9 MAGNETIC FIELD LINES OF A PAIR OF WIRES (Fig. 4-15). The equipotentials of the pair of line charges in Exercise 4-8 happen to be the magnetic field lines of two long wires located at $x = \mp\rho_0$ and carrying opposite currents $\pm I$. They are also the field lines of two perfectly conducting infinite cylinders of radius a separated by $2b$ between centers and carrying the same currents. Show first that

$$b^2 - a^2 = \rho_0^2$$

where a and b are defined in Exercise 4-8. The field in the plane of the wires is

$$B_y = \frac{\mu_0 I}{2\pi}\left(\frac{1}{\rho_0 - x} + \frac{1}{\rho_0 + x}\right)$$

Show that the flux per unit length through the strip between the infinite cylinders is

$$\Phi = \mu_0 I \frac{\cosh^{-1}(b/a)}{\pi}$$

The inductance per unit length is thus

$$L = \mu_0 \frac{\cosh^{-1}(b/a)}{\pi}$$

This result can be generalized: the inductance per unit length of any pair of conductors is a few time smaller than μ_0, that is, about 250 nH/m.

Observe that $LC = 1/c^2$, where C is the capacitance per unit length obtained in Exercise 4-8. In Chapter 11 you will see that this result is valid for any two-conductor transmission line in a medium with permittivity ε_0.

4-10 MAGNETIC FIELD OF AN INFINITE SOLENOID. Consider a closed curve c of arbitrary shape in the x–y plane, as shown schematically in Fig. 4-16, and let c be the generator of an infinite cylinder aligned along the z-axis. Assume now that the cylinder carries a surface current density that is parallel to the x–y plane and has magnitude K. If $d\mathbf{l}$ is a line element of c,

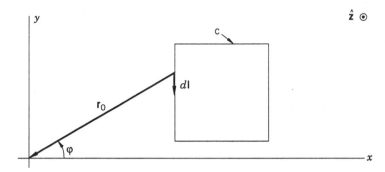

Figure 4-16. Closed curve c of arbitrary shape in the x–y plane, shown for simplicity as a square (Exercise 4-10).

the magnetic field $d\mathbf{B}$ due to the infinite strip generated by $d\mathbf{l}$ is

$$d\mathbf{B} = \frac{\mu_0 K}{4\pi} \int_{-\infty}^{+\infty} \frac{d\mathbf{l} \times \mathbf{r}}{r^3} \, dz$$

where $\mathbf{r}$ is the vector from a source point at a height z on the strip to the field point at the origin. If you decompose $\mathbf{r}$ into the sum of a vector $\mathbf{r}_0$ in the x–y plane and a vector along the z-direction, the contribution from the latter drops out by symmetry and you are left with

$$d\mathbf{B} = \frac{\mu_0 K}{4\pi} \int_{-\infty}^{+\infty} \frac{d\mathbf{l} \times \mathbf{r}_0}{r^3} \, dz$$

Now substitute $r^2 = r_0^2 + z^2$ and $z = \xi r_0$ to get

$$d\mathbf{B} = \frac{\mu_0 K}{4\pi} \frac{d\mathbf{l} \times \mathbf{r}_0}{r_0^2} \int_{-\infty}^{+\infty} \frac{d\xi}{\left(1+\xi^2\right)^{3/2}} = \frac{\mu_0 K}{2\pi} \frac{d\mathbf{l} \times \mathbf{r}_0}{r_0^2}$$

In the x–y plane you have $d\mathbf{l} \times \mathbf{r}_0 = \hat{\mathbf{z}} r_0^2 \, d\varphi$, so that

$$\mathbf{B} = \frac{\mu_0 K}{2\pi} \hat{\mathbf{z}} \oint d\varphi = \begin{cases} \mu_0 K \hat{\mathbf{z}} & \text{inside the cylinder} \\ 0 & \text{outside} \end{cases}$$

4-11 OHM'S LAW IN THE DRUDE–LORENTZ MODEL. Recall that the time T between collisions of the electrons in a metal follows a Poisson distribution with probability density

$$\frac{1}{\tau_F} \theta(T) \exp\left(-\frac{T}{\tau_F}\right)$$

The average value of a function $\psi(T)$ is given by

$$\langle \psi(T) \rangle = \frac{1}{\tau_F} \int_0^\infty \psi(T) \exp\left(-\frac{T}{\tau_F}\right) dT$$

Show that the average value of $\exp(-i\omega T)$, which you will need below, is

$$\langle \exp(-i\omega T)\rangle = \frac{1}{1+i\omega\tau_F}$$

Now consider a metal in which there is an oscillating electric field $\mathbf{E}$ of amplitude $\mathbf{E}_0$:

$$\mathbf{E} = \mathbf{E}_0 \exp(i\omega t)$$

The acceleration of the electrons (of charge $-e$ and mass m) is given by

$$\frac{d\mathbf{v}}{dt} = -\frac{e\mathbf{E}_0}{m}\exp(i\omega t)$$

Consider the situation at a given time t for an electron that had its last collision a time T earlier. The excess (field-induced) velocity at time t is

$$\mathbf{v}(t) = -\frac{e\mathbf{E}_0}{m}\int_{t-T}^{t}\exp(i\omega u)\,du = -\frac{e\mathbf{E}_0}{m}\frac{1-\exp(-i\omega T)}{i\omega}\exp(i\omega t)$$

The drift velocity $\langle \mathbf{v}\rangle$ is the average value of $\mathbf{v}(t)$. Since T follows a Poisson distribution, you get

$$\langle \mathbf{v}\rangle = -\frac{e\tau_F/m}{1+i\omega\tau_F}\mathbf{E}_0\exp(i\omega t)$$

The current density $\mathbf{J}$ is therefore as given in (4-9),

$$\mathbf{J} = -en\langle \mathbf{v}\rangle = \frac{ne^2\tau_F/m}{1+i\omega\tau_F}\mathbf{E} = \frac{\sigma}{1+i\omega\tau_F}\mathbf{E}$$

4-12 THE TOLMAN–STEWART EXPERIMENT (Tolman and Stewart, 1916, 1917). Consider a very long metal rod subjected to a longitudinal acceleration a. Think in terms of the Drude–Lorentz model and convince yourself that at any given time the (excess) velocity of an electron relative to the lattice is $-aT$, where T is the time since the last collision with the lattice. It follows that electrons drift relative to the lattice with an average velocity $-a\tau_F$ and generate a current density

$$J = ne\tau_F a = \sigma(m/e)a$$

In other words, you can make believe that the acceleration gives rise to an electric field of magnitude $(m/e)a$ along the rod. Assuming now that there is an electric field E along the rod (it might arise from charge accumulation at the ends of the rod), convince yourself that the current density becomes

$$J = \sigma(m/e)a + \sigma E$$

Take $a = 100$ m/s^2 (10g) and use $mc^2/e = 5 \cdot 10^5$ V and $c = 3 \cdot 10^8$ m/s to obtain

$$(m/e)a = (mc^2/e)(a/c^2) \approx 5 \cdot 10^{-10} \text{ V/m}$$

It is clear that the effect is small. Tolman and Stewart were nonetheless able to assign a value to the charge-to-mass ratio of electrons as conceived in the Drude–Lorentz theory of metals. In their experiment, a coil was rotated around its axis at high angular velocity and then decelerated in a small fraction of a second. The charge was measured with a ballistic galvanometer; to make the coil resistance comparable to the galvanometer resistance, they used a coil with many turns of thin wire. Realizing that sliding contacts were too erratic, they fixed the wires leading to the galvanometer at a point high above the coil and let the wires twist up when the coil rotated (Fig. 4-17a).

If V, I, r, and a are the coil voltage, current, resistance, and rim acceleration, and l and s are the wire length and cross-sectional area, and R and L are the resistance and inductance of the ballistic galvanometer, obtain

$$I = \sigma s(m/e)a - \sigma s\frac{V}{l} = I_{SC} - \frac{V}{r}$$

$$V = RI + L\frac{dI}{dt}$$

where $I_{SC} = \sigma s(m/e)a$ is the coil short-circuit current, and convince yourself that these equations can be represented by the circuit of Fig. 4-17b. Obtain further that

$$(r+R)I + L\frac{dI}{dt} = l(m/e)a = V_{OC}$$

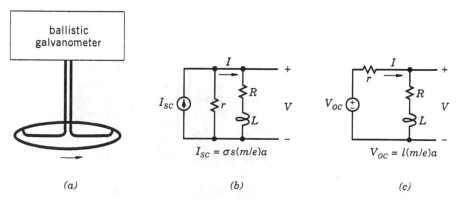

Figure 4-17. (a) Sketch of the Tolman–Stewart experiment described in Exercise 4-12. (b) Norton equivalent circuit. (c) Thévenin equivalent circuit.

where $V_{OC} = l(m/e)a$ is the coil open-circuit voltage, and convince yourself that this equation can be represented by the circuit of Fig. 4-17c. Note that in accordance with Thévenin's theorem,

$$V_{OC} = rI_{SC}$$

Integrate the differential equation for the current and obtain the charge Q that flows when the coil is suddenly brought to rest from a steady rim velocity v:

$$Q = \frac{l(m/e)v}{R+r}$$

Taking $l = 500$ m, $v = 10\pi$ m/s, and $R+r = 40\Omega$ (values close to those used), obtain $Q \simeq 2$ nC. Since the galvanometer sensitivity was about 0.2 nC, Tolman and Stewart were able to get a fair measure of the value of e/m. Incidentally, the papers in which they describe their equipment and measurement procedures are informative, and delightful to read.

4-13   PLASMA OSCILLATIONS. Consider again the infinite slab of copper of Exercise 4-5. Assume that the top and bottom faces have oscillating surface charge densities $\mp\Sigma_0 \cos \omega t$ (in this case power is involved, so that real fields will be more convenient). Obtain the electric field and the current density:

$$E_z = \frac{\Sigma_0}{\varepsilon_0} \cos \omega t$$

$$J_z = \omega\Sigma_0 \sin \omega t$$

On the assumption that the conductivity is infinite (no collisions with the lattice), the electrons all have the same excess velocity v_z. Calculate v_z and J_z. You should get

$$v_z = -\frac{e\Sigma_0}{m\varepsilon_0\omega} \sin \omega t$$

$$J_z = \frac{ne^2}{m\varepsilon_0\omega}\Sigma_0 \sin \omega t$$

Comparison with the preceding expression for J_z indicates that the oscillation is possible only if ω is equal to the plasma frequency:

$$\omega^2 = \omega_P^2 = \frac{ne^2}{m\varepsilon_0}$$

Obtain the kinetic energy density u_K and electric field energy density u_E:

$$u_K = \frac{nmv^2}{2} = \frac{\Sigma_0^2}{2\varepsilon_0} \sin^2 \omega_p t$$

$$u_E = \frac{\varepsilon_0 E^2}{2} = \frac{\Sigma_0^2}{2\varepsilon_0} \cos^2 \omega_p t$$

The total energy density is thus

$$u = \frac{\Sigma_0^2}{2\varepsilon_0}$$

Use Ohm's law (4-9), written as a differential equation, to obtain

$$\mathbf{E} \cdot \mathbf{J} = \frac{\mathbf{J} \cdot \mathbf{J}}{\sigma} + \frac{\tau_F}{\sigma} \frac{d\mathbf{J}}{dt} \cdot \mathbf{J}$$

The average power dissipated per unit volume assuming that $\mathbf{E}$ and $\mathbf{J}$ are sinusoidal is

$$\langle \mathbf{E} \cdot \mathbf{J} \rangle = \frac{\langle \mathbf{J} \cdot \mathbf{J} \rangle}{\sigma} = \frac{\Sigma_0^2}{2\varepsilon_0 \tau_F}$$

You can now obtain a differential equation for the energy density,

$$\frac{du}{dt} = -\frac{u}{\tau_F}$$

You then obtain

$$u(t) = u_0 \exp\left(-\frac{t}{\tau_F}\right)$$

and finally,

$$\Sigma(t) = \Sigma_0 \exp\left(-\frac{t}{2\tau_F}\right) \cos \omega_p t$$

This last expression agrees with (4-11).

4-14 ELECTRIC FIELDS IN INDUCTORS. Consider the two-dimensional circuit of Fig. 4-18. It consists of a one-turn inductor that is connected to a distant source by means of a parallel-plate transmission line of negligible thickness. Assume that this circuit carries a linearly increasing surface current density of magnitude $K_0 \lambda t$. As shown in Exercise 4-10, the magnetic

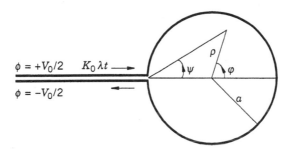

Figure 4-18. Two-dimensional one-turn inductor fed by a parallel-plate transmission line of negligible thickness (Exercise 4-14).

field outside the inductor is zero, whereas inside it is

$$B_z = -\mu_0 K_0 \lambda t$$

In the limit of zero thickness, and in accordance with the results of Section 4-6, neglect the contribution of the current in the transmission line to the electric field and obtain that the electric field due to the surface current is, as shown in Fig. 4-19a,

$$E_\varphi(\rho < a) = \frac{\mu_0 K_0 \lambda \rho}{2} = \frac{V_0 \rho}{2\pi a^2}$$

$$E_\varphi(\rho > a) = \frac{\mu_0 K_0 \lambda a^2}{2\rho} = \frac{V_0}{2\pi\rho}$$

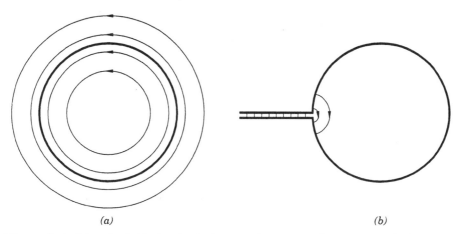

(a)   (b)

Figure 4-19. Electric fields in the inductor of Fig. 4-18 (Exercise 4-14). (a) Field induced by the current. (b) Total field.

where

$$V_0 = \pi a^2 \mu_0 K_0 \lambda$$

This electric field is clearly nonzero in regions where there are conductors; to make the electric field zero inside the conductors and normal to the surface just outside the conductors, charges are reshuffled, and the resulting surface charge distribution generates a potential

$$\phi(\rho < a) = \frac{V_0 \psi}{\pi}$$

$$\phi(\rho > a) = \frac{V_0 \varphi}{2\pi}$$

In particular, the plates of the transmission line acquire potentials

$$\phi(\rho > a, \varphi = \pm \pi) = \pm \frac{V_0}{2}$$

In self-fulfilling notation, the voltage obtained from these potentials agrees with the voltage obtained from Faraday's flux law,

$$-\Phi' = -\pi a^2 B_z' = \pi a^2 \mu_0 K_0 \lambda = V_0$$

For $\rho = a$ you have $\psi = \varphi/2$, and the interior and exterior potentials coincide. For $\rho \geq a$ the electric field from the scalar potential is

$$E_\varphi = -\frac{1}{\rho} \frac{\partial \phi}{\partial \varphi} = -\frac{V_0}{2\pi\rho}$$

and it exactly cancels the field due to the current. The continuity of the tangential component of the electric field implies that the total field is normal to the interior surface of the inductor.

Using

$$\psi = \tan^{-1} \frac{\rho \sin \varphi}{a + \rho \cos \varphi}$$

show that the surface charge on the interior of the inductor is concentrated toward the junction with the transmission line:

$$\Sigma = \varepsilon_0 \frac{\partial \phi}{\partial \rho}\bigg|_{\rho = a} = \frac{\varepsilon_0 V_0}{2\pi a} \tan \frac{\varphi}{2}$$

Figure 4-19b shows the total electric field. In the limit of zero thickness, the electric field in the transmission line is due only to charges and is thus irrotational; that is, it has vanishing curl.

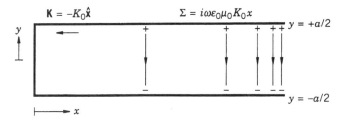

Figure 4-20. Parallel-plate inductor (Exercise 4-15).

4-15 SECOND-ORDER CURRENTS: PARASITIC CAPACITANCE OF AN INDUCTOR. Consider the parallel-plate inductor of Fig. 4-20, and assume that it carries a sinusoidal surface current density of amplitude K_0 and angular frequency ω,

$$\mathbf{K}(y = \pm a/2) = \mp K_0 \hat{\mathbf{x}}$$

Following the lines of Section 4-6, show that

$$\mathbf{B} = \mu_0 K_0 \hat{\mathbf{z}}$$

$$\mathbf{E} = -i\omega\mu_0 K_0 x\hat{\mathbf{y}}$$

and that the surface charge density on the top conductor is

$$\Sigma(y = a) = i\omega\varepsilon_0\mu_0 K_0 x$$

Conservation of charge implies that there must be an additional surface current density $K'\hat{\mathbf{x}}$ that accounts for the surface-charge buildup,

$$i\omega\Sigma = -\frac{\partial K'}{\partial x}$$

Show that K' is of second order in $i\omega x/c$,

$$K' = -\frac{1}{2}\left(\frac{i\omega x}{c}\right)^2 K_0$$

If the width of the inductor is w, the current I is

$$I = \left[1 + \frac{1}{2}\left(\frac{i\omega x}{c}\right)^2\right]K_0 w$$

The voltage at any plane is

$$V = i\omega\mu_0 K_0 ax$$

so that the admittance at that plane is

$$\frac{I}{V} = \frac{w}{i\omega\mu_0 ax} + \frac{i\omega\varepsilon_0 wx}{2a} = \frac{1}{i\omega L} + i\omega C$$

where $L = \mu_0 ax/w$ and $C = \varepsilon_0 wx/2a$.

Once again, inductors unavoidably have parasitic capacitances. In this case, as in most, the capacitance is small: show that $1/i\omega L$ and $i\omega C$ are equal in magnitude at a frequency

$$f_0 = \frac{c}{\pi\sqrt{2}\,x}$$

and that $f_0 \approx 700$ MHz for $x = 10$ cm.

4-16   INTERNAL INDUCTANCE OF A RESISTOR. Consider a cylindrical resistor of radius a and length l, and assume that it is carrying a sinusoidal current of amplitude I. From (4-12), the current density at low frequencies is

$$J_z(\rho) = J_z(0)\left(1 + i\omega\frac{\sigma\mu_0\rho^2}{4}\right)$$

Integrate $J_z(\rho)$ to obtain

$$I = \pi a^2 J_z(0)\left(1 + i\omega\frac{\sigma\mu_0\rho^2}{8}\right)$$

The voltage between the ends, taken along a path just outside the resistor, is

$$V = lE_z(a) = l\frac{J_z(a)}{\sigma} = \frac{lJ_z(0)}{\sigma}\left(1 + i\omega\frac{\sigma\mu_0 a^2}{4}\right)$$

or in terms of the current,

$$V = \frac{lI}{\sigma\pi a^2}\frac{1 + i\omega\dfrac{\sigma\mu_0 a^2}{4}}{1 + i\omega\dfrac{\sigma\mu_0 a^2}{8}} \approx \frac{lI}{\sigma\pi a^2}\left(1 + i\omega\frac{\sigma\mu_0 a^2}{8}\right) = I\left(\frac{l}{\sigma\pi a^2} + i\omega\frac{\mu_0 l}{8\pi}\right)$$

You thus have

$$V = (R + i\omega L)I$$

where R is the static resistance and L is the (parasitic) *internal* inductance, given by

$$L = \frac{\mu_0 l}{8\pi}$$

For $l = 1$ cm you have $L \approx 0.5$ nH, which is generally much smaller than any wiring inductance.

4-17 ELECTRON KINETIC ENERGY. The kinetic energy of electrons is seldom taken into account in energy balances; the energy U in a capacitor, for example, is expressed entirely in terms of electromagnetic quantities:

$$U = \tfrac{1}{2}CV^2$$

The point of this exercise is to show that this neglect is justified at sufficiently low frequencies; it is not justified, for example, in the case of plasma oscillations in metals.

(a) Assume a conductor in the Drude–Lorentz model. At any given time, the excess velocity of an electron that had its last collision a time T earlier is

$$\mathbf{v} = -\frac{e\mathbf{E}T}{m}$$

The average excess kinetic energy density is therefore

$$u_K = \frac{nm\langle v^2 \rangle}{2} = \frac{ne^2E^2 \langle T^2 \rangle}{2m}$$

From (4-7) you have $\langle T^2 \rangle = 2\tau_F^2$. In terms of the conductivity σ given in (4-8), you then have

$$u_K = \sigma E^2 \tau_F = \frac{J^2}{\sigma}\tau_F$$

Since the mean free time τ_F is of order 10^{-13} seconds, u_K is utterly negligible compared with the energy dissipated in any reasonable time.

(b) Consider a leaky planar capacitor with a material of conductivity σ between the plates, and assume that the capacitor is discharging with a time constant $RC = 1$ ns. Show first that in accordance with the discussion in Section 4-5, $RC = \varepsilon_0/\sigma$. Then, taking $\tau_F = 10^{-13}$ s, show that the ratio of the electron kinetic energy density u_K to the electric energy density u_E is

$$\frac{u_K}{u_E} = 2\frac{\sigma}{\varepsilon_0}\tau_F \approx 10^{-4}$$

You can draw the tentative conclusion that the electron kinetic energy is negligible in the case of phenomena that happen on time scales much longer than the mean free time τ_F.

4-18 MUTUAL INDUCTANCE. Consider the transformer discussed in Section 3-4 and shown in Fig. 3-15, and assume that the current in winding 1 is zero. The flux in winding 1 due to winding 2 can be expressed in terms of the vector potential,

$$\Phi_{12} = \int \mathbf{B} \cdot d\mathbf{a}_1 = \oint \mathbf{A} \cdot d\mathbf{l}_1$$

Now use (4-6) to obtain the clearly symmetrical *Neumann formula* for the mutual inductance M,

$$M = \frac{\mu_0}{4\pi} \oint \oint \frac{d\mathbf{l}_1 \cdot d\mathbf{l}_2}{r_{12}}$$

For arbitrary currents the voltages are given by

$$v_1 = L_1 i'_1 - M i'_2$$
$$v_2 = M i'_1 - L_2 i'_2$$

Show that the power into the transformer is

$$P = v_1 i_1 - v_2 i_2 = \left(\tfrac{1}{2} L_1 i_1^2 + \tfrac{1}{2} L_2 i_2^2 - M i_1 i_2 \right)'$$

The energy in the transformer is thus

$$U = \tfrac{1}{2} L_1 i_1^2 + \tfrac{1}{2} L_2 i_2^2 - M i_1 i_2$$

Rewrite the energy as

$$U = \frac{1}{2} L_1 \left(i_1 - \frac{M}{L_1} i_2 \right)^2 + \frac{1}{2} L_2 \left(1 - \frac{M^2}{L_1 L_2} \right) i_2^2$$

Making $i_1 = (M/L_1) i_2$ and demanding that the energy be nonnegative then gives you

$$M^2 \le L_1 L_2$$

5

OPERATIONAL AMPLIFIERS

5-1 INTRODUCTION

Electronics is unthinkable without amplifiers, so the sooner we address the subject, the better. The internal structure of any amplifier has its complications, but all amplifiers can be viewed largely as dependent sources—*active* circuit elements in contrast to *passive* elements such as resistors or transformers—that can deliver substantial power to a load while drawing virtually none from their controlling inputs.

By accepting such a functional description for now, leaving questions of physical structure for later chapters, and focusing on the particularly versatile class of *operational amplifiers*, we will go far toward acquiring a language applicable to all amplifiers and at the same time acquaint ourselves with circuits of great practical value.

Integrated operational amplifiers are exceedingly easy to use and have gained in speed and precision to the point that they are what first comes to mind when contemplating a new design. Focusing on operational amplifiers will thus have the further advantage of leading us into the mainstream of electronics.

A discussion of operational amplifiers necessarily involves *feedback*, a notion that has many aspects and is thus hard to define in few words without limiting its scope. We will therefore introduce feedback gradually by means of specific examples, with the intent of achieving enough familiarity to suit our purposes while implicitly acknowledging that much remains to be explored.

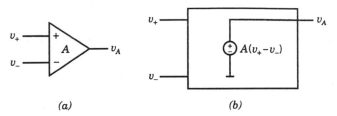

(a)   (b)

Figure 5-1. Operational amplifier. (*a*) Symbol. (*b*) Structure.

5-2 IDEAL OPERATIONAL AMPLIFIERS

An *ideal operational amplifier*, depicted in Fig. 5-1, is a dependent voltage source whose voltage v_A relative to ground is proportional to the difference between its two input voltages v_+ and v_-:

$$v_A = A(v_+ - v_-) \qquad (5\text{-}1)$$

where A is the *open-loop gain* or, more simply, the *gain*. The inputs draw no current from the sources of v_+ and v_-.

We will not specify the gain fully at this time, although we will assume that it is frequency dependent and, as indicated in Fig. 5-2, that its magnitude $A(\omega)$ is constant and large at low frequencies and falls off to zero at high frequencies. What we mean by *large* will depend on the circuit we have in mind: seldom less than 100, typically 10^5, and occasionally as high as 10^8.

Like ideal passive elements, ideal operational amplifiers are useful abstractions from real objects, in this case the operational amplifiers that we will meet in Chapter 7 after we have become familiar with transistor amplifiers; what is more, we will see that deviations from ideality can often be obviated by an appropriately expensive choice of hardware or by minor additional circuitry.

Ideal operational amplifiers apparently violate conservation of energy, but this is simply an artifact of incomplete representation: real operational

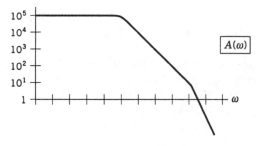

Figure 5-2. Representative frequency response of an operational amplifier.

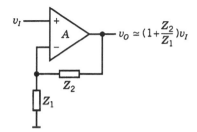

$$v_O \simeq (1 + \frac{Z_2}{Z_1})v_I$$

Figure 5-3. Direct amplifier. Z_1 and Z_2 are almost without exception resistors.

amplifiers are connected to power supplies from which they obtain the energy that they deliver. This point will also become clear in Chapter 7.

Operational amplifiers are best understood by seeing them in action, so let us proceed immediately to the *direct* (or *noninverting*) amplifier shown in Fig. 5-3 (Black, 1934, 1977). The (*feedback*) *loop* has been closed by connecting a *feedback network* $Z_1 | Z_2$ between the output and the *negative* (*inverting*) input of the operational amplifier, and a voltage v_I is applied to the *positive* (*noninverting*) input.

Keeping in mind that the inputs draw no current, we have

$$v_- = \frac{Z_1}{Z_1 + Z_2} v_O$$

Use of the operational-amplifier gain equation (5-1) yields

$$\left(v_I - \frac{Z_1}{Z_1 + Z_2} v_O\right) A = v_O \tag{5-2}$$

and we obtain the closed-loop gain $A_C = v_O/v_I$:

$$A_C = \frac{1 + \dfrac{Z_2}{Z_1}}{1 + \left(\dfrac{Z_1}{Z_1 + Z_2} A\right)^{-1}}$$

The closed-loop gain can be rewritten as

$$A_C = \frac{A_I}{1 + A_L^{-1}} \tag{5-3}$$

where A_I is the *ideal gain*,

$$A_I = 1 + \frac{Z_2}{Z_1}$$

and A_L is the *loop gain*,

$$A_L = \frac{Z_1}{Z_1 + Z_2} A \qquad (5\text{-}4)$$

If A_L is large enough, A_C approaches A_I; in the limit, A_C depends only on the feedback network $Z_1|Z_2$ and not at all on A. This is fundamental, because passive components can be extremely stable: 0.1% over a 10 K range in temperature is standard in film capacitors and metal-film resistors, and better than 0.01% is available in resistors if cost is not an issue. On the other hand, we will soon see that the gain of *open-loop* amplifiers, that is, of amplifiers without feedback, can easily vary by several percent over a 10 K range.

Let us consider some particular cases. If we take $Z_1 = R$ and $Z_2 = 99R$, and assume that the resistors have 0.1% precision and that the low-frequency gain of the operational amplifier is 10^5, we obtain

$$A_I = 100 \qquad A_L(\omega \to 0) = 10^3 \qquad A_C(\omega \to 0) = 99.9 \pm 0.2$$

This speaks for itself: a ×100 voltage amplifier that (ideally) has infinite input impedance and zero output impedance, and has excellent precision at low frequencies, can be built with two inexpensive resistors and, as is usually the case, an almost equally inexpensive operational amplifier.

If we set $Z_1 = \infty$ and $Z_2 = 0$, we obtain the *buffer* or *voltage follower* of Fig. 5-4, which has a gain of 1 to several significant figures. The buffer is used here to isolate a divider from a load resistance that might be several orders of magnitude smaller than the divider output resistance.

Let us now proceed to the *inverting* amplifier shown in Fig. 5-5. Summing currents at the inverting input, we obtain

$$\frac{v_I - v_-}{Z_1} = \frac{v_- - v_O}{Z_2}$$

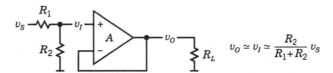

Figure 5-4. Unity-gain buffer used to isolate a divider from its load.

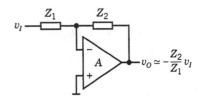

Figure 5-5. Inverting amplifier.

Substituting

$$v_O = -Av_-$$

we obtain the closed-loop gain $A_C = v_O/v_I$,

$$A_C = \frac{-\dfrac{Z_2}{Z_1}}{1 + \left(\dfrac{Z_1}{Z_1 + Z_2} A\right)^{-1}}$$

The closed-loop gain of an inverting amplifier can also be rewritten in terms of an ideal gain and a loop gain in the form (5-3). The loop gain is the same as the loop gain (5-4) of a direct amplifier, whereas the ideal gain is

$$A_I = -\frac{Z_2}{Z_1}$$

As in the case of a direct amplifier, the closed-loop gain A_C approaches the ideal gain A_I if the loop gain A_L is sufficiently large.

Direct and inverting amplifiers with the same feedback network have the same loop gain; as we will see in Section 5-4, this is merely a consequence of the fact that they are structurally identical circuits that differ only in the placement of the (independent) input voltage source. This is a point that we want to stress because for reasons discussed in Section 5-8, a potentially confusing terminology is often used in the literature: direct amplifiers are said to be in a *series feedback* configuration, whereas inverting amplifiers are said to be in a *shunt feedback* configuration. Our point of view will be that although there are differences in the ideal gain and in the impedance seen by the source, there is no difference as far as feedback is concerned.

We might be tempted to think that an increase in the loop gain always leads to a better amplifier, but as we will see in due time, the fact that the loop gain is necessarily frequency dependent implies that making it too large can result in an unstable circuit. This delicate question is the main subject of Chapter 9; for now we will assume that it has been taken care of.

The loop gain goes to zero at high frequencies, and depending on the feedback network, it can also go to zero at low frequencies; the closed-loop gain thus approaches ideal behavior only over a limited bandwidth. This question is also addressed in Chapter 9.

5-3 THE INFINITE-GAIN APPROXIMATION

At frequencies such that the loop gain is large, the *error voltage* $v_+ - v_-$ in a direct amplifier is given by

$$A(v_+ - v_-) = v_O \simeq v_I A_I$$

For typical values $A = 10^5$ and $A_I = 100$, it is clear that the error voltage is much smaller in magnitude than either v_I or v_O. This observation applies to any well-designed operational-amplifier circuit, and loosely speaking, we can say that operational amplifiers in such circuits do whatever is necessary to null their error voltage. By assuming that

$$v_+ = v_-$$

—this does *not* mean *short-circuit the positive and negative inputs*—we eliminate one unknown, and as we will show in Section 5-4, we obtain the ideal gain correctly and with relatively little effort. We will extend the use of the term *ideal* to describe any gain calculated assuming that the error voltage is zero; thus in the direct amplifier of Fig. 5-3, the ideal gain from v_I to v_- is unity.

Since setting $v_+ = v_-$ is formally equivalent to making $A = \infty$, we will refer to this procedure as the *infinite-gain approximation*. By using this approximation we lose all information about stability and bandwidth, so we must assume that proper dynamical behavior has been established separately.

As a first example, let us obtain the ideal gain of the adder shown in Fig. 5-6. Since $v_- = v_+ = 0$ by assumption, summing currents at the negative input yields

$$\frac{v_1}{R_1} + \frac{v_2}{R_1} = -\frac{v_O}{R_2}$$

and we obtain

$$v_O = -\frac{R_2}{R_1}(v_1 + v_2)$$

This scheme can obviously be extended to more than two inputs, and weighted sums can be obtained by adjusting the relative values of the input resistors. There is a price, however: as we will see in Chapter 9, multiple inputs make the response time longer because they reduce the loop gain.

Voltages can of course be summed in purely passive circuits like the one shown in Fig. 5-7, in which

$$v_O = \tfrac{1}{3}(v_1 + v_2)$$

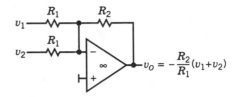

Figure 5-6. Operational-amplifier adder in the infinite-gain approximation.

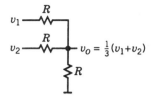

Figure 5-7. Resistive adder.

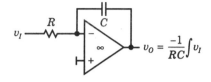

Figure 5-8. Operational-amplifier RC integrator.

The advantage of the operational-amplifier adder is that it can conserve or amplify signal levels and that it has output drive capability.

Let us next take a particular case of an inverting amplifier, the RC *integrator* of Fig. 5-8. Using the procedure of the preceding example, we obtain

$$\frac{v_I}{R} = -C\frac{dv_O}{dt}$$

or, assuming that $v_O(0) = 0$,

$$v_O(t) = -\frac{1}{RC}\int_0^t v_I(u)\,du$$

Operational amplifiers have this name precisely because they were first used to generate electrical analogs of the mathematical operations of addition and integration. An example of their use to solve linear differential equations in *analog computers* is given in Exercise 5-13.

As our next example we will consider the Howland current pump of Fig. 5-9. In the infinite-gain approximation we have

$$v_+ = v_- = \tfrac{1}{2}v_O$$

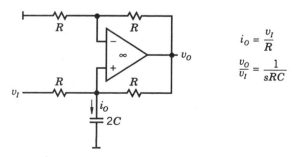

$$i_O = \frac{v_I}{R}$$

$$\frac{v_O}{v_I} = \frac{1}{sRC}$$

Figure 5-9. Howland current pump.

Summing currents at the positive input, we obtain

$$i_O = \frac{v_I - \frac{1}{2}v_O}{R} + \frac{v_O - \frac{1}{2}v_O}{R} = \frac{v_I}{R} \tag{5-5}$$

That is, i_O does not depend on whatever it is driving, and we have our first example of a current source. In the particular case considered here, i_O drives a capacitor, and we get an integrator:

$$\left(\tfrac{1}{2}v_O\right)2sC = i_O = \frac{v_I}{R}$$

or

$$\frac{v_O}{v_I} = \frac{1}{sRC} \tag{5-6}$$

The Howland current pump is not used in practice because its characteristics depend crucially on tight matching of the resistors. It is presented because it exemplifies the somewhat baffling circuits that result from feeding back to the positive as well as the negative input. Another surprising circuit in the same vein is the *negative-impedance converter* shown in Fig. 5-10. Since $R_1 = R_2 = R$ and $v_+ = v_-$, the currents in R_1, R_2, and Z must be equal to the independent source current i, and we get

$$v = v_- = v_+ = -iZ$$

The input impedance Z_I is therefore equal to $-Z$:

$$Z_I = \frac{v}{i} = -Z$$

The use of negative-impedance converters to construct a *gyrator* that converts an impedance into its reciprocal, and can thus make a capacitor look like an inductor, is considered in Exercise 5-10.

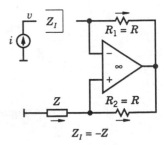

Figure 5-10. Negative-impedance converter.

5-4 IDEAL GAIN AND LOOP GAIN

We have seen in Section 5-2 that direct and inverting amplifiers with the same feedback network have the same loop gain. This is simply a reflection of the general rule that circuit structure is not affected by an independent voltage source in series with a branch or by an independent current source in parallel with a branch. Loop gain is an aspect of circuit structure and is therefore indifferent to input source location, as we will now show.

Let us consider the arbitrary operational-amplifier circuit containing one independent voltage source shown in Fig. 5-11. The feedback network is driven by the amplifier output v_A and by the independent source v_I, so that the amplifier input voltages are linear combinations of v_A and v_I:

$$v_+ = \beta_+ v_A + \alpha_+ v_I$$
$$v_- = \beta_- v_A + \alpha_- v_I$$

(5-7)

The α-factors $\alpha_\pm$ and the β-factors $\beta_\pm$ depend only on the feedback network because the operational-amplifier inputs draw no current. Figure 5-12 shows two examples of β-factor calculations. Calculations of α-factors are taken up in Exercises 5-11 and 5-17.

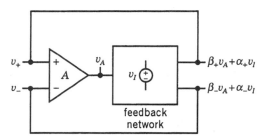

Figure 5-11. Arbitrary operational-amplifier circuit with one independent voltage source.

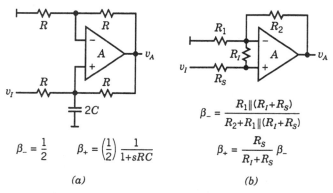

Figure 5-12. Examples of β-factor calculations. (*a*) Howland integrator. (*b*) Direct amplifier with a finite input resistance.

Subtracting equations (5-7) and using the operational-amplifier equation, we obtain

$$(\beta_+ - \beta_-)v_A + (\alpha_+ - \alpha_-)v_I = v_+ - v_- = v_A/A$$

This equation can be recast in the now familiar form (5-3),

$$\frac{v_A}{v_I} = A_C = \frac{A_I}{1 + A_L^{-1}}$$

in which the ideal gain A_I is given by

$$A_I = \frac{\alpha_+ - \alpha_-}{\beta_- - \beta_+} \tag{5-8}$$

and the loop gain A_L by

$$A_L = (\beta_- - \beta_+)A = \beta A \tag{5-9}$$

where $\beta = \beta_- - \beta_+$ is the (net) β-factor. Since β does not depend on where v_I is placed, it is clear that A_L, as promised above, is indifferent to input source location.

The ideal gain can be obtained directly by demanding that $v_+ = v_-$. From (5-7) we have

$$\beta_+ v_A + \alpha_+ v_I = \beta_- v_A + \alpha_- v_I$$

and we immediately recover expression (5-8) for the ideal gain. It is worth pointing out that (5-8) is seldom used since it is usually easier to calculate the ideal gain by assuming that $v_+ = v_-$ and then using whatever method is expedient.

The loop gain (5-9) can also be obtained by imagining that the feedback loop is cut at the output of the amplifier and, as shown in Fig. 5-13a, that an independent voltage source v_I is inserted on the downstream side of the cut

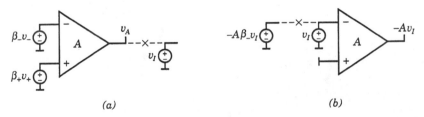

(a)   (b)

Figure 5-13. The loop gain can be obtained by imagining a cut in the feedback loop (a) at the operational-amplifier output in the general case or (b) at the inverting input if $\beta_+ = 0$.

to drive the feedback network, represented here by dependent sources placed at the amplifier inputs. The loop gain is then given by the negative of the gain from the downstream side of the cut to the upstream side:

$$A_L = -\frac{v_A}{v_I} = (\beta_- - \beta_+)A$$

In the case $\beta_+ = 0$ it is frequently seen in the literature that the loop is cut at the negative input as shown in Fig. 5-13b, clearly an equivalent procedure in this case.

Obtaining the loop gain by making an appropriate cut might look like a restatement of the obvious at this point, but it will become fruitful when calculating the loop gain of transistor amplifiers, in which the distinction between gain and β-factors is less sharp than we have shown so far.

Once we have calculated the closed-loop gain from the independent source v_I to the operational-amplifier output v_A, we can calculate the closed-loop gain to any other point in the circuit. We must emphasize, however, that although ideal gains in the infinite-gain approximation are meaningful, the closed-loop gain to points other than the operational-amplifier output will not in general be in the form given in (5-3).

All arguments presented so far are valid, *mutatis mutandis*, if the circuit of Fig. 5-11 is driven by an independent current source. Going further, we can apply such arguments to the dependent sources shown in Fig. 5-14; these are generalized forms of operational amplifiers in which the inputs are sensed without affecting the sources that generate them. We note that although gains and α-factors and β-factors can have dimensions of resistance or conductance, the loop gain βA remains dimensionless. Figure 5-15 presents two examples of gain calculations in circuits that use generalized operational amplifiers.

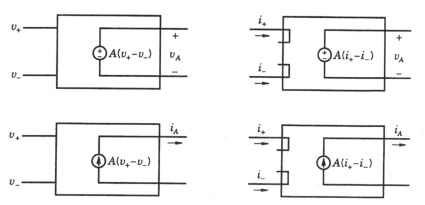

Figure 5-14. Generalized operational amplifiers. The inputs sense either currents or voltages, and their difference controls either a current source or a voltage source.

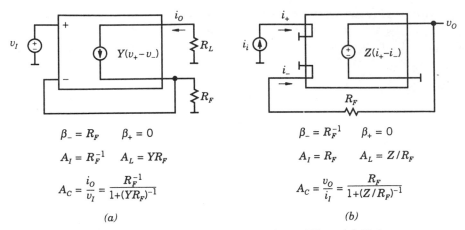

$$\beta_- = R_F \qquad \beta_+ = 0$$

$$A_I = R_F^{-1} \qquad A_L = YR_F$$

$$A_C = \frac{i_O}{v_I} = \frac{R_F^{-1}}{1+(YR_F)^{-1}}$$

(a)

$$\beta_- = R_F^{-1} \qquad \beta_+ = 0$$

$$A_I = R_F \qquad A_L = Z/R_F$$

$$A_C = \frac{v_O}{i_I} = \frac{R_F}{1+(Z/R_F)^{-1}}$$

(b)

Figure 5-15. Circuits with generalized operational amplifiers. (a) Voltage-to-current converter. (b) Current-to-voltage converter. In both cases the loop gain is, as it should be, dimensionless.

5-5   FURTHER BENEFITS FROM FEEDBACK

Real operational amplifiers deviate from ideality in many small ways; in the laboratory proposed in Exercise 5-1 we will see, for example, that the inputs source or sink a finite current and that the output voltage is not zero when the error voltage is zero. The origin of these and other imperfections will become clear after we have become familiar with transistors. What we want to show now is that feedback reduces the effect of certain small imperfections by a factor that is equal or approximately equal to $1+A_L$.

Let us first consider, as shown in Fig. 5-16, a direct amplifier in which the operational amplifier has an input resistance R_I much larger than $R_1 \| R_2$. The current through R_I is zero when the error voltage is zero, so that R_I does not affect the ideal gain. Since v_- follows v_I very closely—we will say that v_- is *bootstrapped* to v_I—it is clear that the current through R_I is much smaller than it would be if R_I were connected to ground and that the magnitude of the closed-loop input impedance Z_{IC} is therefore much larger

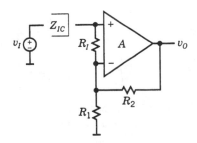

Figure 5-16. Direct amplifier based on an operational amplifier with input resistance R_I.

than R_I. Quantitatively, we have

$$Z_{IC} = \frac{v_I}{(v_I-v_-)/R_I} = \frac{v_I R_I}{v_O/A} = \frac{R_I A}{A_C} = R_I \frac{A}{A_I A_L}(1+A_L)$$

On the other hand, since $R_I \gg R_1 \| R_2$ and $A_I = (R_1+R_2)/R_1$, we also have

$$A_L = \frac{R_1 \| R_I}{R_2 + R_1 \| R_I} A = \frac{R_1}{R_1+R_2} \frac{R_I}{R_1 \| R_2 + R_I} A \simeq \frac{R_1}{R_1+R_2} A = \frac{A}{A_I}$$

It follows that

$$Z_{IC} \simeq R_I(1+A_L) \tag{5-10}$$

Let us next consider, as shown in Fig. 5-17, an amplifier (either direct or inverting, it makes no difference) in which the operational amplifier has an output resistance R_O much smaller than R_1+R_2. We will calculate the closed-loop output impedance Z_{OC} by nulling the input source and applying a voltage source v to the output. The presence of the feedback network results in a voltage v_A at the output of the operational amplifier that by design is much larger in magnitude than v:

$$v_A = -Av_- = \frac{-AR_1}{R_1+R_2} v$$

The current i delivered by the source is thus much larger than it would be if R_O were connected to ground,

$$i \simeq \frac{1}{R_O}\left(v - \frac{-AR_1}{R_1+R_2} v\right) \simeq \frac{1+A_L}{R_O} v$$

Here we have used the fact that $R_O \ll R_1+R_2$ in neglecting the current in the feedback network and in approximating the loop gain. For the closed-loop

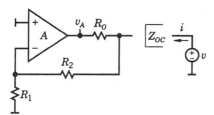

Figure 5-17. Circuit used to determine the closed-loop output impedance Z_{OC} of a direct or inverting amplifier based on an operational amplifier with output resistance R_O.

output impedance we then obtain

$$Z_{OC} \cong \frac{R_O}{1+A_L}$$

In Section 5-6 we will develop a general procedure—Blackman's theorem—for calculating impedances in feedback circuits; the calculations presented above are meant to highlight the mechanisms involved.

Finally, let us quickly see how variations in the open-loop gain affect the closed-loop gain. Setting $A_L = \beta A$ in (5-3), we obtain

$$\frac{dA_C}{dA} = \frac{\beta A_I}{(1+A_L)^2} = \frac{A_C/A}{1+A_L}$$

and for a small variation ΔA,

$$\frac{\Delta A_C}{A_C} \cong \frac{\Delta A}{A} \frac{1}{1+A_L}$$

That is, relative variations in the open-loop gain are suppressed in the closed-loop gain by a factor $1+A_L$. The reduction of nonlinearities in the gain by the same factor is the subject of Exercise 5-7.

5-6   BLACKMAN'S THEOREM

This interesting theorem will allow us to calculate the (closed-loop) impedance between two nodes of an operational-amplifier circuit, using quantities that are normally quite easy to obtain. Figure 5-18 represents an operational-amplifier circuit in which all independent sources but v_I have been nulled and a resistor R of arbitrary value has been inserted between the two nodes

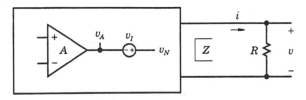

Figure 5-18. Circuit used in the proof of Blackman's theorem. Resistor R is in parallel with the closed-loop impedance Z that is to be determined.

in question; the loop gain is then a function of R. Since v_I is in series with v_A, we have

$$\alpha_+ = \beta_+ \quad \text{and} \quad \alpha_- = \beta_-$$

According to (5-8), the ideal gain from v_I to v_A is -1; the closed-loop gain is therefore

$$\frac{v_A}{v_I} = \frac{-A_L(R)}{1+A_L(R)}$$

and the voltage v_N that drives the feedback network is

$$v_N = v_I + v_A = \frac{1}{1+A_L(R)} v_I$$

The resistor voltage and current are related to v_N by transfer functions $A_T(R)$ and $Y_T(R)$, so that

$$v = \frac{A_T(R)}{1+A_L(R)} v_I \quad \text{and} \quad i = \frac{Y_T(R)}{1+A_L(R)} v_I$$

where

$$Y_T(R) = \frac{A_T(R)}{R}$$

The open-circuit voltage v_{OC} and the short-circuit current i_{SC} are obtained by setting $R = \infty$ and $R = 0$:

$$v_{OC} = \frac{A_T(\infty)}{1+A_L(\infty)} v_I \quad \text{and} \quad i_{SC} = \frac{Y_T(0)}{1+A_L(0)} v_I$$

The impedance Z between the nodes is therefore

$$Z = \frac{v_{OC}}{i_{SC}} = A_T(\infty) Z_T(0) \frac{1+A_L(0)}{1+A_L(\infty)}$$

On the other hand, letting $A \to 0$, we get

$$Z(A = 0) = A_T(\infty) Z_T(0)$$

and we obtain Blackman's theorem,

$$Z = Z(A = 0) \frac{1+A_L^{SC}}{1+A_L^{OC}}$$

where $A_L^{SC} \equiv A_L(0)$ and $A_L^{OC} \equiv A_L(\infty)$ are the short-circuit and open-circuit loop gains. Like ideal gain and loop gain, Blackman's theorem can be extended to circuits containing the generalized operational amplifiers shown in Fig. 5-14.

We will find Blackman's theorem particularly useful when we come to transistor circuits. For now let us take a familiar example, the direct amplifier of Fig. 5-16, and let us calculate its input impedance assuming that the operational amplifier has an input resistance $R_I \gg R_1 \| R_2$. The input impedance for $A = 0$ is

$$Z(A = 0) = R_I + R_1 \| R_2 \simeq R_I$$

The short-circuit loop gain is precisely the loop gain of the amplifier when it is driven as usual by a voltage source,

$$A_L^{SC} = A_L$$

No current flows in R_I when we open the circuit; this implies that $\beta_+ = \beta_-$ in this configuration and therefore that

$$A_L^{OC} = 0$$

We can now recover the result for the closed-loop input impedance given in (5-10):

$$Z_{IC} = Z(A = 0)(1 + A_L^{SC}) \simeq R_I(1 + A_L)$$

5-7 VOLTAGE FEEDBACK AND CURRENT FEEDBACK

The theory of feedback applied to linear servomechanisms takes as its starting point the scheme presented in Fig. 5-19. A sample HX of the output X is picked off by the *feedback block* and subtracted from the input Y at the

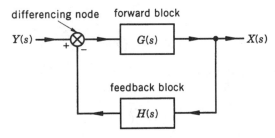

Figure 5-19. Block diagram of a basic linear servomechanism.

differencing node; the *error* $Y - HX$ is applied to the *forward block*, where it is multiplied by the forward gain G to yield X:

$$(Y - HX)G = X \qquad (5\text{-}11)$$

In most servomechanisms, the forward gain G is subject to variations, whereas the gain H of the feedback block is stable by design; the intent in feeding back is that the (large) gain of the forward block should drive X so as to null the error $Y - HX$, and that X should therefore be determined principally by the feedback block. Solving for X in (5-11), we obtain the closed-loop gain,

$$\frac{X}{Y} = \frac{G}{1 + HG}$$

It follows, at least formally, that making HG large implies

$$\frac{X}{Y} \simeq \frac{1}{H}$$

This all looks familiar, as indeed it should: if we compare (5-11) with (5-2), we recognize immediately that the direct amplifier of Fig. 5-3 is in fact a servomechanism that can be mapped into Fig. 5-19 if for Y, X, G, and H we substitute v_I, v_O, A, and $Z_1/(Z_1 + Z_2)$. In addition, we recognize $1/H$ as the ideal gain A_I and HG as the loop gain A_L.

Many electronic circuits are similar to servomechanisms in that an output variable, either a voltage or a current, is intentionally *slaved* to an input variable by means of feedback, in which case we speak of *voltage feedback* or of *current feedback*. Electronic circuits, however, cannot always be reduced to the relatively simple diagram of Fig. 5-19 or, for that matter, to the generalizations discussed in Section 5-8; we therefore need to adopt a broader definition of what we mean by a slaved variable.

With reference to the circuit shown in Fig. 5-20, we will consider that we are dealing with *voltage feedback* if the output voltage v_O is independent of the load impedance Z_L in the infinite-gain approximation, and that we are dealing with *current feedback* if the output current i_O is independent of Z_L in the infinite-gain approximation. These definitions also apply if the circuit is driven by an independent current source rather than an independent voltage source, or if it contains generalized operational amplifiers.

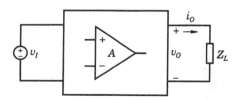

Figure 5-20. Circuit used to determine whether the output voltage v_O or the output current i_O is slaved to the input voltage v_I.

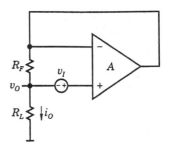

Figure 5-21. Example of current feedback: output current i_O is independent of load resistor R_L in the infinite-gain approximation.

Equivalently, we can say that we have voltage (current) feedback from a pair of terminals if the output impedance in the infinite-gain approximation is zero (infinite). From this point of view, feedback in direct and inverting amplifiers is clearly voltage feedback, even when the operational amplifiers are not completely ideal in that they have finite input and output impedances.

The current source of Fig. 5-21 is an example of current feedback in which it is easy to recognize the output current i_O as the slaved variable according to (5-11) and Fig. 5-19. We have

$$(v_I - R_F i_O) A = (R_L + R_F) i_O$$

or

$$(v_I - R_F i_O) \frac{A}{R_L + R_F} = i_O$$

and we can identify v_I, i_O, R_F, and $A/(R_L + R_F)$ with Y, X, H, and G. We are also dealing with current feedback according to our wider definition because the ideal gain A_I from the input voltage v_I to the output current i_O is independent of the load resistance R_L:

$$A_I = \frac{1}{R_F}$$

In contrast, the Howland current pump of Fig. 5-9 is a circuit in which it is hard to see that a sample of the output current is subtracted from the input voltage to obtain an error signal, yet it is clearly a case of current feedback according to our criterion because the output current in the infinite-gain approximation, given in (5-5), is independent of the load impedance.

5-8 FEEDBACK TERMINOLOGY

So far we have managed to understand operational-amplifier circuits in terms of ideal gain, loop gain, and Blackman's theorem, and we will continue in this style. But this has not been the usual approach, and we will therefore make a

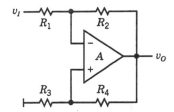

Figure 5-22. Operational-amplifier circuit with both positive and negative feedback.

short digression in order to connect with terminology that for historical reasons is firmly established in the literature.

The terms *negative feedback* and *positive feedback* are widely and somewhat loosely used in the literature, so that it will be convenient at this point to introduce a note of caution. For the time being we will interpret these terms simply as describing feedback to the negative (inverting) or to the positive (noninverting) terminal of an operational amplifier. Refinements will have to wait until Chapter 9, but we can advance the notions that negative feedback is *meant* to be stabilizing and that positive feedback *tends* to be destabilizing; some positive feedback can be tolerated, but too much will get us into trouble unless we explicitly intend to build an oscillator.

As an example, the circuit of Fig. 5-22 is usually unstable if $\beta_+ > \beta_-$, that is, if

$$\frac{R_3}{R_4} > \frac{R_1}{R_2}$$

even if it is stable when R_4 is removed. On the other hand, we have already mentioned that a direct amplifier can become unstable even though the feedback is nominally negative; this is precisely what makes *positive* and *negative* slippery terms. Again, we will withhold further discussion until Chapter 9.

The terms *series feedback* and *shunt feedback*, also widely used, arise from the feedback models shown in simplified form in Fig. 5-23 and discussed in more detail in Exercise 5-15. These models emulate the scheme of Fig. 5-19 by replacing the forward block with a generalized operational amplifier and the feedback block with a *unilateral* two-port. The resulting structures are then classified according to whether voltages (series) or currents (shunt) are subtracted at the input and whether voltage (shunt) or current (series) is sampled at the output.

In terms of these feedback models, the direct amplifier is in a *series–shunt* (briefly *series*) configuration, whereas the inverting amplifier is in a *shunt–shunt* (briefly *shunt*) configuration, because as shown in Exercise 5-16, we can imagine in the latter case that a current proportional to the output voltage is subtracted from an (equivalent) input current source to generate the error signal. As we indicated in Section 5-2, we will accept these distinctions nominally without attaching any deep meaning to them.

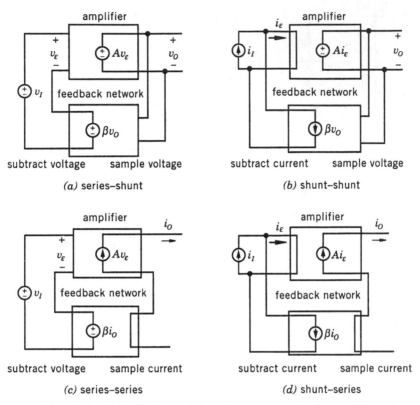

Figure 5-23. Traditional feedback models in the servomechanism mold of Fig. 5-19, based on the generalized operational amplifiers of Fig. 5-14 and on unilateral feedback blocks.

Feedback networks are mostly bilateral; this already makes one-to-one maps of electronic circuits into feedback models at least awkward. One exception—probably the only clean one—is the direct amplifier. Furthermore, although operational-amplifier circuits are almost without exception feedback circuits, we have seen in the example of the Howland current pump that what is being fed back is often hard to visualize, or even—such is the case of a negative-impedance converter—irrelevant, in the sense that we might not be interested in slaving an output to an input as we are in the case of a direct amplifier.

EXERCISES

5-1 LABORATORY I: IMPERFECTIONS OF REAL OPERATIONAL AMPLIFIERS. The equivalent circuit of Fig. 5-24 shows how some imperfections of real operational amplifiers can be represented by external circuit

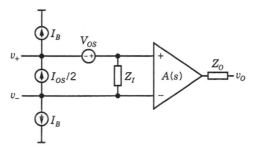

Figure 5-24. Representation of operational-amplifier imperfections by means of external circuit elements (Exercise 5-1).

elements. In Section 5-5 you have already become familiar with the fact that operational amplifiers have a (generally large) input impedance Z_I and a (generally small) output impedance Z_O. The additional elements are the *input offset voltage* V_{OS}, the *input bias current* I_B, and the *input offset current* I_{OS}.

Get hold of an operational amplifier with the following approximate values according to the data sheet:

Low-frequency gain	A	$> 10^5$
Input bias current	I_B	100 nA
Input offset current	I_{OS}	10 nA
Input offset voltage	V_{OS}	1 mV

Given these numbers, convince yourself that the circuits of Fig. 5-25 can be used to extract the indicated quantities. Assemble the circuits and make the measurements.

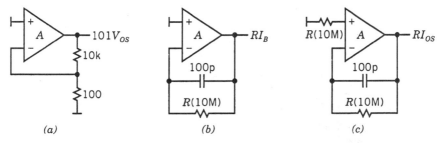

Figure 5-25. Circuits used to measure (*a*) the input offset voltage V_{OS}, (*b*) the input bias current I_B, and (*c*) the input offset current I_{OS} (Exercise 5-1).

5-2 LABORATORY II: GAIN LINEARITY. Assemble the amplifier shown in Fig. 5-26. Use metal-film resistors, and measure them to 0.1%. Vary v_I from 0 to 10V; make 20 measurements of v_O versus v_I with 4-digit voltmeters. Plot the results to make sure that no gross errors have been made, and make a least-squares fit to the gain v_O/v_I. The fit should exhibit deviations under 0.1% and agree with theory within 0.3%. Look at the output with an oscilloscope to make sure that noise pickup or oscillations are not a problem.

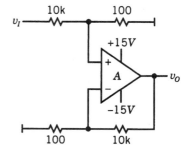

Figure 5-26. Circuit used to measure the gain linearity of a direct amplifier in Exercise 5.2.

5-3 COMMON-MODE REJECTION RATIO OF A DIFFERENTIAL AMPLIFIER. Consider the *differential amplifier* of Fig. 5-27. Assuming that

$$R_4 = R_2 \quad \text{and} \quad R_3 = R_1$$

use superposition to obtain that the output is proportional to the difference of the input voltages,

$$v_O = \frac{R_2}{R_1}(v_2 - v_1)$$

Now assume that there is a slight mismatch in the resistors:

$$R_4 = R_2 \quad \text{and} \quad R_3 = R_1(1 + \varepsilon)$$

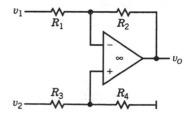

Figure 5-27. Differential amplifier (Exercise 5-3).

Show that

$$v_O \simeq \frac{R_2}{R_1}(v_{DM} - \gamma v_{CM})$$

where v_{DM} and v_{CM} are the *differential-mode* and *common-mode* voltages,

$$v_{DM} = v_2 - v_1 \quad \text{and} \quad v_{CM} = \frac{v_1 + v_2}{2}$$

and γ is the reciprocal of the *common-mode rejection ratio* (CMRR),

$$\gamma = \frac{\varepsilon R_1}{R_1 + R_2} = \frac{1}{\text{CMRR}}$$

Observe that the CMRR improves as the gain increases. For a gain of 100, it is 100 dB (10^5) if the resistors are matched to 0.1%. At this level you should worry about the CMRR of the operational amplifier itself, which will seldom be better than 120 dB (10^6).

5-4 INSTRUMENTATION AMPLIFIERS. Instrumentation amplifiers, a common version of which is shown in Fig. 5-28, are high-impedance differential amplifiers that are used to measure a voltage difference in the presence of a common-mode voltage; they can be used, for example, to measure the offset in a Wheatstone bridge or, when measuring the voltage of a distant source, to reduce common-mode noise injected along the way and eliminate the normally present voltage difference between the remote and local grounds.

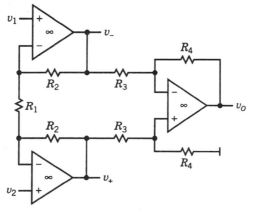

Figure 5-28. Instrumentation amplifier (Exercise 5-4).

Use superposition. First make $v_2 = 0$ and obtain

$$v_- = \left(1 + \frac{R_2}{R_1}\right)v_1 \quad \text{and} \quad v_+ = -\frac{R_2}{R_1}v_1$$

Using superposition again with v_+ and v_-, obtain

$$v_O = -\frac{R_4}{R_3}v_- + \frac{R_4}{R_3+R_4}\left(1+\frac{R_4}{R_3}\right)v_+ = -\frac{R_4}{R_3}\left(1+\frac{2R_2}{R_1}\right)v_1$$

Now make $v_1 = 0$, obtain the same expression for v_O in terms of v_2 except for the sign, and finally obtain

$$v_O = \frac{R_4}{R_3}\left(1+\frac{2R_2}{R_1}\right)(v_2 - v_1)$$

Convince yourself that resistor mismatches in the first stage (R_2) do not affect the common-mode rejection ratio defined in Exercise 5-3.

5-5   SECOND-ORDER LOW-PASS FILTER. Consider the filter shown in Fig. 5-29. The infinite-gain approximation says that $v_+ = v_-$ and therefore that

$$v_O = \frac{v}{1+sR_2C_2}$$

Sum currents at node v to obtain

$$\frac{v_I - v}{R_1} = sC_1(v - v_O) + \frac{v - v_O}{R_2}$$

Conclude that the filter transfer function $H(s) = v_O/v_I$ is

$$H(s) = \frac{1}{1+sC_2(R_1+R_2)+s^2C_1C_2R_1R_2}$$

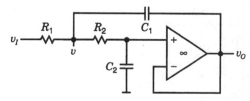

Figure 5-29. Second-order low-pass filter (Exercise 5-5).

Setting $R_1 = R_2 = R$ and $C_2 = C_1/2 = C$ yields

$$H(s) = \frac{1}{1 + 2CRs + 2C^2R^2s^2}$$

This is a second-order system with $\tau_0 = \sqrt{2}\,CR$ and $\zeta = 1/\sqrt{2}$. Show that it is case $n = 2$ of the family of (maximally flat) Butterworth filters that have transfer functions with magnitude

$$H(\omega) = \frac{1}{\left[1 + (\omega\tau_0)^{2n}\right]^{1/2}}$$

5-6 SMALL RESISTORS, LARGE TIME CONSTANT. In the low-pass filter of Fig. 5-30, show that

$$\frac{v_O}{v_I} = \frac{1}{1 + sRC}$$

where

$$R = \frac{R_1 R_2 + R_1 R_3 + R_2 R_3}{R_3}$$

If you have doubts follow the outline of Exercise 5-5, which is quite similar. Consider the case $R_1 = R_2 = R_0$ and $R_3 \ll R_0$, in which

$$R \simeq \frac{R_0^2}{R_3} \gg R_0$$

You then obtain a time constant that is much larger than the time constant of a phase lag formed by R_0 and C. Try to form a physical picture of what is going on; in particular, try to see that R_3 denies to R_2 most of the current through R_1, which is why C charges up so slowly.

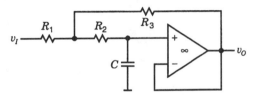

Figure 5-30. Low-pass with a long time constant obtained by positive feedback (Exercise 5-6).

5-7 FEEDBACK REDUCES NONLINEARITIES. Consider a direct amplifier in which the otherwise ideal operational amplifier has the nonlinear transfer characteristic shown in Fig. 5-31,

$$v_O = V_S \tanh\left[\frac{A(v_+ - v_-)}{V_S}\right]$$

The choice is not arbitrary: in Chapter 7 you will see that this is the gain of a transistor differential amplifier. For $|v_+ - v_-| \ll V_S/A$ you have

$$v_O \simeq A(v_+ - v_-)$$

This is the usual case: the output is proportional to the input for small signals and saturates for large signals. Calculate nonlinearities as a function of the output level: for a given level compare the actual input with the input that would give the same output if the gain were linear, that is, if $V_S = \infty$.
    The open-loop nonlinearity for an output level $|v_O| \le V_S/2$ is

$$\frac{v_+ - v_-}{v_O/A} - 1 = \frac{V_S}{v_O}\tanh^{-1}\left(\frac{v_O}{V_S}\right) - 1 \simeq \frac{v_O^2}{3V_S^2}$$

When you close the loop you get

$$v_O = V_S \tanh\left[\frac{A}{V_S}\left(v_I - \frac{R_1}{R_1 + R_2}v_O\right)\right]$$

and you can then obtain

$$v_I = \frac{R_1}{R_1 + R_2}v_O + \frac{V_S}{A}\tanh^{-1}\frac{v_O}{V_S}$$

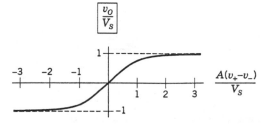

Figure 5-31. Nonlinear open-loop voltage again (Exercise 5-7).

and

$$\frac{v_I}{v_O/A_C} = \frac{A_C R_1}{R_1+R_2} + \frac{A_C}{A} + \frac{A_C}{A}\left[\frac{V_S}{v_O}\tanh^{-1}\left(\frac{v_O}{V_S}\right)-1\right]$$

In a direct amplifier the first two terms on the right-hand side add up to 1 and you also have $A_C/A = 1/(1+A_L)$; the closed-loop nonlinearity is therefore

$$\frac{v_I}{v_O/A_C} - 1 = \frac{A_C}{A}\left[\frac{V_S}{v_O}\tanh^{-1}\left(\frac{v_O}{V_S}\right)-1\right] \approx \frac{1}{1+A_L}\frac{v_O^2}{3V_S^2}$$

The nonlinearity is thus reduced by a factor $1+A_L$.

5-8 INVERSE-FUNCTION GENERATORS. In the circuit of Fig. 5-32 assume that the voltage fed back to the negative input is a function of the output voltage:

$$v_- = V_S g(v_O/V_S)$$

where V_S is a scale factor. Show that the output voltage is given by the inverse function:

$$v_O = V_S g^{-1}(v_I/V_S)$$

In the case of a direct amplifier (Fig. 5-3), g is simply multiplication by $R_1/(R_1+R_2)$, so that as you already know, $v_O = (1+R_2/R_1)v_I$.
    Now take an analog squaring circuit (you will soon meet a real one), in which

$$v_- = v_O^2/V_S$$

The output is then proportional to the square root of the input:

$$v_O = \sqrt{v_I V_S}$$

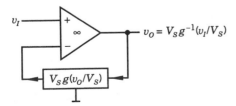

Figure 5-32. Inverse-function generator (Exercise 5-8).

Finally, take an analog multiplier (again, you will soon meet one), in which

$$g(v_O/V_S) = v_O v_I'/V_S^2$$

where v_I' is an independent input. You then get an analog divider:

$$v_O = V_S v_I / v_I'$$

5-9 GYRATOR INDUCTOR. First prove the general result for the totem-pole gyrator of Fig. 5-33a. By making the choices indicated in Fig. 5-33b you can make capacitor C look like an inductor $L = rRC$. Integrated filters make good use of this possibility, since inductors of any significant value cannot be made in integrated form.

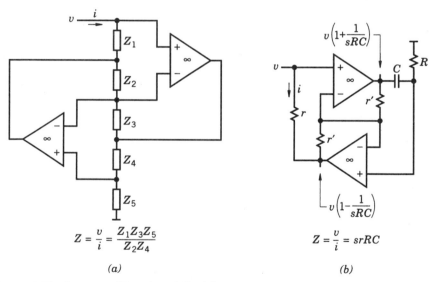

$$Z = \frac{v}{i} = \frac{Z_1 Z_3 Z_5}{Z_2 Z_4}$$

$$Z = \frac{v}{i} = srRC$$

(a)   (b)

Figure 5-33. Gyrators (Exercise 5-9). (a) General case of a totem-pole gyrator. (b) Gyrator inductor.

5-10 NEGATIVE-IMPEDANCE CONVERTERS AND GYRATORS. First prove that Fig. 5-34a represents a negative-impedance converter (the choice between this version and the one in Fig. 5-10 is dictated by stability, as discussed in Section 5-8). Then prove that Fig. 5-34b represents a gyrator.

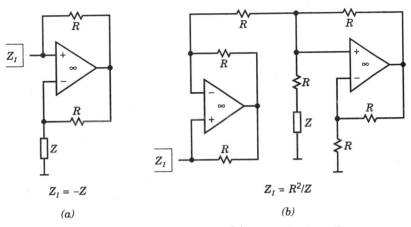

Figure 5-34. Circuits for Exercise 5.10. (*a*) Negative-impedance converter. (*b*) Gyrator.

5-11   IDEAL GAINS THE HARD WAY. Show that the α-factors in the direct amplifier of Fig. 5-12*b* are

$$\alpha_+ = \frac{R_I + R_1 \| R_2}{R_S + R_I + R_1 \| R_2} \quad \text{and} \quad \alpha_- = \frac{R_1 \| R_2}{R_S + R_I + R_1 \| R_2}$$

Use these expressions and the β-factors given in Fig. 5-12*b* to obtain that the ideal gain is given correctly by (5-8),

$$A_I = 1 + \frac{R_2}{R_1}$$

You will agree that it is preferable to obtain A_I by assuming that $v_+ = v_-$ and then using whatever method you find convenient.

5-12   ACTIVE FILTERS. In the circuit of Fig. 5-35, show that

$$\left(-\frac{v_B}{sR_F^2 C} + \frac{v_I}{R_G} \right) \frac{R_B}{1 + sR_B C} = v_B$$

and therefore that

$$\frac{v_B}{v_I} = \frac{R_F}{R_G} \frac{sCR_F}{1 + sC\dfrac{R_F^2}{R_B} + s^2 C^2 R_F^2}$$

and

$$\frac{v_L}{v_I} = \frac{R_F}{R_G} \frac{-1}{1 + sC\dfrac{R_F^2}{R_B} + s^2 C^2 R_F^2}$$

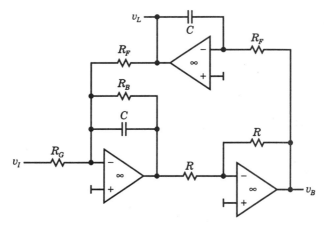

Figure 5-35. Active filter (Exercise 5-12).

These are second-order responses with parameters

$$\omega_0 = \frac{1}{CR_F} \quad \text{and} \quad 2\zeta = \frac{R_F}{R_B}$$

By choosing the resistors appropriately you can set the gain, the bandwidth, and the center frequency for v_B/v_I (bandpass filter), or the gain, the cutoff frequency, and the damping factor for v_L/v_I (low-pass filter).

5-13   ACTIVE FILTERS AS ANALOG COMPUTERS. Convince yourself that differential equation

$$RCv_O' + v_O = -v_I$$

describes the low-pass filter of Fig. 5-36a and, assuming that $v_O(0) = 0$, that it can be rewritten as

$$v_O(t) = -\frac{1}{RC} \int_0^t [v_I(u) + v_O(u)]\, du$$

and represented by the block diagram of Fig. 5-36b.

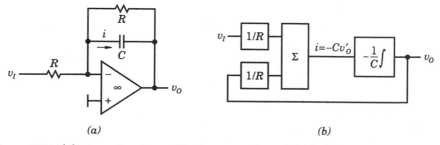

(a)   (b)

Figure 5-36. (a) Operational-amplifier low-pass filter; (b) block-diagram representation of its differential equation (Exercise 5-13).

Now rewrite differential equation

$$C^2 R_F^2 v_L'' + C \frac{R_F^2}{R_B} v_L' + v_L = - \frac{R_F}{R_G} v_I$$

as

$$- C^2 R_F^2 v_L'' = \frac{R_F}{R_G} v_I + \frac{R_F}{R_B} C R_F v_L' + v_L$$

and convince yourself that it is represented by the block diagram of Fig. 5-37, and that the active filter of Exercise 5-12 is a possible hardware implementation, that is, an *analog computer* that solves this differential equation.

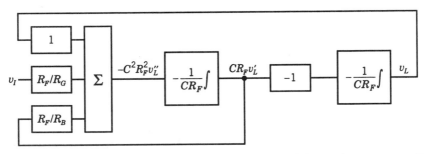

Figure 5-37. Block-diagram representation of the differential equation corresponding to the active filter of Fig. 5-35 (Exercise 5-13).

5-14 MOTOR-SPEED CONTROLLER. The steady-state characteristics of a permanent-magnet dc motor are given by

$$v_A - K\omega = R_A i_A$$

where v_A, i_A, and R_A are the armature voltage, current, and resistance, ω is the motor angular velocity or "speed", and K is a constant. The armature current is proportional to the shaft torque required to drive the load. The purpose of the circuit of Fig. 5-38 is to keep the (steady-state) motor speed constant in the face of variations in the load torque. (In Chapter 9 you will find out that capacitor C is needed to make the circuit stable.)

In the infinite-gain approximation, show that

$$v_O \frac{R_1}{R_1 + R_2} - v_I \frac{R_2}{R_1 + R_2} = i_A R_F$$

and

$$v_O = v_A + i_A R_F$$

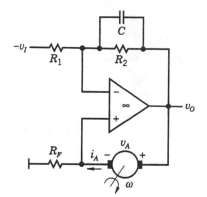

Figure 5-38. Operational-amplifier speed controller for a dc motor (Exercise 5-14).

and therefore that

$$v_A - i_A R_F \frac{R_2}{R_1} = v_I \frac{R_2}{R_1}$$

If you choose R_F such that

$$R_A = R_F \frac{R_2}{R_1}$$

the motor speed will be constant and given by

$$K\omega = v_I \frac{R_2}{R_1}$$

5-15 FEEDBACK MODELS. As suggested in Section 5-8, the feedback models of Fig. 5-39 extend the servomechanism scheme of Fig. 5-19. Although unwieldly in their application to specific feedback circuits, they are theoretically useful in indicating the effects of feedback, particularly on impedances.

Verify a few of the entries for input and output impedances, keeping in mind, for example, that in calculating an input impedance you must open-circuit or short-circuit the output, whichever is appropriate.

5-16 FEEDBACK MODELS: THE INVERTING AMPLIFIER. Consider the inverting amplifier of Fig. 5-5. For $Z_1 = R_1$ and $Z_2 = R_2$, show that

$$\left(\frac{R_2}{R_1 + R_2} v_I + \frac{R_1}{R_1 + R_2} v_O \right) A = -v_O$$

and therefore that

$$\left(\frac{v_I}{R_1} - \frac{-v_O}{R_2} \right) (R_1 \| R_2) A = -v_O$$

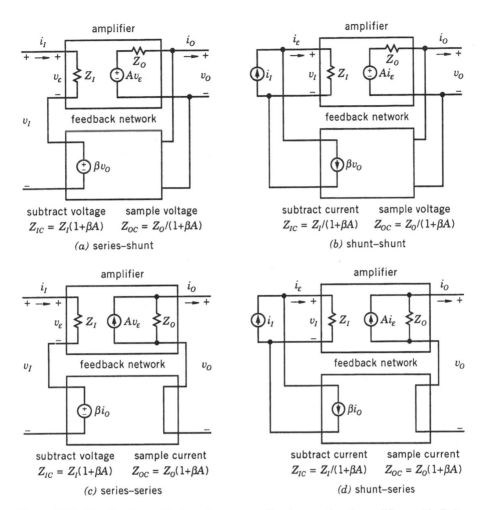

Figure 5-39. Feedback models based on generalized operational amplifiers with finite input and output impedances (Exercise 5-15).

Convince yourself that Fig. 5-40 represents this amplifier in terms of the feedback model of Fig. 5-39b. This is the origin of the term *shunt–shunt* (briefly *shunt*), used in the literature to describe the feedback configuration of inverting amplifiers.

The input impedance calculated according to the prescription of Exercise 5-15 is

$$Z = \frac{R_1 \| R_2}{1 + A_L}$$

where $A_L = AR_1/(R_1 + R_2)$ is the loop gain. Z is *not* the impedance seen by the input voltage source v_I, but rather the impedance at the inverting

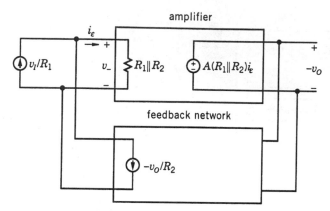

Figure 5-40. Inverting amplifier as viewed in the traditional shunt–shunt feedback model of Fig. 5-39*b* (Exercise 5-16).

terminal of the operational amplifier. Use Blackman's theorem to obtain Z without fuss, and show that

$$Z = R_1 \| \frac{R_2}{1+A}$$

Finally, show that the impedance seen by v_I (which is the one you normally care about) is

$$Z_{IC} = R_1 + \frac{R_2}{1+A}$$

5-17 THE HOWLAND CURRENT PUMP.

(**a**) Show that the α-factors in Fig. 5-12*a* are

$$\alpha_- = 0 \quad \text{and} \quad \alpha_+ = \frac{1}{2}\frac{1}{1+sRC}$$

and, using (5-8), that the ideal gain from v_I to v_A is

$$A_I = \frac{1}{sRC}$$

This result agrees with (5-6), which was obtained more directly by proceeding on the assumption that the error voltage is zero.

(**b**) Use Blackman's theorem to show that the closed-loop output impedance of the current source that feeds the capacitor is

$$Z_{OC} = \frac{R}{2}\left(1 + \frac{A}{2}\right)$$

and that it is thus infinite in the infinite-gain approximation. This is another indication of the fact that you are dealing with current feedback according to the definition given in Section 5-7.

5-18   VOLTAGE FEEDBACK DESPITE APPEARANCES.

(a) Show that the ideal gain from v_I to v_O in the circuit of Fig. 5-41 is

$$A_I = \frac{R_1 + R_2}{R_1}$$

and that this is therefore a case of voltage feedback despite the fact that the generalized operational amplifier is a current source.

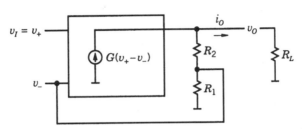

Figure 5-41. Circuit for Exercise 5-18.

(b) Use Blackman's theorem to show that the closed-loop output impedance is

$$Z_{OC} = \frac{R_1 + R_2}{1 + GR_1}$$

In the infinite-gain approximation Z_{OC} is zero, a confirmation of the fact that this is a case of voltage feedback.

5-19   ACCURATE HALF-WAVE RECTIFIER. You will find out about diodes in more detail in Chapter 6, but for certain purposes their description is so simple that a preview is worthwhile. A diode, whose symbol is shown in Fig. 5-42a, is a two-terminal device that conducts current easily in one

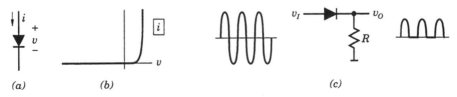

Figure 5-42. (a) Symbol for a diode. (b) Representative voltage-current.

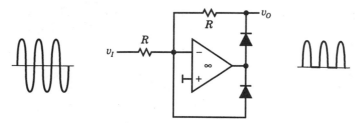

Figure 5-43. Operational-amplifier half-wave rectifier (Exercise 5-19).

direction (*forward*) and not at all in the other (*reverse*); the arrow indicates the forward direction. A typical voltage–current characteristic is shown in Fig. 5-42*b*; at the design current, the forward voltage drop is generally around 800 mV.

The circuit of Fig. 5-42*c* is a *half-wave rectifier*; its output is ideally the positive part of the input waveform. You should have no difficulty in seeing that the nonzero and nonlinear forward voltage drop in the diode makes this circuit nonideal, particularly for low-level signals. At low frequencies, however, there is a solution to this problem. Convince yourself that the circuit of Fig. 5-43 is an ideal half-wave rectifier in the measure that the operational amplifier is ideal.

6

MODELS OF SEMICONDUCTOR DEVICES

6-1 SEMICONDUCTORS

If electronics is unthinkable without amplifiers, it is nowadays virtually unthinkable without semiconductor devices such as transistors. Our purpose here is to gain sufficient insight into the physics of semiconductor devices that we can accept the simple models used to represent them.

We will concentrate on silicon because it is the most widely used semiconductor. Germanium is important for historical reasons, and other semiconductors are increasingly used for special purposes—high speed in the case of gallium arsenide—but devices based on these materials can be readily understood using the concepts we will develop.

6-1-1 Intrinsic Silicon

Pure crystalline silicon (*intrinsic* silicon) is a three-dimensional lattice in which each atom is at the center of a regular tetrahedron formed by its four nearest neighbors. Silicon is element 14 in the periodic table; like carbon, which is in the same column of the table and is thus chemically similar, it has four valence electrons. Each atom in intrinsic silicon forms four covalent bonds with its nearest neighbors. For our purposes, it will be sufficient to think in terms of the two-dimensional lattice shown in Fig. 6-1, in which each atom is also linked to four neighbors.

The energy required to break a covalent bond in silicon is the *bandgap* energy $\mathcal{E}_G = 1.1$ eV (expressed in *electron-volts*, which is the energy gained by an electron of charge $-e = -1.6 \cdot 10^{-19}$ C when it climbs through 1 V. At room temperature (293 K), $\mathcal{E}_G$ is much larger than the average thermal

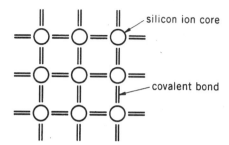

Figure 6-1. Two-dimensional model of intrinsic silicon at zero absolute temperature.

energy of the lattice atoms, which is of order $kT = 0.025$ eV, where $k = 86.17$ μeV/K is Boltzmann's constant and T is the absolute temperature in kelvin. We should therefore expect that compared with the number of atoms, there are few broken bonds in intrinsic silicon at room temperature.

When a covalent bond is broken, the liberated electron is free to roam the lattice and contribute to conduction, much as it might do in a metal. But in semiconductors there is another mechanism for conduction: as illustrated in Fig. 6-2, a broken bond or *hole* can also move quite freely from one site to another, and it turns out that we can think of it as a positive charge carrier with charge e and an (effective) mass of the order of the electron mass.

The density of charge carriers in intrinsic silicon is the result of two opposing processes: conduction electrons and holes are continually generated in pairs by absorption of thermal energy from the lattice, but they also continually recombine pairwise to form covalent bonds. In thermal equilibrium, the electron density n_0 and the hole density p_0 are equal and depend only on the temperature,

$$n_0 = p_0 = n_i(T)$$

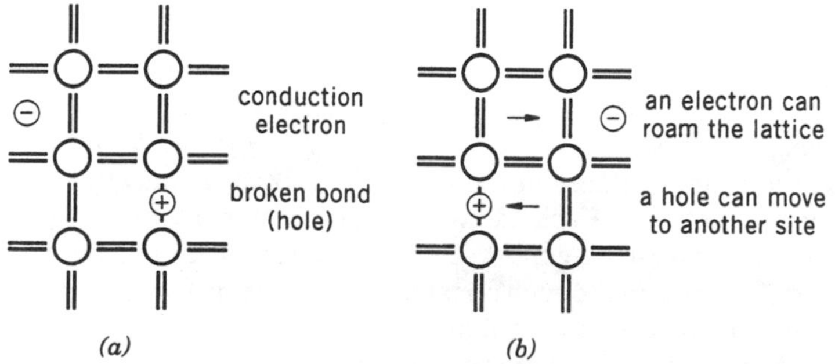

<table>
<tr><td>=O=O=
⊖ ‖ ‖
=O=O=
‖ ⊕ ‖
=O=O=
‖ ‖</td><td>conduction
electron

broken bond
(hole)</td><td>=O=O=
‖ → ‖ ⊖
=O=O=
⊕ ← ‖
=O=O=
‖ ‖</td><td>an electron can
roam the lattice

a hole can move
to another site</td></tr>
<tr><td align="center">*(a)*</td><td></td><td align="center">*(b)*</td><td></td></tr>
</table>

Figure 6-2. Intrinsic silicon at a finite absolute temperature. (*a*) An electron–hole pair. (*b*) A hole can move quite freely from one site to another, and can be imagined as a positive charge carrier with a mass of the order of the electron mass.

where $n_i(T)$ is the *intrinsic carrier density*. Explicitly, we have

$$n_i(T) \sim T^{3/2} \exp\left(-\frac{\mathcal{E}_G}{2kT}\right) \qquad (6\text{-}1)$$

The intrinsic carrier density is a strong function of the temperature: around room temperature it doubles every 10 K. At room temperature we have

$$n_i \simeq 10^{10} \text{ cm}^{-3}$$

As argued above, n_i is many orders of magnitude smaller than the density of silicon atoms, which is about $5 \cdot 10^{22}$ cm^{-3}.

6-1-2   Extrinsic Silicon: Donors and Acceptors

We have looked at intrinsic silicon because it is a necessary starting point for a discussion of semiconductor devices. Things become interesting, however, when the density of charge carriers is varied in a controlled way, and we will concentrate on the fact that this can be done by adding impurities called *dopants* to intrinsic silicon. We will initially consider *uniform* samples, in which the density of dopants is independent of position.

Let us first consider the case illustrated in Fig. 6-3a, in which some of the silicon atoms have been replaced by pentavalent *donor* atoms, usually arsenic (As), phosphorus (P) or antimony (Sb). Four of the valence electrons in these dopants form strong covalent bonds with surrounding silicon atoms; the fifth, however, is weakly bound (its ionization energy is about 50 meV). Between 200 and 400 K, essentially 100% of the donor atoms are ionized once and have thus acquired a *positive* charge and made a conduction electron available. Since the density of donors N_D is usually in the range from 10^{14} to 10^{18} cm^{-3}, and thus much larger than the intrinsic density n_i, we expect the equilibrium density of conduction electrons n_0 to be almost equal to N_D on the basis of a (valid) analogy with strong acids in water.

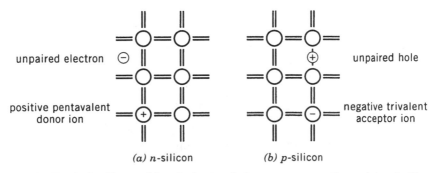

(a) n-silicon   *(b) p*-silicon

Figure 6-3. Extrinsic silicon with substitutional dopants at a small number of silicon sites.

Now let us now turn to the case illustrated in Fig. 6-3b, in which some of the silicon atoms have been replaced by trivalent *acceptor* atoms, usually boron (B). The valence electrons in these dopants can form only three covalent bonds with surrounding silicon atoms. However, it requires very little energy (again about 50 meV) for an acceptor atom to capture a valence electron from another site in the lattice and complete a four-bond structure, leaving behind a hole. Between 200 and 400 K, essentially 100% of the acceptor atoms have completed their bond structure in this way, thus acquiring a *negative* charge and making a hole available for conduction. The density of acceptors N_A is also usually in the range from 10^{14} to 10^{18} cm^{-3}, and we therefore expect the equilibrium hole density p_0 to be almost equal to N_A.

Most semiconductor samples actually contain both donors and acceptors. We will speak of *n-silicon* if $N_D \gg N_A$, and of *p-silicon* if $N_A \gg N_D$. In both cases we will say that the samples are *extrinsic*, and we will extend the use of the term *intrinsic* to cover the case $N_D = N_A$. For simplicity, we will also use N_D and N_A to denote the *net* doping density in extrinsic samples when no confusion can arise.

A uniform sample in thermal equilibrium can have no net charge density; that is, it must obey *charge neutrality*. If we assume that all donors have released a conduction electron and all acceptors have completed a four-bond structure, we must have

$$p_0 - n_0 = N_A - N_D \tag{6-2}$$

We obtain a second equation involving the equilibrium carrier densities by requiring that the rate at which carriers are generated be equal to the rate at which they recombine. We will focus on *direct* processes in which electrons recombine with holes at silicon sites. For these processes, the recombination rate R_D is proportional to the electron density n and to the hole density p,

$$R_D = r_D(T)np \tag{6-3}$$

The addition of small amounts of dopants (or other impurities) does not affect $r_D(T)$ because the environment around most silicon sites remains essentially what it is in intrinsic silicon; that is, it consists of other silicon atoms. The preponderance of silicon atoms over dopants also implies that the density of unbroken bonds at silicon sites is essentially what it is in intrinsic silicon, and that the generation rate G_D for the processes we are considering depends only on the temperature,

$$G_D = g_D(T) \tag{6-4}$$

The principle of detailed balance requires that the generation rate from *any* process be in equilibrium with its inverse in order not to violate the second law of thermodynamics. In equilibrium, it follows from (6-3) and (6-4) that

$$g_D(T) = r_D(T)n_0 p_0$$

and therefore that the product $n_0 p_0$ depends only on the temperature. This holds in particular for intrinsic silicon, in which $n_0 = p_0 = n_i$, and we have

$$n_0 p_0 = n_i^2(T) \tag{6-5}$$

We can now calculate the equilibrium carrier densities in uniform samples of extrinsic silicon when—this is almost always the case—the net density of dopants is much larger than the intrinsic carrier density. Using (6-2) and (6-5) we obtain to a high degree of accuracy that

in n-silicon,   in p-silicon,   (6-6)

$$n_0 = N_D \qquad p_0 = \frac{n_i^2}{N_D} \qquad\qquad p_0 = N_A \qquad n_0 = \frac{n_i^2}{N_A}$$

In n-silicon, electrons are *majority carriers* and holes are *minority carriers*; these roles reverse in p-silicon.

6-1-3   Charge-Carrier Densities, Currents, and Fields

In order to describe semiconductor devices we must learn how to calculate the charge-carrier densities, the current densities, and the electric field in semiconductor samples that are not in equilibrium. We will limit ourselves to a one-dimensional description since it is adequate for our purposes.

We will start by obtaining equations for the current densities. From Section 4-4 (or from elementary physics) we are familiar with drift currents, which are proportional to the electric field E, and we might expect that if we apply a voltage to a long bar of uniform silicon, either extrinsic or intrinsic, the hole drift current density would be

$$J_p^E = \sigma_p E$$

where σ_p is the hole conductivity (in S/cm). This is in fact true for electric fields less than a critical value E_C of order 10^4 V/cm. In more detail, the electric field induces an average drift velocity v_p on the holes,

$$v_p = \mu_p E$$

where μ_p is the *hole mobility* (in cm^2/V-s). The current density is proportional to the hole charge e, the hole density p_0, and the drift velocity v_p,

$$J_p^E = e p_0 v_p$$

and we can write

$$\sigma_p = e \mu_p p_0$$

For fields above E_C these considerations no longer hold: the drift velocity v_p *saturates*; that is, it becomes constant at a value near the average thermal velocity ($5 \cdot 10^6$ cm/s), and the current density thus also becomes constant.

Drift is the dominant conduction mechanism when the density of charge carriers is constant. This is not always the case in semiconductor devices, and we must consider an additional mechanism. Statistical mechanics tells us that when the density of particles of a given type in a medium is not constant, there must be a flow of particles away from regions of higher density toward regions of lower density; this flow or *diffusion* takes place whether or not the particles are charged. In the lowest-order approximation, diffusion is proportional to the gradient of the particle density, and for the hole diffusion current density we can write

$$J_p^D = -eD_p \frac{\partial p}{\partial x}$$

where D_p is the *diffusion constant* (in cm^2/s). In the presence of both fields and density gradients, the hole current density J_p is

$$J_p = J_p^E + J_p^D$$

Analogous considerations apply to electrons, and for the hole and electron current densities we can write

$$J_p = e\mu_p pE - eD_p \frac{\partial p}{\partial x}$$
$$J_n = e\mu_n nE + eD_n \frac{\partial n}{\partial x}$$

(6-7)

Values of D_n in p-silicon and of D_p in n-silicon are given in Fig. 6-4. As we will soon see, $\mu_n = (e/kT)D_n$ and $\mu_p = (e/kT)D_p$; mobilities thus need not

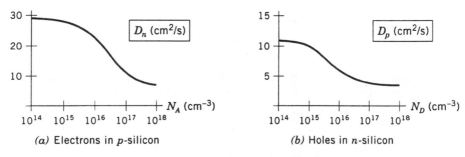

(a) Electrons in p-silicon

(b) Holes in n-silicon

Figure 6-4. Diffusion constants at 290 K as a function of the density of dopants. (Adapted from Conwell, 1958.)

be given separately. Typical values are $\mu_n = 1000$ cm^2/V-s and $\mu_p = 300$ cm^2/V-s.

We obtain another pair of equations by observing that in a fixed volume the number of carriers of a given sign can change only through generation and recombination or because carriers enter and leave the volume. Applying this requirement to an infinitesimal volume yields the *continuity equations*,

$$\frac{\partial p}{\partial t} = -\frac{1}{e}\frac{\partial J_p}{\partial x} - R$$

$$\frac{\partial n}{\partial t} = \frac{1}{e}\frac{\partial J_n}{\partial x} - R$$

(6-8)

where R is the *net* recombination rate for both holes and electrons.

Taking R to be the same for holes and electrons implicitly requires that holes and electrons appear and disappear in pairs; this requirement is satisfied automatically if the processes involved are direct. As discussed in Exercise 6-7, generation and recombination are in general dominated not by direct processes but by indirect processes involving *recombination centers*. But if the density of charge carriers is large, which will almost invariably be the case for us, recombination of a carrier of one sign is followed so quickly by recombination of one of the opposite sign (or regeneration of one of the same sign) that we can consider that holes and electrons appear and disappear in pairs.

The final equation we need is Gauss's law in terms of the carrier and dopant densities,

$$\varepsilon\frac{\partial E}{\partial x} = e(p - n + N_D - N_A)$$

(6-9)

where $\varepsilon = 11.7\varepsilon_0 \simeq 1$ pF/cm is the permittivity of silicon.

At this point we are faced with a set of five coupled nonlinear differential equations (two for the currents, two from continuity, and one from Gauss's law) that cannot be solved in general. In the cases that interest us, however, these equations can be simplified considerably, and we will arrive at results that are useful in themselves and will serve as guides in more general cases.

6-1-4 Excess Carrier Densities in Uniform Samples

In a uniform sample the equilibrium electron and hole densities n_0 and p_0 are constant and given in terms of the density of dopants by (6-6). Since the density of dopants in such samples is assumed to be known, we need only determine the *excess* carrier densities p' and n', given by

$$p' = p - p_0 \quad \text{and} \quad n' = n - n_0$$

In terms of the excess carrier densities, Gauss's law (6-9) becomes

$$\varepsilon \frac{\partial E}{\partial x} = e(p' - n')$$ (6-10)

because $p_0 - n_0 + N_D - N_A = 0$ according to (6-2), and therefore,

$$p - n + N_D - N_A = p' - n' + p_0 - n_0 + N_D - N_A = p' - n'$$

An important property of uniform samples is that they obey *quasineutrality*, that is, that the excess carrier densities as well as their gradients are *almost* equal,

$$p' \simeq n' \quad \text{and} \quad \frac{\partial p'}{\partial x} \simeq \frac{\partial n'}{\partial x}$$

We say "*almost* equal" because true equality implies a zero net charge density, and from Gauss's law (6-10), a constant electric field. As we will see, this is not in general the case, and minute charge imbalances must therefore exist.

We will justify quasineutrality after developing more machinery; for now we will use the physical argument that a charge imbalance generates an electric field that moves charge carriers so as to null the imbalance. We encountered a similar situation when we studied relaxation times in conductors in Section 4-5; based on what we learned there, we should expect that the readjustment in the carrier densities is virtually instantaneous.

6-1-5   Low-Level Injection in Uniform Samples

The equations obtained in Section 6-1-3 can be simplified considerably in the case of *low-level injection* in uniform samples, that is, if the excess carrier densities are much smaller than the net density of dopants and therefore much smaller than the majority-carrier density, which remains almost equal to its equilibrium value. In symbols, we have

in n-silicon,   in p-silicon,

$$p', n' \ll n \simeq n_0 = N_D \qquad \Big| \qquad p', n' \ll p \simeq p_0 = N_A$$

Drift currents are proportional to carrier densities; it follows that if the injection level is low, the minority-carrier drift current is much smaller than the majority-carrier drift current. Since charge imbalances are small in uniform samples because of quasineutrality, we expect only moderate electric fields, and we are led to surmise that minority carriers in uniform samples flow predominantly by diffusion when their density gradient is significant and the injection level is low. If we assume (and later verify) that this will be the case for us, the equations for the current densities in (6-7) become

in n-silicon,   in p-silicon,   (6-11)

$$J_p = -eD_p \frac{\partial p'}{\partial x}$$

$$J_n = e\mu_n n_0 E + eD_n \frac{\partial n'}{\partial x}$$

$$J_p = e\mu_p p_0 E - eD_p \frac{\partial p'}{\partial x}$$

$$J_n = eD_n \frac{\partial n'}{\partial x}$$

The fact that the majority-carrier density remains virtually constant when the injection level is low implies that the (low-level) recombination rate R_L in a given sample, whatever its mechanism, depends only on the minority-carrier density. Since charge carriers act independently of one another, recombination becomes a Poisson process in which the probability per unit time that a minority carrier will recombine is constant, and R_L is therefore proportional to the minority-carrier density. For n-silicon we can then write

$$R_L = \frac{p}{\tau_p}$$

where τ_p is the *minority-carrier lifetime* for holes in n-silicon. As its name correctly implies, τ_p is the average time it takes for a hole to recombine in n-silicon when the injection level is low. The generation rate G_L is equal to the equilibrium recombination rate,

$$G_L = \frac{p_0}{\tau_p}$$

and the net recombination rate $R = R_L - G_L$ is therefore given by

$$R = \frac{p - p_0}{\tau_p} = \frac{p'}{\tau_p}$$

Similar considerations apply to electrons in p-silicon, and we have

in n-silicon,   in p-silicon,   (6-12)

$$R = \frac{p'}{\tau_p}$$

$$R = \frac{n'}{\tau_n}$$

Minority-carrier lifetimes depend dramatically on the metallurgical condition of a sample and can vary by orders of magnitude for a given doping level; the usual range is from 1 ns to 1 ms.

For reference, we note that the recombination rate obtained in Exercise 6-7 is

$$R = \frac{np - n_i^2}{(n + n_i)\tau_p + (p + n_i)\tau_n} \qquad (6\text{-}13)$$

This expression is based on a crude model, but it is shown in the exercises that it reduces to (6-12) when the injection level is low and that it is a useful guide otherwise.

Using expressions (6-12) for R, the continuity equations (6-8) become

in n-silicon,   in p-silicon,   (6-14)

$$\frac{\partial p'}{\partial t} = -\frac{1}{e}\frac{\partial J_p}{\partial x} - \frac{p'}{\tau_p} \qquad\qquad \frac{\partial p'}{\partial t} = -\frac{1}{e}\frac{\partial J_p}{\partial x} - \frac{n'}{\tau_n}$$

$$\frac{\partial n'}{\partial t} = \frac{1}{e}\frac{\partial J_n}{\partial x} - \frac{p'}{\tau_p} \qquad\qquad \frac{\partial n'}{\partial t} = \frac{1}{e}\frac{\partial J_n}{\partial x} - \frac{n'}{\tau_n}$$

We have now linearized and decoupled the equations for the charge-carrier and current densities to the point that solutions in closed form are possible, as we will see next.

6-1-6 Diffusion Equations for Low-Level Injection

If we substitute the minority-carrier current density (6-11) in the minority-carrier continuity equation (6-14) we obtain the *minority-carrier diffusion equation*,

in n-silicon,   in p-silicon,   (6-15)

$$\tau_p\frac{\partial p'}{\partial t} = L_p^2\frac{\partial^2 p'}{\partial x^2} - p' \qquad\qquad \tau_n\frac{\partial n'}{\partial t} = L_n^2\frac{\partial^2 n'}{\partial x^2} - n'$$

where

$$L_p = \sqrt{D_p\tau_p} \quad \text{and} \quad L_n = \sqrt{D_n\tau_n}$$

are the *diffusion lengths* for holes and electrons.

At this time we will solve the minority-carrier diffusion equation (6-15) for the *quasistatic* case in which the time derivative can be neglected. We will do so for holes in n-silicon, for which the diffusion equation becomes

$$p' - L_p^2\frac{d^2 p'}{dx^2} = 0 \qquad (6\text{-}16)$$

In most situations it will turn out that we know the total current density J at some point; in the quasistatic case it follows that we know it everywhere because J is then constant. To see this, we can subtract the continuity equations (6-14) and use $J = J_n + J_p$ to get the continuity equation that expresses conservation of charge,

$$\frac{\partial J}{\partial x} + e\frac{\partial (p' - n')}{\partial t} = 0$$

This equation reduces to $dJ/dx = 0$ in the quasistatic case.[†] The fact that we know J implies that we can obtain all the other unknowns once we have obtained the minority-carrier density. From quasineutrality we have $dn'/dx \simeq dp'/dx$, so that we can calculate the majority-carrier diffusion current density; we can then use (6-11) to obtain the majority-carrier drift current density and the electric field. An example of this procedure is presented in Exercise 6-6.

The boundary conditions for the minority-carrier diffusion equation have standard names in the literature: we speak of an *injecting contact* if the excess minority-carrier density or the minority-carrier current density at the boundary is specified, and of an *ohmic contact* if the excess minority-carrier density at the boundary is zero. These terms will soon acquire a physical interpretation; for now we will accept them as definitions.

Let us consider the situation we will encounter most often, in which we have an injecting contact at $x = 0$ and an ohmic contact at $x = w$. The solution of the quasistatic diffusion equation (6-16) with these boundary conditions in n-silicon, assuming that $p'(0)$ is specified, is

$$p'(x) = p'(0)\frac{\sinh[(w-x)/L_p]}{\sinh(w/L_p)}$$

There are two particular cases of this solution that interest us, for which we will again introduce standard terminology that will soon acquire a meaning; these are the *short-base* solution illustrated in Fig. 6-5a and the *long-base* solution illustrated in Fig. 6-5b.

In the *short-base limit* it is assumed that $w \ll L_p$ (or, equivalently, that the minority-carrier lifetime τ_p is infinite), so that

$$p'(x) = p'(0)\left(1 - \frac{x}{w}\right)$$

The minority-carrier current density in the short-base limit is constant,

$$J_p = \frac{eD_p}{w}p'(0)$$

[†]Actually, in most situations J is independent of x at all frequencies of interest because of quasineutrality. The proof in the short-base limit defined below is proposed in Exercise 6-5.

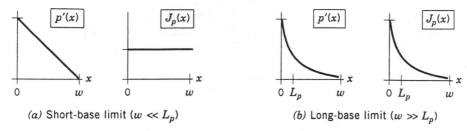

(a) Short-base limit ($w \ll L_p$)   (b) Long-base limit ($w \gg L_p$)

Figure 6-5. Excess hole density $p'(x)$ and hole current density $J_p(x)$ in a uniform sample of n-silicon assuming that the injection level is low and that the contacts are injecting at $x = 0$ and ohmic at $x = w$.

In the *long-base limit* it is assumed that $w \gg L_p$, so that

$$p'(x) = p'(0)\exp\left(-\frac{x}{L_p}\right)$$

In this case the minority-carrier current density also decays exponentially,

$$J_p(x) = \frac{eD_p}{L_p}p'(x)$$

At this point we are in a position to justify that quasineutrality holds in uniform samples when the injection level is low; high-level injection is taken up in Exercise 6-12. We will consider the case of n-silicon. Substituting the expression for the electron current density (6-11) in the electron continuity equation (6-14) yields

$$\frac{\partial n'}{\partial t} = \mu_n n_0 \frac{\partial E}{\partial x} + D_n \frac{\partial^2 n'}{\partial x^2} - \frac{p'}{\tau_p}$$

Using Gauss's law (6-10) we obtain the *majority-carrier diffusion equation*,

$$\tau_D \frac{\partial n'}{\partial t} - L_D^2 \frac{\partial^2 n'}{\partial x^2} + n' = \left(1 - \frac{\tau_D}{\tau_p}\right)p' \qquad (6\text{-}17)$$

where τ_D is the *dielectric relaxation time* (expressed in terms of the electron conductivity $\sigma_n = e\mu_n n_0$),

$$\tau_D = \frac{\varepsilon}{\sigma_n}$$

and L_D is the *extrinsic Debye length*,

$$L_D = \sqrt{D_n \tau_D} \qquad (6\text{-}18)$$

From the form of this differential equation it is clear that the excess minority-carrier density acts as a generator for the excess majority-carrier density and, as expected from the physics, that majority carriers do *all* the rearranging that is required to achieve quasineutrality when the injection level is low. If we take pessimistic values $n_0 = 10^{15}/\text{cm}^{-3}$ and $\mu_n = 600$ cm^2/V-s, we obtain $\sigma_n \simeq 0.1$ S/cm, $\tau_D \simeq 10$ ps, and $L_D \simeq 10^{-5}$ cm. Since minority-carrier lifetimes (τ_p in this case) are seldom under 1 ns, and are thus much larger than the dielectric relaxation time τ_D, it follows that the excess majority-carrier density tracks quasistatic variations of the excess minority-carrier density closely and almost instantaneously.

General statements are somewhat more difficult to make in the case of spatial variations in the minority-carrier density. In high-speed transistors, for example, there are regions in which significant variations occur in a small fraction of a micrometer (10^{-4} cm), and we must determine in such cases whether the Debye length L_D is in fact small enough that quasineutrality holds. On the favorable side, in Exercise 6-6 it is shown in the quasistatic long-base limit that

$$n' - p' \simeq p' \frac{D_n - D_p}{D_n} \frac{L_D^2}{L_p^2}$$

for $x \gg L_D$, whereas in Exercise 6-5 it is shown in the short-base limit that $n' = p'$ for $x \gg L_D$ and $w - x \gg L_D$. Quasineutrality thus holds very well in both cases.

6-1-7   Equilibrium Carrier Densities in Nonuniform Samples

Up to this point we have concentrated on what happens in uniform samples of extrinsic silicon. In a rough approximation, semiconductor devices can be described as one-dimensional stacks of uniform regions with differing dopant densities, and among other things we will need to know how the equilibrium carrier densities vary at the boundaries of these regions.

Let us consider a region of n-silicon with a varying profile of dopants $N_D(x)$. The density of *available* conduction electrons is equal to $N_D(x)$, but electrons diffuse away from regions where N_D is high toward regions where N_D is low. The electric fields resulting from this rearrangement generate drift currents that tend to counterbalance the diffusion currents.

The presence of electric fields implies that the electric potential ϕ—and therefore the potential energy of charge carriers—depends on the position. At the temperatures and dopant levels we are considering, electrons and holes obey Maxwell–Boltzmann statistics; this implies that equilibrium densities in regions that are at potentials ϕ_1 and ϕ_2 are related by Boltzmann factors[†]:

$$\frac{n_0(\phi_1)}{n_0(\phi_2)} = \exp\left[\frac{e(\phi_1-\phi_2)}{kT}\right]$$

$$\frac{p_0(\phi_1)}{p_0(\phi_2)} = \exp\left[-\frac{e(\phi_1-\phi_2)}{kT}\right]$$

If we take the potential in intrinsic silicon to be zero, we can write

$$n_0 = n_i \exp\left(\frac{\phi}{V_T}\right)$$

$$p_0 = n_i \exp\left(-\frac{\phi}{V_T}\right)$$

(6-19)

where

$$V_T = \frac{kT}{e}$$

is the *thermal voltage*, equal to 0.025 V at 290 K. It is clear that even for varying dopant profiles we have

$$n_0 p_0 = n_i^2$$

If we use expressions (6-19) for n_0 and p_0 in Gauss's law (6-9), we obtain the *Poisson–Boltzmann* equation for the potential,

$$\frac{d^2\phi}{dx^2} = -\frac{e}{\varepsilon}\left[N_D(x) - N_A(x) - 2n_i \sinh\frac{\phi(x)}{V_T}\right]$$

(6-20)

We will not do much with this equation; the main reason for exhibiting it is that it shows us that there is a general approach to the problem of obtaining the potential—and therefore all other variables—in nonuniform samples.

In thermal equilibrium the total current density $J = J_n + J_p$ must of course be zero. But the principle of detailed balance further demands that J_n and J_p *separately* be equal to zero. The equilibrium carrier densities and the electric field must therefore be such that the diffusion and drift current densities in (6-7) cancel each other exactly. In terms of the potential ϕ, we must thus have

$$-e\mu_n n_0 \frac{d\phi}{dx} + eD_n \frac{dn_0}{dx} = 0$$

$$-e\mu_p p_0 \frac{d\phi}{dx} - eD_p \frac{dp_0}{dx} = 0$$

[†]Boltzmann factors are discussed briefly and lucidly in Feynman (1964), vol. 1, ch. 40.

If we substitute in these equations the expressions for the carrier densities given in (6-19), we obtain the *Einstein relations*,[†]

$$D_n = \mu_n V_T \quad \text{and} \quad D_p = \mu_p V_T \tag{6-21}$$

In (6-18) we can now substitute $D_n \tau_D = D_n \varepsilon / \sigma_n = D_n \varepsilon / e \mu_n N_D = \varepsilon V_T / e N_D$; we thus obtain alternative expressions for the extrinsic Debye lengths,

in *n*-silicon,   in *p*-silicon,   (6-22)

$$L_D = \sqrt{\frac{\varepsilon V_T}{e N_D}} \qquad\qquad L_A = \sqrt{\frac{\varepsilon V_T}{e N_A}}$$

6-2   THE *p-n* JUNCTION

We have finally acquired enough machinery to describe one of the fundamental structures in semiconductor devices, the junction of a region of *p*-silicon and a region of *n*-silicon, or *p-n junction*. Without great loss of generality, we will limit ourselves to the *abrupt* junction shown in Fig. 6-6, in which over a short distance there is a transition from a region of uniform *p*-silicon with a density of acceptors N_A to a region of uniform *n*-silicon with a density of donors N_D. The distance in question should be much less than the extrinsic Debye length on either side of the junction.

The equilibrium carrier densities in the *p-n* junction can be obtained in principle by solving the Poisson–Boltzmann equation (6-20) and then using (6-19). Rather than do this, we will proceed with a qualitative description that will suggest a productive approximation.

Let us start by retracing the discussion of Section 6-1-7 in the case of the *p-n* junction. Far from the junction, holes are abundant on the *p*-side and scarce on the *n*-side; holes near the junction diffuse from the *p*-side toward the *n*-side and leave behind a *depletion* layer of negative charge due to the fixed acceptor ions. Analogously, electrons diffuse from the *n*-side toward the *p*-side and uncover a depletion layer of positive charge due to the fixed donor ions. Conservation of charge demands that the total charge in the depletion region (or, more generally, *space-charge* region) be zero.

[†]It follows from the equations above that the equilibrium carrier densities depend exponentially on the potential as in (6-19):

$$n_0 = n_i \exp\left(\frac{\mu_0}{D_n}\phi\right)$$

$$p_0 = n_i \exp\left(-\frac{\mu_p}{D_p}\phi\right)$$

Showing that $D_n / \mu_n = D_p / \mu_p = V_T$, however, requires further arguments.

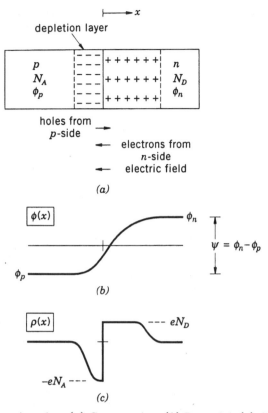

Figure 6-6. The p-n junction. (a) Cross section. (b) Potential. (c) Charge density.

The electric field in the depletion region points from the n-side to the p-side; the ensuing drift currents thus oppose the diffusion currents. In thermal equilibrium, the drift and diffusion currents cancel each other. It follows from what we have said that the equilibrium potential relative to intrinsic silicon has the general aspect shown in Fig. 6-6. In the neutral regions far from the junction, the potential is constant and must reflect the fact that in these regions $n_0 = N_D$ on the n-side and $p_0 = N_A$ on the p-side. Using (6-19), we obtain

on the n-side,

$$\phi_n = V_T \ln \frac{N_D}{n_i}$$

on the p-side

$$\phi_p = -V_T \ln \frac{N_A}{n_i}$$

The height of the potential barrier between the neutral regions or *built-in potential* ψ is thus

$$\psi = \phi_n - \phi_p = V_T \ln \frac{N_A N_D}{n_i^2}$$

The magnitudes of ϕ_n and ϕ_p are typically 350 mV, and thus much larger than the thermal voltage V_T (≈ 25 mV), which gives the scale on which equilibrium carrier densities vary exponentially with the potential. The majority-carrier densities therefore drop to insignificant levels as the junction is approached and give rise to the charge density shown schematically in Fig. 6-6.

If we now assume that the drop in majority-carrier densities near the junction is abrupt and takes place at $-x_A$ on the *p*-side and at x_D on the *n*-side, we can take the charge densities in the depletion layers to be constant, and thus equal to $-eN_A$ on the *p*-side and to eN_D on the *n*-side. This assumption is the essence of the *depletion approximation*, to which we now turn.

6-2-1 The Depletion Approximation

On the basis of the discussion in the preceding section, we will assume that the charge density in the depletion region is, as shown in Fig. 6-7, given by

$$\rho(-x_A \leq x \leq 0) = -eN_A$$
$$\rho(0 \leq x \leq x_D) = eN_D$$

where x_A and x_D are to be determined. The electric field E can be calculated from Gauss's law by observing that E is zero outside the depletion region. We thus obtain

$$E(-x_A \leq x \leq 0) = -\frac{eN_A}{\varepsilon}(x+x_A)$$

$$E(0 \leq x \leq x_D) = \frac{eN_D}{\varepsilon}(x-x_D) \tag{6-23}$$

The boundary conditions on the potential are

$$\phi(-x_A) = \phi_p \quad \text{and} \quad \phi(x_D) = \phi_n$$

The potential is therefore

$$\phi(-x_A \leq x \leq 0) = \phi_p + \frac{eN_A}{2\varepsilon}(x+x_A)^2$$

$$\phi(0 \leq x \leq x_D) = \phi_n - \frac{eN_D}{2\varepsilon}(x-x_D)^2 \tag{6-24}$$

Continuity of the potential at $x = 0$ requires that

$$\phi_p + \frac{eN_A x_A^2}{2\varepsilon} = \phi_n - \frac{eN_D x_D^2}{2\varepsilon}$$

We can now express the built-in potential ψ in terms of x_A and x_D:

$$\psi = \phi_n - \phi_p = \phi(x_D) - \phi(-x_A) = \frac{e}{2\varepsilon}(N_A x_A^2 + N_D x_D^2)$$

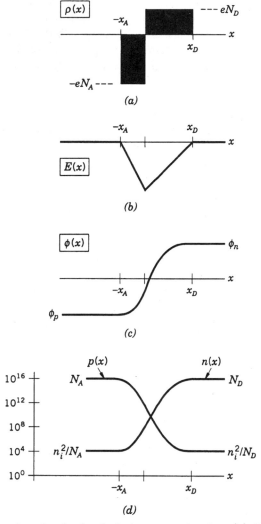

Figure 6-7. The *p-n* junction in the depletion approximation. (*a*) Charge density. (*b*) Electric field. (*c*) Potential. (*d*) Charge-carrier densities.

We get another equation involving x_A and x_D from the continuity of E at $x = 0$ in (6-23),

$$N_A x_A = N_D x_D$$

From the last two equations we finally obtain

$$N_A x_A = N_D x_D = \sqrt{\frac{2\varepsilon\psi N_A N_D}{e(N_A + N_D)}} \tag{6-25}$$

At this point we can check how good the depletion approximation is. Let us first rewrite the potential (6-24) in terms of the extrinsic Debye lengths $L_A = \sqrt{\varepsilon V_T/eN_A}$ and $L_D = \sqrt{\varepsilon V_T/eN_D}$ given in (6-22). The potential then reads

$$\phi(-x_A \leq x \leq 0) = \phi_p + V_T \frac{(x+x_A)^2}{2L_A^2}$$

$$\phi(0 \leq x \leq x_D) = \phi_n - V_T \frac{(x-x_D)^2}{2L_D^2}$$

On either side of the junction, the potential departs by $V_T/2$ from its value in the neutral region in one Debye length and the majority-carrier density becomes insignificant in several Debye lengths. We conclude that the depletion approximation is good as long as $x_A \gg L_A$ and $x_D \gg L_D$. Using (6-25) we get

$$\frac{x_A}{L_A} = \sqrt{\frac{\psi}{V_T} \frac{2N_D}{N_D+N_A}} \quad \text{and} \quad \frac{x_D}{L_D} = \sqrt{\frac{\psi}{V_T} \frac{2N_A}{N_A+N_D}}$$

The built-in potential ψ is typically 700 mV, so that $\psi/V_T \simeq 28$. If N_A and N_D are comparable, x_A/L_A and x_D/L_D are both about 5, and the depletion approximation is reasonable. If as is more usually the case, the density of dopants is much larger on one side than on the other, the approximation is very good on the lightly doped side and barely adequate on the heavily doped side. Nonetheless, the depletion approximation is extremely useful because the charges in the depletion layers are always very well given by $-eN_A x_A$ and $eN_D x_D$, and so therefore is the (maximum) electric field at $x = 0$,

$$E(0) = -\frac{eN_A x_A}{\varepsilon} = -\frac{eN_D x_D}{\varepsilon}$$

6-2-2 Junction Diodes in the Shockley Model

A *diode*, also called a *rectifier*, is a two-terminal device that conducts current in one direction (the *forward* direction) but is virtually an open circuit in the other direction (the *reverse* direction); this property is reflected by the arrow in the symbol for a diode shown in Fig. 6-8. For historical reasons, the terminals of a diode are called *anode* and *cathode*; the forward direction is from anode to cathode. In this section we will become familiar with *junction diodes*, which consist essentially of a *p-n* junction with an ohmic contact some distance away on either side of the junction.

One possible structure of a junction diode is shown in Fig. 6-9. The *p*-side of the junction is joined directly to a metal contact (the anode terminal), and the *n*-side is joined to a metal contact (the cathode terminal) through a thin

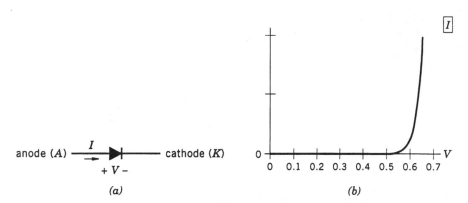

Figure 6-8. (*a*) Symbol and sign conventions for a diode; (*b*) representative *I–V* characteristic.

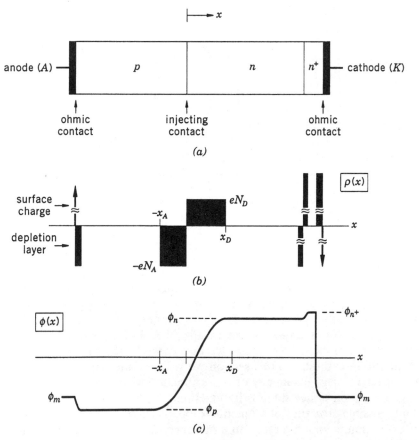

Figure 6-9. Junction diode. (*a*) Cross section. (*b*) Equilibrium charge density. (*c*) Equilibrium potential.

intermediate layer of heavily doped n^+-silicon. Figure 6-9 also shows the charge density and the potential when the diode is open circuited and thus in equilibrium. There are space-charge regions at all junctions; as in the *p-n* junction, these space-charge regions are the result of redistributions of charges, as required to establish potential barriers that null the electron and hole current densities. For our purposes we can describe a metal as having a (uniform) potential ϕ_m relative to intrinsic silicon, and we will assume that the space-charge layer at the junction with a silicon sample is a surface charge across which the potential is continuous. It follows that the potential in the silicon at the junction is ϕ_m. We will always assume that contacts are made of aluminum, in which $\phi_m = -300$ mV.

We are interested in what happens when the diode has a *forward bias*, that is, when a positive voltage V is applied from anode to cathode. We advance the idea that holes are injected from the *p*-side into the *n*-side and that electrons are injected from the *n*-side into the *p*-side. We will assume low-level injection for now, and we therefore know what to do if we are given boundary conditions.

The semiconductor–metal junctions are ohmic contacts as defined in Section 6-1-6 because, roughly speaking, excess minority carriers at the edge of the depletion layer on the semiconductor side roll off the potential barrier at a rate high enough that the excess minority-carrier density can be taken to be zero. The n-n^+ junction introduces a minor complication, but it does not affect the essence of our discussion; we will therefore ignore it for now and offer comments later. To proceed we must make two central assumptions.

The first assumption is that the full applied voltage appears across the *p-n* junction. This requires that the other junctions be ohmic in a further sense, namely, that their built-in potential change negligibly when current flows or, equivalently, that they behave like small resistors. We will accept this assumption for now and discuss it in more detail after we understand what happens at the *p-n* junction; we will thus accept that if we apply a forward bias V to the diode, the potential across the *p-n* junction is equal to $\psi-V$.

The second assumption is that electrons and holes in the depletion region are in *quasi-equilibrium* with the potential barrier $\psi-V$, that is, that densities of carriers at different potentials can be related by Boltzmann factors. We are not in a position to justify this assumption fully, but we will at least make it sound reasonable.

One partial justification of quasi-equilibrium is based on the observation that the carrier density gradients and the electric field in the depletion region are very large and, as discussed in Exercise 6-16, that over most of the depletion region the opposing drift and diffusion currents are therefore much larger than the diffusion-limited currents in the neutral regions. If we make the approximation that the electron and hole currents in the depletion region are zero, we obtain from (6-7) that

$$\mu_n nE + D_n \frac{dn}{dx} = 0 \quad \text{and} \quad \mu_p pE - D_p \frac{dp}{dx} = 0$$

Using the Einstein relations (6-21) and expressing the electric field in terms of the (nonequilibrium) potential ϕ, we can then write

$$-n\frac{d\phi}{dx}+V_T\frac{dn}{dx}\simeq 0 \quad \text{and} \quad -p\frac{d\phi}{dx}-V_T\frac{dp}{dx}\simeq 0$$

It follows that the carrier densities depend exponentially on the potential,

$$n\simeq n_V\exp\left(\frac{\phi}{V_T}\right) \quad \text{and} \quad p\simeq p_V\exp\left(-\frac{\phi}{V_T}\right)$$

where n_V and p_V are functions of the bias voltage V. It also follows that the np product in the depletion region is constant for a given bias voltage,

$$np = n_V p_V$$

Unfortunately, the expressions for the current densities in (6-7) apply to averages over a distance of order 10^{-5} cm, in which charge carriers make many collisions, and this distance is close to the width of the depletion region in most cases (about $3\cdot 10^{-5}$ cm for a density of 10^{16} cm^{-3} on the lightly doped side). Our argument is thus somewhat flawed, and we will have to rely on the fact that the assumption of quasi-equilibrium is experimentally verified.

In the neutral regions, the hole density on the p-side p_p and the electron density on the n-side n_n are essentially equal to their equilibrium values when the injection level is low,

$$n_n\simeq n_{n0} \quad \text{and} \quad p_p\simeq p_{p0}$$

At the edges of the depletion region, the hole density on the n-side p_n and the electron density on the p-side n_p are therefore

$$p_n(+x_D)=p_p(-x_A)\exp\left(-\frac{\psi-V}{V_T}\right)\simeq p_{p0}\exp\left(-\frac{\psi-V}{V_T}\right)=p_{n0}\exp\left(\frac{V}{V_T}\right)$$

$$n_p(-x_A)=n_n(+x_D)\exp\left(-\frac{\psi-V}{V_T}\right)\simeq n_{n0}\exp\left(-\frac{\psi-V}{V_T}\right)=n_{p0}\exp\left(\frac{V}{V_T}\right)$$

For reference we note that the np product at the n-edge of the depletion region is

$$np = n_n(x_D)p_n(x_D)\simeq n_{n0}p_{n0}\exp\left(\frac{V}{V_T}\right)$$

and that the np product in the depletion region is therefore

$$np\simeq n_i^2\exp\left(\frac{V}{V_T}\right) \tag{6-26}$$

The excess minority-carrier densities at the edges of the depletion region are

$$p'_n(+x_D) = p_{n0}\left[\exp\left(\frac{V}{V_T}\right) - 1\right]$$

$$n'_p(-x_A) = n_{p0}\left[\exp\left(\frac{V}{V_T}\right) - 1\right]$$

(6-27)

With these boundary conditions we can now use the results obtained in Section 6-1-6. Setting $p_{n0} = n_i^2/N_D$ and $n_{p0} = n_i^2/N_A$, we obtain that the current densities at the edges of the depletion region are

$$J_p(+x_D) = \frac{en_i^2 D_p}{N_D w_D}\left[\exp\left(\frac{V}{V_T}\right) - 1\right]$$

$$J_n(-x_A) = \frac{en_i^2 D_n}{N_A w_A}\left[\exp\left(\frac{V}{V_T}\right) - 1\right]$$

(6-28)

where w_D and w_A are the *effective widths* of the neutral regions, equal to L_p and L_n in the long-base limit, and equal to the real widths in the short-base limit.

We must make one more assumption: that recombination in the depletion region is negligible. This is *not* a good assumption in most real diodes, but it is an excellent assumption in transistors, which is primarily what we have in mind. We then have

$$J_p(-x_A) = J_p(x_D)$$
$$J_n(-x_A) = J_n(x_D)$$

The total current density J is constant and equal to $J_p + J_n$; using either of the conditions above, we then obtain

$$J = en_i^2\left(\frac{D_p}{N_D w_D} + \frac{D_n}{N_A w_A}\right)\left[\exp\left(\frac{V}{V_T}\right) - 1\right]$$

(6-29)

The carrier densities, the excess carrier densities, and the current densities are shown in Fig. 6-10 in the long-base limit. We observe in this limit that the current density far from the junction on either side is due only to majority carriers.

For a diode of cross-sectional area A, we obtain from (6-29) that in the Shockley model the current I for a bias V is

$$I = I_S\left[\exp\left(\frac{V}{V_T}\right) - 1\right]$$

(6-30)

where I_S is the *saturation current*,

$$I_S = en_i^2 A\left(\frac{D_p}{N_D w_D} + \frac{D_n}{N_A w_A}\right)$$

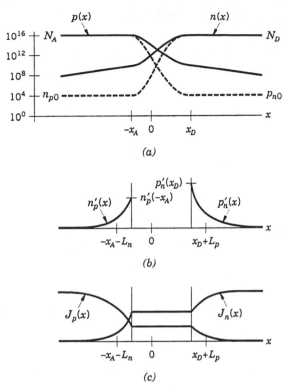

Figure 6-10. Forward-biased junction diode in the long-base limit of the Shockley model. The width of the depletion region is grossly exaggerated. (*a*) Charge-carrier densities. (*b*) Excess minority-carrier densities. (*c*) Current densities.

The saturation current is typically in the range 10^{-15}–10^{-12} A, and the working forward bias is around 600 mV. With these values, currents in the range $1 \mu A$–1A are obtained.

We can now take up the question of the n-n^+ junction in Fig. 6-9. In long-base diodes the current is purely electron current at the n-n^+ junction; electrons are abundant on either side of the junction, and there is thus no obstacle to current flow, so we need not modify what we have said. In short-base diodes the n-n^+ junction increases the effective width of the n-side and thus reduces the hole current (this point is addressed in Exercise 6-17), but it is still no obstacle for the electron current. Thus, in the worst of cases, the saturation current I_S is reduced but the Shockley model (6-30) remains valid.

6-2-3   Deviations from the Shockley Model

To obtain the Shockley model of a diode we assumed that there was no significant recombination in the depletion region, and we also assumed that the injection level was low. In addition, we said nothing about what happens when a diode is reverse biased. We will now look into these questions.

Let us first deal with recombination. From the continuity equation for electrons in (6-8), we obtain

$$J_n(x_D) = J_n(-x_A) + J_r$$

where J_r is the recombination current density,

$$J_r = e \int_{-x_A}^{x_D} R(x)\, dx$$

The current density J is therefore

$$J = J_p(x_D) + J_n(x_D) = J_p(x_D) + J_n(-x_A) + J_r$$

But $J_p(x_D)$ and $J_n(-x_A)$ continue to be given by (6-28) because recombination does not affect quasi-equilibrium in the depletion region; their sum is thus proportional to $\exp(V/V_T)$ if V is larger than 100 mV. On the other hand, as discussed in Exercise 6-9, the recombination current density is of the form

$$J_r = J_0 \exp\left(\frac{V}{mV_T}\right)$$

where J_0 is a constant and m is near 2, and thus varies much more slowly with the bias voltage. As shown in Fig. 6-11, J_r becomes dominant at some point below 300 mV.

At high current densities the current grows more slowly than in the Shockley model. One reason is that the level of injection becomes high, so that as shown in the case of transistors in Exercise 6-12, the current varies like $\exp(V/2V_T)$. Another reason is that the voltage drops in the neutral regions are no longer negligible: at very high current densities, the current varies linearly with the voltage. These effects are also shown in Fig. 6-11.

If a reverse bias larger than several times V_T is applied to a junction diode, the excess minority-carrier densities at the edges of the depletion

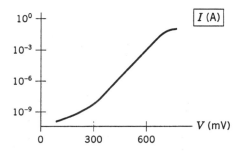

Figure 6-11. Current as a function of voltage in a forward-biased junction diode. The Shockley model is valid over a large range of currents, but the recombination current finally wins out at low currents, and high-level injection— among other things—becomes important at high currents.

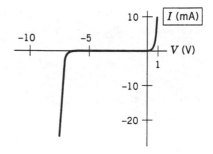

Figure 6-12. I–V characteristic of a real diode, showing that reverse breakdown sets in at some point.

Figure 6-13. Symbol for a Zener diode.

region, given in (6-27), become equal and opposite to the equilibrium minority-carrier densities. The diffusion currents in the neutral regions thus become constant and very small according to the Shockley model (6-30): the current for a large reverse bias is $-I_S$. As discussed in Exercise 6-10, the generation of carriers in the depletion region makes the reverse current much larger than I_S, but nonetheless small, of order 1 nA in a small-signal diode.

At some reverse bias, however, the reverse current grows dramatically due to field emission (Zener effect) or avalanche multiplication. We will not say anything further about these *reverse-breakdown* effects; they tell us, however, that we should check the specifications of any diode we use to see what maximum reverse voltage it can sustain. Taking reverse breakdown into account, the I–V curve of a junction diode is as shown in Fig. 6-12.

On the positive side, reverse breakdown is not destructive if the current is kept low enough that the diode does not overheat. What is more, the voltage at constant reverse current can be quite independent of temperature, particularly for reverse breakdown voltages around 6 V. Voltage reference diodes that are operated in the third quadrant of Fig. 6-12 are called *Zener diodes*, and they are represented as shown in Fig. 6-13.

6-2-4   Metal–Semiconductor Junctions: Ohmic Contacts and Schottky Diodes

Now that we have seen what a *p-n* junction looks like, we can make some brief comments on the nature of metal–semiconductor junctions. Let us consider the *m-n* junction shown in Fig. 6-14. In the simplest picture, there is a negative surface charge in the metal and a depletion layer with a built-in potential $\psi = \phi_n - \phi_m$ in the *n*-silicon. If a forward bias V (metal positive) is applied to the junction, electrons flow from the *n*-silicon into the metal. The I–V curve, shown in Fig. 6-15a, is similar to that of a *p-n* junction,

$$I = I_S\left[\exp\left(\frac{V}{V_T}\right) - 1\right]$$

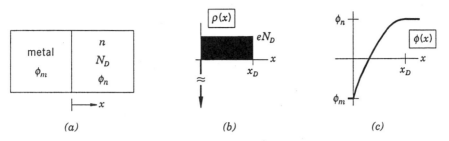

Figure 6-14. Metal–semiconductor junction. (*a*) Cross section. (*b*) Charge density. (*c*) Potential.

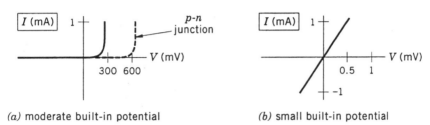

(*a*) moderate built-in potential   (*b*) small built-in potential

Figure 6-15. *I–V* characteristics of metal–semiconductor junctions.

There are, however, important differences that merit at least a qualitative discussion. The current is not diffusion limited as it is in a *p-n* junction, and depends mainly on the height $\phi_n - \phi_m - V$ of the potential barrier. As a result, the saturation current I_S is orders of magnitude larger than in a *p-n* junction with equal doping on the lightly doped side and depends weakly on *V*, and at equal currents, the forward bias is much smaller than in a *p-n* junction.

If the built-in potential is small, of order 100 mV, I_S is enormous. As shown in Fig. 6-15*b*, the *m-n* junction then looks like a small resistor rather than a diode at currents that are typical in junction diodes; in other words, it is ohmic. Similar considerations hold for *m-p* junctions such as the one at the anode of the junction diode in Fig. 6-9.

If the *n*-silicon is heavily doped ($N_D \approx 10^{19}$ cm^{-3}), the depletion layer is very thin, of order 10^{-6} cm, and electrons can tunnel through the potential barrier with relative ease. For moderate built-in potentials, the junction is ohmic. This is the case of the n^+-*m* junction at the cathode of the junction diode in Fig. 6-9.

The *Schottky diode* shown together with its symbol and its *I–V* curve in Fig. 6-16 has an *m-n-n$^+$-m* structure. In agreement with what we have just seen, the *m-n* junction is rectifying and the n^+-*m* junction is ohmic. Small-signal Schottky diodes typically have a forward bias of 300 mV at 1 mA; in Chapter 8 we will make good use of this particularity, and we will also see that Schottky diodes are fast.

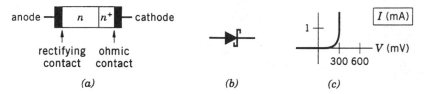

Figure 6-16. Schottky diode. (*a*) Cross section. (*b*) Symbol. (*c*) $I-V$ characteristic.

6-3   BIPOLAR JUNCTION TRANSISTORS

A *bipolar junction transistor* (briefly *transistor*, or *bipolar transistor* when necessary) can be roughly described as a one-dimensional structure with three layers of extrinsic silicon that form two closely spaced *p-n* junctions. There are thus two kinds of transistors, *npn* and *pnp*, depending on whether the outer layers are *n*-type or *p*-type.

The essential aspects of the physical structure of an *npn* transistor are shown in Fig. 6-17. From top to bottom, the layers are the *emitter*, the *base*, and the *collector*; the corresponding densities of dopants are N_{DE}, N_{AB}, and N_{DC}. The base width w_B is usually in the range 0.1–1 μm; the emitter width w_E is several times w_B, and the collector width w_C is usually larger than w_E. All layers are connected to the outside world by means of aluminum contacts. In the case of the collector, an additional heavily doped n^+ layer is required to make the contact ohmic; the emitter is always heavily doped and can thus be connected directly to its contact.

Let us consider the situation when the transistor is *active*, that is, when the base–emitter diode is forward biased and the base–collector diode is reverse biased. We will focus first on the base–emitter junction. Since the emitter is much more heavily doped than the base, the excess electron density at the base edge of the base–emitter depletion region is much larger than the excess hole density at the emitter edge. In addition, the electron diffusion constant is larger than the hole diffusion constant, and the emitter width is larger than the base width. The current in the base–emitter junction is thus predominantly electron current, as we will see in more detail shortly. [This conclusion can also be drawn by comparing the current densities in (6-28).]

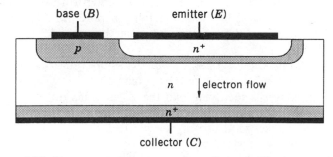

Figure 6-17. Representative cross section of an *npn* bipolar transistor.

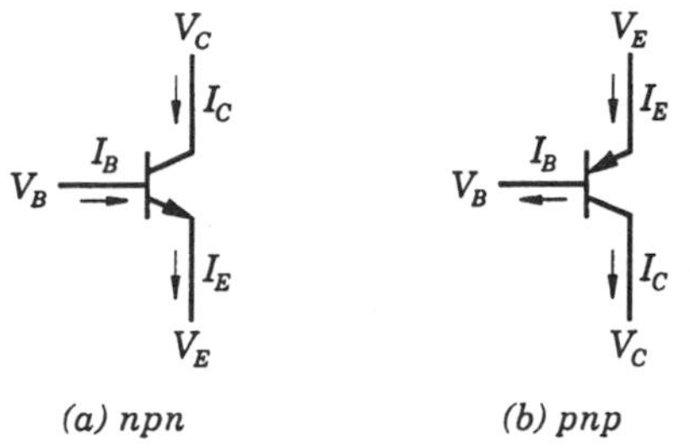

(a) npn   *(b) pnp*

Figure 6-18. Symbols and sign conventions for bipolar transistors.

Electrons injected from the emitter into the base diffuse over to the reverse-biased base–collector junction, where they are swept into the collector by the favorable potential they encounter. The collector current is thus essentially equal to the emitter current, and it is *prima facie* independent of the reverse bias on the base–collector junction. The base current is equal to the hole component of the current at the base–emitter junction and is therefore much smaller than the collector current.

Briefly stated, the base–emitter voltage controls the collector current; from another point of view, which we will use occasionally, it is the base current that controls the collector current. From either point of view, we are dealing with a dependent current source; according to what we have seen, amplifiers should be possible, and we will soon see that this is the case. Symbols and sign conventions for *npn* and *pnp* transistors are shown in Fig. 6-18.

6-3-1 The Collector Current

Since current flow in transistors is mostly one-dimensional, we will continue our discussion using the model with constant cross-sectional area A shown in Fig. 6-19. For convenience, the depletion regions are represented as having zero width. We will make the assumptions that the doping in all regions is uniform, that the junctions are abrupt, and that minority-carrier lifetimes are infinite, so that we will always be in the short-base limit. There are many exceptions to these rules, but they will not affect our main conclusions. We will also assume that *forward bias* means a bias much larger than V_T.

Figure 6-19 shows the minority-carrier profiles in the base and in the emitter when the base–emitter junction has a forward bias V_{BE} and the base–collector junction is reverse biased. The excess electron density at the base edge of the base–emitter junction is

$$n'_B(0) = \frac{n_i^2}{N_{AB}} \exp \frac{V_{BE}}{V_T} \tag{6-31}$$

whereas it is essentially zero at the base edge of the base–collector junction.

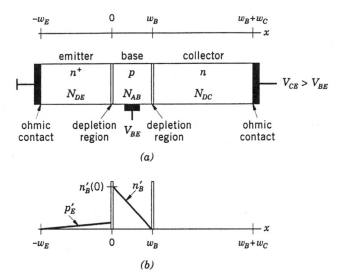

Figure 6-19. One-dimensional model of an *npn* transistor. (*a*) Cross section. (*b*) Excess minority-carrier densities.

The collector current I_C is equal to the electron current in the base,

$$I_C = \frac{e n'_B(0) D_n A}{w_B} \tag{6-32}$$

Using the expression for $n'_B(0)$ in (6-31), we obtain

$$I_C = I_S \exp\frac{V_{BE}}{V_T} \tag{6-33}$$

where

$$I_S = \frac{A e n_i^2 D_n}{w_B N_{AB}} \tag{6-34}$$

The collector current thus varies exponentially with the base–emitter voltage; a change by a factor of 10 in I_C requires a change by $V_T \ln 10 \approx 60$ mV in V_{BE}. In Chapter 7 we will see that in addition to their applications in amplifiers, transistors can be used to execute mathematical operations on voltages and currents; we will also encounter devices that use the exponential variation of the collector current in other ingenious ways.

6-3-2 The Emitter and Base Currents

The excess hole density in the emitter is zero at the ohmic contact; at the emitter edge of the base–emitter junction, it is

$$p_E'(0) = \frac{n_i^2}{N_{DE}} \exp \frac{V_{BE}}{V_T}$$

The emitter current is equal to the sum of the electron and hole currents at the base–emitter junction,

$$I_E = Aen_i^2 \left(\frac{D_n}{w_B N_{AB}} + \frac{D_p}{w_E N_{DE}} \right) \exp \frac{V_{BE}}{V_T}$$

We can rewrite this expression as

$$I_E = \frac{Aen_i^2 D_n}{w_B N_{AB}} (1 + \delta_E) \exp \frac{V_{BE}}{V_T}$$

where δ_E is the *emitter defect*, equal to the ratio of the hole and electron currents at the base–emitter junction,

$$\delta_E = \frac{D_p w_B N_{AB}}{D_n w_E N_{DE}}$$

From (6-33) and (6-34) we have $I_E = I_C(1 + \delta_E)$ if recombination in the base is negligible, as assumed. For completeness, however, we will define the *base defect* δ_B as the fractional loss in collector current due to recombination in the base. We can then write

$$I_C = I_E \frac{1 - \delta_B}{1 + \delta_E}$$

The recombination rate in the base is equal to Q_B/τ_n, where $-Q_B$ is the excess electron charge in the base. Since δ_B is small, $n_B(x)$ retains its triangular shape; using (6-32) we then get

$$Q_B \simeq \frac{en_B'(0) Aw_B}{2} = \frac{w_B^2}{2D_n} I_C \qquad (6\text{-}35)$$

and substituting $D_n = L_n^2/\tau_n$,

$$\frac{Q_B}{\tau_n} \simeq \frac{w_B^2}{2L_n^2} I_C$$

The base defect is thus

$$\delta_B \simeq \frac{w_B^2}{2L_n^2} \tag{6-36}$$

Taking $w_E \approx 3w_B$ and $N_{DE} \approx 10N_{AB}$, and using the fact that $D_n \simeq 3D_p$, we get $\delta_E \approx 0.01$; δ_B is usually an order of magnitude smaller.

We can now obtain the base current,

$$I_B = I_E - I_C \simeq (\delta_E + \delta_B)I_C \tag{6-37}$$

The base current is thus composed of two small components: the hole current at the emitter junction and the (negligible) recombination current in the base. We will refer to the fact that I_C is less than I_E as the *base loss*. Turning things around, we can write

$$I_C = \beta I_B$$

where

$$\beta = \frac{1}{\delta_E + \delta_B}$$

is the (forward) *current gain* from the base to the collector. As indicated in Section 6-3, we can look at a transistor as a dependent current source in which the base current controls the collector current. In most transistors, β is in the range 50–200. On the exceptional side, it can be as low as 10 in power transistors and as high as 5000 in narrow-base *superbeta* transistors.

6-3-3   Deviations at Low and High Currents

In Section 6-2-3 we saw that diodes deviate from the Shockley model at low and high currents; this is of course also true for transistors. The results of Section 6-3-2 are nonetheless valid over a wide range; in small-signal transistors such as the standard 2N3904, the exponential dependence of I_C on V_{BE} holds for 10 nA $\leq I_C \leq$ 10 mA, that is, over six decades in collector current. Figure 6-20 shows a *Gummel plot* of the collector and base currents as a function of the base–emitter voltage. In this coarse view both currents depend exponentially on the base–emitter voltage over large ranges, and their ratio β is constant over a somewhat smaller range.

The injection level first becomes high in the base because the emitter is much more heavily doped than the base; a high level thus sets in when $n_B'(0) \approx N_{AB}$ or, using (6-31), when

$$V_{BE} \approx V_T \ln \frac{N_{AB}^2}{n_i^2}$$

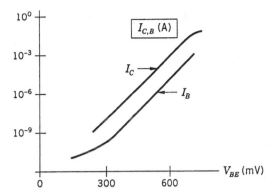

Figure 6-20. Gummel plot of an active transistor, showing that the current gain $\beta = I_C/I_B$ is roughly constant over a large range of collector currents.

The corresponding collector current is given by (6-32),

$$I_C \approx \frac{AeD_n N_{AB}}{w_B}$$

For a typical value $N_{AB} = 10^{16}$ cm^{-3}, the injection level becomes high for V_{BE} in the vicinity of 700 mV; in small-signal transistors, the corresponding value of I_C is around 10 mA. Since the emitter injection level remains low, the base current continues to grow like $\exp(V_{BE}/V_T)$. As shown in Fig. 6-20, β therefore decreases at high currents. There are actually other effects that come into play at high currents, but they do so approximately at the current at which high-level injection sets in, and they have similar consequences.

At low currents, recombination in the base–emitter junction first starts to have an effect on the base current, whereas the collector current continues to fall like $\exp(V_{BE}/V_T)$ because it is much larger. As shown in Fig. 6-20, β also decreases at low currents.

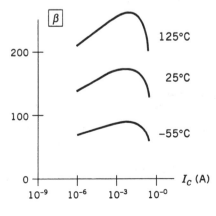

Figure 6-21. Current gain as a function of the collector current at several temperatures.

Figure 6-21 shows the behavior of β in more detail and also shows how β varies as a function of temperature. The explanation of these details is complicated, and not necessary for our purposes, which will only require that β be "large enough". Around room temperature, β increases by about 0.7% K^{-1}.

6-3-4   Depletion and Diffusion Capacitances

Transistors and diodes are in some aspects capacitors because they can store charges; the excess electron charge in the base of the active transistor shown in Fig. 6-19 is a clear example. The amounts of stored charges depend on the state of a device; when the device changes state, the external terminals must provide the charges required to establish the new state.

Let us reconsider the depletion approximation presented in Section 6-2-1, but let us now assume that the junction is subjected to a bias voltage V. There will still be a depletion region, but the charges will be different. Retracing our arguments, we find that the widths of the depletion layers are given by

$$N_A x_A = N_D x_D = \sqrt{\frac{2\varepsilon(\psi-V)N_A N_D}{e(N_A+N_D)}} \tag{6-38}$$

The (negative) charge Q_T in the depletion layer on the p-side is thus given by

$$-Q_T = eAN_A x_A = A\sqrt{\frac{2e\varepsilon(\psi-V)N_A N_D}{N_A+N_D}}$$

If V changes by a small amount ΔV, the charge will change by an amount $C_T \Delta V$, where $C_T = dQ_T/dV$ is the (incremental) *depletion capacitance* of the junction, given by

$$C_T = A\sqrt{\frac{e\varepsilon N_A N_D}{2(N_A+N_D)(\psi-V)}} \tag{6-39}$$

The depletion capacitance for an abrupt junction thus varies like $1/\sqrt{\psi-V}$. In junctions with other doping profiles, the functional dependence is different (a cube root rather than a square root if the doping profile is linear), but the general idea is still valid: the depletion capacitance is small if the reverse bias is large, but it can become large if the forward bias is large. We will denote the depletion capacitances in transistors by C_{T_π} (base–emitter junction) and C_{T_μ} (base–collector junction). In small-signal transistors, both these capacitances are of order 10 pF for a forward bias and 1 pF for a reverse bias.

In an active *npn* transistor, the excess electron charge in the base as given in (6-35) is

$$-Q_B = -\frac{w_B^2}{2D_n} I_C$$

From charge neutrality, there is also an excess hole charge Q_B in the base. We observe that these charges are proportional to the collector current and independent of the cross-sectional area of the transistor. We also observe, in contrast with a capacitor, that they occupy the same physical space. The excess charges in the emitter are negligible because the emitter is much more heavily doped than the base and only several times wider. If we define the *base charging time* τ_B by

$$\tau_B = \frac{w_B^2}{2D_n} \tag{6-40}$$

we can write

$$Q_B = \tau_B I_C \tag{6-41}$$

The base charging time is generally in the range 10–500 ps. We will see below that it is the determining factor in the speed of a transistor.

For small signals, we will define the (incremental) *diffusion capacitance* $C_{D\pi}$ by

$$C_{D\pi} = \frac{dQ_B}{dV_{BE}}$$

From the chain rule we have

$$\frac{dQ_B}{dV_{BE}} = \frac{dQ_B}{dI_C} \frac{dI_C}{dV_{BE}}$$

From (6-33) and (6-41) we also have

$$\frac{dI_C}{dV_{BE}} = \frac{I_C}{V_T} \quad \text{and} \quad \frac{dQ_B}{dI_C} = \tau_B$$

We then obtain

$$C_{D\pi} = \tau_B \frac{I_C}{V_T} \tag{6-42}$$

Like the excess charges in the base, the diffusion capacitance is proportional to the collector current and independent of the cross-sectional area of the transistor. When the diffusion capacitance is charged up, the base provides the hole charge and the emitter provides the electron charge; it thus acts like a capacitor between the base and the emitter.

6-3-5 The Large-Signal Model

For positive collector–emitter voltages, an *npn* transistor can be described as being in one of three states. It is at *cutoff* if the base–emitter voltage is such that the collector current is insignificant. A transistor is thus certainly at cutoff if its base–emitter voltage is zero or negative. How positive a base–emitter voltage makes a transistor come out of cutoff is of course a relative question: it depends on the current level we consider to be significant, and must be determined in each case. We are already familiar with the second state, the *active* state, defined up to now by a reverse-biased base–collector diode and a forward-biased base–emitter diode. The third state is *saturation*, in which both the base–emitter and the base–collector diodes are forward biased.

A transistor at cutoff does nothing; except for the effect of its depletion capacitances, it could well be removed from a circuit. We will therefore first consider in more detail the large-signal behavior of an active transistor and then take up saturation.

Up to now we have required that the base–collector diode in an active transistor have a fixed reverse bias. We will now see that the collector current depends to some extent on the collector–base voltage; this dependence is known as the *Early effect*.

The effective width w_B of the base is equal to the metallurgical width x_M of the base minus the widths x_E and x_C of the depletion layers on the base side at the base–emitter and base–collector junctions,

$$w_B = x_M - x_E - x_C$$

According to (6-38), x_E is proportional to $\sqrt{\psi_{BE} - V_{BE}}$, where ψ_{BE} is the built-in potential of the base–emitter junction, and x_C is proportional to $\sqrt{\psi_{BC} + V_{CB}}$, where ψ_{BC} is the built-in potential of the base–collector junction; w_B thus depends on V_{BE} and on V_{CB}.

The collector current is given by (6-32),

$$I_C = \frac{eD_n A}{w_B} n'_B(0)$$

The variations in w_B due to V_{BE} are swamped by the exponential variations in $n'_B(0)$ also due to V_{BE} and can therefore be neglected. In contrast, $n'_B(0)$ is independent of V_{CB}, and we have

$$\frac{\partial I_C}{\partial V_{CB}} = -\frac{I_C}{w_B} \frac{\partial w_B}{\partial V_{CB}}$$

If we now define the *Early voltage V_A* by

$$\frac{1}{V_A} = -\frac{1}{w_B} \frac{\partial w_B}{\partial V_{CB}}$$

we can write

$$\frac{\partial I_C}{\partial V_{CB}} = \frac{I_C}{V_A} \tag{6-43}$$

V_A depends linearly on x_M through w_B and ranges from 15 V in narrow-base transistors with transition frequencies in the region of several gigahertz to 200 V in medium-frequency transistors such as the 2N3904. For a given transistor, V_A is usually a slowly increasing function of V_{CB}.

If the base defect δ_B is nonzero, the component $\delta_B I_C$ of the base current given in (6-37) is also modified by V_{CB} because according to (6-36), $\delta_B = w_B^2/2L_n^2$. In the circuits we will encounter in the following chapters, δ_B is negligible if it is at least 10 times smaller than the emitter defect δ_E. This is always a safe assumption; we will therefore continue to ignore δ_B and consider that I_B is independent of V_{CB}.

If we assume that the Early voltage is constant, we can integrate $\partial I_C/\partial V_{CB}$ in (6-43). We then obtain the *large-signal model* for an *npn* transistor in the active region,

$$I_C = I_S \exp\left(\frac{V_{BE}}{V_T}\right) \exp\left(\frac{V_{CB}}{V_A}\right) \tag{6-44}$$

$$\beta I_B = I_S \exp\left(\frac{V_{BE}}{V_T}\right) \tag{6-45}$$

where I_S is evaluated at $V_{CB} = 0$. The base–collector diode does not start to conduct significantly until V_{CB} is about −400 mV; the expression for I_C can thus be extrapolated to this point. Generally speaking, transistors are operated at values of V_{CB} that are a small fraction of V_A; in addition, V_A grows with V_{CB}. As a result, I_C depends quite linearly on V_{CB} in the active region. With the conventions given in Fig. 6-18, (6-44) and (6-45) are valid for a *pnp* transistor if V_{BE} and V_{CB} are replaced by V_{EB} and V_{BC}.

Transistor characteristics are usually presented in terms of V_{CE} rather than V_{CB}, and we will follow this usage. Figure 6-22*a* shows I_C as a function of V_{CE} for several values of I_B; we note that we can take I_B as the parameter because it depends only on V_{BE}.

Let us now see what happens in the saturation region when the base current is constant and the collector–emitter voltage is varied. For V_{CE} above 500 mV or so, the transistor is active: the base–emitter voltage is V_{BE0}, the excess electron density at the emitter end of the base is $n'_{B0}(0)$, and neglecting the small influence of the Early effect, the collector current is I_{C0}.

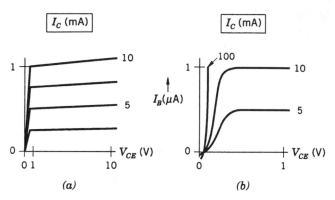

Figure 6-22. (*a*) Characteristics of an *npn* transistor; (*b*) details of the saturation region.

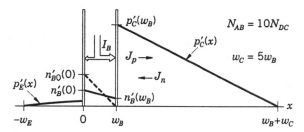

Figure 6-23. Excess minority-carrier densities in a saturated *npn* transistor.

Figure 6-23 shows the excess minority-carrier densities in the collector and base neutral regions when V_{CE} is low enough that the base–collector diode has turned on. The excess electron density $n'_B(w_B)$ at the collector end of the base and the excess hole density $p'_C(w_B)$ at the base end of the collector are now significant; the excess hole density at the collector contact is of course still essentially zero. In addition, part of the base current has been diverted toward the collector, so that V_{BE} and $n'_B(0)$ have dropped below V_{BE0} and $n'_{B0}(0)$.

The net result is that the electron current in the collector drops below I_{C0} and that there is in addition a hole current in the opposite direction. When V_{CE} is low enough, in the vicinity of 50 mV, the opposing electron and hole currents cancel, and the collector current becomes zero. When V_{CE} becomes zero, $n'_B(0)$ and $n'_B(w_B)$ become equal, and the electron current becomes zero; the collector current is all hole current and, ignoring the minute current into the emitter, equal to $-I_B$. Figure 6-22*b* shows I_C as a function of V_{CE} for several values of I_B.

For a transistor of cross-sectional area A, the excess hole charge in the collector is

$$Q_C = \frac{eAw_C}{2}p'_C(w_B)$$

In order to minimize the Early effect, the collector in most transistors is much more lightly doped than the base, and $p'_C(w_B)$ is therefore much larger than $n'_B(w_B)$. In addition, the collector is usually several times wider than the base. As a result, the excess charges in a saturated transistor are mostly in the collector and are much larger than the excess charges stored in the base when the transistor is active. When the transistor is strongly saturated, that is, when I_C is below I_{C0} by a factor of 10 or so, the base current is essentially equal to minus the collector hole current,

$$I_B = \frac{eAD_p}{w_C} p'_C(w_B)$$

If we now define the *collector charging time* τ_C by

$$\tau_C = \frac{w_C^2}{2D_p}$$

we can write

$$Q_C \simeq \tau_C I_B$$

Since the hole diffusion constant is several times smaller than the electron diffusion constant and the collector is usually several times wider than the base, the collector charging time in *npn* transistors can be much larger than the base charging time, as much as three orders of magnitude in high-voltage transistors with wide collectors. In a 2N3904, for example, $\tau_B \simeq 300$ ps and $\tau_C \simeq 150$ ns.

   If a saturated transistor is pulled out of saturation by reversing the base current, the base–emitter voltage stays essentially constant until the excess charges in the collector have been removed; the time involved is of order τ_C and can thus be quite long. We will study this behavior in more detail in Chapter 8; the laboratory in Exercise 6-3 deals with similar behavior in the simpler case of diodes and is meant to give a preview of what to expect.

6-3-6   The Small-Signal Model

In a large class of electronic devices, amplifiers in particular, the signals are small enough that we can work with the incremental values of the voltages and the currents around an *operating point* defined by the static values of the collector current and the collector–emitter voltage. If we set $V_{CB} = V_{CE} - V_{BE}$ in (6-44), we obtain

$$I_C = I_S \exp\left(\frac{V_{BE}}{V_T}\right) \exp\left(\frac{V_{CE}}{V_A}\right) \exp\left(-\frac{V_{BE}}{V_A}\right)$$

Since V_A is much larger than V_T, we can ignore the last exponential in this expression, and instead of (6-44) and (6-45) we can write

$$I_C \simeq I_S \exp\left(\frac{V_{BE}}{V_T}\right) \exp\left(\frac{V_{CE}}{V_A}\right)$$

$$\beta I_B = I_S \exp\left(\frac{V_{BE}}{V_T}\right)$$

We will now define the *transconductance* g_m, the *base input conductance* g_π, and the *collector output conductance* g_o by

$$g_m = \frac{\partial I_C}{\partial V_{BE}} \qquad g_\pi = \frac{dI_B}{dV_{BE}} \qquad g_o = \frac{\partial I_C}{\partial V_{CE}}$$

We then obtain

$$g_m = \frac{I_C}{V_T} \qquad g_\pi = \frac{I_B}{V_T} \qquad g_o = \frac{I_C}{V_A}$$

Corresponding to these conductances, we have the *emitter resistance* r_m, the *base input resistance* r_π, and the *collector output resistance* r_o, given by

$$r_m = \frac{1}{g_m} \qquad r_\pi = \frac{1}{g_\pi} \qquad r_o = \frac{1}{g_o}$$

For static incremental signals (denoted by small letters) we then have

$$i_C = g_m v_{BE} + g_o v_{CE} \qquad\qquad (6\text{-}46)$$

$$i_B = g_\pi v_{BE} \qquad\qquad (6\text{-}47)$$

The circuit shown in Fig. 6-24 is the *small-signal model* of a transistor in the active region. As is easily verified, this model is a representation of (6-46) and

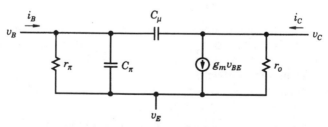

Figure 6-24. Small-signal model of a bipolar transistor.

(6-47) if we ignore the capacitors. The collector current I_C is always a given, and we can estimate the base current I_B quite well by taking

$$I_B \simeq \frac{I_C}{\beta}$$

The parameters in the small-signal model are thus usually expressed in the form

$$g_m = \frac{I_C}{V_T} \qquad r_\pi = \beta r_m \qquad r_o = \frac{V_A}{I_C}$$

For reference, we note that for $I_C = 1$ mA, $\beta = 100$, and $V_A = 100$ V we have

$$g_m = 40 \text{ mA/V} \qquad r_\pi = 2.5 \text{ k}\Omega \qquad r_o = 100 \text{ k}\Omega$$

Following standard usage in the literature, the transconductance is expressed in mA/V rather than in siemens. We also have $r_m = 25$ Ω at $I_C = 1$ mA. If we keep these values in mind, we can quickly estimate the resistive parameters at any collector current because g_m is proportional to the collector current and r_π and r_o are inversely proportional to the collector current.

For time-varying signals, we must take into account the stored charges discussed in Section 6-3-4. The small-signal base–emitter capacitance C_π is the sum of the depletion and diffusion capacitances given in (6-39) and (6-42),

$$C_\pi = C_{T\pi} + C_{D\pi}$$

Since the base–collector junction is reverse biased, the only contribution to the base–collector capacitance comes from the depletion capacitance,

$$C_\mu = C_{T\mu}$$

In Chapter 8 we will see that the speed of a transistor for small signals is measured by the *transition frequency* f_T, defined by

$$f_T = \frac{1}{2\pi(C_\pi + C_\mu)r_m}$$

From (6-42) we have $C_{D\pi} = \tau_B(I_C/V_T)$, and we can write

$$C_\pi + C_\mu = C_{D\pi} + C_{T\pi} + C_{T\mu} = \tau_B \frac{I_C}{V_T} + C_{T\pi} + C_{T\mu}$$

According to (6-39), the base–emitter depletion capacitance $C_{T\pi}$ depends on the base–emitter voltage; it therefore depends logarithmically on the collector current, and we can take it to be roughly constant. The transition

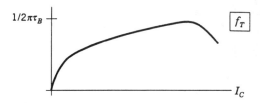

Figure 6-25. Representative variation of the transition frequency as a function of the collector current.

frequency is measured at constant collector–emitter voltage, so that we can also take the base–collector depletion capacitance $C_{T\mu}$ to be constant. Substituting $r_m = V_T/I_C$ we obtain

$$f_T = \frac{1}{2\pi} \frac{I_C}{\tau_B I_C + (C_{T\pi} + C_{T\mu}) V_T}$$

According to this expression, f_T grows linearly at low collector currents and tends to $1/2\pi\tau_B$ at infinite collector current. At high collector currents, however, r_m becomes larger than V_T/I_C, and f_T starts to fall, but its maximum value is nonetheless usually well given by $1/2\pi\tau_B$. A representative plot of f_T as a function of I_C is shown in Fig. 6-25. At $I_C = 1$ mA, a 2N3904 has $C_\pi \approx 30$ pF and $C_\mu \approx 3$ pF, and thus $f_T \approx 200$ MHz. In contrast, microwave transistors typically have $C_\pi = 2$ pF and $C_\mu = 0.5$ pF, and thus $f_T = 2.5$ GHz.

As an example, let us use the small-signal model to obtain the incremental behavior of a simple circuit, the diode-connected transistor shown in Fig. 6-26a. The base is tied to the collector, so that $V_{BE} = V_{CE}$, and we of course have $I_E = I_C + I_B$. From (6-44) and (6-45) we then get

$$I_E = \left(1 + \frac{1}{\beta}\right) I_S \exp\left(\frac{V_{CE}}{V_T}\right)$$

so that as shown in Fig. 6-26b, the transistor acts like a diode in this

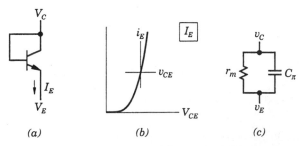

(a)   (b)   (c)

Figure 6-26. Diode-connected transistor. (a) Circuit. (b) I–V characteristic. (c) Incremental circuit.

configuration. For small signals we have $v_{BE} = v_{CE}$; the current in C_μ is zero, and the incremental emitter current is

$$i_E = v_{CE}(g_m + g_\pi + g_o + sC_\pi)$$

Neglecting g_π and g_o compared with g_m, we obtain

$$i_E \simeq v_{CE}(g_m + sC_\pi)$$

This equation is represented by the incremental circuit of Fig. 6-26c.

6-3-7 Temperature Dependence of Parameters

In order to fix the operating point of a transistor, we need to know how the collector current and the base–emitter voltage vary with temperature. According to (6-1), n_i^2 varies like $T^3 \exp(-\mathcal{E}_G/kT)$. If we ignore the relatively slow T^3 dependence in n_i^2 and the also relatively slow temperature variations in the other ingredients of I_S, we can write

$$I_S \simeq I_0 \exp\left(-\frac{V_G}{V_T}\right)$$

where I_0 is a constant and $V_G = \mathcal{E}_G/e = 1.1\text{V}$ is the bandgap energy expressed in volts. The collector current is thus given approximately by

$$I_C \simeq I_0 \exp\left(\frac{V_{BE} - V_G}{V_T}\right)$$

At constant base–emitter voltage we have

$$\frac{\partial I_C}{\partial T} \simeq \frac{V_G - V_{BE}}{V_T} \frac{I_C}{T}$$

For $V_{BE} = 0.6$ V and $T = 290$ K we get $(1/I_C)\partial I_C/\partial T \simeq 0.07$ K^{-1}, which implies that the collector current doubles every 10 K. We conclude that holding the base–emitter voltage constant is *not* the way to set the operating point of a transistor and that other means will have to be found. At constant collector current we have

$$\frac{\partial V_{BE}}{\partial T} \simeq \frac{V_{BE} - V_G}{T}$$

With the values we used above, we get $\partial V_{BE}/\partial T \simeq -1.7$ mV/K. To complete the picture, we will recall that β increases by about 0.7% K^{-1} at room temperature.

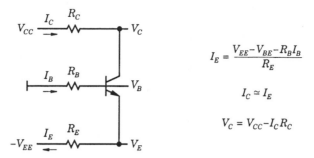

$$I_E = \frac{V_{EE} - V_{BE} - R_B I_B}{R_E}$$

$$I_C \simeq I_E$$

$$V_C = V_{CC} - I_C R_C$$

Figure 6-27. Classic circuit used to set the operating point of a transistor.

As an example, Fig. 6-27 shows a classic method of setting the operating point of an *npn* transistor so that it remains stable for large variations in temperature, say ± 50 K. The base is connected to ground through a resistor R_B, the emitter is connected through a resistor R_E to a constant negative voltage $-V_{EE}$ a few times larger in magnitude than V_{BE}, and the collector is connected through a resistor R_C to a constant positive voltage V_{CC}. (That this arrangement is still useful will become apparent in Chapter 7.) The emitter current is

$$I_E = \frac{V_{EE} - V_{BE} - R_B I_B}{R_E}$$

Since V_{BE} varies logarithmically with I_C, and $I_C \simeq I_E$, I_C is well determined if $R_B I_B$ can be neglected. The variation in V_{BE} due to temperature is about 100 mV in 50K, and thus quite tolerable if $V_{EE} - V_{BE} \geq 1$V. The collector–emitter voltage can then be set by a suitable choice of V_{CC} and R_C.

6-4   FIELD-EFFECT TRANSISTORS

Current flow in bipolar transistors is mainly longitudinal, that is, normal to the junctions, and its mechanism is diffusion of minority carriers in the base. The current level is regulated by the density of excess minority carriers at the emitter end of the base, and thus by the forward bias on the base–emitter junction.

In a *field-effect transistor* (FET), current flow takes place in a transverse *channel* whose *source* and *drain* terminals are analogs of the emitter and collector terminals in a bipolar transistor, and the mechanism is drift of majority carriers in the channel. The voltage on a *gate* electrode adjacent to the channel modulates the channel conductivity and consequently controls the amount of current that flows when a voltage is applied between the drain and the source; the gate terminal is thus the analog of the base terminal in a bipolar transistor. The gate and the channel form a capacitorlike structure in which a voltage translates into an electric field. We can thus say equally well that the electric field modulates the channel conductivity. This is the historical origin of the term *field effect*.

An outstanding feature of field-effect transistors is that their gate current is extremely low, of order 10 pA at room temperature. Correspondingly, the resistance analogous to r_π in bipolar transistors is typically 10^{12} Ω and can be as high as 10^{14} Ω. Another invaluable feature of field-effect transistors is that they can be used to construct *linear gates*, that is, two-state devices that look like an open circuit in one state and like a small resistor in the other.

On the negative side, the transconductance in a field-effect transistor is more than an order of magnitude smaller than it is in a bipolar transistor at the same current; as we will see in Chapter 7, elementary FET amplifiers consequently have lower gains than their bipolar analogs. This disadvantage is not absolute, however, and can be overcome by making FET amplifiers somewhat more elaborate.

FETs are classified according to the nature of the gate. We will look at two representative types, the *junction* FET or *JFET*, and the *metal–oxide–silicon* FET or *MOSFET*. FETs are also classified according to the majority carriers in the channel; there are thus *p*-channel FETS and *n*-channel FETs. MOSFETs can further be *depletion mode* or *enhancement mode*, depending on whether the channel exists when all terminals are at the same potential or whether it has to be induced by applying a voltage between the gate and the source.

We will deal first with junction FETs because they can be described quite straightforwardly in terms of what we already know. We will then look at a structure that is as fundamental as the *p-n* junction, the *MOS capacitor*. Finally, we will complicate the MOS capacitor slightly and obtain MOSFETs.

6-4-1   JFETs

Let us consider the *n*-channel JFET shown in cross section in Fig. 6-28. The moderately doped *n*-silicon channel is sandwiched between two *p*-regions. The top or *gate p*-region is heavily doped, whereas the bottom or *substrate*

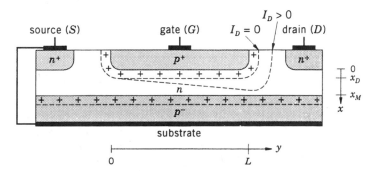

Figure 6-28. Cross section of a junction field-effect transistor.

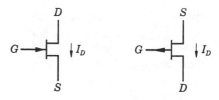

Figure 6-29. Symbols and sign conventions for JFETs.

(a) n-channel

(b) p-channel

p-region is lightly doped. The gate–channel depletion region therefore lies mostly in the channel, whereas the substrate–channel depletion region lies mostly in the substrate, and we can ignore the depletion layer on the channel side. Ohmic aluminum contacts connect the gate and the substrate to the outside world. At its ends, the channel has aluminum *source* and *drain* contacts that are made ohmic by means of n^+-regions. The source and substrate terminals are tied together. Symbols and sign conventions for n-channel and p-channel JFETs are shown in Fig. 6-29.

Let us assume initially that the drain is open circuited and thus at the same potential as the source, and let us apply a (negative) voltage V_{GS} between the gate and the source; this is equivalent to applying a bias voltage $\psi_{GC} = V_{GS}$ to the gate–channel junction. The width of the depletion layer in the channel is given by (6-38),

$$x_D = \sqrt{\frac{2\varepsilon}{eN_D}} \sqrt{\psi - \psi_{GC}} \qquad (6\text{-}48)$$

where ψ is the built-in potential of the gate–channel junction and we have used the fact that the density N_A of acceptors in the gate is much larger than the density N_D of donors in the channel.

As ψ_{GC} is made more negative, the width of the depletion layer in the channel increases until it becomes equal to the metallurgical width x_M of the channel. At this point, the channel is *pinched off*; that is, the density of charge carriers in the channel is everywhere negligible. The value of $\psi_{GC} = V_{GS}$ at which this occurs is the *pinch-off voltage* V_P, generally between -1 and -5 V and given by

$$x_M = \sqrt{\frac{2\varepsilon}{eN_D}} \sqrt{\psi - V_P}$$

If V_{GS} is larger (less in magnitude) than V_P, there are charge carriers in the undepleted part of the channel between x_D and x_M, and current will flow if a voltage V_{DS} is applied between the drain and the source. For small values of V_{DS}, the channel is *ohmic*; that is, it looks like a resistor. Let us now consider only *positive* values of V_{DS}. At any point along the channel the voltage ψ_{CS} relative to the source increases the reverse bias on the gate–channel junction and widens the depletion layer. At a high enough value of V_{DS}, the width of the depletion layer at the drain end of the channel becomes equal to x_M, and the channel is thus pinched off at this end. For higher values of V_{DS}, the drain current *saturates*; that is, it levels off and remains essentially equal to its value at pinch-off.

To quantify the behavior of a junction FET, we will use the *gradual-channel approximation*; that is, we will assume that the width of the depletion layer in the channel changes slowly enough that we can calculate it using the local value of the bias voltage ψ_{GC} in (6-48). In the same spirit, we will assume that current flow is purely transverse, that is, that the x-component of the current density is negligible. The metallurgical width x_M of the channel is typically around 1 μm; since the gradual-channel approximation requires a channel length L that is not too small, we can neglect the voltage drops in the much wider regions between the source and drain contacts and the channel and take the voltage drop in the channel to be equal to V_{DS}.

The width of the depletion layer in the gradual-channel approximation is

$$x_D(y) = \sqrt{\frac{2\varepsilon}{eN_D}} \sqrt{\psi - \psi_{GC}(y)}$$

In terms of the pinch-off voltage, the width of the channel is thus given by

$$x_C(y) = x_M - \frac{\sqrt{\psi - \psi_{GC}(y)}}{\sqrt{\psi - V_P}} x_M$$

To simplify our calculations, we will use the empirical fact that this expression for x_C can be replaced with minimal error (20% at worst) by an expression that is independent of ψ and linear in ψ_{GC},

$$x_C(y) \simeq \frac{x_M}{2}\left[1 - \frac{\psi_{GC}(y)}{V_P}\right]$$

Substituting $\psi_{GC}(y) = V_{GS} - \psi_{CS}(y)$, we obtain x_C in terms of the channel–source voltage ψ_{CS},

$$x_C(y) = \frac{x_M}{2} \frac{-V_P + V_{GS} - \psi_{CS}(y)}{-V_P} \tag{6-49}$$

If the width of the FET (in the z-direction in Fig. 6-28) is W, the (constant) drain current I_D is given in terms of the transverse electron current density J by

$$I_D = -Wx_C(y)J(y)$$

If we now use Ohm's law $J = \sigma E$ and express E in terms of ψ_{CS}, we have

$$I_D = \sigma W x_C(y) \frac{d\psi_{CS}}{dy} \tag{6-50}$$

Integrating I_D between 0 and L and changing variables, we obtain

$$I_D L = \sigma W \int_0^L x_C \frac{d\psi_{CS}}{dy} dy = \sigma W \int_0^{V_{DS}} x_C \, d\psi_{CS}$$

$$= -\frac{\sigma W x_M}{2V_P} \int_0^{V_{DS}} [-V_P + V_{GS} - \psi_{CS}(y)] \, d\psi_{CS}$$

Performing the integral, we finally obtain

$$I_D = -\frac{\sigma W x_M}{2V_P L}\left[(-V_P + V_{GS})V_{DS} - \frac{V_{DS}^2}{2}\right]$$

This expression is valid in the ohmic region, that is, for $V_{DS} \leq V_{GS} - V_P$. For $V_{DS} \geq V_{GS} - V_P$, that is, in the saturation region, the drain current is equal to its value in the ohmic region for $V_{DS} = V_{GS} - V_P$, and we get

$$I_D = -\frac{\sigma W x_M}{4V_P L}(-V_P + V_{GS})^2 \tag{6-51}$$

If we now define

$$I_{DSS} = -\frac{\sigma W x_M V_P}{4L} \tag{6-52}$$

as the saturated drain current for zero gate–source voltage, in the ohmic region we have

$$I_D = \frac{2I_{DSS}}{V_P^2}\left[(V_{GS} - V_P)V_{DS} - \frac{V_{DS}^2}{2}\right] \tag{6-53}$$

whereas in the saturation region we have

$$I_D = \frac{I_{DSS}}{V_P^2}(V_{GS} - V_P)^2 \tag{6-54}$$

As we will see in Section 6-4-3, (6-53) and (6-54) are of the same form as the equations that describe MOSFETs, and they are therefore a particular case of the *universal FET characteristics*.

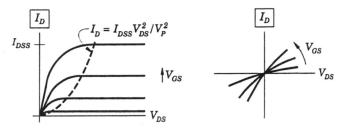

Figure 6-30. (a) Characteristics of an n-channel JFET; (b) blow-up in the ohmic region.

Figure 6-30a shows I_D as a function of V_{DS} for several values of V_{GS}. The ohmic region is shown in more detail for both positive and negative values of V_{DS} in Fig. 6-30b. For $|V_{DS}| \ll V_{GS} - V_P$ we have

$$I_D \simeq \frac{2 I_{DSS}}{V_P^2} (V_{GS} - V_P) V_{DS}$$

Near the origin, a JFET thus acts like a resistor whose value is controlled by the gate–source voltage.

In the saturation region, the effective length of the channel decreases slowly with the drain–source voltage and the drain current increases correspondingly. As in bipolar transistors, this behavior is reasonably well described by an Early voltage V_A of order 100 V. For the large-signal model of a JFET in the saturation region we then obtain

$$I_D = \frac{I_{DSS}}{V_P^2} (V_{GS} - V_P)^2 \exp \frac{V_{DS}}{V_A} \tag{6-55}$$

The small-signal model of a JFET in the saturation region is shown in Fig. 6-31. As is usually done in practice, we will ignore the Early effect when calculating the transconductance g_m; we will thus obtain g_m from (6-54) rather than (6-55):

$$g_m = \frac{\partial I_D}{\partial V_{GS}} = \frac{2 I_{DSS}}{V_P^2} (V_{GS} - V_P) \tag{6-56}$$

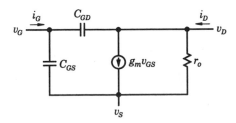

Figure 6-31. Small-signal model of a JFET.

Using (6-54) again, we see, in contrast with what we saw in transistors, that g_m is proportional to the square root of I_D and, in addition, that it depends on device parameters,

$$g_m = \frac{2\sqrt{I_{DSS}I_D}}{-V_P}$$

Since I_D is of order I_{DSS} and $-V_P$ is of order 1 V and thus much larger than $V_T = 25$ mV, it is clear that the transconductance of JFETs is much smaller than the transconductance of bipolar transistors. To calculate the output conductance g_o we must of course use (6-55), and we obtain

$$g_o = \frac{I_D}{V_A}$$

The gate–source capacitance C_{GS} is the capacitance of a reverse-biased junction and thus generally in the 10 pF range. As shown in Exercise 6-15, C_{GS} is independent of the drain current in the saturation region and given by

$$C_{GS} = \frac{eN_D WLx_M}{-3V_P} \tag{6-57}$$

In principle, the gate–drain capacitance C_{GD} is zero in the saturation region because the gate–drain voltage does not affect the charge in the channel. The existence of the Early effect tells that this is not quite so, but C_{GD} is nonetheless generally much smaller than C_{GS}.

The most notable feature in the small-signal model of a JFET compared with the small-signal model of a bipolar transistor is of course the absence of r_π. At high frequencies, however, the capacitors can conduct significantly, and the difference starts to fade. As in bipolar transistors, the figure of merit for speed is the transition frequency f_T given by

$$f_T = \frac{1}{2\pi r_m(C_{GS} + C_{GD})}$$

As a rule, JFETs designed for amplification cannot achieve the transition frequencies in the region of several GHz available in bipolar transistors.

6-4-2   The MOS Capacitor

Before proceeding to MOSFETS, we will take a good look at the MOS capacitor because it has a simpler structure but it nonetheless clearly exhibits the mechanisms of channel formation and modulation that characterize MOSFETs.

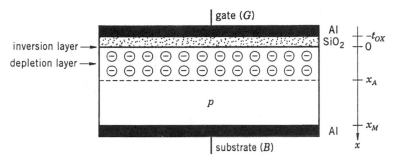

Figure 6-32. Cross section of a MOS capacitor.

The structure of an *n*-channel MOS capacitor is shown in cross section in Fig. 6-32. From top to bottom, there are layers of aluminum, silicon oxide (SiO_2), *p*-silicon, and aluminum; all junctions are assumed to be abrupt. The top aluminum layer is connected to the *gate* terminal, and the bottom aluminum layer is connected to the *substrate* or *body* terminal. We will assume that silicon oxide has zero conductivity and that the capacitor is thus in equilibrium even if there is a voltage V_{GB} between the gate and the substrate. The oxide layer is characterized only by its permittivity $\varepsilon_{OX} = 3.9\varepsilon_0$ and by its thickness t_{OX}, which we will take to be about 25 nm.

The potential ϕ relative to intrinsic silicon and the charge density ρ are shown for small values of V_{GB} in Fig. 6-33*a*. The potential ϕ_p in the neutral region of the *p*-silicon is given by

$$\phi_p = -V_T \ln\frac{N_A}{n_i}$$

As discussed in Section 6-2-2, the potential in the *p*-silicon at the junction with the aluminum layer is equal to the potential ϕ_m of aluminum relative to intrinsic silicon.

We observe first that if V_{GB} is made equal to the *flat-band voltage*

$$\psi_{FB} = \phi_p - \phi_m$$

there is no voltage across the oxide and no depletion layer next to the oxide in the *p*-silicon. If V_{GS} is larger than the flat-band voltage, there is a charge on the gate electrode and a corresponding depletion layer in the *p*-silicon. If the width of the depletion layer is x_A, the potential for $0 \le x \le x_A$ is

$$\phi(x) = \phi_p + \frac{eN_A}{2\varepsilon}(x - x_A)^2$$

The width of the depletion layer is therefore

$$x_A = \sqrt{\frac{2\varepsilon[\phi(0) - \phi_p]}{eN_A}}$$

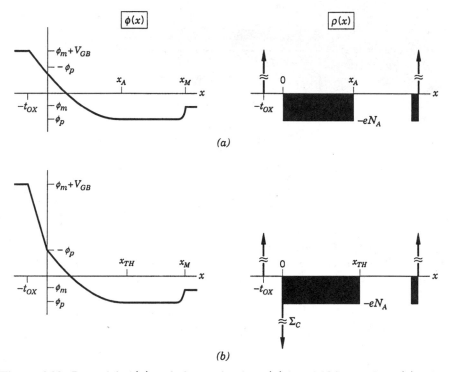

Figure 6-33. Potential $\phi(x)$ and charge density $\rho(x)$ in a MOS capacitor: (a) below threshold; (b) above threshold.

The electron density in the depletion layer is

$$n(x) = n_i \exp\frac{\phi(x)}{V_T}$$

and it increases rapidly as V_{GB}—and therefore $\phi(x)$—increases. We will define the *threshold voltage* V_{TH} as the value of V_{GB} such that

$$\phi(0) = -\phi_p = V_T \ln\frac{N_A}{n_i}$$

At threshold we have

$$n(0) = N_A$$

The threshold voltage is thus the voltage at which the electron density at the oxide–semiconductor interface becomes significant. As indicated below, however, the electron density falls off in a very short distance, so that the electron charge is negligible at threshold even though $n(0) = N_A$.

For $V_{GB} = V_{TH}$, the voltage across the depletion layer is $\psi_{TH} = -2\phi_p$, and the width x_{TH} of the depletion layer is

$$x_{TH} = \sqrt{\frac{2\varepsilon\psi_{TH}}{eN_A}}$$

The charge on the gate electrode is equal and opposite to the charge in the depletion layer, and we thus have

$$\frac{\varepsilon_{OX}}{t_{OX}}(V_{TH} + \phi_m + \phi_p) = eN_A x_{TH} = \sqrt{2e\varepsilon N_A \psi_{TH}}$$

Solving for V_{TH}, we obtain

$$V_{TH} = \psi_{FB} + \psi_{TH} + \sqrt{\psi_{BE}\,\psi_{TH}} \qquad (6\text{-}58)$$

where ψ_{BE} (typically 300 mV) is the *body-effect* voltage,

$$\psi_{BE} = 2e\varepsilon N_A \frac{t_{OX}^2}{\varepsilon_{OX}^2}$$

Above threshold, electrons accumulate in a thin *inversion layer* or *channel* at the oxide interface; in Exercise 6-13 it is shown that the inversion layer is only a few nanometers thick. Since the electron density grows exponentially with the potential, only a small increase in the width of the depletion layer and therefore of the potential at the oxide interface is needed to obtain a large increase in the inversion-layer charge; the quantitative details are worked out in Exercise 6-13.

We will make the simplifying assumption that the width of the depletion layer remains constant above threshold, so that $\phi(0)$ remains at its threshold value $-\phi_p$. The situation above threshold is then as shown in Fig. 6-33b, and it is clear that the electron surface charge density Σ_C in the channel is equal and opposite to the additional surface charge density on the gate electrode,

$$\Sigma_C = -\frac{\varepsilon_{OX}}{t_{OX}}(V_{GB} - V_{TH}) \qquad (6\text{-}59)$$

The electron channel will of course turn out to be what its name implies when we come to MOSFETs. In the present case we are dealing with an enhancement process in which a voltage must be applied between the gate and the substrate in order to induce the channel. There is another possibility, however. Let us assume that there is an immobile *interface* surface charge density Σ_I in the p-silicon at the oxide interface. Looking back at Fig. 6-33a, we observe that the only effect of Σ_I is to shift the flat-band voltage to a new value

$$\psi'_{FB} = \phi_p - \phi_m - \frac{t_{OX}}{\varepsilon_{OX}}\Sigma_I$$

and that the rest of our presentation goes through unmodified. By making Σ_I positive and sufficiently large, the threshold voltage in (6-58) can be made negative; that is, the channel can exist even for $V_{GB} = 0$. This is the basis on which depletion-mode MOSFETs are made. Positive interface surface charges due to crystal discontinuities are actually always present in a MOS capacitor; the net interface surface charge density, however, can be established by implanting additional p-type or n-type impurities in the channel region.

6-4-3 MOSFETs

If we now add ohmic source and drain contacts to an n-channel MOS capacitor, we obtain the n-channel MOSFET shown in Fig. 6-34. Symbols and sign conventions for n-channel and p-channel MOSFETs are shown in Fig. 6-35.

The MOSFET is still an equilibrium structure if we leave the drain and source terminals open circuited, in which case we have $V_{SB} = V_{DB} = 0$. Assuming that V_{GB} is above threshold, electrons are abundant in the channel, so that the junctions of the channel with the source and drain n^+-regions are ohmic. Let us now make $V_{SB} = V_{DB} = \psi_{CB}$; the channel potential relative to the substrate is then also raised by ψ_{CB}, and the situation is as shown in Fig. 6-36. The width of the depletion layer in this case is

$$x_A = \sqrt{\frac{2\varepsilon(\psi_{TH} + \psi_{CB})}{eN_A}}$$

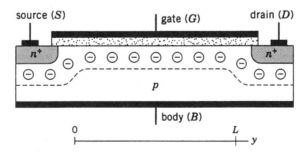

Figure 6-34. Cross section of an n-channel MOSFET.

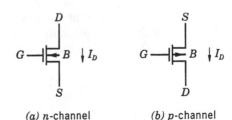

Figure 6-35. Symbols and sign conventions for MOSFETs.

(a) n-channel   (b) p-channel

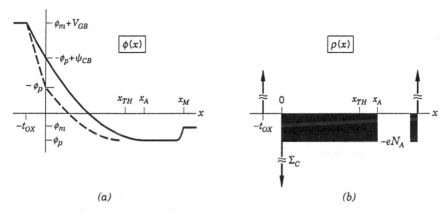

Figure 6-36. (*a*) Potential and (*b*) charge density in a MOSFET above threshold when the channel-substrate voltage is raised by an amount ψ_{CB} above its equilibrium value.

and the surface charge density in the depletion layer is

$$\Sigma_A = -eN_A x_A = -\sqrt{2e\varepsilon N_A(\psi_{TH}+\psi_{CB})}$$

But from Gauss's law we have

$$\Sigma_C+\Sigma_A = -\frac{\varepsilon_{OX}}{t_{OX}}(V_{GB}+\phi_m+\phi_p-\psi_{CB}) = -\frac{\varepsilon_{OX}}{t_{OX}}(V_{GB}-\psi_{FB}-\psi_{TH}-\psi_{CB})$$

and we then obtain

$$\Sigma_C = -\frac{\varepsilon_{OX}}{t_{OX}}\left(V_{GB}-\psi_{FB}-\psi_{TH}-\psi_{CB}-\sqrt{\psi_{BE}(\psi_{TH}+\psi_{CB})}\right)$$

The channel charge thus decreases in magnitude as ψ_{CB} increases. To a lesser extent, the charge in the depletion layer increases in magnitude.

As we did in the case of JFETs, we will use the gradual-channel approximation to calculate the drain current. To do so, we will first rewrite the channel surface charge density in terms of voltages relative to the source:

$$\Sigma_C = -\frac{\varepsilon_{OX}}{t_{OX}}\left[V_{GS}-\psi_{FB}-\psi_{TH}-\psi_{CS}-\sqrt{\psi_{BE}(\psi_{TH}+\psi_{CS}-V_{BS})}\right] \quad (6\text{-}60)$$

We will next ignore the transverse variation in the depletion layer by setting $\psi_{CS}=0$ in the square-root term, so that we are left with

$$\Sigma_C = -\frac{\varepsilon_{OX}}{t_{OX}}\left[V_{GS}-\psi_{FB}-\psi_{TH}-\psi_{CS}-\sqrt{\psi_{BE}(\psi_{TH}-V_{BS})}\right]$$

Finally, we will redefine the threshold voltage V_{TH} as

$$V_{TH} = \psi_{FB} + \psi_{TH} + \sqrt{\psi_{BE}(\psi_{TH} - V_{BS})} \tag{6-61}$$

The expression for the channel surface charge density then becomes

$$\Sigma_C = -\frac{\varepsilon_{OX}}{t_{OX}}(V_{GS} - V_{TH} - \psi_{CS}) \tag{6-62}$$

Let us now assume that $V_{DS} > 0$, so that ψ_{CS} varies from 0 to V_{DS}. At any point along the channel the (constant) drain current is

$$I_D = -W\mu_n \Sigma_C \frac{d\psi_{CS}}{dy} \tag{6-63}$$

where W is the channel width. If the channel length is L, we have

$$I_D = -\frac{W\mu_n}{L}\int_0^L \Sigma_C \frac{d\psi_{CS}}{dy}\,dy = -\frac{W\mu_n}{L}\int_0^{V_{DS}} \Sigma_C\,d\psi_{CS}$$

For $0 \le V_{DS} \le V_{GS} - V_{TH}$ we then have

$$I_D = K\left[(V_{GS} - V_{TH})V_{DS} - \frac{V_{DS}^2}{2}\right]$$

where

$$K = \frac{\mu_n W}{L}\frac{\varepsilon_{OX}}{t_{OX}}$$

When V_{DS} becomes equal to $V_{GS} - V_{TH}$, the channel surface charge density in (6-62) becomes zero at the drain end of the channel, and I_D levels off at higher values of V_{DS}. As in the case of JFETs, we will say that the channel is pinched off and that we are in the saturation region. For most purposes we will assume that the drain current is constant above pinch-off. For $0 \le V_{GS} - V_{TH} < V_{DS}$ we then have

$$I_D = \frac{K}{2}(V_{GS} - V_{TH})^2 \tag{6-64}$$

Figure 6-37a shows I_D as a function of V_{DS} for several values of $V_{GS} - V_{TH}$. For small values of V_{DS}, MOSFETs, like JFETs, act like resistors whose value is controlled by the gate–source voltage; their behavior near the origin is thus as shown in Fig. 6-30b. Enhancement MOSFETs can be diode connected by shorting the gate and the drain; the I–V curve is then as shown in Fig. 6-37b.

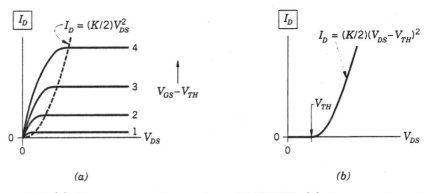

(b)

Figure 6-37. (a) Characteristics of an n-channel MOSFET; (b) characteristic in the diode connection.

The effective length of the channel becomes smaller as V_{DS} grows, and the drain current thus increases slowly with V_{DS} above pinch-off. Once again, this behavior can be described in terms of an Early voltage V_A; for the large-signal model of a MOSFET in the saturation region we then obtain

$$I_D = \frac{K}{2}(V_{GS}-V_{TH})^2 \exp\frac{V_{DS}}{V_A} \tag{6-65}$$

The small-signal model of a MOSFET in the saturation region is shown in Fig. 6-38. Taking the source as the reference, for static incremental signals we have

$$i_D = g_m v_{GS} + g'_m v_{BS} + g_o v_{DS} \tag{6-66}$$

where

$$g_m = \frac{\partial I_D}{\partial V_{GS}} \qquad g'_m = \frac{\partial I_D}{\partial V_{BS}} \qquad g_o = \frac{\partial I_D}{\partial V_{DS}}$$

In contrast with JFETs, in which the substrate is usually tied to the source, there is a transconductance g'_m associated with the substrate–source voltage.

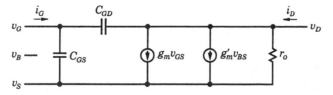

Figure 6-38. Small-signal model of a MOSFET.

Ignoring the Early effect, from (6-64) we obtain

$$g_m = \frac{\partial I_D}{\partial V_{GS}} = K(V_{GS} - V_{TH}) = \sqrt{2KI_D}$$

and

$$g'_m = \frac{\partial I_D}{\partial V_{BS}} = -K(V_{GS} - V_{TH})\frac{\partial V_{TH}}{\partial V_{BS}} \equiv \chi g_m \qquad (6\text{-}67)$$

From (6-61) we have

$$\chi = -\frac{\partial V_{TH}}{\partial V_{BS}} = \frac{1}{2}\sqrt{\frac{\psi_{BE}}{(\psi_{TH} - V_{BS})}}$$

For $V_{BS} = 0$, χ is generally between 0.1 and 0.3. The output conductance g_o is obtained from (6-65),

$$g_o = \frac{I_D}{V_A}$$

The gate–source capacitance C_{GS} is determined by the charge Q_G on the gate electrode, which is essentially equal and opposite to the channel charge. We thus have

$$Q_G = -W\int_0^L \Sigma_C \, dy$$

Substituting $dy = -(W\mu_n/I_D)\Sigma_C \, d\psi_{CS}$ from (6-63), we have

$$Q_G = \frac{W^2\mu_n}{I_D}\int_0^{V_{GS}-V_{TH}} \Sigma_C^2 \, d\psi_{CS}$$

Substituting Σ_C from (6-62) and then using (6-64) on the result, we obtain

$$Q_G = \frac{2}{3}WL\frac{\varepsilon_{OX}}{t_{OX}}(V_{GS} - V_{TH})$$

The gate–source capacitance is thus independent of the drain current, as it is in JFETs,

$$C_{GS} = \frac{2}{3}WL\frac{\varepsilon_{OX}}{t_{OX}}$$

Also as in JFETs, the gate–drain capacitance C_{GD} is in principle zero because the drain and the channel are decoupled in the saturation region.

C_{GD} thus depends on secondary effects, mainly the overlap of the gate electrode and the drain region, and is generally much smaller than C_{GS}. There are other interelectrode capacitances (between the gate and the substrate, for example), but they are generally small compared with C_{GS}, and in complete circuits they are either shorted out or in parallel with substantially larger capacitances.

Assuming then that C_{GS} is the main contributor to the transition frequency $f_T = \omega_T/2\pi$, and using the Einstein relation, we obtain

$$\omega_T^{-1} \simeq \frac{C_{GS}}{g_m} = \frac{2L^2}{3D_n} \frac{V_T}{V_{GS}-V_{TH}}$$

We note, however, that for L below 1 μm, the electric field in the channel is large enough that the drift velocity saturates, and ω_T^{-1} becomes proportional to L rather than L^2. In *npn* bipolar transistors, the minimum value of ω_T^{-1} is approximately equal to the base charging time τ_B given in (6-40),

$$\tau_B = \frac{w_B^2}{2D_n}$$

In both cases, device speed is determined by the square of a characteristic length; in MOSFETs, in addition, speed increases with $V_{GS}-V_{TH}$. Speed is also determined by a diffusion constant, which favors *npn* transistors and *n*-channel MOSFETs. For $V_{GS}-V_{TH} = 1$ V, $L = 1$ μm, and $D_n = 15$ cm^2/s, we obtain $\omega_T^{-1} = 11$ ps, comparable to what we obtain in an *npn* bipolar transistor with $w_B = 0.2$ μm.

EXERCISES

6-1 LABORATORY I: TRANSISTOR CHARACTERISTICS. Assemble the circuit of Fig. 6-39 using small-signal transistors, 2N3904 or similar. I_C is measured on a 10 kΩ resistor in order to retain 1% accuracy at currents down to 1 μA using easily available 3-digit voltmeters.

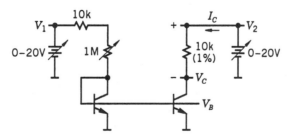

Figure 6-39. Circuit used in Exercise 6-1 to obtain the parameters of an *npn* transistor.

(a) Vary V_1 and measure I_C and V_B at 10 points for I_C between 1 μA and 1 mA. Make the ratio between successive currents about 2 and adjust V_2 so as to keep V_C equal to 1V. Plot I_C versus V_B on a log-lin scale and extract I_S ($\approx$ 5 fA) and V_T (25 mV).

(b) Set $I_C = 1$ μA and estimate V_A ($\approx$ 100V) by increasing V_C from 1V to 10V. Repeat for 10 μA, 100 μA, and 1 mA. Verify that V_A is reasonably independent of I_C.

6-2 LABORATORY II: MOSFET CHARACTERISTICS.

(a) Assemble the circuit of Fig. 6-40a using a low-power MOSFET with $V_{TH} \approx 1$V. Measure I_D as a function of V_G and extract K and V_{TH} by fitting to the formula for the drain current in the saturation region,

$$I_D = \frac{K}{2}(V_G - V_{TH})^2$$

(b) Use the circuit of Fig. 6-40b to measure I_D and V_D in the ohmic region for several values of V_G and see to what point

$$I_D = K(V_G - V_{TH})V_D$$

You might need to adjust K to make things come out right.

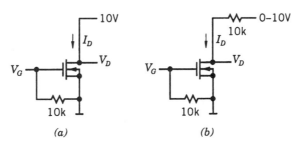

(a)   (b)

Figure 6-40. Circuits used in Exercise 6-2 to obtain the parameters of an n-channel enhancement MOSFET.

6-3 LABORATORY III: STORED CHARGE IN DIODES. Assemble the circuit of Fig. 6-41 and verify that the outputs are as shown on a time scale of several microseconds. With fast equipment you should be able to observe that a 1N4148 also has a finite *reverse recovery time* t_{rr}, but that t_{rr} is much shorter than in a 1N4006, nanoseconds rather than microseconds.

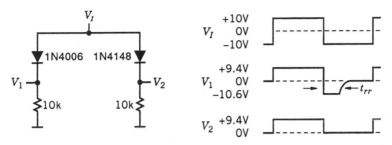

Figure 6-41. Circuit for Exercise 6-3.

6-4 DIFFUSION CURRENTS. Consider a sample of n-silicon whose electron density increases linearly in the z-direction,

$$n = n(0) + z\frac{dn}{dz}(0)$$

Assume that the electrons collide occasionally with the lattice ions and that the mean free time between collisions is τ_F. Assume further that collisions are elastic, and uniform in angular distribution. Now consider an element of area dA located at the origin in the x–y plane. In polar coordinates, the average number of collisions per unit time of electrons with velocities between v and $v+dv$ in a ring of volume $2\pi r^2 \sin\theta\, d\theta\, dr$ is

$$\frac{np(v)}{\tau_F}2\pi r^2 \sin\theta\, d\theta\, dr\, dv$$

where $p(v)$ is the probability density of the velocity. Of the electrons that have suffered collisions, the fraction aimed at dA is

$$\frac{dA\cos\theta}{4\pi r^2}$$

and the probability that they cross dA without further collisions is

$$\exp\left(-\frac{r}{v\tau_F}\right)$$

Do the integrals, and show that the rates per unit area at which electrons cross the x–y plane in the positive and negative z-directions are

$$\frac{dN^{\pm}}{dt} = \frac{n(0)}{4}\langle v\rangle \mp \frac{\tau_F}{6}\frac{dn}{dz}(0)\langle v^2\rangle$$

where

$$\langle v \rangle = \int_0^\infty v p(v)\, dv \quad \text{and} \quad \langle v^2 \rangle = \int_0^\infty v^2 p(v)\, dv$$

The electron current density is therefore

$$J_n = eD \frac{dn}{dz}$$

where D is the diffusion constant,

$$D = \frac{\tau_F}{3} \langle v^2 \rangle$$

Note incidentally that the rate per unit area at which electrons cross any plane in one direction is $n\langle v \rangle / 4$.

6-5 QUASINEUTRALITY IN THE SHORT-BASE LIMIT.

Consider a uniform region of n-silicon with an injecting contact at $x = 0$ and an ohmic contact at $x = w$, and assume that w is much larger than the extrinsic Debye length L_D and that $\tau_p = \infty$, so that you are in the short-base limit. In the quasistatic case, the majority-carrier diffusion equation in (6-17) then reads

$$-L_D^2 \frac{d^2 n'}{dx^2} + n' = p'$$

The solutions $\exp(-x/L_D)$ and $\exp[(x - w)/L_D]$ of the homogeneous equation take care of whatever boundary conditions are imposed on n' and become negligible a few Debye lengths away from $x = 0$ and $x = w$. Over most of the region you are then left with the particular solution

$$n' = p'$$

Quasineutrality thus holds exactly except near the boundaries. From Gauss's law (6-10) you have $dE/dx = 0$, so that E is constant, and so therefore is the electron drift current. The total current density J is thus also constant, as required by conservation of charge in the quasistatic case.

If you wish to go further, consider showing—again far from the boundaries—that $e\partial(p' - n')/\partial t = -\partial J/\partial x$ is much smaller than $J(0)/w$ at frequencies such that $[\omega^3 \tau_D^2 (w^2/2D_p)]^{1/2} \ll 1$. This condition is essentially always satisfied, so that J is known everywhere if it is known at one point. To simplify matters, assume that the electron current is zero at $x = 0$.

6-6 QUASINEUTRALITY IS SELF-CONSISTENT. Consider the quasi-static long-base solution of the minority-carrier diffusion equation (6-15) in n-silicon when the injection level is low. You then have

$$p' = p(0) \exp\left(-\frac{x}{L_p}\right)$$

For $x \gg L_D = \sqrt{\varepsilon V_T/eN_D}$, that is, assuming that the homogeneous solution has died out, use the majority-carrier diffusion equation (6-17) to show that

$$n' = \frac{1-\left(L_D^2/L_p^2\right)\left(D_p/D_n\right)}{1-\left(L_D^2/L_p^2\right)} p'$$

and that

$$p'-n' \simeq -\frac{D_n-D_p}{D_n}\frac{L_D^2}{L_p^2} p'$$

Now *assume* quasineutrality, that is, that

$$n' \simeq p' = p(0) \exp\left(-\frac{x}{L_p}\right)$$

The electron and hole density gradients are also approximately equal, so that

$$J_p^D = e\frac{D_p}{L_p}p' \quad \text{and} \quad J_n^D \simeq -e\frac{D_n}{L_p}p'$$

The total current density is constant,

$$J = J_n + J_p$$

and the electron drift-current density J_n^E is

$$J_n^E = J_n - J_n^D$$

The electric field is then given by

$$eN_D \mu_n E \simeq J + e\frac{D_n-D_p}{L_p}p'$$

Use Gauss's law and the Einstein relation to show that you recover the result you got above,

$$p' - n' = \frac{\varepsilon}{e} \frac{dE}{dx} \simeq -\frac{D_n - D_p}{D_n} \frac{L_D^2}{L_p^2} p'$$

6-7   RECOMBINATION (Shockley and Read, 1952; Hall, 1952; Sah, Noyce, and Shockley, 1957). If recombination is dominated by direct processes in which only a hole and an electron take part, the (net) recombination rate is of the form

$$R_D = (np - n_0 p_0) r_D(T)$$

Assuming further that holes and electrons in the depletion region of a diode are in equilibrium with the potential barrier, the np product in this region is constant and given by (6-26),

$$np = n_i^2 \exp \frac{V}{V_T}$$

The recombination rate is therefore also constant,

$$R_D = n_i^2 \left(\exp \frac{V}{V_T} - 1 \right) r_D(T)$$

At low forward bias ($V \ll \psi$) the width of the depletion region varies little, and the recombination current should thus be proportional to the diffusion current. This is contrary to experiment, which shows that the recombination current varies roughly like $n_i \exp(V/mV_T)$, where m is near 2.

In the Shockley–Read–Hall theory, generation and recombination proceed through *recombination centers* or *traps*, which in the simplest case can be thought of as impurities with an activation energy (minimum ionization energy) $\mathcal{E}_T$ that is a fraction of the bandgap energy $\mathcal{E}_G$. Further simplification is achieved by assuming that second ionization of a trap is negligible: traps are either *occupied* (not ionized) or *empty* (ionized once).

Traps can capture and emit conduction electrons, and you have the processes

empty trap + conduction electron ⇌ occupied trap

The process to the right is (conduction) *electron capture*, and its inverse is (conduction) *electron emission*. Traps can also capture and emit valence electrons, and you have the processes (with activation energy $\mathcal{E}_G - \mathcal{E}_T$)

empty trap + valence electron ⇌ occupied trap

which can also be described as

empty trap ⇌ occupied trap + hole

The process to the right is *hole emission*, and its inverse is *hole capture*.

In this picture, recombination is a two-step process,

(empty trap + electron) + hole → occupied trap + hole → empty trap

or

(occupied trap + hole) + electron → empty trap + electron → occupied trap

If f is the fraction of occupied traps, the electron capture rate (assumed to be independent of the doping level) is proportional to n and to $1-f$ and can be written as

$$C_n = \frac{n(1-f)}{\tau_n}$$

where τ_n involves, among other things, the density of traps and the capture cross section for electrons. The electron emission rate is proportional to f and can be written as

$$E_n = \frac{n^*f}{\tau_n}$$

where n^* is a constant to be determined. The (net) recombination rate for electrons R_n is therefore

$$R_n = \frac{n(1-f) - n^*f}{\tau_n}$$

For hole emission and capture the roles of f and $1-f$ are reversed, and the (net) recombination rate for holes is

$$R_p = \frac{pf - p^*(1-f)}{\tau_p}$$

You can determine n^* and p^* partially by applying the principle of detailed balance and requiring that $R_n = R_p = 0$ in thermal equilibrium, that is, that

$$n_0(1-f) = n^*f$$

and

$$p_0 f = p^*(1-f)$$

Multiplying these equalities yields

$$n^*p^* = n_0 p_0 = n_i^2$$

It should not be hard to accept that $n^* = p^* = n_i$ if $\mathcal{E}_t = \mathcal{E}_G/2$, that is, if electron emission and hole emission have the same activation energy. Assume that this is the case, so that

$$R_n = \frac{n(1-f) - n_i f}{\tau_n}$$

$$R_p = \frac{pf - n_i(1-f)}{\tau_p}$$

In a steady state you have $R_n = R_p = R$. Show that this leads to

$$f = \frac{n\tau_p + n_i \tau_n}{(n+n_i)\tau_p + (p+n_i)\tau_n}$$

You can now obtain the expression for R in (6-13),

$$R = \frac{np - n_i^2}{(n+n_i)\tau_p + (p+n_i)\tau_n}$$

Under low-level injection conditions in extrinsic material you have

$$n' \simeq p' \ll n_0 + p_0 \simeq |N_D - N_A|$$

Substitute $n_0 + n'$ and $p = p_0 + p'$, and show that in p-type material

$$R \simeq \frac{n - n_0}{\tau_n} = \frac{n'}{\tau_n}$$

If you think about it a bit, you can see why: holes are abundant, and you can expect that as soon as a trap is occupied it is emptied by a hole; most traps are thus empty, and the recombination rate is limited by the much slower rate of electron capture into empty traps. From symmetry it is clear that in n-type material you have

$$R \simeq \frac{p - p_0}{\tau_p} = \frac{p'}{\tau_p}$$

This and the preceding expression are the ones given in (6-12).

6-8 DEPLETION CAPACITANCE. Show that the depletion capacitance of a p-n junction can be written as

$$C_T = \frac{\varepsilon A}{x_A + x_D}$$

Figure 6-42. Incremental charges in a *p-n* junction for a small increase in the forward bias (Exercise 6-8).

With the aid of Fig. 6-42 convince yourself that this is what you should expect since the charge between $-x_A$ and x_D does not change when you increase the voltage across the junction.

6-9 RECOMBINATION CURRENT IN A FORWARD-BIASED JUNCTION. Consider the idealized case of an abrupt *p-n* junction in which $N_A = N_D = N$, $D_n = D_p = D$, $\tau_n = \tau_p = \tau$, and $L_n = L_p = L$. If electrons and holes are in equilibrium with the potential barrier, the *np* product in the depletion layer when the junction is forward biased by more than a few times V_T is

$$np = n_i^2 \exp\frac{V}{V_T} \gg n_i^2$$

In this case, the recombination rate R given in (6-13) becomes

$$R = \frac{np}{(n+p)\tau}$$

Refer to Fig. 6-10 and convince yourself that R has a sharp peak at

$$n = p = n_i \exp\frac{V}{2V_T}$$

The peak has a maximum value

$$\frac{n_i}{2\tau}\exp\frac{V}{2V_T}$$

and it decreases by a factor $\exp(1) = 2.72$ when the potential changes by V_T on either side of the maximum. The width of the peak is thus approximately

$$\frac{2V_T}{E}$$

where E is the magnitude of the electric field at the maximum,

$$E = \frac{eN_A x_A}{\varepsilon} = \sqrt{\frac{e(\psi-V)N}{\varepsilon}}$$

The recombination current density in the depletion layer is therefore

$$J_r = \frac{eV_T n_i}{\tau\sqrt{e(\psi-V)N/\varepsilon}} \exp\frac{V}{2V_T}$$

On the other hand, the diffusion current density is

$$J_d = \frac{2en_i^2 D}{Nw} \exp\frac{V}{V_T}$$

where w is the effective width of the neutral regions. For $V \ll \psi$ you can now obtain

$$\frac{J_r}{J_d} = \frac{N}{2n_i}\frac{w}{L^2}\frac{V_T}{\sqrt{e\psi N/\varepsilon}}\exp\left(-\frac{V}{2V_T}\right) = K\exp\left(-\frac{V}{2V_T}\right)$$

Take $N = 10^{16}$ cm^{-3}, $\tau = 1$ μs, and $D = 25$ cm^2/s ($L = 5 \cdot 10^{-3}$ cm and $\psi = 0.7$ V). In the case of the base–emitter junction of a medium-frequency transistor, you have $w = 1$ μm and $K = 1.5$; the recombination current thus does not affect the collector current. In the base current, however, the diffusion component is β times smaller and K is of order β; the recombination current therefore becomes dominant for base–emitter voltages under $10V_T$ or so.

Minority-carrier lifetimes in fast diodes are small by design, and recombination is dominant; you should thus expect the slope in the I–V characteristic to be closer to 120 mV/decade than to the 60 mV/decade predicted by the Shockley model, and this is in fact usually the case.

6-10   REVERSE CURRENT IN A p-n JUNCTION. Consider a long-base diode with $N_A = N_D = N$, $\tau_n = \tau_p = \tau$, $D_n = D_p = D$, and $L_n = L_p = L$. In the Shockley diffusion model, the current in such a diode is

$$I = I_S\left[\exp\frac{V}{V_T} - 1\right]$$

The current for a reverse bias of a few V_T is thus constant and very small,

$$I_R = I_S = \frac{2en_i^2 DA}{NL}$$

You will now see that electron–hole pair generation in the depletion region gives rise to much larger reverse currents. Over most of the depletion region of a junction with a large reverse bias you have

$$n, p \ll n_i$$

Use (6-13) to show that the generation rate for electron−hole pairs in the depletion region is

$$-R = \frac{n_i}{2\tau}$$

For every pair generated, one electron must circulate in the external circuit; the reverse current is thus

$$I_R = \frac{e n_i A x}{2\tau}$$

where x is the width of the depletion region,

$$x = 2\sqrt{\frac{\varepsilon(\psi - V)}{eN}}$$

You then have

$$\frac{I_R}{I_S} = \frac{N}{2n_i} \frac{\sqrt{\varepsilon V_T / eN}}{L} \sqrt{\frac{(\psi - V)}{V_T}} = K\sqrt{\frac{(\psi - V)}{V_T}}$$

Take $N = 10^{16}$ cm^{-3}, $\tau = 1$ μs, and $D = 25$ cm^2/s; you then have $K \approx 400$. I_R is thus much larger than I_S; in addition, it grows slowly with reverse bias.

6-11   QUASINEUTRALITY IN BIPOLAR TRANSISTORS. If recombination is negligible, the hole current density in the base of an *npn* transistor is negligible for quasistatic excitations, whatever the injection level. Use (6-11) to show that in this case

$$E = \frac{V_T}{p} \frac{dp'}{dx}$$

Use this expression in Gauss's law (6-10) to obtain the diffusion equation for holes at any injection level,

$$p' - \frac{d}{dx}\left(\frac{\varepsilon V_T}{ep} \frac{dp'}{dx} \right) = n'$$

If the base width is w_B, convince yourself that $p' \simeq n'$ if $L_A^2 / w_B^2 \ll 1$, where $L_A = \sqrt{\varepsilon V_T / e N_{AB}}$ is the extrinsic Debye length. *Suggestion:* Integrate $p' - n'$.

6-12   HIGH-LEVEL INJECTION IN AN *npn* TRANSISTOR. The injection level in an *npn* transistor is considered to be high when the electron density in the base is of the order of the acceptor density N_{AB}. The carrier densities in the case $n(0) = 3N_{AB}$ are sketched in Fig. 6-43. When the

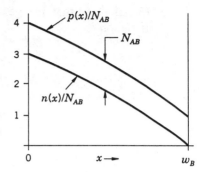

Figure 6-43. Charge-carrier densities in the base of an *npn* transistor when the injection level is high (Exercise 6-12). In the case illustrated, $n(0) = 3N_{AB}$ and $p(0) \simeq 4N_{AB}$.

injection level is high, it is no longer true that electrons flow by diffusion only, and you have

$$J_n = en\frac{D_n}{V_T}E + eD_n\frac{dn}{dx}$$

The quasistatic hole current density in the base is negligible; this allows you to calculate the electric field if you assume that the base is uniformly doped and therefore quasineutral, so that $p \simeq n + N_{AB}$. In this case you have

$$p\frac{D_p}{V_T}E = D_p\frac{dp}{dx} \simeq D_p\frac{dn}{dx}$$

or

$$E = \frac{V_T}{p}\frac{dn}{dx}$$

The electron current density is then

$$J_n = eD_n\left(1 + \frac{n}{p}\right)\frac{dn}{dx} = eD_n\frac{2n + N_{AB}}{n + N_{AB}}\frac{dn}{dx}$$

If you integrate from 0 to w_B you obtain

$$J_n w_B = -eD_n\left[2n(0) - N_{AB}\ln\frac{n(0) + N_{AB}}{N_{AB}}\right]$$

From (6-26) you have

$$n(0)\left[n(0) + N_{AB}\right] = n_i^2 \exp\frac{V}{V_T}$$

For $n(0) \ll N_{AB}$, that is, for low-level injection, show that you recover the Shockley model,

$$J_n \simeq -\frac{eD_n n_i^2}{w_B N_{AB}} \exp \frac{V}{V_T}$$

For $n(0) \gg N_{AB}$, show that

$$J_n \simeq -\frac{2eD_n n_i}{w_B} \exp \frac{V}{2V_T}$$

This last result explains in part the deviations from the Shockley model that you observe in Fig. 6-20.

6-13   THE INVERSION LAYER IN MOS STRUCTURES IS THIN. The objective of this exercise is to show, as suggested in Fig. 6-44, that the inversion layer in MOS structures is a few nanometers thick, and to justify that the width of the depletion layer remains almost constant above threshold.

Start by calculating some useful numbers for an n-channel MOS capacitor. Take $N_A = 2 \cdot 10^{16}$ cm^{-3}, $\varepsilon = 1$ pF/cm, $\varepsilon_{OX} = 0.35$ pF/cm, and $t_{OX} = 25$ nm, and obtain

$$\psi_{TH} = 2V_T \ln \frac{N_A}{n_i} = 725 \text{ mV}$$

$$x_{TH} = \sqrt{\frac{2\varepsilon\psi_{TH}}{eN_A}} = 2.1 \cdot 10^{-5} \text{ cm}$$

$$L_A = \sqrt{\frac{\varepsilon V_T}{eN_A}} = 2.8 \cdot 10^{-6} \text{ cm}$$

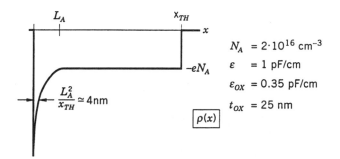

$$
\begin{array}{ll}
N_A & = 2 \cdot 10^{16} \text{ cm}^{-3} \\
\varepsilon & = 1 \text{ pF/cm} \\
\varepsilon_{OX} & = 0.35 \text{ pF/cm} \\
t_{OX} & = 25 \text{ nm}
\end{array}
$$

Figure 6-44. Charge density in an n-channel MOS capacitor.

and from these,

$$\frac{x_{TH}^2}{2L_A^2} = \frac{\psi_{TH}}{V_T} = 29$$

$$\frac{L_A^2}{x_{TH}} = 3.7 \cdot 10^{-7} \text{ cm}$$

The potential due to the depletion layer is

$$\phi(x) = \phi_p + \frac{eN_A}{2\varepsilon}(x-x_A)^2 = \phi_p + V_T \frac{(x-x_A)^2}{2L_A^2}$$

You can now obtain the built-in potential

$$\psi = \phi(0) - \phi_p = V_T \frac{x_A^2}{2L_A^2}$$

and the electron density

$$n(x) = n_i \exp\left[\frac{\phi(x)}{V_T}\right] = \frac{n_i^2}{N_A} \exp\left[\frac{(x-x_A)^2}{2L_A^2}\right]$$

Convince yourself that for $x \ll L_A$,

$$n(x) \simeq \frac{n_i^2}{N_A} \exp\left(\frac{x_A^2}{2L_A^2}\right) \exp\left(-\frac{x_A}{L_A^2}x\right)$$

The decay constant is $L_A^2/x_A \ll L_A$, which is consistent with the assumption. The channel charge per unit area is

$$\Sigma_C = -e\int_0^{x_A} n(x)\, dx \simeq -e\int_0^\infty n(x)\, dx = -e\frac{n_i^2}{N_A}\frac{L_A^2}{x_A}\exp\left(\frac{x_A^2}{2L_A^2}\right)$$

On the other hand, according to (6-59), which assumes that $x_A = x_{TH}$ above threshold,

$$\Sigma_C \simeq -\frac{\varepsilon_{OX}}{t_{OX}}(V_{GS} - V_{TH})$$

and you can estimate the value of x_A from

$$\exp\left(\frac{x_A^2}{2L_A^2}\right) \simeq -\frac{\Sigma_C}{e}\frac{N_A}{n_i^2}\frac{x_{TH}}{L_A^2}$$

At $V_{GS}-V_{TH}=1$V you have

$$\Sigma_C \simeq -1.4\cdot10^{-7}\,\text{C/cm}^2$$

and

$$\frac{x_A^2}{2L_A^2} \simeq 33.8$$

This represents a small (8%) increase in x_A, consistent with the assumption that x_A remains approximately constant above threshold, and an increase in the built-in potential that is small compared with $V_{GS}-V_{TH}=1$V,

$$\Delta\psi = V_T\frac{x_A^2-x_{TH}^2}{2L_A^2} = 120\,\text{mV}$$

So far you have implicitly assumed that the potential $\phi_C(x)$ due to the channel charge is negligible. Integrate the channel charge density $-en(x)$ twice using the conditions $\phi_C'(0)=\Sigma_C/\varepsilon$ and $\phi_C(\infty)=0$, and obtain

$$\phi_C(x) = -\frac{\Sigma_C}{\varepsilon}\frac{L_A^2}{x_A}\exp\left(-\frac{x_A}{L_A^2}x\right)$$

For $V_{GS}-V_{TH}=1$V you have

$$-\frac{\Sigma_C}{\varepsilon}\frac{L_A^2}{x_A} \simeq 50\,\text{mV} = 2V_T$$

The total potential thus drops by $3V_T$ rather than V_T in a distance L_A^2/x_A, which means that the inversion layer is even thinner than you estimated above. At this point you might consider using an integrating factor $d\phi/dx$ in the Poisson–Boltzmann equation (6-20) and integrating from $x=0$ to $x=\infty$ to obtain $d\phi/dx$ at $x=0$, and therefore a more accurate estimate of the width of the inversion layer. For the record, the result is

$$\left.\frac{d\phi}{dx}\right|_{x=0} = -\sqrt{2}\,\frac{V_T}{L_A}\sqrt{\frac{\phi(0)-\phi_p}{V_T}+\exp\frac{\phi(0)+\phi_p}{V_T}-1}$$

6-14   SUBTHRESHOLD CONDUCTION IN MOSFETS. In Exercise 6-13 you obtained that the channel charge per unit area in a MOS capacitor is

$$\Sigma_C = -e\frac{n_i^2}{N_A}\frac{L_A^2}{x_A}\exp\left(\frac{x_A^2}{2L_A^2}\right) = -e\frac{n_i^2}{N_A}\frac{L_A^2}{x_A}\exp\left(\frac{\psi}{V_T}\right)$$

where with reference to Fig. 6-33b, $\psi=\phi(0)-\phi_p$ is the built-in potential.

The channel charge does not suddenly drop to zero below threshold, and you should therefore expect the same behavior in the drain current of a MOSFET.

Using the data in Exercise 6-13, show that the channel charge per unit area at threshold is

$$\Sigma_C(V_{TH}) = -1.1 \cdot 10^{-9} \ \text{C/cm}^2$$

and thus small compared with the charge per unit area in the depletion layer,

$$\Sigma_A(V_{TH}) = -eN_A x_{TH} = -\sqrt{2\varepsilon\psi_{TH}eN_A} = -6.8 \cdot 10^{-8} \ \text{C}$$

Use this fact and Gauss's law to show that

$$[V_{GB} + \phi_m - \phi(0)]\frac{\varepsilon_{OX}}{t_{OX}} = eN_A x_A = \sqrt{2\varepsilon eN_A[\phi(0) - \phi_p]}$$

and that you therefore get a generalization of (6-58),

$$V_{GB} = \psi_{FB} + \psi + \sqrt{\psi_{BE}\psi}$$

Using $\psi_{TH} = 725$ mV and $\psi_{BE} = 300$ mV, plot $V_{GB} - V_{TH}$ for a few values of ψ between zero and ψ_{TH} and convince yourself that for $-\psi_{TH} \leq V_{GB} - V_{TH} \leq 0$,

$$\psi \simeq \psi_{TH} + \frac{V_{GB} - V_{TH}}{m}$$

where $m \simeq 1.4$. Since x_A is a slowly varying function of ψ, it follows that Σ_C decreases exponentially for a few hundred millivolts below threshold; this result also holds for the saturation drain current in MOSFETs.

6-15 CHANNEL WIDTH AND GATE–SOURCE CAPACITANCE IN JFETS. From (6-50) and (6-49) you have

$$\frac{d\psi_{CS}}{dy} = \frac{I_D}{\sigma W x_C(y)} = \frac{-2V_P I_D}{\sigma W x_M(-V_P + V_{GS} - \psi_{CS})}$$

Integrate this differential equation subject to the boundary condition that $\psi_{CS} = 0$ at $y = 0$, and obtain

$$(-V_P + V_{GS})^2 - (-V_P + V_{GS} - \psi_{CS})^2 = -\frac{4V_P I_D L}{\sigma W x_M}\frac{y}{L}$$

Using (6-49) and (6-51), now obtain

$$(-V_P + V_{GS})^2 - \left(-2V_P\frac{x_C}{x_M}\right)^2 = (-V_P + V_{GS})^2\frac{y}{L}$$

Conclude that

$$x_C(y) = \frac{x_M}{2} \frac{-V_P+V_{GS}}{-V_P} \sqrt{1-\frac{y}{L}}$$

The electron charge in the channel is

$$Q = -eN_DW\int_0^L x_C(y)\,dy = \frac{eN_DWx_M L}{3V_P}(-V_P+V_{GS})$$

It follows that the gate–source capacitance C_{GS} is as given in (6-57),

$$C_{GS} = \frac{eN_DWLx_M}{-3V_P}$$

6-16 QUASI-EQUILIBRIUM IN THE POTENTIAL BARRIER. The point of this exercise is to show that the drift currents in the depletion region of a *p-n* junction are much larger than the diffusion currents in the neutral regions. Consider a junction with $N_A = N_D = N$ and $D_n = D_p = D$. In the short-base limit, the hole diffusion current density on the *n*-side for a large forward bias is

$$J^D = \frac{en_i^2 D}{Nw}\exp\frac{V}{V_T}$$

where w is the width of the neutral regions. From Gauss's law and (6-25), the electric field at the center of the depletion region (of width $2x$) is

$$-E_C = \frac{e}{\varepsilon}Nx = \frac{e}{\varepsilon}\sqrt{\frac{(\psi-V)\varepsilon N}{e}} = \sqrt{\frac{(\psi-V)eN}{\varepsilon}}$$

Assume a forward bias large enough that $\psi - V \le 50$ mV. For $N = 10^{16}$ cm^{-3}, check that $-E_C$ is under 10^4 V/cm, so that the drift velocity can be considered to be unsaturated. The hole drift current density at the center of the depletion region is thus

$$-J_C^E = -e\frac{D}{V_T}n_i\exp\left(\frac{V}{2V_T}\right)E_C = e\frac{D}{V_T}n_i\exp\left(\frac{V}{2V_T}\right)\sqrt{\frac{(\psi-V)eN}{\varepsilon}}$$

You then have

$$-\frac{J_C^E}{J^D} = \frac{N}{n_i}\frac{w}{V_T}\sqrt{\frac{(\psi-V)eN}{\varepsilon}}\exp\left(-\frac{V}{2V_T}\right)$$

Take $V = \psi - V_T$. Show that the ratio in this case is

$$-\frac{J_C^E}{J^D} = \frac{w}{L_D}\exp\frac{1}{2}$$

where L_D is the Debye length,

$$L_D = \sqrt{\frac{\varepsilon V_T}{eN}}$$

For $N = 10^{16}$ cm^{-3} and $w = 1$ μm (10^{-4} cm), the ratio is about 40. To simulate the effect of drift-velocity saturation at a lower forward bias, keep $\psi - V = V_T$ in the expression for E_C. You then have

$$-\frac{J_C^E}{J^D} = \frac{w}{L_D} \exp\frac{\psi - V}{2V_T}$$

The ratio is thus enormous at a low forward bias, and not only at the center of the depletion region.

6-17 THE HOLE CURRENT IN SHORT-BASE JUNCTION DIODES IS SMALL. Consider the adjacent regions of n-silicon depicted in Fig. 6-45, and assume that the region from $x = 0$ to $x = w_-$ is lightly doped ($N_D = N_-$) and that the region from $x = w_-$ to $x = w_- + w_+$ is heavily doped ($N_D = N_+$ $\gg N_-$). Assume further that there is an injecting contact at $x = 0$ ($p = p_f$) and an ohmic contact at $x = w_- + w_+$ ($p \approx 0$). This is essentially what you have on the n-side of the junction diode of Fig. 6-9 when the forward bias is substantial.

Neglect the width of the space-charge region around the junction at $x = w_-$, and assume that the potential across this region changes negligibly when current flows, so that the ratio of the hole densities at its edges is equal

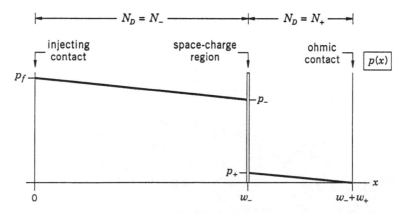

Figure 6-45. Hole density on the n-side of a junction diode in the short-base limit (Exercise 6-17).

to the equilibrium value:

$$\frac{p_+}{p_-} = \frac{N_-}{N_+}$$

For simplicity, assume that the diffusion constant D is independent of the doping level. In the short base limit you then have

$$D\frac{p_f - p_-}{w_-} = J_p = D\frac{p_+}{w_+}$$

If w_- and w_+ are of the same order, deduce that $p_- \simeq p_f$ and that

$$J_p \simeq \frac{D}{w}p_f$$

where w is the effective width,

$$w \simeq w_-\frac{w_+N_+}{w_-N_-} \gg w_-$$

The hole current is thus much smaller than it would be if the junction at $x = w_-$ were ohmic.

7

LOW-FREQUENCY
CIRCUIT FRAGMENTS

7-1 INTRODUCTION

Transistor circuits are multifarious in appearance, but most of them can be reduced to combinations of easily recognizable substructures or *fragments* with remarkably simple properties. These fragments, in turn, can be understood in terms of three basic one-transistor configurations, characterized by which terminal of a transistor is held at constant voltage.

Our purpose in this chapter is to become thoroughly acquainted with the low-frequency properties of the basic configurations and construct the fragments most commonly found in integrated circuits. With these elements in hand, we will see that the low-frequency behavior of circuits with dozens of transistors can often be predicted with just a few calculations. Much of our effort will go into calculating the small-signal input and output impedances of the basic configurations and of the fragments we construct with them; this will allow us to see how fragments affect one another.

In examples and exercises we will present a variety of the elegant circuits that become feasible if it can be assumed—this is one of the great advantages of integrated circuits—that certain critical transistors have identical characteristics and, implicitly, the same temperature.

We will concentrate on bipolar transistors because unique circuits can be constructed by exploiting the exponential dependence of the collector current on the base–emitter voltage in these devices. Another reason for this choice is that JFET or MOSFET versions of bipolar transistor circuits, when they exist, have incremental circuits that with appropriate labels and parameter values are identical to the ones we will obtain for the bipolar versions.

286

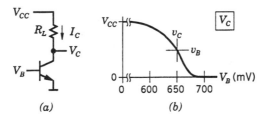

Figure 7-1. Common-emitter amplifier. (*a*) Structure. (*b*) Voltage transfer characteristic.

7-2 COMMON-EMITTER CONFIGURATION

In the *common-emitter* configuration, the emitter voltage of a transistor is constant and the input and output terminals are the base and the collector; the output signal is generated on a load connected to the collector. The low-frequency prototype of this configuration is the *common-emitter amplifier*, shown for an *npn* transistor in Fig. 7-1. The emitter is grounded and the collector load is a resistor R_L connected to a (constant) positive *supply* voltage V_{CC}.

If the base voltage V_B is negative or only slightly positive, the transistor is cut off and the collector voltage V_C is equal to V_{CC}. At higher values of V_B (600 mV or so), the collector current I_C becomes important and we have

$$V_C = V_{CC} - I_C R_L$$

When V_C approaches zero, the transistor goes into saturation and I_C remains essentially constant at a value $I_C \simeq V_{CC}/R_L$.

If R_L is not too large, we can ignore the Early effect and write

$$I_C = I_S \exp \frac{V_B}{V_T}$$

The slope of the voltage transfer curve is

$$\frac{dV_C}{dV_B} = -\frac{I_C}{V_T} R_L = -g_m R_L$$

For small variations around a given bias current I_C, we have

$$v_C = -g_m R_L v_B$$

We thus have a voltage amplifier with an (inverting) gain

$$A = -g_m R_L$$

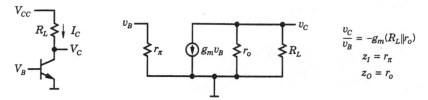

Figure 7-2. Incremental circuit of a common-emitter amplifier.

The voltage gain of a common-emitter amplifier can be quite large, even with the simple resistive load we are assuming. Let us take $V_{CC} = 10$ V, $R_L = 5$ kΩ, and $I_C = 1$ mA; this makes $V_C = 5$ V, so that the transistor is unquestionably in the active region. We then have $g_m = I_C/V_T = 40$ mA/V, and therefore $A = -200$. We will soon see that even larger gains are obtainable in more sophisticated circuits.

In a good design, the operating point of the transistor in a common-emitter amplifier must be made independent of variations due to temperature or, in discrete versions, to replacement by another transistor in case of a failure. In addition, of course, the operating point has to be set to its design value. At this time we will not worry about this problem: its solution will become apparent when we construct complete circuits.

The incremental circuit of a common-emitter amplifier is shown in Fig. 7-2; the presence of r_o indicates that the Early effect is taken into account. We can now obtain the small-signal voltage gain for arbitrary load resistances,

$$\frac{v_C}{v_B} = -g_m\left(R_L\|r_o\right) \tag{7-1}$$

The input impedance does not depend on the load resistance,

$$z_I = r_\pi$$

and the output impedance does not depend on the source resistance (which for this reason is not even shown in the incremental circuit),

$$z_O = r_o$$

We note that R_L is considered to be external to the amplifier proper and that it therefore does not contribute to z_O.

The maximum possible gain of a common-emitter amplifier is the *amplification factor* μ; from (7-1) is clear that μ is the gain for $R_L = \infty$,

$$\mu = g_m r_o$$

Substituting $g_m = I_C/V_T$ and $r_o = V_A/I_C$, we obtain

$$\mu = \frac{V_A}{V_T}$$

Transistors designed for high amplification typically have $V_A = 100$ V, and thus $\mu = 4000$. In Section 7-7 we will see that gains as high as μ can be obtained without resorting to large resistors by using active collector loads, that is, transistor current sources with a large incremental output resistance.

For small collector loads, the gain is linear within 1% if the magnitude of v_B is less than 0.5 mV. Returning briefly to the large-signal model, we have

$$i_C = I_C\left[\exp\left(\frac{v_B}{V_T}\right) - 1\right] \simeq I_C\left(\frac{v_B}{V_T} + \frac{v_B^2}{2V_T^2}\right) = g_m v_B\left(1 + \frac{v_B}{2V_T}\right)$$

For 1% linearity we therefore require $|v_B| < 2V_T/100 = 0.5$ mV.

Calculating nonlinearities for large R_L is involved and model dependent; this is not usually a problem, however, since large collector load resistances are normally found in devices such as operational amplifiers, in which nonlinearities are strongly suppressed by feedback.

7-3   COMMON-COLLECTOR CONFIGURATION

In the *common-collector* configuration, the collector of a transistor is held at constant voltage, and the input and output terminals are the base and the emitter. The low-frequency prototype of this configuration is the *emitter follower*, shown for an *npn* transistor in Fig. 7-3. The collector is connected to a positive supply voltage V_{CC} and the emitter load is a resistor R_L connected to ground.

If V_B is negative or small, we have $V_E = 0$. As V_B increases, the transistor enters the active region at a certain point, and V_{BE} thereafter increases only slightly if R_L is not too small. In other words, the emitter voltage *follows* the

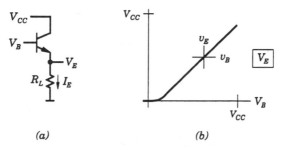

(a)   (b)

Figure 7-3. Emitter follower. (*a*) Structure. (*b*) Voltage transfer characteristic.

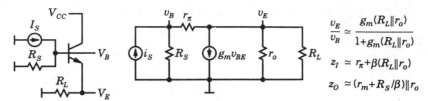

$$\frac{v_E}{v_B} \simeq \frac{g_m(R_L\|r_o)}{1+g_m(R_L\|r_o)}$$

$$z_I \simeq r_\pi+\beta(R_L\|r_o)$$

$$z_O \simeq (r_m+R_S/\beta)\|r_o$$

Figure 7-4. Incremental circuit of an emitter follower driven by a source with a finite output resistance.

base voltage. For small variations around a given operating point the gain is somewhat less than 1,

$$v_E \simeq v_B$$

This might seem uninteresting until we observe that the base current i_B is approximately β times smaller than the emitter current i_E and that the source of v_B can thus drive R_L as if it were a resistor β times larger.

Figure 7-4 shows the incremental circuit of an emitter follower driven by a source with output resistance R_S. We have included R_S in the circuit because as it does in most circuits—the common-emitter amplifier happens to be one of the few exceptions—it contributes to the output impedance.

Summing currents at the emitter node yields

$$(v_B-v_E)(g_m+g_\pi) = v_E(g_o+G_L)$$

Solving this equation for the gain and neglecting g_π compared with g_m, we get

$$\frac{v_E}{v_B} \simeq \frac{g_m(R_L\|r_o)}{1+g_m(R_L\|r_o)} \tag{7-2}$$

The input impedance $z_I = v_B/i_B$ can be obtained by observing that

$$r_\pi i_B = v_{BE} = v_B\frac{1}{1+g_m(R_L\|r_o)}$$

and therefore that

$$z_I \simeq r_\pi+\beta(R_L\|r_o) \tag{7-3}$$

The output impedance z_O can be obtained by removing r_o, calculating the output impedance of the remainder, and then putting the result in parallel with r_o. The output impedance of the remainder can be obtained in turn by calculating Thévenin equivalents v_T and i_T at the emitter node. For these we get

$$v_T = i_S R_S$$

and again neglecting g_π compared with g_m,

$$i_T \simeq i_S g_m (R_S \| r_\pi) = i_S \frac{\beta R_S}{R_S + \beta r_m}$$

Putting v_T/i_T in parallel with r_o, we obtain

$$z_O \simeq \left(r_m + \frac{R_S}{\beta} \right) \| r_o \qquad (7\text{-}4)$$

Roughly speaking, we can say that an emitter follower has a gain near 1, and that it is an *impedance converter* that makes load impedances look β times larger and source impedances β times smaller than they actually are.

Emitter followers are often required to drive ground-referenced loads to large positive and negative voltages; in operational amplifiers, for example, the voltage swing is typically ± 12V. The basic emitter follower is inadequate for this task, which is usually given to the *push–pull follower* shown in Fig. 7-5. This variant can drive a ground-referenced load to voltages within one diode drop from the supply voltages (600 mV or so) while drawing a small

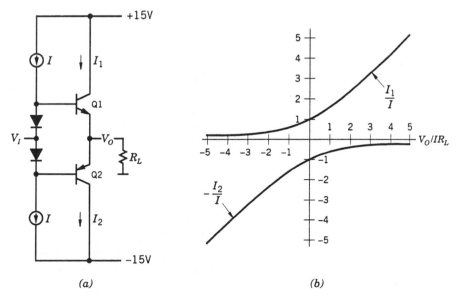

(a)   (b)

Figure 7-5. Push–pull follower. (*a*) Structure. (*b*) Collector currents as a function of the output current.

quiescent current. For positive input voltages most of the current is *pushed* to the load by Q1, whereas for negative input voltages it is *pulled* from the load by Q2. We will assume that the diodes are diode-connected transistors and that all transistors have identical characteristics.

The difference of the base–emitter voltage drops in Q1 and Q2 is constant, and the product of the collector currents is therefore also constant. We have

$$0 = \Delta V_{BE1} + \Delta V_{EB2} = V_T \ln\frac{I_1}{I} + V_T \ln\frac{I_2}{I} = V_T \ln\frac{I_1 I_2}{I^2}$$

and therefore

$$I_1 I_2 = I^2$$

Summing currents at the output node and neglecting base losses results in a quadratic equation for I_1,

$$I_1 = I_2 + \frac{V_O}{R_L} = \frac{I^2}{I_1} + \frac{V_O}{R_L}$$

Solving for I_1, we get

$$\frac{I_1}{I} = \frac{V_O}{2 I R_L} + \sqrt{\left(\frac{V_O}{2 I R_L}\right)^2 + 1}$$

We can now obtain the input voltage:

$$V_I = V_O + V_T \ln\frac{I_1}{I} = V_O + V_T \sinh^{-1}\frac{V_O}{2 I R_L}$$

For small signals we have

$$v_I \simeq v_O + v_O \frac{V_T}{2 I R_L}$$

and thus

$$\frac{v_O}{v_I} \simeq \frac{2 g_m R_L}{1 + 2 g_m R_L}$$

This is as it should be, since we have two transistors in parallel and the equivalent transconductance is therefore $2 g_m$. If we take $R_L = 2$ kΩ and $I = 1$ mA (these are typical values in operational amplifiers), we get $V_O/V_I = 0.995$ within 0.2% for $|V_I| < 10$ V. With the same values of R_L and I we get $v_O/v_I = 0.994$ for small signals.

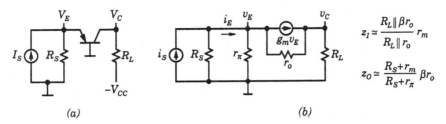

Figure 7-6. Common-base amplifier driven by a source with a finite output resistance. (*a*) Structure. (*b*) Incremental circuit.

7-4 COMMON-BASE CONFIGURATION

In the *common-base* configuration, exemplified by the *pnp common-base amplifier* shown in Fig. 7-6, the base voltage of a transistor is constant, and the input and output terminals are the emitter and the collector.

Let us first obtain output Thévenin equivalents. If the output current is zero, the current in r_o is $g_m v_E$ and the current in $r_\pi \| R_S$ is i_S; the open-circuit voltage is therefore

$$v_T = i_S(r_\pi \| R_S)(1 + g_m r_o) \simeq i_S \frac{\beta r_o R_S}{R_S + r_\pi}$$

The short-circuit current is obtained by current division:

$$i_T = i_S \frac{g_m + g_o}{g_m + g_\pi + g_o + G_S} \simeq i_S \frac{R_S}{R_S + r_m}$$

The output impedance is then obtained as the ratio of v_T and i_T,

$$z_O \simeq \frac{R_S + r_m}{R_S + r_\pi} \beta r_o \qquad (7\text{-}5)$$

If $R_S \geq r_\pi$, the output impedance is very large, of order βr_o.

For $R_S = \infty$, we have $i_T = i_S = i_E$ and $z_O = \beta r_o$; the gain from i_E to v_C is thus

$$\frac{v_C}{i_E} \simeq R_L \| \beta r_o$$

On the other hand, summing currents at the collector node yields

$$g_m v_E = v_C G_L + (v_C - v_E) g_o$$

The following equations appear at the right of the figure:

$$z_I \simeq \frac{R_L \| \beta r_o}{R_L \| r_o} r_m$$

$$z_O \simeq \frac{R_S + r_m}{R_S + r_\pi} \beta r_o$$

and thus

$$v_C(g_o+G_L) = v_E(g_m+g_o) \simeq v_E g_m$$

The voltage gain is therefore, as we might expect, equal to the gain (7-1) of a common-emitter amplifier,

$$\frac{v_C}{v_E} \simeq g_m(R_L \| r_o)$$

We can now calculate the input impedance $z_I = v_E/i_E = (v_C/i_E)/(v_C/v_E)$, and we get

$$z_I = \frac{R_L \| \beta r_o}{R_L \| r_o} r_m \qquad (7\text{-}6)$$

The input impedance is small, of order r_m if $R_L \leq r_o$, and always less than r_π.

We can now construct an important fragment, the *cascode* amplifier shown in Fig. 7-7, which consists of a common-emitter amplifier (Q1) whose load is a common-base amplifier (Q2). There are two versions: the *totem* cascode with stacked transistors of the same type, and the *folded* cascode with transistors of opposite types. The constant voltage V_{BB} should be high enough to keep Q1 out of saturation.

The output resistance of Q1 (r_o) is much larger than the input resistance of Q2 ($\leq r_\pi$). This implies that the output impedance z_O is βr_o and, ignoring base losses, that the short-circuit current is $g_m v_B$. The cascode amplifier thus has the incremental circuit of a common-emitter amplifier except for the value of the equivalent of the collector output resistance, which is βr_o rather

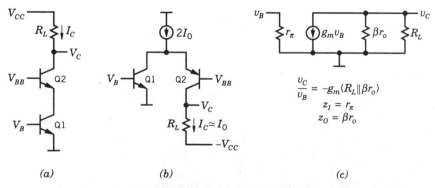

Figure 7-7. Cascode amplifiers. (*a*) Totem. (*b*) Folded. (*c*) Incremental circuit indicating that a cascode amplifier is essentially a transistor with a large collector output resistance.

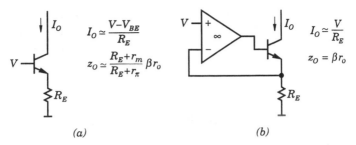

(a)   *(b)*

Figure 7-8. (*a*) Basic transistor current source. (*b*) Current reference obtained by slaving the emitter voltage to the input voltage.

than r_o. In other words, we can look at the cascode fragment as if it were a transistor with a very large Early voltage.

Another important fragment is the voltage-controlled current source shown in Fig. 7-8*a*, in which the base is connected to a voltage V. Ignoring base losses and assuming that $V_{CB} > 0$, the output current is

$$I_O \simeq \frac{V - V_{BE}}{R_E}$$

If R_E is sufficiently large, the output impedance is βr_o. A precision version of this current source is the current reference shown in Fig. 7-8*b*. The operational amplifier keeps the emitter current constant, so that

$$I_O \simeq I_E = \frac{V}{R_E}$$

As far as the transistor is concerned, it is in a common-base configuration with infinite emitter resistance, and the output impedance is therefore βr_o.

Finally, let us turn to the common-emitter amplifier with *emitter degeneration* shown in Fig. 7-9*a*. This fragment receives its name because it can be

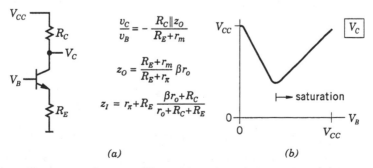

(a)   *(b)*

Figure 7-9. Common-emitter amplifier with emitter degeneration. (*a*) Structure. (*b*) Voltage transfer characteristic indicating that the collector as well as the emitter follows the base when the transistor saturates.

viewed as a circuit with *degenerative* (negative) feedback[†] in which the difference between the base voltage v_B and a sample $i_C R_E$ of the collector current i_C is multiplied by the transconductance g_m to obtain i_C; for $R_C \ll r_o$ and $\beta = \infty$, we have

$$(v_B - i_C R_E) g_m = i_C$$

and thus

$$i_C = v_B \frac{g_m}{1 + g_m R_E}$$

The collector voltage is $v_C = -R_C i_C$, and the voltage gain is therefore

$$\frac{v_C}{v_B} = -\frac{R_C}{R_E + r_m}$$

We can thus think of a transistor with a resistance R_E in series with the emitter as if it were a transistor with a transconductance $1/(R_E + r_m)$.

As shown in Fig. 7-9b, a problem with this fragment is that the collector as well as the emitter will follow the base when the transistor saturates if $R_E \| R_C$ is large enough that it does not load down the source of V_B; the slope of the voltage transfer curve is then positive for large V_B. In a feedback circuit this inversion can result in an internal latch-up condition in which the circuit does not respond to its inputs.

According to Exercise 7-10, the short-circuit current is

$$i_T = -\frac{v_B}{R_E + r_m}$$

if $R_E \leq r_o$. With this minimal restriction, which is insignificant in practice, the voltage gain for arbitrary R_C is

$$\frac{v_C}{v_B} = -\frac{R_C \| z_O}{R_E + r_m} \tag{7-7}$$

where z_O is the output impedance of the amplifier, clearly equal to the output impedance of a common-base amplifier with source resistance R_E and thus given by (7-5),

$$z_O = \frac{R_E + r_m}{R_E + r_\pi} \beta r_o$$

The expression for the input impedance is straightforward to obtain; since it does not involve anything really new, we will quote it without proof:

$$z_I = r_\pi + R_E \frac{\beta r_o + R_C}{r_o + R_C + R_E} \tag{7-8}$$

For $R_C \ll r_o$, this expression reduces, as expected, to the emitter-follower input impedance given in (7-3).

[†]Positive feedback is correspondingly *regenerative*.

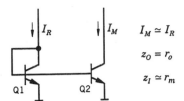

$$I_M \simeq I_R$$

$$z_O = r_o$$

$$z_I \simeq r_m$$

Figure 7-10. Basic current mirror.

7-5 CURRENT MIRRORS

Integrated circuits make extensive use of *current mirrors* like the one shown in Fig. 7-10. Q1 and Q2 are identical and have equal base–emitter voltages; their collector currents are thus also equal. If we ignore base losses and the Early effect, we have

$$I_M \simeq I_R$$

In other words, the reference current I_R is reflected by the mirror into an (almost) equal image current I_M. The input impedance is approximately equal to the small-signal resistance of diode-connected transistor Q1,

$$z_I \simeq r_m$$

Since Q2 is a common-emitter amplifier, the output impedance is

$$z_O = r_o$$

As shown in Fig. 7-11, other current ratios are obtained by adding transistors to either the input or the output stage or, equivalently, by using transistors with different values of the saturation current I_S.

An improvement over the basic current mirror is the *Wilson mirror* shown in Fig. 7-12a. Its advantages are that the relative difference between its image and reference currents is of order $1/\beta^2$ rather than $1/\beta$ (this point is taken up in Exercise 7-13) and that its output impedance is $\beta r_o/2$ rather than r_o, as we will now see.

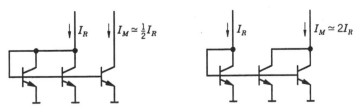

Figure 7-11. Current mirrors with nonunity current ratios.

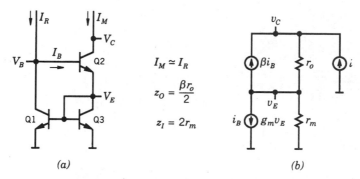

Figure 7-12. Wilson current mirror. (*a*) Structure. (*b*) Incremental circuit used to calculate the output impedance.

Q1 and Q3 have the same base–emitter voltage and therefore the same collector currents. We thus have

$$I_M \simeq I_R$$

Since $V_B = 2V_E$, the input impedance is

$$z_I = 2r_m$$

For small signals, Q2 is a common-emitter amplifier with emitter degeneration, provided in this case by the resistance r_m of diode-connected transistor Q3. According to (7-8), the input resistance at the base of Q2 is at most $2r_\pi$. In the incremental circuit, shown in Fig. 7-12*b*, we have taken advantage of this fact to ignore the collector output resistance of Q1 because it is in parallel with the input resistance of Q2|Q3. If we drive the output with a current source i, we have

$$2v_E g_m = i = (v_C - v_E)g_o - \beta g_m v_E$$

and thus

$$v_E(2g_m + g_o + \beta g_m) = v_C g_o$$

We then obtain

$$v_C \simeq v_E \beta g_m r_o = i\frac{\beta r_o}{2}$$

The output impedance $z_O = v_C/i$ is therefore very large,

$$z_O = \frac{\beta r_o}{2} \qquad\qquad (7\text{-}9)$$

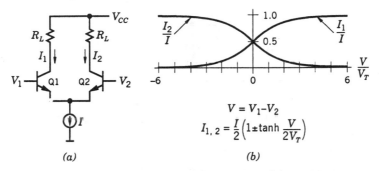

Figure 7-13. Basic differential amplifier. (a) Structure. (b) Collector currents as a function of the differential voltage $V = V_1 - V_2$.

7-6 DIFFERENTIAL AMPLIFIER

The accurate measurement of voltage differences is a fundamental problem in a large class of electronic devices that includes operational amplifiers as paramount examples. The circuit that addresses the problem is the *differential amplifier*, whose basic version with *npn* transistors is shown in Fig. 7-13. We will assume that R_L is much smaller than r_o; the interesting cases in which this assumption does not hold are covered in Section 7-7 and Exercise 7-8.

Let V be the differential voltage between the bases of Q1 and Q2,

$$V = V_1 - V_2$$

We then have

$$V = V_{BE1} - V_{BE2} = V_T \ln \frac{I_1}{I_2}$$

If β is sufficiently large we can ignore base losses; the sum of the collector currents must then be equal to the bias current I:

$$I_1 + I_2 = I$$

We can now obtain the collector currents (also graphed in Fig. 7-13),

$$I_{1,2} = \frac{I}{2}\left(1 \pm \tanh \frac{V}{2V_T}\right) \tag{7-10}$$

With a differential voltage of about 120 mV ($4.8V_T$), one of the transistors carries 99% of the current, whereas the other is almost at cutoff: we have a current switch, of which we will make use when we come to digital circuits.

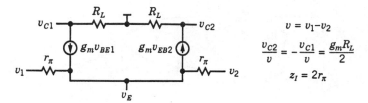

Figure 7-14. Incremental circuit of the differential amplifier of Fig. 7-13 in the case $I_1 = I_2 = I/2$.

A bipolar differential amplifier is linear within 1% if $|V|$ is less than 9 mV. It is easy to see physically why there is a large improvement relative to a common-emitter amplifier, in which 1% linearity requires a signal level under 0.5 mV: the increase of the transconductance in one transistor is compensated by a reduction in the other. The nonlinearity is given by

$$\frac{I_1 - \dfrac{I}{2}}{\dfrac{I}{4V_T}V} - 1 = \frac{\tanh\dfrac{V}{2V_T}}{\dfrac{V}{2V_T}} - 1 \simeq -\frac{1}{3}\left(\frac{V}{2V_T}\right)^2$$

For 1% linearity we must thus have $|V| \leq 0.35 V_T \approx 9$ mV.

The incremental circuit for $I_1 = I_2 = I/2$ is shown in Fig. 7-14. At node v_E we have

$$(v_1 - v_E)(g_m + g_\pi) = (v_E - v_2)(g_m + g_\pi)$$

so that

$$v_E = \frac{v_1 + v_2}{2}$$

and therefore

$$v_{EB2} = v_{BE1} = \frac{v_1 - v_2}{2} = \frac{v}{2}$$

and

$$-i_{C2} = i_{C1} = g_m \frac{v}{2}$$

The voltage gains are

$$\frac{v_{C2}}{v} = -\frac{v_{C1}}{v} = \frac{g_m R_L}{2}$$

and the differential input impedance is

$$z_I = \frac{v}{i_{B1}} = 2r_\pi$$

These results can also be obtained by observing that each transistor is an emitter follower with a load r_m equal to its output impedance (the collectors can be considered to be at constant voltage because R_L is assumed to be much smaller than r_o). Grounding the base of Q2 and applying a voltage v to the base of Q1 then leads to $v_{BE1} = v_{EB2} = v/2$, and the result for the gain follows; the input impedance is obtained from (7-3).

The input voltages to a differential amplifier can be written in the form

$$v_{1,2} = v_{CM} \pm \frac{v_{DM}}{2}$$

where v_{CM} is the *common-mode* voltage and v_{DM} is the *differential-mode* voltage,

$$v_{CM} = \frac{v_1 + v_2}{2} \quad \text{and} \quad v_{DM} = v_1 - v_2$$

In an ideal differential amplifier the outputs depend only on v_{DM}; this is of course not true for real differential amplifiers, which are characterized by their *common-mode rejection ratio* (CMRR), equal to the ratio of the responses to differential-mode and common-mode inputs.

Figure 7-15 shows the incremental circuit of a differential amplifier when the collectors are grounded; r_E represents the output resistance of the bias-current source that feeds the emitters. Assuming identical transistors, we will use this simple circuit to set upper limits on the CMRR we can expect in practice.

Let us first assume that $r_E = \infty$. If we apply a common-mode voltage v_{CM} to both inputs, we obtain

$$(v_{CM} - v_E)(2g_m + 2g_\pi) = v_E(2g_o)$$

and, setting $v_{BE1} = v_{BE2} = v_{BE}$,

$$v_E = \frac{g_m r_o / \alpha}{1 + g_m r_o / \alpha} v_{CM} \quad \text{and} \quad v_{BE} = \frac{1}{1 + g_m r_o / \alpha} v_{CM}$$

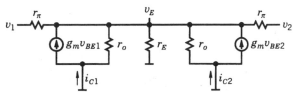

Figure 7-15. Incremental circuit of the differential amplifier of Fig. 7-13 when the collectors are grounded. The collector output resistances and the output resistance r_E of the bias-current source are included in order to calculate the common-mode rejection ratio.

where $\alpha = \beta/(\beta+1)$; we note that we do *not* neglect g_π compared with g_m. The collector currents are

$$-i_{C1} = -i_{C2} = v_E g_o - v_{BE} g_m = v_{CM} \frac{g_m(1/\alpha-1)}{1+g_m r_o/\alpha} \simeq \frac{v_{CM}}{\beta r_o}$$

For a differential voltage v_{DM} we have $-i_{C2} = i_{C1} = g_m v_{DM}/2$, so that

$$\text{CMRR} \simeq \frac{\beta g_m r_o}{2}$$

For $V_A = 100$ V and $\beta = 100$, we have CMRR $\simeq 2 \cdot 10^5$, that is, about 106 dB. We can improve this figure by measuring the differential output current $i_{C2} - i_{C1}$ rather than the single-ended currents i_{C1} or i_{C2}. In operational amplifiers, where this is in fact done, the CMRR can in practice be as high as 120 dB.

If r_E is of order βr_o (as it is in Wilson mirrors), it adds a current of order $v_{CM}/2\beta r_o$ to the common-mode currents, so that the CMRR is not much affected. On the other hand, if r_E is of order r_o (as it is in basic current mirrors), its effect is dominant, and the common-mode currents are of order $v_{CM}/2r_o$, so that

$$\text{CMRR} \approx g_m r_o$$

which, again for $V_A = 100$ V, is 4000 (72 dB). As in the preceding case, the CMRR can be raised, this time to about 85 dB, by measuring the differential output current.

7-7   DIFFERENTIAL AMPLIFIER WITH AN ACTIVE LOAD

Bipolar transistor amplifiers can achieve very high gains if the collector load is large. The differential amplifier with an active load, shown together with its incremental circuit in Fig. 7-16, is an example of this technique. Q1 feeds diode-connected transistor Q3 of *pnp* current mirror Q3|Q4, whereas Q2 is connected to mirror output transistor Q4, which has output impedance r_o. If the external load is made large by using emitter followers, we can expect the gain to be of order $g_m r_o$. In the incremental circuit we have used the fact that Q1 has a collector load r_m much smaller than its output resistance r_o to set $v_{EB4} = -v_{C1} = v_{BE1}$.

This fragment and its variants—some of which are quite elegant—are at the heart of virtually every existing operational amplifier. In addition to high gain, it has other valuable properties. It converts differential voltage $v_1 - v_2$ to a single-ended output voltage v_O by subtracting collector currents i_{C1} (or, more properly, its image i_{C4}) and i_{C2}, that is, by measuring the differential collector current in Q1 and Q2. It thus simultaneously doubles the gain and improves the common-mode rejection ratio.

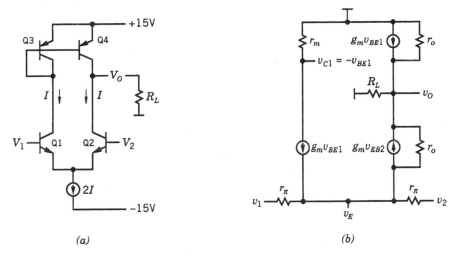

Figure 7-16. Differential amplifier with an active load. (*a*) Structure. (*b*) Incremental circuit.

We will solve this circuit for differential-mode inputs $v_1 = -v_2 = v/2$. To do so we will calculate Thévenin equivalents v_T and i_T at the output. When the output is grounded, Q1, Q2, and Q4 have zero or negligible collector loads; according to Section 7-6, we then have $v_E = 0$ and therefore $v_{BE1} = v_{EB2} = v/2$, and the short-circuit current is

$$i_T \simeq g_m v$$

The open-circuit equations at nodes v_E and v_O are

$$\left(\frac{v}{2} - v_E\right)(g_m + g_\pi) + (v_O - v_E)g_o = \left(v_E + \frac{v}{2}\right)(g_m + g_\pi)$$

$$\left(\frac{v}{2} - v_E\right)g_m + \left(v_E + \frac{v}{2}\right)g_m = v_O g_o + (v_O - v_E)g_o$$

Neglecting g_π and g_o compared with g_m and observing that v_E is negligible compared with v_O, we have

$$g_o v_O = (2g_m + 2g_\pi + g_o)v_E \simeq 2g_m v_E$$

$$g_m v = -g_o v_E + 2g_o v_O \simeq 2g_o v_O$$

(7-11)

and thus

$$v_T = v_O \simeq \frac{g_m r_o}{2} v$$

$$v_E \simeq \frac{v}{4}$$

The output impedance is the ratio of v_T and i_T,

$$z_O \simeq \frac{r_o}{2}$$

and the gain is therefore

$$\frac{v_O}{v} = g_m \left(R_L \| \frac{r_o}{2} \right)$$

From the first equation in (7-11), which is for node v_E and is thus valid for arbitrary values of R_L, we have

$$v_E = \frac{g_o}{2g_m} v_O = \frac{g_o}{2} \left(R_L \| \frac{r_o}{2} \right) v$$

If $R_L \gg r_o$ we have $v_E = v/4$ and $i_{B2} = -3i_{B1}$, so that the differential input impedance is not well defined. If $R_L \leq r_o/5$, however, v_E is reasonably close to zero and we can use the incremental circuit shown in Fig. 7-17.

We can obtain a well-defined differential input impedance for arbitrary values of R_L by replacing Q1 and Q2 with cascode amplifiers; in this case, the output impedance becomes r_o. In Section 7-9 we will see that the output impedance becomes $\beta r_o/3$ if we further make the active load a Wilson mirror, in which case we can obtain very high gains *and* a well-defined differential input impedance.

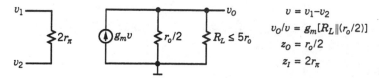

Figure 7-17. Simplified incremental circuit of the differential amplifier of Fig. 7-16, valid for $R_L \leq 5r_o$.

7-8   CIRCUITS WITH JFETs AND MOSFETs

For small signals, all transistors can be described in a first approximation as voltage-controlled current sources. We should therefore expect that what we have learned about bipolar transistors is applicable in a large measure to field-effect transistors. Most of the bipolar transistor circuits that we have seen have FET versions. This certainly holds for the basic configurations, and there are thus common-source amplifiers, source followers, and common-gate amplifiers. In addition, it holds for important fragments such as differential and cascode amplifiers and, in the case of enhancement MOSFETs, for current mirrors.

JFETs as well as MOSFETs with the substrate tied to the source have the same small-signal model as bipolar transistors. The expressions for the gains and impedances in the FET analogs are therefore also the same; to use them we need only set $\beta = \infty$.

In contrast, n-channel (p-channel) MOSFETs in integrated circuits often have a common substrate tied to the most negative (positive) supply voltage; since the drain current depends on V_{BS} as well as on V_{GS}, common-gate amplifiers and source followers in such circuits will differ somewhat in behavior from their bipolar analogs. Ignoring the Early effect, the incremental drain current is given by (6-66) and (6-67),

$$i_D = g_m v_{GS} + \chi g_m v_{BS}$$

A common-gate amplifier has $v_G = 0$. In this circuit, a MOSFET with $v_B = 0$ acts as if it had a transconductance $g_m(1+\chi)$; with this modification, we can use the results we obtained for a common-base amplifier. In source followers, such as the one shown together with its incremental circuit in Fig. 7-18, the effect of a constant substrate voltage is more noticeable. Summing currents at the source node, we get

$$g_m(v_G - v_S) - \chi g_m v_S = G_L v_S$$

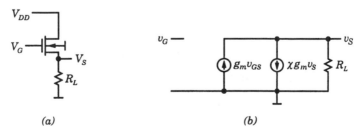

(a)   (b)

Figure 7-18. MOSFET source follower. (a) Structure. (b) Incremental circuit.

and thus

$$\frac{v_S}{v_G} = \frac{g_m R_L}{1+(1+\chi)g_m R_L}$$

The gain cannot exceed $1/(1+\chi)$ even for large R_L; since χ is of order 1, the MOSFET source follower is inferior in this case to its bipolar analog, whose gain can be quite close to 1.

Despite their conceptual similarity, JFETs, MOSFETs, and bipolar transistors are of course not interchangeable; each has its advantages and drawbacks, which is why they are often found side by side, notably so in integrated circuits. The great virtue of JFETs (and MOSFETs) is that their gate current at room temperature is orders of magnitude less than the base current of bipolar transistors with comparable transconductances; JFETs are consequently often used only in the input stage of circuits that require a high input impedance, whereas all other functions are assigned to bipolar transistors because these are, generally speaking, easier to handle.

A consideration based on device physics comes into play in amplifiers with field-effect transistors. The transconductance g_m of FETs is proportional to the square root of the drain current, whereas the drain output resistance r_o is inversely proportional to the drain current; the amplification factor $\mu = g_m r_o$ is thus inversely proportional to the square root of the drain current. This is why amplifier FETs are often operated at relatively low drain currents.

The unique properties of MOSFETs makes them supreme for certain types of mixed analog and digital circuits and for large-scale integrated circuits. We will look into this domain when we come to digital circuits in Chapter 10.

That analogs exist does not always mean that they are used in practice. Take for instance the cascode current mirrors shown in Fig. 7-19a. Both have a high output impedance, and we might expect them to be used with equal

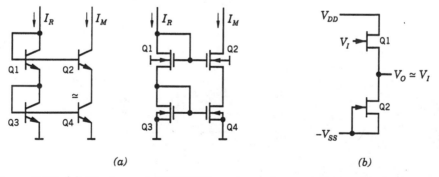

(a)   *(b)*

Figure 7-19. (*a*) Bipolar and MOSFET versions of a cascode current mirror; the former is usually bypassed in favor of a Wilson mirror. (*b*) JFET zero-offset follower.

frequency; the bipolar version is seldom seen, however, because it is usually replaced by a Wilson mirror, which has an equally high output impedance but much smaller base losses.

The most striking aspects of circuits are often due to peculiarities in the large-signal behavior of transistors. Consider for instance the zero-offset source follower with identical JFETS illustrated in Fig. 7-19b. The drain currents are equal; since current source Q2 has $V_{DS2} = 0$, we must also have $V_{DS1} = 0$ in the measure that we can ignore the Early effect. Thus, even for large signals, we have $V_O \simeq V_I$.

7-9 EXAMPLE: HIGH-SPEED OPERATIONAL AMPLIFIER

We are finally in a position to assemble what we have learned and construct a complete circuit, in this case the operational amplifier shown in Fig. 7-20, a stripped-down version of the National Semiconductor LM6165. This circuit is basically a folded-cascode differential amplifier with an active load and an output push–pull follower. We will assume that all transistors have identical characteristics and that diodes are diode-connected transistors. As usual, we will ignore base losses.

The n-channel JFET is a constant-current source that sets currents $2I$ in diodes D1 and D3 and thus also in mirror transistors Q3, Q4, and Q5. The

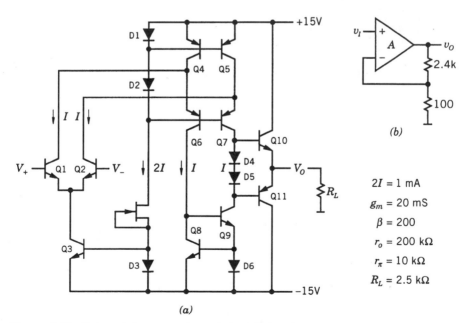

Figure 7-20. Stripped-down version of the National Semiconductor LM6165 high-speed operational amplifier. (*a*) Structure. (*b*) Connection as a ×25 direct amplifier.

currents in Q1, Q2, Q6, and Q7 are therefore equal to I. Q8|Q9|D6 is a Wilson mirror that makes the current in Q9 also equal to I.

Q1|Q2|Q6|Q7 is a differential amplifier like the one in Fig. 7-16, except that the transistors have been replaced by cascode amplifiers Q1|Q6 and Q2|Q7, which have output resistance βr_o (40 MΩ), and that the active load is a Wilson mirror with output resistance $\beta r_o/2$ (20 MΩ). The output resistance of the differential amplifier is therefore $(\beta r_o)\|(\beta r_o/2) = \beta r_o/3$ (13 MΩ).

D4|D5|Q10|Q11 is a push-pull follower; its input resistance βR_L (500 kΩ) is much smaller than the output resistance of the differential amplifier; the gain A of the operational amplifier is therefore $g_m \beta R_L = 10{,}000$.

The follower sees a source resistance $\beta r_o/3$, and its effective collector output resistance is $r_o/2$ because Q10 and Q11 are in parallel; the output impedance z_O of the operational amplifier is therefore $(r_o/3)\|(r_o/2) = r_o/5$ (40 kΩ).

The collector loads on Q1 and Q2 are much smaller than r_o; the differential input impedance z_I is therefore well defined and equal to $2r_\pi$ (20 kΩ).

The figures for the gain and the impedances are not impressive unless we are told that a direct amplifier with operational amplifiers of this type can have a 30 MHz bandwidth at a closed-loop gain $A_C = 25$. The loop gain in this case is $A_L = A/25 = 400$; the closed-loop input and output impedances are therefore $z_{IC} \simeq A_L z_I = 8$ MΩ and $z_{OC} \simeq z_O/A_L = 100$ Ω. The closed-loop figures turn out to be quite reasonable, and they can be improved by at least an order magnitude by buffering the operational amplifier with an additional push–pull follower.

At this point we can give a partial answer to the question, posed in Section 7-2, of how to set the operating point in transistor circuits. The answer in this case (and in many others) is quite straightforward: the feedback network that sets the gain implicitly ensures that the operating point is right.

7-10 EXAMPLE: CURRENT-FEEDBACK INSTRUMENTATION AMPLIFIER

The presence of operational-amplifier blocks often makes the low-frequency analysis of apparently formidable circuits actually quite easy, as we will now see. Figure 7-21 shows a current-feedback instrumentation amplifier that can achieve superb performance in sophisticated versions such as the Analog Devices AMP-01.[†] We will assume that the operational amplifiers are ideal and that the transistors are identical and have $\beta = \infty$.

Output operational amplifier A0 keeps the currents in input transistors Q3 and Q4 equal and, since these currents add up to $2I_0$, equal to I_0. In addition, the collector voltages of Q3 and Q4 are constant (that these transistors have a constant collector current *and* a constant collector voltage is one of the

---

[†]Originally introduced by Precision Monolithics.

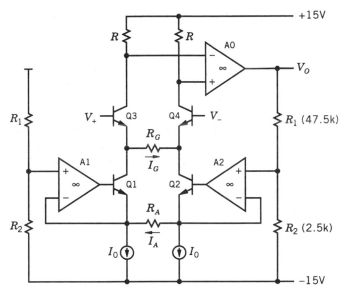

Figure 7-21. Simplified version of the Analog Devices AMP-01 current-feedback instrumentation amplifier.

marvels of operational-amplifier circuits). If we use these constraints in the large-signal model (6-44), we get

$$I_0 = I_0 \exp\left(\frac{v_{BE3}}{V_T}\right)\exp\left(\frac{v_{CB3}}{V_A}\right)$$

where v_{BE3} and v_{CB3} are the deviations in V_{BE3} and V_{CB3} from their values for $V_+ = 0$. Since V_{C3} is constant, we have $v_{CB3} = -V_+$, and

$$v_{BE3} = \frac{V_T}{V_A}V_+$$

An analogous equation holds for v_{BE4}. It follows that

$$V_{E3} - V_{E4} \approx V_+ - V_-$$

because $V_T/V_A \ll 1$, and therefore that

$$I_G \approx \frac{V_+ - V_-}{R_G}$$

The fragments formed by operational amplifiers A1 and A2 and their followers Q1 and Q2 have unity gain, and zero output impedance at the emitters of Q1 and Q2, so that

$$I_A = \frac{R_2}{R_1 + R_2}\frac{V_O}{R_A}$$

The collector and emitter currents are equal in Q1 and Q2, which implies that

$$I_A = I_G$$

We can now calculate the gain:

$$V_O \simeq \left(1 + \frac{R_1}{R_2}\right) \frac{R_A}{R_G} (V_+ - V_-)$$

In most designs, the feedback resistors R_1 and R_2 and the load resistances R on Q3 and Q4 are fixed, and the gain is set with R_A and R_G. We observe in this example, as we did in the preceding one, that feedback loops set the operating points of transistors.

Operational amplifier A0 is of course not ideal and must thus have a finite open-loop gain A_0. Setting $V_+ = V_- = 0$, and ignoring the current in R_G because R_G is presumably much larger than r_m, we obtain that the loop gain involving A0 is

$$A_L = \frac{R_2}{R_1 + R_2} \frac{2R}{R_A} A_0$$

The loop gain depends inversely on R_A, which must therefore be made as small as possible. There is a lower limit, however, because too much loop gain will surely result in an unstable circuit.

7-11   EXAMPLE: BANDGAP VOLTAGE REFERENCE

The exponential dependence of the collector current on the base–emitter voltage in bipolar transistors is put to good use in the Analog Devices AD580 *bandgap voltage reference* shown in simplified form in Fig. 7-22 (Brokaw, 1974;

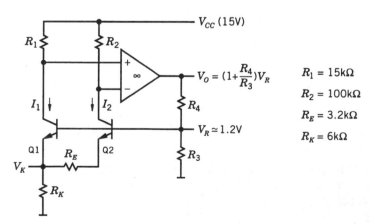

Figure 7-22. Simplified version of the Analog Devices AD580 bandgap voltage reference.

Widlar, 1971). As we will now see, this circuit generates a reference voltage V_R and a voltage V_K *proportional to absolute temperature* (PTAT). The output voltage V_O is proportional to V_R.

The operational-amplifier inputs are at the same voltage, which implies that Q1 and Q2 have the same collector–base voltage and that the ratio of their collector currents is constant. If Q1 and Q2 are identical, the difference of their base–emitter voltages is PTAT:

$$V_{BE1} - V_{BE2} = V_T \ln \frac{I_1}{I_2} = V_T \ln \frac{R_2}{R_1}$$

We then have

$$I_2 = \frac{V_T}{R_E} \ln \frac{R_2}{R_1}$$

$$I_1 = \frac{R_2}{R_1} I_2$$

(7-12)

and, ignoring base losses,

$$V_K = (I_1 + I_2) R_K = \chi V_T$$

where

$$\chi = \left(1 + \frac{R_2}{R_1}\right) \frac{R_K}{R_E} \ln \frac{R_2}{R_1}$$

We can now obtain the reference voltage,

$$V_R = V_{BE1} + V_K = V_{BE1} + \chi V_T$$

The temperature coefficients of V_T (+86.17 μV/K) and V_{BE1} (≈ -2.5 mV/K) have opposite signs; with an appropriate choice of χ we can null the temperature coefficient of V_R at a given reference temperature T_0. For T_0 near 300 K, χ is about 30; V_R is quite close to the bandgap voltage $V_G = \mathcal{E}_G/e \approx 1.1$ V (hence the name) and V_K has a temperature coefficient of about 2.5 mV/K. The variation of V_R with temperature is explored in Exercise 7-6.

With similar methods we can obtain current sources that are PTAT and voltage reference diodes that are far superior to Zener diodes.

7-12 EXAMPLE: THE GILBERT MULTIPLIER

Another interesting circuit that makes use of the exponential characteristics of bipolar transistors is the *Gilbert multiplier* shown in Fig. 7-23. We will define

$$V_X = V_X^+ - V_X^- \quad \text{and} \quad V_Y = V_Y^+ - V_Y^-$$

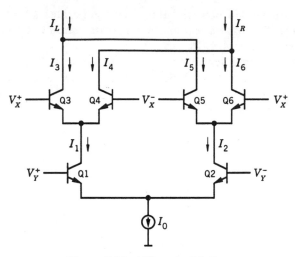

Figure 7-23. Gilbert multiplier.

According to (7-10), the currents in the differential amplifiers are

$$I_{1,2} = \frac{I_0}{2}\left(1 \pm \tanh\frac{V_Y}{2V_T}\right)$$

$$I_{3,4} = \frac{I_1}{2}\left(1 \pm \tanh\frac{V_X}{2V_T}\right)$$

$$I_{5,6} = \frac{I_2}{2}\left(1 \mp \tanh\frac{V_X}{2V_T}\right)$$

The output currents I_L and I_R are therefore

$$I_L = I_3 + I_5 = \frac{I_1 + I_2}{2} + \frac{I_1 - I_2}{2}\tanh\frac{V_X}{2V_T}$$

$$I_R = I_4 + I_6 = \frac{I_1 + I_2}{2} - \frac{I_1 - I_2}{2}\tanh\frac{V_X}{2V_T}$$

and the differential output current is

$$I_L - I_R = I_0 \tanh\frac{V_X}{2V_T}\tanh\frac{V_Y}{2V_T}$$

For small signals we thus have

$$i_L - i_R \simeq \frac{I_0}{4V_T^2}v_X v_Y$$

A large-signal voltage multiplier based on similar principles is discussed in Exercise 7-15. Once we can multiply voltages, we can also divide voltages by putting a multiplier in a feedback loop, as shown in Exercise 5-8. We have seen adders and subtractors in Chapter 5; the four arithmetic operations can thus be performed with electronic circuits.

EXERCISES

7-1 LABORATORY: CURRENT MIRRORS (Fig. 7-24).

(a) In configuration A, measure I_2 versus I_1 at $V_{C2} = 1V$ for 10 $\mu A \leq I_1 \leq 1$ mA. The relation should be linear, although it might not have unity slope.

(b) In configuration A, estimate z_O at $I_2 = 1$ mA by increasing V_{C2} by several volts; it should be of order 100 kΩ.

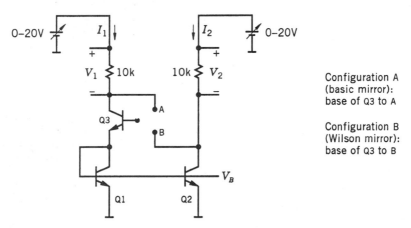

Figure 7-24. Circuit used in Exercise 7-1 to study the properties of current mirrors.

(c) In configuration B, measure I_1 versus I_2 at $V_{C3} = 2V$ for 10 $\mu A \leq I_2 \leq 1$ mA. The slope should be the reciprocal of the slope in part (a).

(d) In configuration B, set a lower limit on z_O at $I_1 = 1$ mA by increasing V_{C3} from 2V to 12V. It should be barely detectable, of order 10 MΩ.

7-2 THE CLASSIC COMMON-EMITTER AMPLIFIER. This exercise shows you how to design a biasing circuit that guarantees that the operating point of a bipolar transistor is well defined. To be specific, consider the classic common-emitter amplifier shown in Fig. 7-25.

   Choose R_1 and R_2 so that $I_B(R_1 \| R_2) \leq 50$ mV and $V_E \geq 1V$; this guarantees that variations in V_{BE} and β due to temperature fluctuations or replacement of the transistor will be negligible and that $I_C \approx I_E$ is deter-

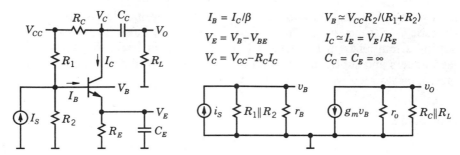

$$I_B = I_C/\beta$$
$$V_E = V_B - V_{BE}$$
$$V_C = V_{CC} - R_C I_C$$

$$V_B \simeq V_{CC} R_2/(R_1 + R_2)$$
$$I_C \simeq I_E = V_E/R_E$$
$$C_C = C_E = \infty$$

Figure 7-25. The classic common-emitter amplifier (Exercise 7-2).

mined mainly by R_E. Set V_C with R_C and check that the transistor remains in the active region for all admissible values of V_{CC} and V_{BE}.

V_{CC} is fixed by definition, so that $v_{CC} = 0$, and if blocking capacitor C_C and *bypass capacitor* C_E are sufficiently large, you can assume that $v_O = v_C$ and that $v_E = 0$. You can then obtain the incremental circuit of the amplifier by using the small-signal model for the transistor, grounding the power-supply and emitter nodes, and short-circuiting the collector and output nodes.

7-3 DARLINGTON AND SZIKLAI FOLLOWERS. In the two-stage followers shown in Fig. 7-26, transistors Q1 and Q2 act quite like a transistor with current gain β^2. Obtain the indicated results assuming that $R_L \ll r_{o2}$ and that $\beta_1 = \beta_2 = \beta$; use the fact that $I_{C2} \simeq \beta I_{C1}$ implies that $g_{m2} \simeq \beta g_{m1}$. You will save a lot of work if you make use of what you know about emitter followers; thus, for example, you know that the emitter load on Q1 in the Darlington follower is $\beta(r_{m2} + R_L)$.

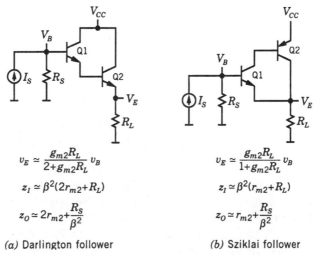

$$v_E \simeq \frac{g_{m2}R_L}{2 + g_{m2}R_L} v_B$$

$$z_I \simeq \beta^2(2r_{m2} + R_L)$$

$$z_O \simeq 2r_{m2} + \frac{R_S}{\beta^2}$$

$$v_E \simeq \frac{g_{m2}R_L}{1 + g_{m2}R_L} v_B$$

$$z_I \simeq \beta^2(r_{m2} + R_L)$$

$$z_O \simeq r_{m2} + \frac{R_S}{\beta^2}$$

(a) Darlington follower

(b) Sziklai follower

Figure 7-26. Two-stage followers (Exercise 7-3).

7-4 POWER-SUPPLY REJECTION RATIOS. Power-supply variations modify the output voltage of operational amplifiers. The main contribution is usually from the input stage because contributions from other stages are reduced by the gain of preceding stages. Take a good look at the differential amplifier of Fig. 7-13 and convince yourself that power-supply variations are akin to common-mode voltage variations, and that power-supply rejection ratios (PSRR) and common-mode rejection ratios (CMRR) should therefore be of the same order. Now go look at an operational-amplifier data book and verify that this is generally the case.

7-5 OUTPUT IMPEDANCE OF A CURRENT REFERENCE. Figure 7-27 shows the incremental circuit of the current reference discussed in Section 7-4 and shown in Fig. 7-8b. Calculate the output impedance by injecting a current into the collector node (this involves less algebra than putting a voltage source at the input of the operational amplifier). Use the fact that the operational amplifier holds the emitter node at ground and that all the collector current must therefore flow through r_π:

$$v_{EB} = i r_\pi$$

Now sum currents at the collector node to obtain

$$z_O \approx \beta r_o$$

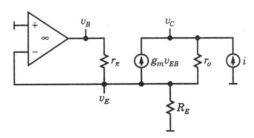

Figure 7-27. Circuit used in Exercise 7-5 to calculate the output impedance of the current reference of Fig. 7-8b.

7-6 TEMPERATURE VARIATIONS IN A BANDGAP REFERENCE. The temperature coefficient of the bandgap reference presented in Section 7-11 can be nulled at a given reference temperature T_0. In this exercise you will see how the reference voltage V_R varies for temperatures near T_0; to do so you need to know how the base–emitter voltage V_{BE} of transistor Q1 in Fig. 7-22 varies with temperature.

For simplicity, ignore the Early effect. You then have

$$V_{BE} = \frac{kT}{e} \ln \frac{I_C}{I_S}$$

where I_S is the saturation current given in (6-34),

$$I_S = \frac{eAD_n n_i^2}{w_B N_{AB}}$$

According to (7-12), the collector current I_C is proportional to absolute temperature (PTAT), so that

$$I_C = I_{C0} \frac{T}{T_0}$$

where I_{C0} is the collector current at T_0. Accept for now that the main contributors to variations in I_S as a function of temperature are D_n and n_i^2; w_B and N_{AB} will be dealt with later.

At the relatively low doping levels (10^{16} cm^{-3}) typically found in the base of silicon transistors, D_n falls like a reciprocal power of T for T near 300 K. For the purpose of this exercise, assume that D_n varies like $1/\sqrt{T}$.

According to (6-1), n_i^2 varies like $T^3 \exp(-eV_G/kT)$, where V_G is the bandgap voltage $\mathcal{E}_G/e$; for $T_0 \simeq 300$ K, V_G is given to sufficient accuracy by

$$V_G = V_{G0} + TV_G'(T_0)$$

where V_{G0} is the *extrapolated* bandgap voltage at $T = 0$, about 1.2 V in silicon for $T_0 \simeq 300$ K, and $V_G' = dV_G/dT$. The term linear in T gives rise to a factor that is independent of T (but does, however, depend on the choice of T_0), so that n_i^2 varies like $T^3 \exp(-eV_{G0}/kT)$.

The base–emitter voltage of Q1 can now be written as

$$V_{BE} = \frac{T}{T_0} V_{BE0} + \left(1 - \frac{T}{T_0}\right) V_{G0} - \eta \frac{kT_0}{e} \frac{T}{T_0} \ln \frac{T}{T_0}$$

where $\eta = \frac{3}{2}$ and V_{BE0} is V_{BE} at T_0. The reference voltage is of the form

$$V_R = \chi \frac{kT}{e} + V_{BE}$$

Show that the temperature coefficient of V_R is zero at T_0 if

$$\chi = \eta + \frac{e}{kT_0}(V_{G0} - V_{BE0})$$

Inserting this value of χ in the expression for V_R, you can now obtain

$$V_R = V_{G0} + \eta \frac{kT_0}{e} \frac{T}{T_0} \left(1 - \ln \frac{T}{T_0}\right)$$

At the reference temperature T_0 you have

$$V_{R0} = V_{G0} + \eta \frac{kT_0}{e}$$

Since $\eta(kT_0/e) \ll V_{G0}$, you have $V_{R0} \simeq V_{G0}$, which explains why bandgap references are so named. The derivative of V_R with respect to temperature is

$$\frac{dV_R}{dT} = -\eta \frac{k}{e} \ln \frac{T}{T_0}$$

which is, as it should be, zero at T_0. For $\Delta T / T_0 \ll 1$ you then have

$$\Delta V_R \simeq -\frac{\eta}{2} \frac{kT_0}{e} \left(\frac{\Delta T}{T_0} \right)^2$$

ΔV_R is thus an inverted parabola, symmetric around T_0; the drift is 21 μV for $\Delta T = 10$ K and $\eta = \frac{3}{2}$. In practice, the drift can be several times larger, but the parabolic shape remains.

To finish up, you might consider the variations in w_B and in N_{AB}. Assuming an abrupt junction, the width x_A of the depletion layer at the emitter end of the base varies like $\sqrt{\psi - V_{BE}}$. See if you can show that x_A varies like T/T_0 to a power near $\frac{1}{2}$, and that w_B varies like T/T_0 to a much smaller power since it is large compared with x_A. For $N_{AB} \simeq 10^{16}$ cm^{-3}, the acceptors are about 99% ionized; the remainder is reasonably well described by a Boltzmann factor, and N_{AB} thus varies like $1 - 0.01 \exp(-\mathcal{E}_I / kT_0) \exp(\mathcal{E}_I / kT)$, where $\mathcal{E}_I \simeq 50$ meV is the energy required for an acceptor to capture a valence electron. It follows that N_{AB} varies like T/T_0 to a power much smaller than 1.

7-7   SMALL-SIGNAL RESPONSE OF A RECTIFIER. Consider the full-wave rectifier of Fig. 7-28. Assume that the diodes can be described by the Shockley model and that $V_I \gg V_T$. Assume also that $I_L / \omega C_L \ll V_T$; since C_L

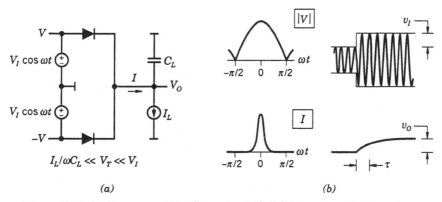

Figure 7-28. Full-wave rectifier (Exercise 7-7). (a) Structure. (b) Waveforms.

is discharged only by I_L, it follows that in a steady state the variation of V_O during a half-cycle is much smaller than V_T and that V_O can be taken to be constant when C_L is being charged by the diode current I. In a steady state the average current $\langle I \rangle$ must of course be equal to I_L.

In terms of $\varphi = \omega t$, the current in one half-cycle is

$$I = I_S \exp\left(\frac{V_I \cos\varphi - V_O}{V_T}\right) = I_S \exp\left(\frac{V_I - V_O}{V_T}\right) \exp\left[\frac{V_I}{V_T}(\cos\varphi - 1)\right]$$

Since $V_I \gg V_T$, the current is important only for φ near zero, and you have approximately

$$I = I_S \exp\left(\frac{V_I - V_O}{V_T}\right) \exp\left(-\frac{V_I}{2V_T}\varphi^2\right)$$

Use $\int_{-\infty}^{+\infty} \exp(-u^2/2)\, du = \sqrt{2\pi}$ to obtain that

$$\langle I \rangle = \frac{1}{\pi}\int_{-\pi/2}^{\pi/2} I\, d\varphi \simeq \frac{1}{\pi}\int_{-\infty}^{+\infty} I\, d\varphi = I_S \sqrt{\frac{2V_T}{\pi V_I}} \exp\left(\frac{V_I - V_O}{V_T}\right)$$

The square root varies slowly compared with the exponential, so that for variations v_I and v_O much smaller than V_T, the variation i of $\langle I \rangle$ is related to $v_I - v_O$ by the small-signal conductance of either diode at a current I_L:

$$i \simeq \frac{I_L}{V_T}(v_I - v_O)$$

For an input v_I, the output v_O is given by

$$C_L \frac{dv_O}{dt} = i = \frac{I_L}{V_T}(v_I - v_O)$$

and you thus have

$$\frac{C_L V_T}{I_L}\frac{dv_O}{dt} + v_O = v_I$$

The conclusion is clear: for small variations in the amplitude of the input sine wave, the rectifier responds like a linear first-order lag with a time constant $\tau = C_L V_T / I_L$.

Take $\omega = 2\pi \cdot 15$ MHz, $V_I = 600$ mV, $I_L = 10$ μA, and $C_L = 0.1$ μF. These values are for the rectifier in the automatic gain control (AGC) circuit of a 15 MHz sine-wave oscillator. The AGC circuit senses the amplitude of the sine wave and feeds it back to the oscillator in order to keep the amplitude constant. In this case, $V_I/V_T \simeq 24$ and $I_L/\omega C_L \simeq 1$ μV, so that the required conditions are amply satisfied. In addition, time constant τ is 250 μs, which gives the AGC circuit plenty of time to apply its corrections to the oscillator.

7-8 DIFFERENTIAL AMPLIFIERS WITH LARGE LOAD RESIS-TANCES. The gain of a differential amplifier with load resistances much smaller than r_o was obtained in Section 7-6. In practice, there are two cases of interest in which this condition is not satisfied: in the first case, one of the load resistances is zero; in the second case, common in operational amplifiers with a very large gain, the two load resistances are equal.

Consider the first case, shown in Fig. 7-29a. The short-circuit current is easy to obtain because $v_E = 0$ when $R_L = 0$,

$$i_T = g_m \frac{v}{2}$$

The impedance looking into the emitter of Q1 is r_m. Use (7-5) to obtain that the impedance looking into the collector of Q2 is

$$z_T = 2r_o$$

The differential gain is therefore

$$\frac{v_O}{v} = \frac{g_m}{2}(R_L \| 2r_o)$$

Now consider the second case, shown in Fig. 7-29b. The differential gain is easy to obtain in this case because $v_E = 0$, so that each transistor has the gain

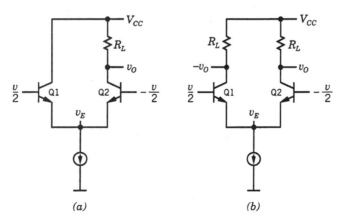

Figure 7-29. Differential amplifiers for Exercise 7-8. (a) Single-ended output. (b) Differential output.

of a common-emitter amplifier,

$$\frac{v_O}{v} = \frac{g_m}{2}(R_L \| r_o)$$

The impedance looking into either emitter is given by (7-6), and the output impedance z_T is given by (7-5). Check that $z_T \simeq 2r_o$ for $R_L \ll r_o$ (this is, as it should be, the result obtained in the first case). Also check that $z_T \simeq 3r_o$ for $R_L = r_o$ and that $z_T \simeq \beta r_o/3$ for $R_L = \beta r_o$.

In the same vein, use superposition to show that the differential amplifier with an active load shown in Fig. 7-16 has output impedance $r_o/2$.

7-9   HIGH-GAIN ONE-STAGE JFET AMPLIFIER. Junction FETs have a relatively small amplification factor $\mu = g_m r_o$ compared with bipolar transistors. It is nonetheless possible to obtain large gains in a single stage by going to a cascode configuration with a large drain load.

Consider the integrated JFET amplifier shown in Fig. 7-30. The transistors are identical except for their widths, which can have relative values 1 or 25. All transistors thus have the same pinchoff voltage $V_P = -1V$, and because they are of the same length, the same Early voltage $V_A = 30V$. Narrow transistors have $I_{DSS} = 1$ mA, and wide transistors have $I_{DSS} = 25$ mA.

The quiescent drain currents are set by Q3 and Q8, which have $V_{GS} = 0$; all transistors thus have $I_D = 1$ mA. Ignoring the Early effect for a moment, the drain current is given by

$$I_D = \frac{I_{DSS}}{V_P^2}(V_{GS} - V_P)^2$$

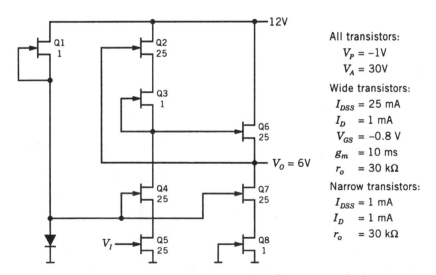

Figure 7-30. High-gain one-stage JFET amplifier (Exercise 7-9). (After Radeka, 1991.)

and the transconductance by

$$g_m = 2\frac{I_{DSS}}{V_P^2}(V_{GS} - V_P) = 2\frac{\sqrt{I_{DSS}I_D}}{-V_P}$$

For the wide transistors, show that $I_D = 1$ mA and $g_m = 10$ mS when $V_{GS} = -0.8$ V. You can now check that $V_{DS} > 1$ V for all transistors (Q3 in particular), so that all transistors are in the saturation region.

The large drain load on Q4 is obtained by bootstrapping the drain of current source Q3, that is, by driving it with source followers Q6 and Q2 so that v_{D3} follows v_{S3} very closely. From (7-5) obtain that the output impedance of current source Q7|Q8 is

$$r_o(1 + g_m r_o) \gg r_o$$

Source follower Q6 is therefore loaded only by its own drain output resistance r_o, so that its gain, given by (7-2), is

$$\frac{v_{S6}}{v_{G6}} = \frac{g_m r_o}{1 + g_m r_o}$$

Now obtain the output admittance of current source Q2|Q3|Q6 by superposition:

$$-\frac{i_{D3}}{v_{S3}} = \frac{1}{r_m + r_o} - \frac{g_m r_o}{1 + g_m r_o}\frac{g_m r_o/2}{1 + g_m r_o/2}\frac{1}{r_o} = \frac{1}{r_m + r_o}\frac{1}{1 + g_m r_o/2} \approx \frac{2}{g_m r_o^2}$$

(remember that the source load on Q2 is in parallel with r_o) and show that the output impedance of cascode Q4|Q5 is

$$r_o(1 + g_m r_o) \approx g_m r_o^2$$

The effective load resistance for cascode Q4|Q5 is thus

$$\frac{g_m r_o^2}{3}$$

and the small-signal gain $A = v_O/v_I$ is in fact quite high,

$$A \approx -\frac{(g_m r_o)^2}{3} = -30,000$$

7-10   COMMON-EMITTER AMPLIFIER WITH EMITTER DEGENER-ATION. Consider the common-emitter amplifier with emitter degeneration shown in Fig. 7-31, and assume that $R_E \leq r_o$. First take $R_C = 0$ and calculate the short-circuit current i_T. Summing currents at the emitter node, obtain

$$(v_B - v_E)(g_m + g_\pi) = v_E(g_o + G_E)$$

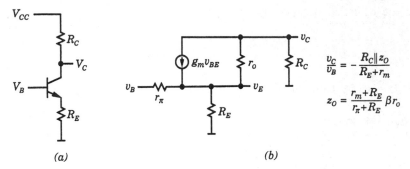

Figure 7-31. Common-emitter amplifier with emitter degeneration (Exercise 7-10). (a) Structure. (b) Incremental circuit.

and

$$i_T = g_o v_E - g_m v_{BE} = v_E \frac{g_\pi g_o - g_m G_E}{g_m + g_\pi} \simeq -\frac{v_E}{R_E}\left(1 - \frac{R_E}{\beta r_o}\right)$$

Since $R_E \le r_o$, you have

$$i_T \simeq -\frac{v_E}{R_E}$$

and, solving for v_E in terms of v_B,

$$i_T \simeq -\frac{v_B}{R_E + r_m}$$

For $R_C \ne 0$ you then obtain the gain given in (7-7),

$$\frac{v_C}{v_B} = -\frac{R_C \| z_O}{R_E + r_m}$$

where z_O is the output impedance of a common-base amplifier, given in (7-5).

7-11 CURRENT SOURCE PROPORTIONAL TO ABSOLUTE TEMPERATURE. The gain of a bipolar differential amplifier is independent of the temperature if the bias-current source that feeds the emitters is proportional to absolute temperature (PTAT). Figure 7-32 shows an example of such a source.

Ignore the diode circuit initially. Q1 and Q2 constitute a *Widlar* current source in which I_1 is the reference current and I_2 is the output current; Q2

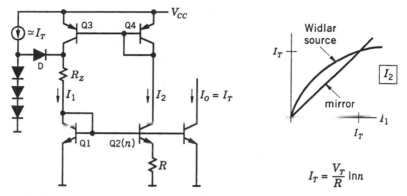

Figure 7-32. Current source proportional to absolute temperature (Exercise 7-11).

has n times the area of Q1, that is, $I_{S2} = nI_{S1}$. With this information you can show that

$$I_2 R = V_{BE1} - V_{BE2} = V_T \ln \frac{nI_1}{I_2}$$

If you now use current mirror Q3|Q4 to force $I_1 = I_2$, you obtain

$$I_O = I_1 = I_2 = I_T \equiv \frac{V_T}{R} \ln n$$

and therefore that I_O is PTAT.

There is a problem, however, because $I_1 = I_2 = 0$ is also a solution. This is the reason for the diode startup circuit. If R_Z is chosen such that $R_Z I_T$ is two diode drops when the circuit is in the nonzero solution, there is no voltage across diode D and the startup circuit has no effect. But if the circuit falls into the zero solution, there is at least one diode drop across R_Z, which is enough to force the circuit into the nonzero solution.

7-12   COMMON-MODE FEEDBACK. The circuit of Fig. 7-33a shows, in simplified form, the first two stages of the National Semiconductor LF156 JFET-bipolar operational amplifier (Russell and Culmer, 1974). The points marked with an A are connected together, as are the points marked with a B. Differential amplifier Q1|Q2 drives differential amplifier Q6|Q7. The collector of Q7 goes to the output stage, which is not shown. JFETs Q3, Q4, and Q8 are identical, as are JFETs Q1 and Q2. In addition, all bipolar transistors have identical characteristics, and the Early voltage V_A is the same for all devices.

The drain loads on Q1 and Q2 are current sources Q3 and Q4, which have the same current I_0 as current source Q8. Inverting amplifier Q5 senses the common-mode voltage V_B at the emitters of Q6|Q7 and forces the common-

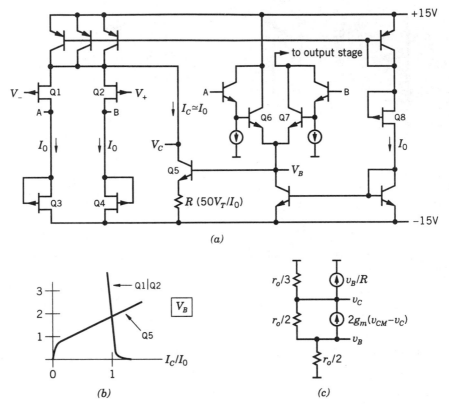

Figure 7-33. Simplified version of the first two stages of the National Semiconductor LF156 JFET–bipolar operational amplifier (Exercise 7-12). Common-mode feedback is used to stabilize the operating point of the input stage. (*a*) Structure. (*b*) Plots of I_C versus V_B for Q5 and of V_B versus I_C for Q1|Q2. (*c*) Incremental circuit used to obtain the common-mode rejection ratio.

mode current in Q1|Q2 to be also equal to I_0. To convince yourself of this fact, sketch I_C as a function of V_B for Q5, and V_B as a function of I_C for Q1|Q2; as shown in Fig. 7-33*b*, the curve for Q1|Q2 is almost vertical at $I_C = I_0$ because of the large gain of Q1|Q2 in the common-gate configuration, and therefore intercepts the curve for Q5 essentially at $I_C = I_0$.

The common-mode voltage at the drains of Q1|Q2 is RI_0 plus three diode drops, and the feedback circuit tends to keep it at this value when the input common-mode voltage changes. The common-mode rejection ratio (CMRR) for Q1|Q2 should therefore be higher than it would be if there were no feedback circuit. To see this, join the drains of Q1 and Q2 and apply a common-mode voltage v_{CM} to the inputs. Obtain the incremental circuit of Fig. 7-33*c* and show that

$$\frac{v_C}{v_{CM}} \simeq \frac{g_m r_o}{1 + g_m r_o} \quad \text{and} \quad \frac{v_B}{v_{CM}} \simeq -\frac{3R}{r_o}$$

With no feedback, you would in principle have $v_B/v_{CM} = -1$, since two rather than three transistors would be used to bias Q1|Q2; the CMRR would thus be a factor $r_o/3R$ lower. At this point you should look carefully at the circuit and convince yourself that feedback is essential because the Q1|Q2 common-mode drain voltage is extremely sensitive to parameter mismatches in its absence.

The value of R should be small, but not so small that it affects the CMRR of differential amplifier Q6|Q7. Think about why, and convince yourself that a reasonable value for R is $50V_T/I_0$ if $\beta = 100$, and that with this choice the CMRR is improved by a factor $V_A/150V_T$, equal to 20 if $V_A = 75$ V.

7-13   GAIN OF A WILSON CURRENT MIRROR. Assuming that $\beta_1 = \beta_2 = \beta_3 = \beta$, show that the image current I_M in the Wilson current mirror of Fig. 7-12 is given by

$$I_M = I_R \frac{1 + \dfrac{2}{\beta}}{1 + \dfrac{2}{\beta} + \dfrac{2}{\beta^2}} \simeq I_R \left(1 - \frac{2}{\beta^2}\right)$$

7-14   BLACKMAN'S THEOREM OPENS UP AN ALTERNATIVE APPROACH. Blackman's theorem can give you an independent check on your results when you are dealing with unfamiliar circuits. Here you will recover known results.

(a) To begin with, calculate the input and output impedances of the common-base amplifier of Fig. 7-6. Take source $g_m v_E$ to be the generalized operational amplifier. For the input impedance you should obtain

$$z_I(g_m = 0) \simeq r_\pi \qquad A_L^{SC} = 0 \qquad A_L^{OC} \simeq \frac{\beta r_o}{r_o + R_L}$$

and from these you should recover (7-6). For the output impedance you should obtain

$$z_O(g_m = 0) \simeq r_o \qquad A_L^{SC} \simeq g_m(R_S \| r_\pi) \qquad A_L^{OC} = 0$$

and from these you should recover (7-5).

(b) Now calculate the output impedance of the Wilson mirror of Fig. 7-12. Take source $g_m v_E$ to be the generalized operational amplifier. You should obtain

$$z_O(g_m = 0) \simeq r_o \qquad A_L^{SC} \simeq \beta \qquad A_L^{OC} = 1$$

from which (7-9) follows.

(c) Finally, calculate the input and output impedances of the emitter follower of Fig. 7-4 and recover (7-3) and (7-4).

7-15   MULTIPLIER FOR LARGE SIGNALS (Gilbert, 1974). The inputs to the Gilbert multiplier presented in Section 7-12 are voltages that are required to be much smaller than the thermal voltage V_T. In this exercise you will analyze a variant that multiplies large-signal currents, the Motorola MC1494 shown in simplified form in Fig. 7-34. A large-signal voltage multiplier is obtained by adding voltage-to-current converters at the input and a current-to-voltage converter at the output.

Assume that the transistors are identical and that $\beta = \infty$. For Q1|Q2 show that

$$V_T \ln\frac{I+I_X}{I-I_X} = V_T \ln\frac{I_2}{I_1}$$

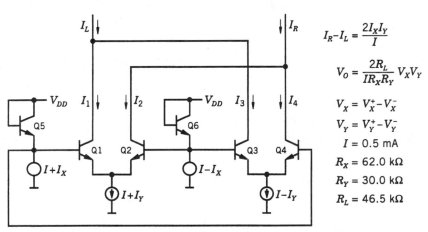

$$I_R - I_L = \frac{2I_X I_Y}{I}$$

$$V_O = \frac{2R_L}{IR_X R_Y} V_X V_Y$$

$$V_X = V_X^+ - V_X^-$$
$$V_Y = V_Y^+ - V_Y^-$$
$$I = 0.5 \text{ mA}$$
$$R_X = 62.0 \text{ k}\Omega$$
$$R_Y = 30.0 \text{ k}\Omega$$
$$R_L = 46.5 \text{ k}\Omega$$

large–signal current multiplier

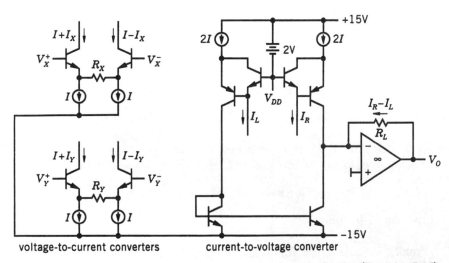

voltage-to-current converters   current-to-voltage converter

Figure 7-34. The Motorola MC1494 large-signal voltage multiplier (Exercise 7-15).

and, using $I_1 + I_2 = I + I_Y$, that

$$(I + I_X)I_1 = (I - I_X)(I + I_Y - I_1)$$

For Q3|Q4 obtain analogously that

$$(I - I_X)I_3 = (I + I_X)(I - I_Y - I_3)$$

Do the algebra in the last two equations and add the results to obtain that

$$I_L = I_1 + I_3 = I - \frac{I_X I_Y}{I}$$

and from continuity, that

$$I_R = I + \frac{I_X I_Y}{I}$$

The differential current is therefore

$$I_R - I_L = \frac{2 I_X I_Y}{I}$$

Expressing all voltages in volts, use the values given in Fig. 7-34 to finally obtain

$$V_O = \frac{V_X V_Y}{10}$$

7-16   LOW-OUTPUT-IMPEDANCE VOLTAGE FOLLOWERS. The zero-offset follower of Fig. 7-35a is much like an operational-amplifier buffer, whereas the JFET-input follower of Fig. 7-35b shows how you can drive a ground-referenced load with a p-channel JFET. Use superposition to show that both circuits have a very low output impedance that is approximately equal to $(r_{m0} + r_{m1})/\beta$; you should therefore expect their gains to be

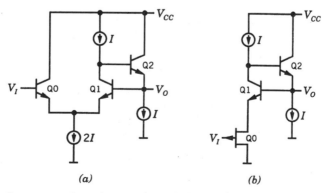

(a)   (b)

Figure 7-35. Low-output-impedance voltage followers (Exercise 7-16). (a) Zero-offset. (b) JFET-input.

very close to 1. The calculation of the gains can be simplified by first considering fragment Q1|Q2, because you know what to do once you have its gain and input impedance. For simplicity, take the load resistance to be infinite; this will not change the spirit of the exercise.

Consider then the circuit shown in Fig. 7-36. Q1 has a large collector load and therefore a large gain. It follows that $|v_{CB}| \gg |v_{EB}|$, even though both are quite small. Looking at the incremental circuit, it also follows that you can remove the resistor (it happens to be $r_{\pi 1}$) between the points marked "×". You can therefore consider that emitter follower Q2 is loaded only by its

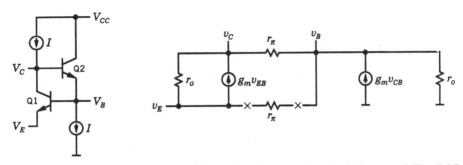

Figure 7-36. Incremental circuit of the Q1|Q2 fragment in the followers of Fig. 7-35 (Exercise 7-16).

own collector output resistance. From expression (7-2) for the emitter-follower gain, you then immediately get

$$v_B \simeq \frac{g_m r_o}{1+g_m r_o} v_C$$

Summing currents at node v_C yields

$$v_E(g_m+g_o) = v_C(g_\pi+g_o) + v_B(g_m-g_\pi)$$

You can now obtain v_B and v_C in terms of v_E (and confirm that $|v_{CB}| \gg |v_{EB}|$):

$$v_B \simeq \frac{\beta g_m r_o}{1+\beta g_m r_o} v_E \quad \text{and} \quad v_C \simeq \frac{1+g_m r_o}{g_m r_o} v_E$$

The input current is

$$i_E \simeq v_{CB} g_\pi \simeq \frac{v_E}{\beta r_o}$$

and the input impedance is therefore

$$z_E = \beta r_o$$

If you substitute $(R_L \| r_o)$ for r_o, the results for v_B/v_E and z_E are valid for $\beta R_L \gg r_o$. With this mild restriction, the excellence of the followers of Fig. 7-35 is limited only by the amplification factor of Q0 because $z_E \gg r_o$. You can verify your result for z_E quite easily using Blackman's theorem and the fact that $v_B \simeq v_C$. You should get

$$z_E(g_{m1} = 0) = \beta r_o \qquad A_L^{SC} = A_L^{OC} = g_m r_o$$

If you are not happy about the idea of taking $r_{\pi 1}$ to be infinite, you can check your result for z_E by using Blackman's theorem again, this time on the complete incremental circuit. In this case you should get

$$z_E(g_{m1} = 0) \simeq r_\pi (\beta + g_m r_o / \beta) \qquad A_L^{SC} \simeq \frac{\beta g_m r_o}{\beta + g_m r_o / \beta} \qquad A_L^{OC} \simeq \beta$$

8

DYNAMIC RESPONSE OF SEMICONDUCTOR CIRCUITS

8-1   INTRODUCTION

In this chapter we will look into the small-signal frequency response of the circuits presented in Chapter 7. One of our principal objectives will be to show that the simplicity of the fragments discussed there is largely conserved and that calculations in terms of such fragments remain quite straightforward. For completeness, we will consider the dynamic response for large signals in a few representative cases.

As we have done so far, we will stick to the basics and avoid special methods. In the exercises, however, we will introduce the y-parameter model for the common-emitter (or common-source) configuration because it is so similar to the small-signal models developed in Chapter 6 that it is no burden on memory, yet it has the advantage of making calculations much easier.

The small-signal models of bipolar and field-effect transistors are formally identical; we will therefore continue to concentrate on bipolar circuits since our calculations can be applied to field-effect versions by making the current gain infinite.

8-2   PRELIMINARY EXERCISE: THE SHORT-CIRCUIT FORWARD CURRENT GAIN

The short-circuit forward current gain $h_{fe}(s)$ of a bipolar transistor is the small-signal current gain from the base to the collector when the emitter is grounded and the collector is held at a fixed voltage, as shown in Fig. 8-1a. It is thus one of the parameters of the obsolete h-parameter model of bipolar

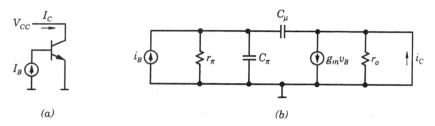

Figure 8-1. Measurement of the short-circuit forward current gain. (*a*) Biasing circuit. (*b*) Incremental circuit.

transistors; we will maintain the traditional nomenclature because it is well installed in the literature.

The incremental circuit is shown in Fig. 8-1*b*; the collector output resistance r_o is included as a formality, since it is short-circuited and does not affect the collector current. Solving this circuit, we obtain

$$h_{fe}(s) = \frac{i_C}{i_B} = \frac{r_\pi}{1+s(C_\pi+C_\mu)r_\pi}(g_m - sC_\mu) = \frac{\beta(1-sC_\mu r_m)}{1+s(C_\pi+C_\mu)r_\pi}$$

Dropping time constant $C_\mu r_m$ for reasons we will see below, we have

$$h_{fe}(s) \simeq \frac{\beta}{1+s(C_\pi+C_\mu)r_\pi}$$

To proceed we need to define a *transition frequency* as a frequency at which the magnitude of a frequency response, much larger than 1 over some range as in the case of h_{fe}, falls to 1. We can now see that the transition frequency of h_{fe} is essentially the f_T we introduced in Section 6-3-6 to characterize the speed of a transistor:

$$f_T = \frac{1}{2\pi(C_\pi+C_\mu)r_m}$$

The small-signal model starts to fail at frequencies near f_T; for this reason, the value of f_T quoted in data sheets is normally extrapolated from a measurement of h_{fe} at a frequency f_M (usually 100 MHz) that is several times lower than f_T but much higher than the cutoff frequency $f_C = f_T/\beta$. The quoted value is then

$$f_T = f_M h_{fe}(f_M)$$

Equivalently, both f_M and $h_{fe}(f_M)$ are sometimes quoted. We note that measuring h_{fe} at a lower frequency justifies our neglect of time constant $C_\mu r_m$. We further note that data sheets also quote C_μ, which allows us to extract C_π.

From now on we will normally drop time constants of order $C_\pi r_m$ and $C_\mu r_m$: they are much smaller than the dominant time constants of the circuits we will be dealing with, and since they correspond to frequencies of order f_T or higher, at which the small-signal model is no longer fully valid, they account for only part of the frequency response. For reference, we note that $C_\pi r_m \approx 1$ ns at $I_C = 1$ mA in a 2N3904 transistor.

8-3   THE COMMON-EMITTER AMPLIFIER AND THE MILLER EFFECT

The incremental circuit of the common-emitter amplifier presented in Section 7-2, including source and load impedances, is shown in Fig. 8-2. With the substitutions

$$v_T = i_S(R_S \| r_\pi) \qquad R_T = R_S \| r_\pi \qquad C_T = C_S + C_\pi + C_\mu$$

we obtain the simplified circuit shown in Fig. 8-3 (why we include a term C_μ in C_T will become clear below).

The presence of the collector–base capacitance C_μ has important consequences. To get an idea of what C_μ does, let us first consider the inverting amplifier shown in Fig. 8-4, and let us assume that the operational amplifier is ideal and has a *constant* gain A. We then have

$$V_O = -AV_I$$

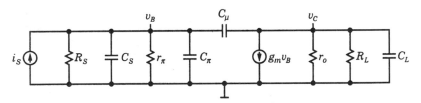

Figure 8-2. Incremental circuit of a common-emitter amplifier, including source and load impedances.

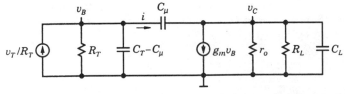

Figure 8-3. Simplified version of the incremental circuit of Fig. 8-2.

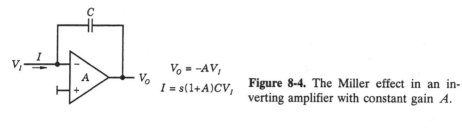

$$V_O = -AV_I$$

$$I = s(1+A)CV_I$$

Figure 8-4. The Miller effect in an inverting amplifier with constant gain A.

and

$$I = sC(V_I - V_O) = s(1+A)CV_I$$

A feedback capacitor C thus results in an input capacitance $(1+A)C$; this is known as the *Miller effect*, and it is quite easy to visualize in this case: a voltage V_I at the input generates a voltage $-AV_I$ at the output, so that the capacitor voltage—and therefore the current—is $1+A$ times larger than it would be if the capacitor were connected from the input to ground. With some complications, this is the effect of C_μ.

We can now continue our analysis of the common-emitter amplifier. Summing currents at the collector node in Fig. 8-3, we have

$$sC_\mu(v_B - v_C) = g_m v_B + (g_o + G_L + sC_L)v_C$$

and thus

$$\frac{v_C}{v_B} = -\frac{g_m(R_L \| r_o)(1 - sr_m C_\mu)}{1 + s(R_L \| r_o)(C_L + C_\mu)}$$

Neglecting time constant $r_m C_\mu$, we obtain the voltage gain from the base to the collector,

$$\frac{v_C}{v_B} = \frac{-g_m(R_L \| r_o)}{1 + s(R_L \| r_o)(C_L + C_\mu)} \qquad (8\text{-}1)$$

The current i through C_μ is $sC_\mu(v_B - v_C)$, so that

$$\frac{i}{v_B} = sC_\mu + \frac{g_m(R_L \| r_o)}{1 + s(R_L \| r_o)(C_L + C_\mu)} sC_\mu$$

The first term in this expression corresponds to a capacitor C_μ to ground. The reciprocal of the second term can be expressed as

$$\frac{1}{sC_M} + R_M$$

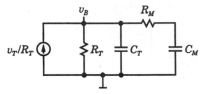

Figure 8-5. Input circuit of a common-emitter amplifier, exhibiting the Miller capacitance C_M and the Miller resistance R_M.

where C_M and R_M are the Miller capacitance and the Miller resistance,

$$C_M = g_m(R_L\|r_o)C_\mu \quad \text{and} \quad R_M = r_m\left(1 + \frac{C_L}{C_\mu}\right) \tag{8-2}$$

We observe that $R_M C_M$ is equal to the time constant in the voltage gain from v_B to v_C in (8-1),

$$R_M C_M = (R_L\|r_o)(C_L + C_\mu) \tag{8-3}$$

As far as the source is concerned, C_μ looks like the parallel combination of a capacitor C_μ (already included in C_T) and the Miller impedance $R_M + 1/sC_M$. The input circuit can thus be represented as shown in Fig. 8-5. According to Exercise 3-8, v_B is given in terms of v_T by

$$\frac{v_B}{v_T} = \frac{1 + sR_M C_M}{1 + s(R_T C_M + R_M C_M + R_T C_T) + s^2 R_M C_M R_T C_T}$$

Using (8-1) and (8-3) we then obtain the voltage gain from v_T to v_C,

$$\frac{v_C}{v_T} = \frac{-g_m(R_L\|r_o)}{1 + s(R_T C_M + R_M C_M + R_T C_T) + s^2 R_M C_M R_T C_T}$$

Let us first consider the case in which $R_T \gg R_M$, which is very important in practice. For simplicity we will assume that C_M is at least several times larger than C_T, as it is in high-gain amplifiers with reasonable source capacitances. Since the maximum value of $R_T = R_S\|r_\pi$ is $r_\pi = \beta r_m$, we obtain from (8-2) that $R_T \gg R_M$ implies $C_L \ll \beta C_\mu$; in most cases this means that C_L can be at most a few tens of picofarads. We now observe that

$$\frac{R_M C_M R_T C_T}{(R_T C_M + R_M C_M + R_T C_T)^2} \leq \frac{R_M C_M R_T C_T}{(R_T C_M + R_T C_T)^2}$$

$$= \frac{C_M C_T}{(C_M + C_T)^2}\frac{R_M}{R_T} \leq \frac{1}{4}\frac{R_M}{R_T} \ll 1$$

Observing further that

$$1+s\tau_1+s^2\tau_2^2 \simeq (1+s\tau_1)\left(1+s\frac{\tau_2^2}{\tau_1}\right) \tag{8-4}$$

if $\tau_2 \ll \tau_1$, we obtain

$$\frac{v_C}{v_T} \simeq \frac{-g_m(R_L\|r_o)}{[1+sR_T(C_M+C_T)](1+sR_MC_T)} \tag{8-5}$$

We note that if $R_T>5R_M$ and $C_M>5C_T$, R_MC_T is at least 30 times smaller than $R_T(C_M+C_T)$. The condition on R_T is not very restrictive: for $C_\mu = 4$ pF, $C_L = 8$ pF, and $I_C = 1$ mA, it is satisfied if $R_T>375$ Ω.

If C_μ were negligible, the base and collector circuits would become uncoupled, and their time constants would be R_TC_T and $(R_L\|r_o)(C_L+C_\mu) = R_MC_M$, which are usually of comparable magnitude. As we will see in Chapter 9, this would be a great inconvenience in a feedback circuit. In the presence of C_μ, and assuming that $C_M \gg C_T$, the time constants in (8-5) are $R_TC_M \gg R_TC_T$ and $R_MC_T \ll R_MC_M$; that is, they depart in opposite directions from the uncoupled values. This behavior is called *pole splitting*, after the fact that the zeros of the denominator of a rational function are called *poles* in complex variable theory.

If we drop the second term in the denominator on the right-hand side in (8-5)—its time constant R_MC_T is often of order r_mC_π—we get

$$\frac{v_C}{v_T} \simeq \frac{-g_m(R_L\|r_o)}{1+sR_T(C_M+C_T)}$$

In this case, v_C/v_T can be obtained from the equivalent circuit shown in Fig. 8-6. This circuit is unidirectional and can thus not be used to calculate the output impedance; it is nonetheless a compelling image of the Miller effect in a common-emitter amplifier, and we will find it quite useful.

The Miller effect makes inverting amplifiers slower and can thus represent a serious limitation; it can be virtually eliminated, however, by using cascode amplifiers, as we will see in the following section. In Section 8-6 we will also

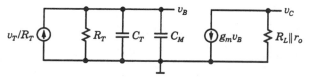

Figure 8-6. Unidirectional equivalent circuit of a common-emitter amplifier with a well-developed Miller effect.

see that it can be used to advantage. In fact, it should already be clear that *RC* integrators—and all operational-amplifier inverting circuits, for that matter—are based on the Miller effect.

Let us now consider the case $C_L \gg \beta C_\mu$. We then have $R_M \gg \beta r_m = r_\pi$, and we can eliminate R_M and C_M from the input circuit in Fig. 8-5. The reason is clear: the voltage gain is negligible at the frequencies at which C_μ can contribute significantly to the input current. The amplifier input impedance is then simply r_π in parallel with $C_\pi + C_\mu$, and the voltage gains are

$$\frac{v_B}{v_T} = \frac{1}{1 + sR_T C_T}$$

and

$$\frac{v_C}{v_T} = \frac{-g_m(R_L \| r_o)}{(1 + sR_T C_T)[1 + s(R_L \| r_o)C_L]}$$

Returning to Fig. 8-3, we will now obtain the output circuit seen by the load. For the short-circuit current we have

$$i_{SC} = -\frac{g_m(1 - sr_m C_\mu)}{1 + sR_T C_T} v_T \simeq -\frac{g_m}{1 + sR_T C_T} v_T$$

To calculate the output impedance we will null the input source, remove the load impedance, and apply a voltage v_C to the collector. Let us first obtain v_B in terms of v_C by summing currents at the base:

$$\left[s(C_T - C_\mu) + G_T \right] v_B = sC_\mu(v_C - v_B)$$

and thus

$$v_B = \frac{sR_T C_\mu}{1 + sR_T C_T} v_C$$

We can now calculate the collector current i_C,

$$i_C = sC_\mu(v_C - v_B) + g_o v_C + g_m v_B$$

$$= (sC_\mu + g_o)v_C + (1 - sC_\mu r_m)g_m v_B \simeq (sC_\mu + g_o)v_C + g_m v_B$$

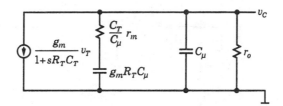

Figure 8-7. Output circuit of a common-emitter amplifier.

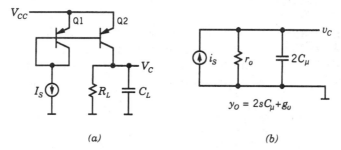

(a)   (b)

Figure 8-8. Current mirror. (a) Viewed as a common-emitter amplifier. (b) Output circuit.

and, expressing v_B in terms of v_C, the output admittance y_O:

$$y_O \approx sC_\mu + g_o + \frac{sg_m R_T C_\mu}{1 + sR_T C_T} = sC_\mu + g_o + \frac{1}{\dfrac{C_T}{C_\mu} r_m + \dfrac{1}{sg_m R_T C_\mu}}$$

Putting everything together, we obtain the output circuit shown in Fig. 8-7.

Let us apply this result to the current mirror of Fig. 8-8a. We first observe that $C_{\pi 1}$, $C_{\pi 2}$, and $r_{\pi 2}$ are in parallel with the emitter resistance r_{m1} of Q1 and can therefore be ignored. Substituting $v_T = -i_s r_m$, $R_T = r_m$, and $C_T = 0$ in Fig. 8-7, we obtain the output circuit shown in Fig. 8-8b. The short-circuit current and the output admittance are thus

$$i_{SC} = i_S$$

and

$$y_O = 2sC_\mu + g_o$$

In Exercise 8-9 it is shown that a Wilson mirror also has output capacitance $2C_\mu$ and that its output admittance is therefore $2sC_\mu + 2g_o/\beta$.

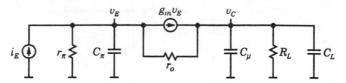

Figure 8-9. Incremental circuit of a loaded common-base amplifier driven by a current source.

8-4 THE COMMON-BASE FRAGMENT

The incremental circuit of a loaded common-base amplifier is shown in Fig. 8-9. We will only consider the case of an infinite source impedance. We note that dependent current source $g_m v_E$ acts like a resistor when the output is grounded. The short-circuit current can thus be obtained by current division:

$$i_{SC} = i_E \frac{g_m + g_o}{g_m + g_o + g_\pi + sC_\pi} \simeq i_E \frac{1}{1 + sr_m C_\pi} \simeq i_E$$

To obtain the output admittance y_O, we will remove C_μ and the load impedance and then calculate the output admittance y_T of the remainder and put it in parallel with C_μ. The short-circuit current i_T of the remainder is also i_E,

$$i_T = i_E$$

Continuity implies that the open-circuit voltage v_T of the remainder is

$$v_T = i_E \frac{r_\pi}{1 + sr_\pi C_\pi} (1 + g_m r_o) \simeq i_E \frac{\beta r_o}{1 + sr_\pi C_\pi}$$

Observing that $C_\pi / g_m r_o \ll C_\mu$, we then get

$$y_T = \frac{i_T}{v_T} = \frac{1}{\beta r_o} + \frac{sC_\pi}{g_m r_o} \simeq \frac{1}{\beta r_o}$$

and, putting y_T in parallel with C_μ,

$$y_O = \frac{1}{\beta r_o} + sC_\mu$$

The gain from the input current i_E to the collector voltage v_C is thus

$$\frac{v_C}{i_E} = \frac{R_L \| \beta r_o}{1 + s(R_L \| \beta r_o)(C_L + C_\mu)} \tag{8-6}$$

To obtain the voltage gain v_C/v_E we will sum currents at the collector node:

$$g_m v_E = (G_L + sC_L + sC_\mu)v_C + g_o(v_C - v_E)$$

or, collecting terms,

$$(G_L + g_o + sC_L + sC_\mu)v_C = (g_m + g_o)v_E \approx g_m v_E$$

The voltage gain is thus

$$\frac{v_C}{v_E} = \frac{g_m(R_L \| r_o)}{1 + s(R_L \| r_o)(C_L + C_\mu)} \tag{8-7}$$

Finally, we can obtain the input impedance z_I using (8-6) and (8-7):

$$z_I = \frac{v_E}{i_E} = r_m \frac{R_L \| \beta r_o}{R_L \| r_o} \frac{1 + s(R_L \| r_o)(C_L + C_\mu)}{1 + s(R_L \| \beta r_o)(C_L + C_\mu)} \tag{8-8}$$

If $\omega(R_L \| r_o)(C_L + C_\mu) \gg 1$, we have $z_I = r_m$. If R_L is of order r_o or larger, $z_I(\omega)$ increases at low frequencies, and for infinite R_L, we have $z_I = r_\pi$ at $\omega = 0$.

The increase in the input impedance at low frequencies, if present at all, has no important consequences. It might increase the Miller capacitance C_M of a preceding stage, but then only at frequencies at which C_M is an open circuit; in addition, we are assuming that the source resistance is much larger than r_π, so that the low-frequency gain is not affected. We could therefore take the input impedance to be simply r_m; however, for consistency with Section 8-5, we will retain input capacitance C_π. We then get the incremental circuit shown in Fig. 8-10. We note that there is no coupling from the collector to the emitter or, equivalently, that the reverse transadmittance is zero.

We can now combine the results we have obtained for common-emitter and common-base amplifiers and apply them to the cascode amplifier shown in Fig. 8-11. As in the low-frequency case of Section 7-4, the cascode

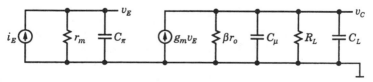

Figure 8-10. Simplified incremental circuit of a common-base amplifier, showing that the reverse transadmittance is zero.

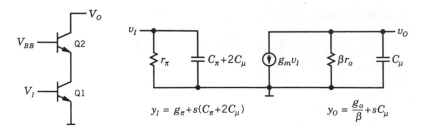

$$y_I = g_\pi + s(C_\pi + 2C_\mu) \qquad y_O = \frac{g_o}{\beta} + sC_\mu.$$

Figure 8-11. The cascode fragment is equivalent to a transistor with a large collector output resistance and essentially zero collector–base capacitance.

amplifier has advantages over a common-emitter amplifier: there is no capacitive coupling between the output and the input, which virtually does away with the Miller effect, and the output resistance is βr_o rather than r_o because the output resistance of Q1 (r_o) is much larger than the input resistance of Q2 $(\leq r_\pi)$.

If the collector load on output transistor Q2 is moderate, the collector load on input transistor Q1 is the input resistance r_m at the emitter of Q2, and the voltage gain of Q1 is therefore $-g_m r_m = -1$. The Miller capacitance C_M is thus equal to C_μ, and the input capacitance is equal to $C_\pi + 2C_\mu$.

In Exercise 8-4 it is shown that $C_M = C_\mu$ irrespective of the load on Q2. The incremental circuit shown in Fig. 8-11 is therefore always valid. We note that in most cases we have $C_\pi \gg C_\mu$, so that the input capacitance is simply C_π.

8-5   THE EMITTER FOLLOWER

The frequency response of emitter followers is somewhat complicated to describe because it seldom exhibits a dominant time constant such as the one that characterizes a common-emitter amplifier with a well-developed Miller effect. In the ensuing discussion we should therefore not lose sight of the fact that an emitter follower is mainly an impedance converter: the load impedance appears larger at the input, and the source impedance appears smaller at the output. Since everything counts, we will not summarily neglect small time constants such as $r_m C_\pi$ as we have done so far, but rather proceed step by step from general results given for reference to particular cases that are of interest to us.

The incremental circuit of an emitter follower, including source and load impedances, is shown in Fig. 8-12. To simplify our results, we have absorbed C_μ in C_S, which is not too risky because C_S is usually much larger than C_μ; we have also absorbed r_o in R_L, so that R_L actually stands for $R_L \| r_o$.

To obtain the gain v_E/v_B we will sum currents at the emitter node:

$$(v_B - v_E)(g_m + y_\pi) = v_E Y_L$$

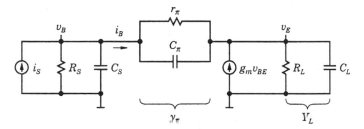

Figure 8-12. Incremental circuit of an emitter follower.

and thus

$$v_E = \frac{g_m + y_\pi}{g_m + y_\pi + Y_L} v_B$$

Neglecting g_π in comparison with g_m, we obtain

$$\frac{v_E}{v_B} = \frac{g_m R_L}{1 + g_m R_L} \frac{1 + sC_\pi r_m}{1 + s(C_L + C_\pi)(r_m \| R_L)} \qquad (8\text{-}9)$$

We can now calculate the base current:

$$z_\pi i_B = v_{BE} = \frac{Y_L}{g_m + y_\pi + Y_L} v_B \simeq \frac{1}{(1 + g_m R_L)} \frac{1 + sC_L R_L}{1 + s(C_L + C_\pi)(r_m \| R_L)} v_B$$

The input impedance $z_I = v_B / i_B$ is thus

$$z_I = \beta(r_m + R_L) \frac{1 + s(C_L + C_\pi)(r_m \| R_L)}{(1 + sC_L R_L)(1 + sC_\pi r_\pi)} \qquad (8\text{-}10)$$

Our first simplifying assumption will be that $g_m R_L \gg 1$. The gain and the input impedance then become

$$\frac{v_E}{v_B} = \frac{1 + sC_\pi r_m}{1 + s(C_L + C_\pi)r_m}$$

$$\qquad (8\text{-}11)$$

$$z_I = \beta R_L \frac{1 + s(C_L + C_\pi)r_m}{(1 + sC_L R_L)(1 + sC_\pi r_\pi)}$$

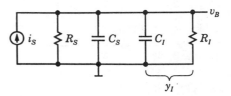

Figure 8-13. Input circuit of an emitter follower assuming that $g_m R_L \gg 1$ and that $C_L r_m$ and $C_\pi r_m$ are negligible. C_I and R_I are defined in (8-13).

If we further assume that $C_\pi r_m$ is negligible, we obtain

$$\frac{v_E}{v_B} = \frac{1}{1 + s C_L r_m}$$

$$z_I = \beta R_L \frac{1 + s C_L r_m}{(1 + s C_L R_L)(1 + s C_\pi r_\pi)} \qquad (8\text{-}12)$$

Simplifying even further, we will assume that C_L, while not negligible, is small enough, let us say of order C_π, that $C_L r_m$ is negligible. As shown in Fig. 8-13, we can then represent y_I as the parallel combination of a capacitance C_I and a frequency-dependent resistance R_I:

$$y_I = \frac{(1 + s C_L R_L)(1 + s C_\pi r_\pi)}{\beta R_L} = s C_I + \frac{1}{R_I} \qquad (8\text{-}13)$$

where C_I and R_I are given by

$$C_I = \frac{C_L}{\beta} + \frac{C_\pi}{g_m R_L}$$

and

$$\frac{1}{R_I} = \frac{1 - \omega^2 C_L R_L C_\pi r_\pi}{\beta R_L}$$

Assuming that C_π is much larger than C_μ, we can rewrite $1/R_I$ in terms of the transition frequency $\omega_T \approx 1/r_m C_\pi$,

$$\frac{\beta R_L}{R_I} = 1 - \frac{\omega^2}{\omega_T^2} \frac{C_L}{C_\pi} \beta g_m R_L$$

Since $C_L \approx C_\pi$ and $\beta g_m R_L \approx 1000$, R_I can become negative at frequencies well below ω_T. This implies that emitter followers (and source followers) are inherently only marginally stable, and it is in fact quite easy to obtain an oscillator by replacing R_S with an inductor, thus converting the input circuit into a tank circuit. The tendency to oscillate vanishes if C_L is sufficiently

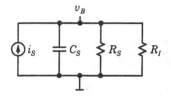

Figure 8-14. Simplified version of the input circuit of Fig. 8-13, obtained by observing that $C_I \ll C_S$.

large: as shown in Exercise 8-5, the real part of the input impedance as given in (8-10) is positive at all frequencies if $C_L > \beta C_\pi$. This extreme condition is of course rarely satisfied, certainly not in the case that we are considering, and as we will see immediately, it is then C_S that plays a crucial stabilizing role.

If we now recall that C_S includes C_μ, we can neglect C_I in Fig. 8-13 and calculate v_B and v_E with the circuit shown in Fig. 8-14. Summing currents at the base node, we obtain

$$i_S = \left(sC_S + \frac{1}{R_S} + \frac{1}{R_I} \right) v_B = \left(sC_S + \frac{1}{R_S} + \frac{1}{\beta R_L} + s^2 C_L C_\pi r_m \right) v_B$$

We can then obtain the gains from i_S to v_B and v_E,

$$\frac{v_E}{i_S} \simeq \frac{v_B}{i_S} \simeq \frac{R_S \| \beta R_L}{1 + s(R_S \| \beta R_L)C_S + s^2(R_S \| \beta R_L)r_m C_\pi C_L} \qquad (8\text{-}14)$$

From this expression we conclude that the source capacitance C_S is in fact stabilizing, and in practice it turns out that quite modest values of C_S result in a damped response. As we will see below, this conclusion remains valid in a more complete treatment. If $(R_S \| \beta R_L)C_S$ is sufficiently large, we can use the approximation given in (8-4) to obtain

$$\frac{v_E}{i_S} \simeq \frac{v_B}{i_S} \simeq \frac{R_S \| \beta R_L}{[1 + s(R_S \| \beta R_L)C_S](1 + sr_m C_\pi C_L/C_S)}$$

If $C_L \approx C_S$ the second term in the denominator can be ignored, and we get

$$\frac{v_E}{i_S} \simeq \frac{v_B}{i_S} \simeq \frac{R_S \| \beta R_L}{1 + s(R_S \| \beta R_L)C_S} \qquad (8\text{-}15)$$

This is the result we would have obtained if we had ignored C_π all along. If we again recall that C_μ was absorbed in C_S, we conclude that for small C_L and large R_S the input impedance of a follower is simply βR_L in parallel with C_μ, and that neither C_π nor C_L contributes to the frequency response.

Using only the assumption that $g_m R_L \gg 1$, it is shown in Exercise 8-6 that

$$\frac{v_E}{i_S} = (R_S \| \beta R_L) \frac{1 + sr_m C_\pi}{1 + s\tau_1 + s^2 \tau_2^2} \qquad (8\text{-}16)$$

where τ_1 and τ_2 are given by

$$\tau_1 = (R_S \| \beta R_L)\left(C_S + \frac{C_\pi + C_L}{g_m R_S} + \frac{C_L}{\beta} + \frac{C_\pi}{g_m R_L}\right)$$

and

$$\tau_2^2 = r_m(R_S \| \beta R_L)(C_L C_\pi + C_S C_\pi + C_S C_L)$$

We deduce that (8-14) is slightly optimistic, although still indicative. Let us take values that are typical in fast operational amplifiers, $C_S = C_\pi$, $C_L = 2C_\pi$, $R_S = r_o$, and $R_L = \infty$. We then have

$$\tau_1 \simeq r_o C_\pi$$
$$\tau_2^2 = 5r_m r_o C_\pi^2$$

whereas (8-14), while agreeing on τ_1, gives $\tau_2^2 = 2r_m r_o C_\pi^2$. Nonetheless, we have

$$\frac{\tau_2^2}{\tau_1} \simeq 5r_m C_\pi$$

so that (8-15) is valid up to frequencies not too far below the transition frequency $\omega_T \simeq 1/r_m C_\pi$. In Exercise 8-11, expression (8-16) is used to study the case in which the source resistance is small, and again it turns out that a modest source capacitance is stabilizing.

Returning to Fig. 8-12, we will now obtain the output impedance of an emitter follower. Making the load impedance infinite, we obtain that the open-circuit output voltage v_T is

$$v_T = \frac{R_S}{1 + sC_S R_S} i_S$$

When the output is grounded we have

$$(sC_S + G_S + sC_\pi + g_\pi)v_B = i_S$$

and the short-circuit current i_T is given in terms of v_B by

$$i_T = (g_m + g_\pi + sC_\pi)v_B \simeq (g_m + sC_\pi)v_B$$

Substituting v_B in terms of i_S, we get

$$i_T = \frac{g_m + sC_\pi}{s(C_S + C_\pi) + G_S + g_\pi} i_S = g_m(R_S \| r_\pi)\frac{1 + sC_\pi r_m}{1 + s(C_S + C_\pi)(R_S \| r_\pi)} i_S$$

Finally, we can obtain $z_O = v_T/i_T$,

$$z_O = \left(r_m + \frac{R_S}{\beta}\right)\frac{1+s(C_S+C_\pi)(R_S\|r_\pi)}{(1+sC_SR_S)(1+sC_\pi r_m)} \qquad (8\text{-}17)$$

The source capacitance C_S is essential in keeping the high-frequency output impedance low when the source resistance R_g is large; in this case R_g/β is also large, and so therefore is the time constant it forms with the load capacitance C_L. Let us assume that we can neglect $C_\pi r_m$. For $\omega C_S R_S \gg 1$ and $R_S \gg r_\pi$ we then have

$$z_O \simeq \frac{1}{s\beta C_S} + r_m\left(1+\frac{C_\pi}{C_S}\right)$$

The output impedance at high frequencies is thus the series combination of a large capacitance and a resistance that is of order r_m and therefore much smaller than R_S/β.

8-6 AMPLIFIERS WITH SMALL EMITTER DEGENERATION

Let us consider the circuit of Fig. 8-15a, and let us assume that the emitter load resistance R_E is of order r_m, that the emitter load capacitance C_E is of order C_π, and that the collector load resistance R_C is only a few times larger than R_E. It follows that we can ignore the collector output resistance r_o and that we can calculate v_E as if the circuit were an emitter follower. From (8-9) we then have

$$\frac{v_E}{v_B} = \frac{g_m R_E}{1+g_m R_E}\frac{1+sC_\pi r_m}{1+s(C_E+C_\pi)(r_m\|R_E)}$$

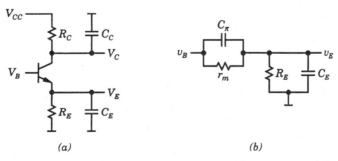

(a) *(b)*

Figure 8-15. Amplifier with emitter degeneration. (*a*) Structure. (*b*) Equivalent circuit at the emitter node.

For the moment we will assume that $C_\mu = 0$. From (8-10) we then also have

$$z_I = \beta(r_m + R_E)\frac{1+s(C_E+C_\pi)(r_m\|R_E)}{(1+sC_E R_E)(1+sC_\pi r_\pi)}$$

Given our assumptions, we could neglect all time constants except $C_\pi r_\pi$. But we can do better. We will choose C_E such that

$$C_E R_E = C_\pi r_m \tag{8-18}$$

or, equivalently, such that

$$(C_E+C_\pi)(r_m\|R_E) = C_\pi r_m$$

With this choice, the gain and the input admittance become

$$\frac{v_E}{v_B} = \frac{g_m R_E}{1+g_m R_E} \tag{8-19}$$

and

$$y_I = \frac{1+sC_\pi r_\pi}{r_\pi(1+g_m R_E)} \tag{8-20}$$

and the collector current is proportional to the input voltage,

$$i_C = g_m v_{BE} = \frac{g_m}{1+g_m R_E}v_B \tag{8-21}$$

If we now take C_μ into account, we obtain the incremental circuit shown in Fig. 8-16. The circuit of Fig. 8-15a therefore behaves like a common-emitter amplifier if $C_E R_E = C_\pi r_m$, except for the fact that the base input impedance is $1+g_m R_E$ times larger and the transconductance is $1+g_m R_E$ times smaller. This implies that we can calculate the Miller effect according to Section 8-3.

From a different point of view, we can set $R_S = 0$ in (8-17) because we are assuming that we know v_B. The output impedance at the emitter is r_m in

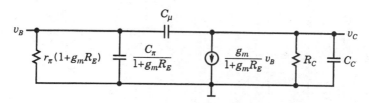

Figure 8-16. A common-emitter amplifier with small emitter degeneration looks like a common-emitter amplifier based on a transistor with a transconductance $1+g_m R_E$ times smaller and a base input impedance $1+g_m R_E$ times larger.

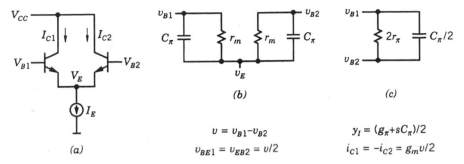

Figure 8-17. Differential amplifier with negligible collector loads. (*a*) Structure. (*b*) Equivalent circuit looking into the emitter node. (*c*) Input circuit.

parallel with C_π in this case, and we can calculate v_E with the incremental circuit shown in Fig. 8-15*b*. If (8-18) is satisfied, we immediately recover (8-19), (8-20), and (8-21).

Let us apply these results to the differential amplifier of Fig. 8-17*a*. We will assume that the collector loads are small enough that we can ignore the Miller effect. For either transistor, the output and load impedances at the emitter are equal to r_m in parallel with C_π, as shown in Fig. 8-17*b*, so that (8-18) holds. For a differential voltage $v = v_{B1} - v_{B2}$ it is then clear that

$$v_{BE1} = v_{EB2} = \frac{v}{2}$$

and, neglecting the effect of C_μ, that the collector currents are

$$i_{C1} = -i_{C2} = g_m \frac{v}{2}$$

As indicated in Fig. 8-17*c*, the input admittance is

$$y_I = \frac{g_\pi + sC_\pi}{2}$$

The input stage of many operational amplifiers is a differential amplifier like the one shown in Fig. 8-18*a*. In almost all cases, the load capacitance C_L is the Miller capacitance of the following stage and is thus much larger than βC_μ. As we saw in Section 8-3, this implies that the Miller effect in Q2 is negligible. The Miller effect is also clearly negligible in Q1 because Q3 is diode connected. From Section 7-7 we know that the output resistance at the collectors of Q2|Q4 is $r_o/2$; since C_L is large, we can ignore the amplifier output capacitance. We then obtain the extremely simple incremental circuit shown in Fig. 8-18*b*.

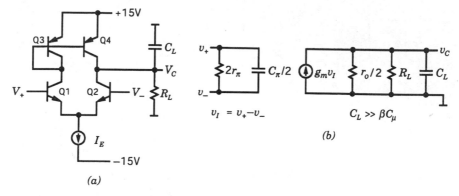

Figure 8-18. Widely used operational-amplifier input circuit. (*a*) Structure. (*b*) Incremental circuit.

8-7 EXAMPLE: VIDEO AMPLIFIER

To see how easy the analysis of a composite circuit can be, let us consider the amplifier shown in Fig. 8-19*a*. It consists of a cascode amplifier (Q1|Q2) with an active load (Q3), and an output emitter follower (Q4) with an infinite load resistance. As usual, we will assume that all transistors have identical characteristics and that diodes are diode-connected transistors.

Emitter follower Q4 sees a high source resistance (r_o); according to Section 8-5, its input capacitance is equal to C_μ if C_L is small.

The load resistance of the cascode amplifier is equal to the output resistance r_o of mirror transistor Q3 because the output resistance of Q2 and the input resistance of Q4 are both equal to βr_o. The load capacitance has contributions from follower Q4 (C_μ), mirror Q3 ($2C_\mu$), and cascode output transistor Q2 (C_μ). Capacitor C_B can be used to reduce the bandwidth without introducing an additional phase lag.

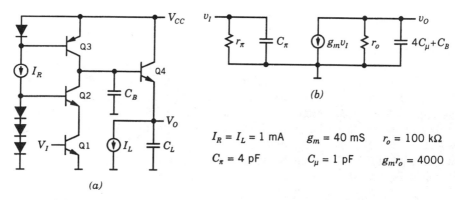

$$I_R = I_L = 1 \text{ mA} \qquad g_m = 40 \text{ mS} \qquad r_o = 100 \text{ k}\Omega$$
$$C_\pi = 4 \text{ pF} \qquad C_\mu = 1 \text{ pF} \qquad g_m r_o = 4000$$

Figure 8-19. Video amplifier. (*a*) Structure. (*b*) Incremental circuit.

The incremental circuit is shown in Fig. 8-19b. With the indicated values we obtain that the low-frequency gain is $g_m r_o = 4000$. At frequencies out to about 50 MHz we also have

$$\frac{v_O}{v_I} = \frac{g_m r_o}{1 + s(4C_\mu + C_B)r_o}$$

Beyond this point more detailed calculations become necessary and, eventually, the small-signal model must be refined.

8-8   EXAMPLE: COMPENSATED OPERATIONAL AMPLIFIER

The circuit shown in Fig. 8-20 is a two-stage operational amplifier. Input stage Q1|Q2|Q3|Q4 is a differential amplifier with an active load, and second stage Q6 is a common-emitter amplifier that is buffered by emitter follower Q5 at its input and by push–pull follower Q7|Q8 at its output. The feedback capacitance $C = 30$ pF in the second stage is much larger than the transistor diffusion and depletion capacitances; it is thus the determining factor in the frequency response, which is essentially an integrator. The only other capacitances we need to consider are $C_{\pi 1} = C_{\pi 2}$, which are quite small, say 4 pF, because Q1 and Q2 have small areas and carry small currents.

The first-stage load resistance R_{L1} (2 MΩ) is the parallel combination of the collector output resistances of Q2 and Q4 (20 MΩ each) and the input resistance of Q5 (2.5 MΩ). The second-stage load resistance R_{L2} (170 kΩ) is

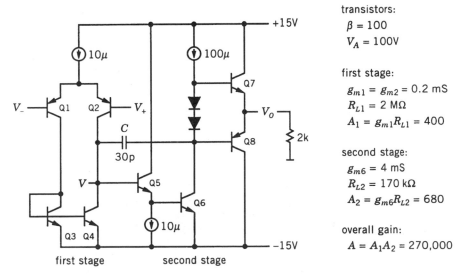

transistors:
$\beta = 100$
$V_A = 100$V

first stage:
$g_{m1} = g_{m2} = 0.2$ mS
$R_{L1} = 2$ MΩ
$A_1 = g_{m1}R_{L1} = 400$

second stage:
$g_{m6} = 4$ mS
$R_{L2} = 170$ kΩ
$A_2 = g_{m6}R_{L2} = 680$

overall gain:
$A = A_1A_2 = 270,000$

Figure 8-20. Compensated operational amplifier.

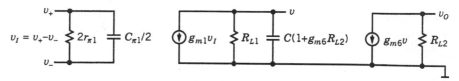

Figure 8-21. Unidirectional equivalent circuit for the compensated operational amplifier of Fig. 8-20.

the parallel combination of the collector output resistance of Q6 (1 MΩ) and the input resistance of the push–pull follower, which is 200 kΩ for $\beta = 100$ and a 2 kΩ load.

The Miller effect is negligible in the input differential amplifier because the load on Q1 is the resistance of diode-connected transistor Q3 and the load on Q2 is essentially the (large) Miller capacitance of the second stage. In the unidirectional equivalent circuit shown in Fig. 8-21, the Miller effect in the second stage has been taken into account in accordance with Fig. 8-6. We can now obtain the gain,

$$\frac{v_O}{v_I} = \frac{g_{m1}R_{L1}g_{m6}R_{L2}}{1 + sCR_{L1}(1 + g_{m6}R_{L2})}$$

For the values indicated in Fig. 8-20, the cutoff frequency f_C is about 4 Hz. Beyond the cutoff frequency, the gain becomes an integrator,

$$\frac{v_O}{v_I} = \frac{g_{m1}}{sC}$$

Since C is so large, the gain is *compensated*, meaning that it continues to integrate all the way out to the transition frequency $f_T = g_{m1}/2\pi C \simeq 1$ MHz.

Summarizing, the low-frequency gain is around 300,000, the input impedance is 1 MΩ in parallel with 2 pF ($C_{\pi 1} = C_{\pi 2} = 4$ pF), the input bias current is about 50 nA, and the transition frequency is 1 MHz. These are roughly the characteristics of the classic μA741 compensated operational amplifier introduced by Fairchild Semiconductor in the 1960s.

8-9   THE TIME-DEPENDENT DIFFUSION EQUATION

In Section 6-3-4 we learned that there are excess charges $\pm Q_B$ stored in the base of an active transistor, and that $Q_B = \tau_B I_C$, where τ_B is the base charging time. We will assume that we are dealing with *pnp* transistors, so that

$$\tau_B = \frac{w_B^2}{2D_p}$$

We also accepted that we could interpret small variations in Q_B in terms of the diffusion capacitance $C_{D\pi}$, where

$$C_{D\pi} = g_m \tau_B$$

The base–emitter capacitance C_π is equal to $C_{D\pi}$ in parallel with the depletion capacitance $C_{T\pi}$. For convenience, we will assume that the current density is high enough that we can ignore $C_{T\pi}$. We will also assume that the base defect θ_B is negligible and that the emitter defect δ_E is small enough that we can ignore the electron current at the emitter–base junction.

Let us consider a common-base amplifier with a negligible collector load. For a small step of amplitude i_0 in the emitter current, the small-signal model predicts that the emitter voltage and the collector current both vary like $1 - \exp(-t/r_m C_\pi)$. Assuming that $C_\pi = C_{D\pi}$, the time constant is $r_m C_{D\pi} = \tau_B$. We will now look more closely at these responses.

We will define p'' as the excess hole density in the base relative to the value in the initial state. For infinite hole lifetime, the time-dependent diffusion equation for p'', obtained from (6-15), is

$$\frac{\partial p''}{\partial t} = D_p \frac{\partial^2 p''}{\partial x^2}$$

Taking Laplace transforms on both sides, we obtain

$$s p'' = D_p \frac{d^2 p''}{dx^2}$$

We will assume, as indicated in Fig. 8-22a, that the emitter–base junction is located at $x = 0$ and that the collector–base junction is located at $x = w_B$.

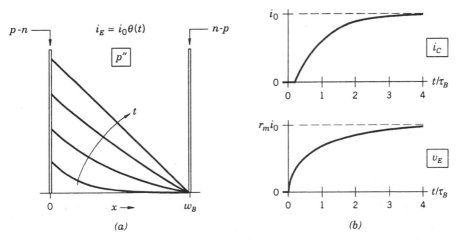

Figure 8-22. Dynamical behavior of a *pnp* common-base amplifier with negligible collector load for a small step of amplitude i_0 in the emitter current. (*a*) Snapshots of p'', the excess hole density in the base relative to the value in the initial state. (*b*) Waveforms of the collector current i_C and the emitter voltage v_E.

Since the transistor is in the active region, we require $p''(w_B) = 0$. To satisfy this boundary condition, p'' must be of the form

$$p''(x) = K \sinh\left[(w_B - x)\sqrt{s/D_p}\right]$$

where K is independent of x. The hole current density $J(x)$ is then

$$J(x) = -eD_p \frac{dp''}{dx} = eK\sqrt{sD_p}\cosh\left[(w_B - x)\sqrt{s/D_p}\right]$$

and, in terms of the emitter current density $J_E = J(0)$, we must have

$$K = \frac{J_E}{e\sqrt{sD_p}\cosh\left[w_B\sqrt{s/D_p}\right]}$$

It follows that the excess hole density $p''_E = p''(0)$ at the emitter end of the base is given by

$$\frac{p''_E}{(w_B J_E/eD_p)} = \frac{\tanh\left(w_B\sqrt{s/D_p}\right)}{w_B\sqrt{s/D_p}} \tag{8-22}$$

For small signals, p''_E is proportional to v_E, so we also get

$$\frac{v_E}{r_m i_E} = \frac{\tanh\left(w_B\sqrt{s/D_p}\right)}{w_B\sqrt{s/D_p}} \tag{8-23}$$

We will use the approximation

$$\frac{p''_E}{(w_B J_E/eD_p)} \simeq \frac{1}{\sqrt{1 + 2\tau_B s}} \tag{8-24}$$

For $s = i\omega$, transfer functions (8-22) and (8-24) have the same asymptotes at low and high frequencies. Approximation (8-24) has a cutoff frequency $0.87/\tau_B$ that is about 40% lower than the cutoff frequency $1.3/\tau_B$ of (8-22); its time response is thus somewhat slower than the true response.

It is easy to show that the convolution of $\theta(t)/\sqrt{\pi t}$ with itself is $\theta(t)$, which has Laplace transform $1/s$. We thus have

$$\mathbf{L}\left[\theta(t)/\sqrt{\pi t}\right] = 1/\sqrt{s}$$

It follows that the impulse response $h(t)$ of a system with transfer function $1/\sqrt{1 + 2\tau_B s}$ is given by

$$h(t) = \frac{\theta(t)}{\sqrt{2\pi\tau_B t}}\exp(-t/2\tau_B)$$

and that for $t \geq 0$ the step response $r(t)$ is given by

$$r(t) = \frac{2}{\sqrt{\pi}} \int_0^{\sqrt{t/2\tau_B}} \exp(-u^2) \, du = \mathrm{erf}\left(\sqrt{t/2\tau_B}\right)$$

For a step of amplitude J_0 in the emitter current density we thus have

$$p_E''(t) \simeq \left(w_B J_0 / e D_p\right) \mathrm{erf}\left(\sqrt{t/2\tau_B}\right)$$

If J_0 is small and corresponds to a current of amplitude i_0, we have

$$v_E(t) \simeq i_0 r_m \, \mathrm{erf}\left(\sqrt{t/2\tau_B}\right) \tag{8-25}$$

The collector current density is equal to the hole current density at $x = w_B$; the transfer function from the emitter current i_E to the collector current i_C is thus

$$\frac{i_C}{i_E} = \frac{1}{\cosh\left(w_B\sqrt{s/D_p}\right)} \tag{8-26}$$

We will use the much simpler approximation

$$\frac{i_C}{i_E} \simeq \frac{\exp(-0.18\tau_B s)}{1 + 0.82\tau_B s} \tag{8-27}$$

For $s = i\omega$ and $\omega\tau_B \leq 10$, (8-26) and (8-27) are virtually indistinguishable. At higher frequencies, approximation (8-27) falls much more slowly than (8-26). According to the approximation, the step response of the collector current is zero for $t < 0.18\tau_B$ and thereafter rises toward 1 with a time constant $0.82\tau_B$. The step response of (8-26) is essentially the same on the time scale that we are considering.

The waveforms of $v_E(t)$ and $i_C(t)$ are shown in Fig. 8-22b; for reference, Fig. 8-22a shows $p''(x)$ at several times. For small times we have from (8-25) that $v_E(t)$ varies like $\sqrt{t/2\tau_B}$, so that it rises sharply and quickly takes on values close to its final value; that is, it looks much like a step function. This observation allows us to turn things around and argue that the collector current in the small-signal model is better given by

$$i_C \simeq g_m v_{BE} \exp(-\kappa\tau_B s) \tag{8-28}$$

where $\kappa \approx \frac{1}{4}$; this argument is strengthened in Exercise 8-12. The delay in the collector current response is important because it results in large phase shifts at high frequencies. As we will see in Chapter 9, such phase shifts can make feedback amplifiers unstable.

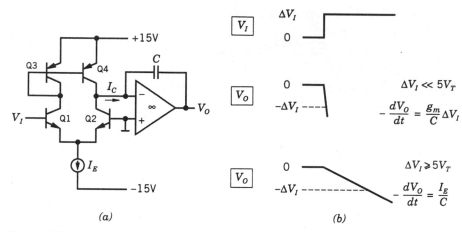

Figure 8-23. Large-signal response of operational amplifiers. (*a*) Simplified structure of a representative operational amplifier. (*b*) Step responses for $\Delta V_I \ll 5 V_T$ (proportional to ΔV_I) and for $\Delta V_I \gtrsim 5 V_T$ (limited by the slew rate).

8-10   LARGE-SIGNAL DYNAMIC RESPONSES

We will look at only a few examples of how circuits respond to large signals, just enough to see that entirely new phenomena appear.

Let us first consider the differential amplifier shown in Fig. 8-23; its load is an integrator with a feedback capacitor C. As we saw in Sections 8-6 and 8-8, this is a typical configuration in operational amplifiers. The response to a small step of amplitude ΔV_I is initially a ramp with a slope proportional to ΔV_I,

$$-\frac{dV_O}{dt} = \frac{g_m}{C} \Delta V_I$$

The time t_R required for the output to change by an amount $\Delta V_O = -\Delta V_I$ is thus independent of ΔV_I,

$$t_R = r_m C$$

The maximum current that the differential amplifier can deliver is the bias current I_E; for steps of amplitude $\Delta V_I \gtrsim 5 V_T$, the slope is limited by the *slew rate*

$$-\frac{dV_O}{dt} = \frac{I_E}{C}$$

The time t'_R in which the output changes by $-\Delta V_I$ is proportional to ΔV_I in this case,

$$t'_R = \frac{C}{I_E} \Delta V_I$$

Using $g_m \simeq I_E/2V_T$, we obtain

$$\frac{t'_R}{t_R} \simeq \frac{\Delta V_I}{2V_T}$$

For $\Delta V_I = 10V = 400V_T$, we have $t'_R/t_R = 200$. This is huge; as a consequence, considerable effort has been invested in designing operational amplifiers with a large slew rate. An example is considered in Exercise 8-13.

From another point of view, we can consider how the amplifier of Fig. 8-23 responds to sinusoidal signals. For an output amplitude of magnitude V_P and phase zero we have

$$i_C = sCV_P$$

The magnitude of i_C cannot exceed I_E; the output amplitude can thus attain a magnitude V_P only below the *power-bandwidth* frequency ω_P given by

$$\omega_P = I_E/CV_P$$

For typical values $I_E = 10$ μA, $C = 30$ pF, and $V_P = 10$ V, we have $f_P \simeq 5$ kHz. In correspondence with what we saw for the slew rate, f_P is much smaller than the small-signal bandwidth in a unity-gain buffer; as we will see in Chapter 9, this bandwidth is equal to $1/2\pi r_m C$, which is about 1 MHz.

As our next example we will consider the emitter follower shown in Fig. 8-24a. If we apply a pulse with short rise and fall times to the base, the emitter will follow the base reasonably well on the positive flank, and the rise time will be limited mostly by parasitic resistances in series with the base and the emitter. On the falling flank, however, the emitter voltage is clamped by capacitor C_L, and the electron density at the emitter end of the base drops almost instantaneously to zero. As shown in Fig. 8-24b, the excess electron profile becomes similar to an inverted parabola, and the emitter current reverses. Based on what we learned in Section 8-9, we can see that the charge

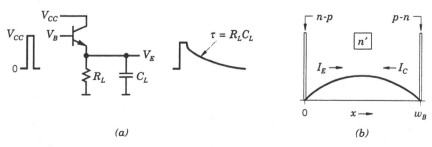

Figure 8-24. Large-signal response of an *npn* emitter follower to a square pulse with zero rise and fall times. (*a*) Waveforms. (*b*) Sketch of the excess electron density in the base immediately after the falling edge of the input pulse.

extracted from C_L is of order $\tau_B V_{CC}/R_L$; to this we must add the charge in the base–emitter depletion capacitance. The emitter voltage V_E thus quickly drops by an amount inversely proportional to C_L, and thereafter decays exponentially with a time constant $R_L C_L$. There are advantages and disadvantages in this behavior. On the negative side, it might force us to change our design and use a push–pull follower. On the positive side, we have a means of defining a time interval starting at the falling edge of the input pulse if we detect where the exponential reaches a given fraction of its initial value.

Finally, we will consider the large-signal inverter of Fig. 8-25. This circuit is the basic building block of the TTL digital circuits that we will see in Chapter 10. The input voltage V_I is either zero or V_{CC}. When V_I is zero, the transistor is cut off, and $V_C = V_{CC}$; when V_I is V_{CC}, the base voltage takes on a value V_{ON} that is approximately 700 mV; we will assume that $V_{CC} \geq 5$ V, so that we can neglect V_{ON} compared with V_{CC}. Assuming $\beta \approx 100$ and $R_B = 10R_C$, the base current V_{CC}/R_B is much larger than $V_{CC}/\beta R_C$, and the transistor is heavily saturated. For simplicity, we will assume that the base–emitter depletion capacitance $C_{T\pi}$ and the base–collector depletion capacitance $C_{T\mu}$ are constant.

When V_I goes from 0 to V_{CC}, the base voltage must rise to a value slightly under V_{ON} before the transistor can turn on. This implies charging up $C_{T\pi}$ and $C_{T\mu}$, and therefore a delay in the response. The *delay time* t_D is

$$t_D \approx \left(\frac{V_{ON}}{V_{CC}}\right) R_B (C_{T\pi} + C_{T\mu})$$

Once the transistor turns on, the base voltage remains almost constant at a value near V_{ON}, and the circuit looks like an RC integrator with input resistance R_B and feedback capacitor $C_{T\mu}$. The output voltage V_O thus goes linearly to zero in a *rise time*

$$t_R \approx R_B C_{T\mu}$$

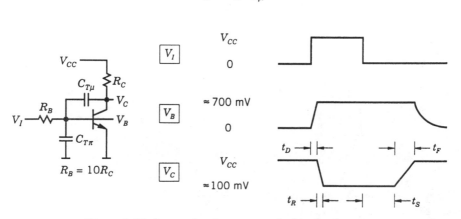

Figure 8-25. Large-signal response of a bipolar inverter.

The rise time t_R is actually a fall time, but the literature has it this way. In standard TTL circuits, t_D and t_R are both about 10 ns.

So far, we have ignored the excess charge Q_B in the base. To see its magnitude, let us assume that the input is adjusted so that V_C is near zero but the transistor is still in the active region. The collector current is then V_{CC}/R_C, and $Q_B \simeq V_{CC}\tau_B/R_C$, which is usually much less than $V_{CC}C_{T\mu}$.

When the transistor saturates, the forward hole current in the base–collector diode is

$$I_{BF} \simeq \frac{V_{CC}}{R_B}$$

The excess hole charge in the collector is thus $Q_C = V_{CC}\tau_C/R_B$. As we saw in Section 6-3-5, τ_C is usually orders of magnitude larger than τ_B. In a 2N3904, for example, we have $\tau_B \simeq 300$ ps and $\tau_C \simeq 150$ ns.

When V_I goes from V_{CC} to zero, the base current reverses and takes on a magnitude

$$I_{BR} \simeq \frac{V_{ON}}{R_B}$$

According to (8-27), the (negative) collector hole current remains constant for a time $0.18\tau_C$ and, in principle, thereafter decays exponentially toward I_{BR} with a time constant $0.82\tau_C$; when it becomes zero, we can assume that all the excess charges have been swept out of the collector and that the transistor has come out of saturation. The *storage time* t_S is thus given roughly by

$$t_S \approx 2\tau_C$$

During the storage time, the base voltage drops only slightly below V_{ON}.

Once the transistor is back in the active region, the situation is similar to what it was during the rise time: V_B remains near V_{ON}, and $C_{T\mu}$ is charged linearly by I_{BR}. The *fall time* t_F is thus

$$t_F \simeq \frac{V_{CC}}{V_{ON}}R_B C_{T\mu}$$

The storage and fall times can be considerable. There are two standard methods for eliminating the storage time. In one method, illustrated in Fig. 8-26a, a Schottky diode with a forward voltage drop around 300 mV is connected from the base to the collector. The Schottky diode turns on before the base–collector diode can become forward biased and diverts the excess base current into the collector. As a result, the transistor never saturates.

Another method, often used in discrete circuits and illustrated in Fig. 8-26b, is to place a small *speed-up* capacitor C_B in parallel with base resistor R_B. If properly chosen (usually around 10 pF), C_B can eliminate the fall time as well as the storage time.

We note incidentally that there are no *excess* stored charges in Schottky diodes as there are in the neutral regions of junction diodes and bipolar

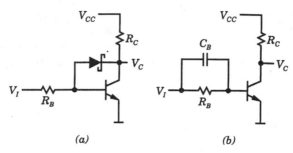

(a)   (b)

Figure 8-26. Elimination of the storage time in a bipolar inverter (a) with a Schottky diode from the base to the collector or (b) with a small capacitor in parallel with the input resistor.

transistors, so that the storage time is zero. Switching transients are determined only by the depletion capacitance (typically 1 pF), and can therefore easily be in the picosecond range rather than in the nanosecond range that obtains in small-signal junction diodes.

EXERCISES

8-1 LABORATORY I: THE MILLER EFFECT AND THE CASCODE AMPLIFIER. The circuit of Fig. 8-27 will allow you to observe the Miller effect in a common-emitter amplifier and to see how it can be suppressed by going to a cascode configuration. In configuration A, the base and the collector of Q2 are tied together, so that Q2 becomes a diode that has essentially no effect on common-emitter amplifier Q1. In configuration B, the base of Q2 is tied to a fixed voltage, and Q2 becomes the common-base

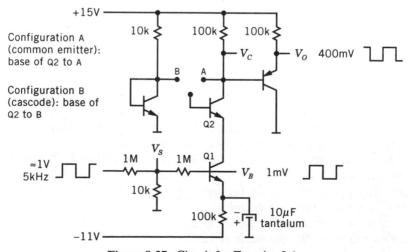

Figure 8-27. Circuit for Exercise 8-1.

amplifier in a cascode amplifier formed by Q1 and Q2. With the values indicated, Q1 and Q2 will have a collector current of 100 μA.

(a) Adjust the 5 kHz square-wave input voltage until the amplitude of V_B is about 1 mV. The response will then be linear within 1%.

(b) Estimate the input resistance r_π at the base of Q1 by comparing v_S and v_B. Observe that it varies little when you switch configurations. This confirms in some measure that the input resistance of a common-emitter amplifier does not depend on the collector load resistance. Calculate β (≈ 100).

(c) In configuration A (common emitter), measure the gain v_O/v_B (≈ 400) and the output rise time; the waveform should be essentially of first order. Estimate C_μ (≈ 4 pF) using $C_\pi = 0$.

(d) In configuration B (cascode), observe the much faster output rise time; the waveform should be of second order. Measure the rise time of v_B and estimate C_π (≈ 20 pF) keeping in mind that the $\times 10$ oscilloscope probe you are presumably using has a 15 pF input capacitance. Switch between configurations, and by comparing the gains, estimate the output resistance r_o in configuration A (≈ 1 MΩ) assuming that it is infinite (βr_o) in configuration B.

8-2   LABORATORY II: STORED CHARGES IN TRANSISTORS

(a) Assemble the circuit of Fig. 8-28 without capacitor C or Schottky diode D. Observe the considerable storage time t_S and fall time (actually, rise time) t_F when the transistor comes out of saturation.

(b) From the base current and the slope of V_C during the fall time, estimate C_μ, which should be about 4 pF.

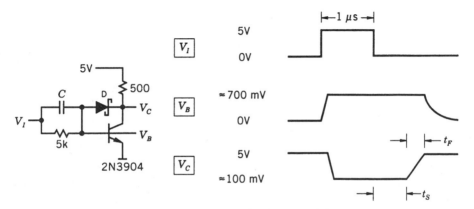

Figure 8-28. Circuit for Exercise 8-2.

(c) Add diode D and observe that the storage time becomes zero.

(d) Remove diode D and add capacitor C (of order 10 pF). Adjust C until the storage time becomes zero. The injected charge should be several times larger than the charge extracted from the base during the storage time in the absence of C because the reverse base current is quite small in the latter case, so that most of the charge stored in the collector is removed at the collector terminal.

(e) Adjust C until the waveform at the base is almost square. The additional injected charge should be just what is necessary to charge up C_μ, and the fall time should be zero.

8-3 THE y-PARAMETER MODEL OF A TRANSISTOR IN THE COMMON-EMITTER CONFIGURATION. The y-parameter model is more convenient for calculations than the standard small-signal model of Fig. 6-24. In addition, it is often used in data sheets to represent high-frequency transistors. The established nomenclature for the parameters is shown in Fig. 8-29.

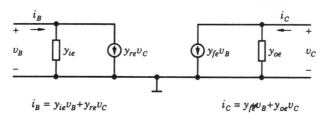

$$i_B = y_{ie}v_B + y_{re}v_C \qquad\qquad i_C = y_{fe}v_B + y_{oe}v_C$$

Figure 8-29. Standard nomenclature for the y-parameters of a transistor in the common-emitter configuration (Exercise 8-3).

Show that the y-parameters in terms of the standard parameters are

$$y_{ie} = g_\pi + s(C_\pi + C_\mu) \qquad y_{fe} = g_m(1 - sC_\mu r_m)$$

$$y_{re} = -sC_\mu \qquad\qquad y_{oe} = g_o + sC_\mu$$

Neglecting time constant $C_\mu r_m$ in y_{fe}, you obtain the model shown in Fig. 8-30, easy to remember and identical to the standard model at low frequencies.

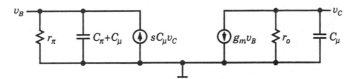

Figure 8-30. The y-parameter model of a transistor in the common-emitter configuration (Exercise 8-3).

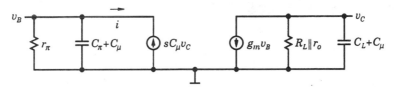

Figure 8-31. Calculation of the Miller effect using the y-parameter model (Exercise 8-3).

To see how useful the y-parameter model is, use the circuit of Fig. 8-31 to calculate the Miller effect with minimal fuss. In two steps, you should obtain i in terms of v_C, and v_C in terms of v_B:

$$i = -sC_\mu v_C = -sC_\mu \frac{-g_m(R_L \| r_o)}{1+s(R_L \| r_o)(C_L+C_\mu)} v_B \equiv \frac{1}{R_M+(sC_M)^{-1}} v_B$$

If this does not convince you, use the circuit of Fig. 8-32 to calculate the output impedance of the current mirror represented in Fig. 8-8. The source resistance is r_m, which allows you to eliminate $C_{\pi 1}$, $C_{\pi 2}$, and $r_{\pi 2}$ from the input circuit. Again in two steps, you should obtain

$$i = g_m v_B = sC_\mu v_C$$

and therefore

$$y_O = g_o + 2sC_\mu$$

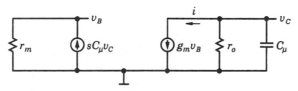

Figure 8-32. Calculation of the output impedance of a current mirror using the y-parameter model (Exercise 8-3).

8-4 INPUT IMPEDANCE OF A CASCODE AMPLIFIER. Consider a bipolar cascode amplifier with input transistor Q1 and output transistor Q2. The input impedance at the emitter of Q2 is as given in (8-8),

$$z_{E2} = r_m \frac{R_L \| \beta r_o}{R_L \| r_o} \frac{1 + s(R_L \| r_o)(C_L + C_\mu)}{1 + s(R_L \| \beta r_o)(C_L + C_\mu)}$$

The worst possible loading condition on Q1, that is, the one that will most enhance the Miller effect, is obtained by making the load impedance and β both infinite. In this case you get

$$z_{E2} = r_m + \frac{1}{sg_m r_o C_\mu}$$

If you further ignore the collector output resistance r_o of Q1, you obtain the incremental circuit shown in Fig. 8-33. Summing currents at node v_{C1} yields

$$sC_\mu(v_{B1} - v_{C1}) = g_m v_{B1} + \frac{sg_m r_o C_\mu}{1 + sr_o C_\mu} v_{C1}$$

Assuming $\omega r_m C_\mu \ll 1$ and $g_m r_o \gg 1$, you can then obtain

$$v_{B1} - v_{C1} = 2v_{B1} + \frac{1}{sr_o C_\mu} v_{B1}$$

and finally,

$$\frac{i}{v_{B1}} = 2sC_\mu + \frac{1}{r_o}$$

Since $r_o \gg r_\pi$, the input impedance of a bipolar cascode amplifier can always be taken to be r_π in parallel with $C_\pi + 2C_\mu$. In the case of FET-input amplifiers, the absence of r_π implies that you have to be more careful.

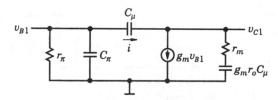

Figure 8-33. Circuit for Exercise 8-4.

8-5 STABILITY OF EMITTER FOLLOWERS. Show that the input admittance $y_I = 1/z_I$ of an emitter follower, as given in (8-10), has a positive real part at all frequencies if

$$C_L > (\beta - 1)C_\pi$$

Emitter followers are generally stable if C_L is either large or small. They are prone to oscillations for C_L typically in the 100 pF range.

8-6 EMITTER-FOLLOWER GAIN. The input impedance of an emitter follower for $g_m R_L \gg 1$ is given in (8-11). Use this expression to calculate v_B / i_S in Fig. 8-12, and show that

$$\frac{v_B}{i_S} = (R_S \| \beta R_L) \frac{1 + s r_m (C_\pi + C_L)}{1 + s \tau_1 + s^2 \tau_2^2}$$

where

$$\tau_1 = (R_S \| \beta R_L) \left(C_S + \frac{C_\pi + C_L}{g_m R_S} + \frac{C_L}{\beta} + \frac{C_\pi}{g_m R_L} \right)$$

and

$$\tau_2^2 = r_m (R_S \| \beta R_L)(C_L C_\pi + C_S C_\pi + C_S C_L)$$

Now use (8-11) to show that the gain v_E / i_S given in (8-16) is correct.

8-7 GAIN INVERSION AT RELATIVELY LOW FREQUENCIES. Consider the amplifier with emitter degeneration shown in Fig. 8-34, and imagine C_μ as an external capacitor. Assume that $R_C \ll r_o$; in this case you can calculate v_E from the expressions for an emitter follower. If you further assume that $g_m R_E \gg 1$, that $\omega r_m C_\pi \ll 1$, and that $\omega r_m C_E \ll 1$, you can see

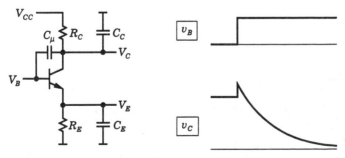

Figure 8-34. The gain of a common-emitter amplifier with emitter degeneration can change sign at relatively low frequencies (Exercise 8-7).

from (8-12) that $v_E \simeq v_B$ and that

$$i_C = (G_E + sC_E) v_B$$

Summing currents at the collector node, you have

$$sC_\mu (v_B - v_C) = (G_E + sC_E) v_B + G_C v_C + sC_C v_C$$

Solve this equation to obtain

$$\frac{v_C}{v_B} = -\frac{R_C}{R_E} \frac{1 + s(C_E - C_\mu) R_E}{1 + s(C_C + C_\mu) R_C}$$

If $C_E = 0$ and $C_C \approx C_\mu$, the gain becomes positive at relatively low frequencies if R_E is large, and the step response is as shown in Fig. 8-34. On the other hand, it is clear that this effect can be tuned out by making $C_E = C_\mu$. What is more, by making $(C_E - C_\mu) R_E = (C_C + C_\mu) R_C$, you can knock out the collector time constant.

8-8 EMITTER FOLLOWERS WITH LARGE CAPACITIVE LOADS ARE STABLE. Consider an emitter follower with $R_L = \infty$ and $C_L = \beta C_\pi$. In the expression for the gain given in (8-16), show that

$$\tau_1 = R_S(C_S + C_\pi) + r_\pi C_\pi + \underline{r_m C_\pi}$$

and

$$\tau_2^2 = r_\pi C_\pi R_S(C_S + C_\pi) + \underline{r_m C_\pi R_S C_S}$$

If $\beta \gg 1$ you can drop the underlined terms, and the denominator of the gain can then be factored:

$$\frac{v_E}{i_S} = R_S \frac{1 + sr_m C_\pi}{[1 + sR_S(C_S + C_\pi)](1 + sr_\pi C_\pi)}$$

The damping factor ζ is thus larger than 1, and the gain is very stable.

8-9 OUTPUT ADMITTANCE OF THE WILSON MIRROR. Consider the Wilson mirror shown in Fig. 8-35a. The incremental circuit of Fig. 8-35b has been simplified by taking into account

1. that r_{o1} is much larger than the input resistance of follower Q2, which achieves its maximum value $2r_\pi$ when the mirror output is grounded.
2. that $C_{\mu 1}$ is in parallel with $C_{\pi 2}$ (it has been absorbed in C_π).
3. that $r_{\pi 1}$, $C_{\pi 1}$, and $C_{\pi 3}$ are in parallel with the resistance r_{m3} of diode-connected transistor Q3.

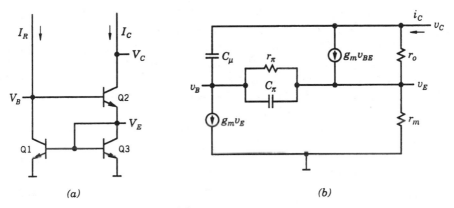

Figure 8-35. (*a*) Wilson mirror; (*b*) incremental circuit used to calculate the output impedance. (Exercise 8-9.)

Remove C_μ initially and drive node v_C with a voltage source. The currents in r_m and $g_m v_E$ are then equal, and from continuity, equal to $\frac{1}{2} i_C$. You thus have

$$v_E = \tfrac{1}{2} r_m i_C$$

and

$$v_{EB} = \frac{\tfrac{1}{2} r_\pi}{1 + s C_\pi r_\pi} i_C$$

Now sum currents at node v_E to obtain

$$v_C \simeq v_C - v_E \simeq \frac{\tfrac{1}{2} \beta r_o}{1 + s C_\pi r_\pi} i_C$$

In the absence of C_μ the output admittance is thus

$$\frac{i_C}{v_C} \simeq \frac{2 g_o}{\beta} + \frac{2 s C_\pi}{g_m r_o} \simeq \frac{2 g_o}{\beta}$$

(The capacitive contribution can be dropped because $g_m r_o \gg 1$.)

Setting $v_C = 0$ and applying a current source i_B to node v_B, observe now that $v_E = \frac{1}{2} v_B$, that $i_C = i_B$, and that the input impedance at node v_B is $2 r_m$ in parallel with $2 r_\pi$ and $\frac{1}{2} C_\pi$, and thus approximately equal to $2 r_m$.

If you now restore C_μ and drive node v_C with a voltage source, C_μ acts like a current source $s C_\mu v_C$ applied to node v_B because $2 \omega C_\mu r_m \ll 1$, and i_C increases by $2 s C_\mu v_C$. The output admittance of the Wilson mirror is thus

$$y_O = \frac{2 g_o}{\beta} + 2 s C_\mu$$

8-10 UNCOMPENSATED OPERATIONAL AMPLIFIER. Convince yourself that the incremental circuit shown in Fig. 8-36 will allow you to calculate the main aspects of the frequency response of the operational amplifier presented in Section 7-9. In particular, with reference to Fig. 7-20, you should agree that the capacitance to ground C_L at the collectors of Q7 and Q9 is the sum of the input capacitance of the push–pull emitter follower

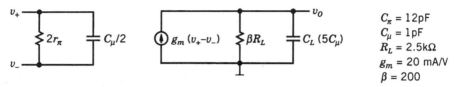

Figure 8-36. Incremental circuit for the operational amplifier of Fig. 7-20 (Exercise 8-10).

$(2C_\mu)$ and the output capacitances of the cascode amplifier (C_μ) and the Wilson mirror $(2C_\mu)$. The open-loop gain A is

$$A = \frac{g_m \beta R_L}{1+s\beta R_L C_L}$$

This gain has a cutoff frequency

$$f_C = \frac{1}{2\pi\beta R_L C_L} \simeq 65 \text{ kHz}$$

and a transition frequency

$$f_T = \frac{1}{2\pi r_m C_L} \simeq 650 \text{ MHz}$$

This is not the full story, however, and at frequencies around 100 MHz you should expect further cutoff frequencies, due in part to time constants that have been ignored, of order $r_m C_\pi$, and in part to the incompleteness of the small-signal model. In addition, you should expect the gain to have a delay of order 1 ns. The Bode plots should then have the general aspect shown in Fig. 8-37. You are thus dealing with an *uncompensated* operational amplifier because the phase of the gain at the transition frequency is well beyond $-135°$.

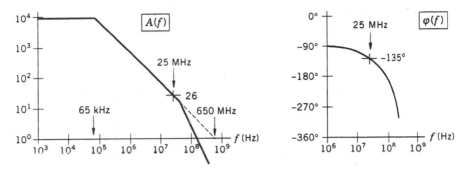

Figure 8-37. Bode plots for the operational amplifier of Fig. 7-20 (Exercise 8-10).

8-11 EMITTER FOLLOWERS WITH MODERATE SOURCE RESISTANCE. Consider the emitter-follower gain given in (8-16), and assume that $\beta R_L \gg R_S$ and that C_L/β and $C_\pi/g_m R_L$ can be neglected compared with C_S. Show that τ_2^2/τ_1^2 is maximum for

$$R_S = r_m \frac{C_\pi + C_L}{C_S}$$

and that the minimum in the damping factor $\zeta = \tau_1/2\tau_2$ is given by

$$\zeta_{min}^{-2} = 1 + \frac{1}{C_S} \frac{C_\pi C_L}{C_\pi + C_L}$$

It is clear that a large source capacitance is stabilizing. Take reasonable values $C_\pi = 50$ pF, $C_L = 10$ pF, $C_S = C_\mu = 4$ pF, $r_m = 25$ Ω, $R_L = 5$ kΩ, and $\beta = 200$. Obtain that the minimum damping factor is 0.57 for $R_S = 375$ Ω. It should now be clear that even a rather small source capacitance is quite stabilizing.

8-12 THE TRANSCONDUCTANCE HAS A SMALL DELAY. Consider the common-base amplifier discussed in Section 8-9. From (8-23) and (8-26) obtain that the emitter voltage and the collector current are related by

$$\frac{i_C}{g_m v_E} = \frac{w_B \sqrt{s/D_p}}{\sinh\left(w_B \sqrt{s/D_p}\right)}$$

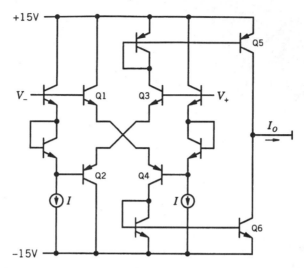

Figure 8-38. Hearn input circuit, used to obtain a large slew rate in operational amplifiers (Exercise 8-13).

Now consider the approximation

$$\frac{i_C}{g_m v_E} = \frac{\exp(-0.13\tau_B s)}{1 + 0.2\tau_B s}$$

For $s = i\omega$ and $\omega\tau_B \leq 5$, check that these two transfer functions have virtually identical phase and that they differ in magnitude by less than 5%. The step response of the approximation is $\theta(t)[1 - \exp(-5t/\tau_B)]$ delayed by $0.13\tau_B$. If you now roughly straighten out this response around the point where its value is $\frac{1}{2}$, you can equally roughly replace it with a delay by $\kappa\tau_B = (0.13 + 0.2\ln 2)\tau_B$. This is the justification for the statement in (8-28).

8-13 OPERATIONAL-AMPLIFIER INPUT CIRCUIT WITH A LARGE SLEW RATE. The slew rate in operational amplifiers can be increased considerably by using the input circuit shown in Fig. 8-38 (Hearn, 1971). Convince yourself first that this circuit is in fact a differential amplifier, that all transistors have a bias current I, and that

$$i_O = g_m(v_+ - v_-)$$

Now convince yourself that the push–pull structure of this circuit makes the slew rate very large.

9

BANDWIDTH AND STABILITY OF OPERATIONAL-AMPLIFIER CIRCUITS

9-1 INTRODUCTION

So far we have dealt with operational amplifiers mostly in the infinite-gain approximation. This procedure is valid if we are interested only in the ideal response of circuits that are well designed and therefore presumably stable. In Chapter 8 we learned that the loop gain of an operational-amplifier circuit necessarily goes to zero at high frequencies; in some circuits—*RC* integrators, for example—it also goes to zero at low frequencies. We will now see that this behavior not only limits the bandwidth over which the response is ideal, but that it can also result in an unstable response.

We will first study direct amplifiers in order to determine which parameters of the loop gain control their dynamical behavior. We will then show that our conclusions can be extended to arbitrary operational-amplifier circuits; in doing so we will develop the *phase-margin criterion*, a simple rule for judging the degree of stability of a circuit. An important conclusion we will draw is that the loop gain of stable circuits looks much like an integrator in the vicinity of the upper transition frequency and, given the nature of things, that the best possible loop gain is an integrator with an infinite transition frequency rather than a (physically impossible) infinite constant gain.

For the purposes of studying their dynamical behavior, most linear feedback systems can be represented as operational-amplifier circuits; we will thus be equipped to go beyond the strict domain of operational amplifiers and analyze electromechanical and hydraulic servomechanisms as well as transistor feedback amplifiers, and we will do so in examples and exercises.

9-2   OPERATIONAL-AMPLIFIER GAIN

We learned in Chapter 8 that the gain of an operational amplifier can be modeled as the product of a constant gain, a small delay, and one or more first-order lags:

$$A(s) = \frac{A_0 \exp(-\tau_D s)}{(1+A_0\tau_1 s)(1+\tau_2 s) \cdots}$$

The magnitude of the gain decreases monotonically with frequency, and we will assume for the time being that the feedback network is such that this is also true of the loop gain; in particular, we will assume that the loop gain has a well-defined upper transition frequency. (In Exercise 9-16 we will see that the presence of a lower transition frequency does not present a problem if the two transition frequencies are well separated, and it will become apparent how to extend our conclusions to this case.)

The delay τ_D is of order 1 nanosecond and therefore important only in wideband circuits. Since the loop gain does not have a lower transition frequency, the closed-loop dynamics are determined by the behavior of $A(s)$ at frequencies well above the cutoff frequency of the first lag (whose time constant $A_0\tau_1$ is much larger than any of the other time constants) and are not modified in any significant way if this lag is replaced by an integrator:

$$\frac{A_0}{1+A_0\tau_1 s} \rightarrow \frac{1}{\tau_1 s}$$

The minor effects of this replacement on the closed-loop dynamics and on the low-frequency closed-loop gain are examined in Exercise 9-12. In compensated operational amplifiers like the one we analyzed in Section 8-8, the first lag is dominant by design, so that $A(s)$ can simply be taken to be an integrator:

$$A(s) \rightarrow \frac{1}{\tau_1 s}$$

9-3   DYNAMICS OF THE DIRECT AMPLIFIER

Although we eventually want to emphasize the role of the loop gain, we will start exploring the dynamic effects of feedback by studying the closed-loop gain of a direct amplifier. This approach is not the most effective, but it will serve us well initially and allow us to introduce language found in the literature and in data sheets.

Let us thus consider the direct amplifier of Fig. 9-1, and let us initially assume that the operational amplifier is compensated and that its gain is an

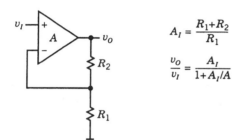

$$A_I = \frac{R_1+R_2}{R_1}$$

$$\frac{v_O}{v_I} = \frac{A_I}{1+A_I/A}$$

Figure 9-1. Direct amplifier with resistive feedback.

integrator with a transition frequency $\omega_T = 1/\tau$:

$$A(s) = \frac{1}{\tau s}$$

The loop gain is

$$A_L(s) = \frac{R_1 A(s)}{R_1+R_2} = \frac{A(s)}{A_I} = \frac{1}{A_I \tau s}$$

and the closed-loop gain is therefore

$$A_C(s) = \frac{A_I}{1+A_I \tau s}$$

This is a first-order lag, the paradigm of stability. The cutoff frequency

$$\omega_C = \frac{1}{A_I \tau} = \frac{\omega_T}{A_I}$$

is inversely proportional to the ideal gain, and it is equal to the bandwidth because the closed-loop gain does not have a low-frequency cutoff; the *gain–bandwidth product* (GBWP)—often simply the *bandwidth*—is therefore constant and equal to the transition frequency of the operational-amplifier gain,

$$A_I \omega_C = \omega_T$$

Bode plots of the open-loop and closed-loop gains are shown superimposed in Fig. 9-2.† From these plots it is clear that the cutoff frequency of the closed-loop gain can be obtained by locating where a horizontal line with ordinate A_I intersects the open-loop plot.

†For simplicity and clarity of representation, wherever possible we will use the stylized Bode plots discussed in Exercise 1-9.

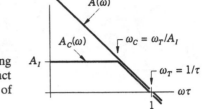

Figure 9-2. Direct amplifier with an integrating open-loop gain. The gain-bandwidth product $A_I \omega_C$ is equal to the transition frequency ω_T of the open-loop gain.

The gain–bandwidth product is strictly constant only in *direct* amplifiers with an *integrating* open-loop gain, but it is sufficiently constant under certain other conditions that the concept is still useful: at large ideal gains, for example, it is clearly applicable to inverting amplifiers with an integrating open-loop gain.

Compensated operational amplifiers are very convenient devices, but we should keep in mind that their simplicity is obtained by degrading wideband devices so that they are stable in direct amplifiers with unity ideal gain; as we will see below, the closed-loop response at higher ideal gains is therefore slower than it needs to be for the purposes of stability.

To obtain large bandwidths at high ideal gains we must turn to uncompensated operational amplifiers; to see what new dynamic effects appear, we will start by considering a second-order open-loop gain, that is, the cascade of an integrator and a first-order lag:

$$A(s) = \frac{1}{\tau_1 s (1 + \tau_2 s)}$$

The closed-loop gain in this case is a second-order lag,

$$A_C(s) = \frac{A_I}{1 + A_I \tau_1 s + A_I \tau_1 \tau_2 s^2}$$

If we rewrite the closed-loop gain in terms of its characteristic time and its damping factor, we obtain

$$A_C(s) = \frac{A_I}{1 + 2\zeta \tau_0 s + \tau_0^2 s^2}$$

where

$$\tau_0 = \sqrt{A_I \tau_1 \tau_2} \quad \text{and} \quad 2\zeta = \sqrt{A_I \tau_1 / \tau_2}$$

Figure 9-3 shows the closed-loop step response for $\tau_2 / \tau_1 = 10$ and several values of A_I. The response is fairly stable $\left(\zeta > \frac{1}{2}\right)$ if $A_I > \tau_2 / \tau_1$ but becomes increasingly oscillatory as A_I is reduced below this value.

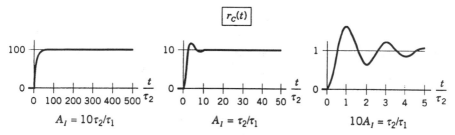

Figure 9-3. Closed-loop step responses of a direct amplifier with a second-order open-loop gain for the case $\tau_2/\tau_1 = 10$.

This is quintessential behavior in feedback circuits: with few exceptions, mostly theoretical, increasing the loop gain uniformly eventually makes the response worse rather than better. To see why this is so in a manner that can be generalized, let us recall that the closed-loop gain A_C is given in terms of the ideal gain A_I and the loop gain A_L by

$$A_C(s) = \frac{A_I}{1+[A_L(s)]^{-1}}$$

and that the magnitude of A_L increases uniformly as A_I is decreased:

$$A_L(\omega) = \frac{A(\omega)}{A_I}$$

The origin of the oscillations can then be understood with the aid of Fig. 9-4 by observing that the phase φ_L of the loop gain tends to $-180°$ at high frequencies, and that as A_I decreases the loop gain approaches -1 in the neighborhood of the transition frequency. The result is a resonance in the closed-loop gain at the frequency at which $[A_L(s)]^{-1}$ comes closest to -1. We note that the resonance is due to the excess phase shift in the loop gain relative to an integrator and not to the change of shape in the magnitude of the loop gain; this point is taken up in Exercise 9-3.

In real uncompensated operational amplifiers the phase of the gain exceeds $-180°$ at high frequencies, and the closed-loop response of a direct amplifier is unstable if A_I is below a certain critical value, usually of order 10. For A_I above this value, however, the closed-loop response behaves roughly as shown in Fig. 9-3 and Fig. 9-4.

The trend toward a resonance as A_I decreases is of course possible only if the closed-loop response is stable for A_I sufficiently large, and although it is characteristic of direct amplifiers and many other feedback systems, it is not always present. To illustrate this point, let us consider a direct amplifier with an integrating open-loop gain and *positive* feedback. We then have

$$A_L(s) = -\frac{1}{A_I\tau s}$$

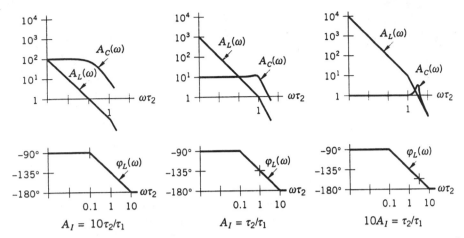

Figure 9-4. Loop gains and closed-loop gains corresponding to the step responses of Fig. 9-3.

and, at least formally,

$$A_C(s) = \frac{A_I}{1 - A_I \tau s}$$

This response is unstable no matter how large A_I is made; its loop gain is purely imaginary and thus never comes close to -1, and there is no resonance at all.

Returning to second-order open-loop gains, we note that for $A_I \geq \tau_2/\tau_1$ we can estimate the closed-loop cutoff frequency by locating where a horizontal line with ordinate A_I intersects the Bode plot of the open-loop gain, as shown in Fig. 9-5. We will see in Section 9-4 that this graphic determination underestimates the cutoff frequency by about 10% for $A_I = 10\tau_2/\tau_1$ and by 30 to 60% for $A_I = \tau_2/\tau_1$, depending on whether we use a stylized or an exact plot of the open-loop gain. To this level the gain–bandwidth product is constant and equal to the *extrapolated* transition frequency $\omega_T' = 1/\tau_1$:

$$A_I \omega_C \simeq \omega_T'$$

Data sheets usually show Bode plots of the open-loop gain or at least give the gain–bandwidth product and the lowest ideal gain at which stability is

Figure 9-5. Direct amplifier with a second-order open-loop gain. The gain-bandwidth product is approximately equal to the extrapolated transition frequency $\omega_T' = 1/\tau_1$ of the open-loop gain if $A_I \geq \tau_2/\tau_1$.

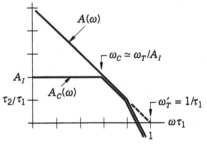

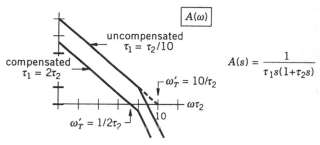

Figure 9-6. The gain–bandwidth product in a compensated operational amplifier is substantially smaller than it is in the uncompensated version.

guaranteed. With this information the closed-loop bandwidth at any ideal gain can be obtained.

Compensated operational amplifiers are usually obtained by uniformly reducing the gain of uncompensated versions until satisfactory stability at unity ideal gain is obtained; as shown in Fig. 9-6, this objective can be attained in the case of a second-order gain by making $\tau_1 = 2\tau_2$, which results in $\zeta = 1/\sqrt{2}$ at unity ideal gain. Since uncompensated versions can easily have the equivalent of $\tau_1 = \tau_2/10$, it is clear that compensation implies a considerable reduction in bandwidth, by a factor of 20 in this example. It should also be clear that further compensation is pointless since it reduces the bandwidth even more without contributing much to stability.

9-4   DYNAMICS AND THE LOOP GAIN

It is now time to bring the loop gain to the fore. In what follows we will show that knowledge of the loop gain is sufficient to obtain the principal characteristics of the dynamic response of operational-amplifier circuits: bandwidth and stability. This is important because the loop gain is easy to calculate or measure and because the redesign that is required if the closed-loop response is not satisfactory is conveniently visualized on Bode plots of the loop gain.

As shown in Fig. 9-7, the closed-loop gain of an operational-amplifier circuit—we specifically mean the gain to the output of the operational amplifier—can be represented by a two-stage amplifier consisting of a unity-gain buffer whose open-loop gain is the circuit loop gain and an output amplifier whose gain is the circuit ideal gain. Since the ideal gain is independent of the operational-amplifier gain, the contribution of the operational amplifier to the dynamic response is entirely in the buffer. We must emphasize that other than the fact that it has the same loop gain, such a representation carries no information about the impedances of the original circuit. As

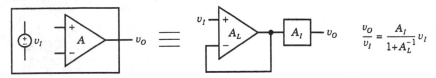

Figure 9-7. An arbitrary operational-amplifier circuit can be represented as the cascade of a unity-gain buffer and an amplifier with gain A_I.

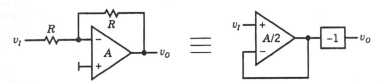

Figure 9-8. Representation of an inverting amplifier according to Fig. 9-7.

an example, the representation of a unity-gain inverting amplifier is as shown in Fig. 9-8 if we assume that the operational amplifier has infinite input impedance and zero output impedance, so that $A_L = A/2$.

The buffer in Fig. 9-7 represents the discrepancy between the closed-loop gain and an ideal gain that is, at least initially, fixed by design. Complete knowledge of the closed-loop response can involve a great deal of work; we will limit ourselves to determining the rise time (bandwidth) and overshoot (stability) of the buffer step response. We will consider our mission complete if we can estimate both of these within a factor of 2, and we will now see that the loop gain can give us this information directly. In the remainder of this chapter we will generally ignore the ideal-gain stage in Fig. 9-7 and concentrate on the buffer; unless otherwise indicated, when we refer to closed-loop gain we will implicitly mean *buffer* closed-loop gain, and ditto for the bandwidth.

Given that the magnitude of the loop gain is large at low frequencies and goes to zero monotonically, and that we are now dealing with a direct amplifier with unity ideal gain, it follows that the magnitude of the closed-loop gain is near 1 for frequencies well below the transition frequency of the loop gain and near zero for frequencies well above the transition frequency:

$$\omega \ll \omega_T \quad \Rightarrow \quad A_L(\omega) \gg 1 \quad \Rightarrow \quad A_C(\omega) \simeq 1$$

$$\omega \gg \omega_T \quad \Rightarrow \quad A_L(\omega) \ll 1 \quad \Rightarrow \quad A_C(\omega) \approx 0$$

Based on what we saw in direct amplifiers, we might expect that the cutoff frequency of the closed-loop gain is approximately equal to the transition frequency of the loop gain:

$$\omega_C \simeq \omega_T$$

Before going on to more complex systems, we will see how good this estimate is in second-order systems (and in first-order systems as a by-product), and we will also introduce the phase margin as a measure of stability.

In a second-order system with unity ideal gain we have

$$A_L(s) = \frac{1}{\tau_1 s(1+\tau_2 s)} \quad \text{and} \quad A_C(s) = \frac{1}{1+\tau_1 s+\tau_1\tau_2 s^2}$$

or in terms of the characteristic time and the damping factor,

$$A_L(s) = \frac{1}{2\zeta\tau_0 s+\tau_0^2 s^2} \quad \text{and} \quad A_C(s) = \frac{1}{1+2\zeta\tau_0 s+\tau_0^2 s^2}$$

where

$$\tau_0 = \sqrt{\tau_1\tau_2} \quad \text{and} \quad 2\zeta = \sqrt{\tau_1/\tau_2}$$

In first-order systems we have $\tau_2 = 0$, so that $\tau_0 = 0$ and $2\zeta\tau_0 = \tau_1$.

At the transition frequency we have $A_L(\omega_T) = 1$, whereas at the cutoff frequency we have $A_C(\omega_C) = 1/\sqrt{2}$; from these conditions we obtain

$$\omega_T^2\tau_0^2 = \sqrt{4\zeta^4+1} - 2\zeta^2$$

and

$$\omega_C^2\tau_0^2 = \sqrt{(1-2\zeta^2)^2 +1} +(1-2\zeta^2)$$

It is easy to show that ω_T underestimates ω_C by a factor of 1.1 for $\tau_1 = 10\tau_2$ ($\zeta = 1.6$) and by a factor of 1.6 for $\tau_1 = \tau_2$ ($\zeta = \frac{1}{2}$). In first-order systems ω_T is equal to ω_C.

We will define the *phase margin* φ_M as the difference between the phase of the loop gain at the transition frequency and $-180°$:

$$\varphi_M = \varphi_L(\omega_T) +\pi = \tan^{-1}\frac{2\zeta}{\omega_T\tau_0}$$

The phase margin is a monotonic function of the damping factor and thus a direct measure of stability. Phase margins above 52° correspond to $\zeta>\frac{1}{2}$, that is, to systems that are fairly stable. First-order systems have a phase margin

TABLE 9-1 Parameters of Second-Order Systems

ζ	φ_M (deg)	$t_R\omega_T$	Overshoot (%)	ω_C/ω_T	Resonance (%)
0.16	18	1.13	60	1.56	220
0.27	30	1.19	42	1.58	93
0.42	45	1.26	23	1.61	31
0.50	52	1.29	16	1.62	16
0.61	60	1.33	9	1.60	3
0.71	66	1.38	4	1.55	0
0.95	75	1.58	0	1.36	0
1.58	84	1.97	0	1.11	0
∞	90	2.21	0	1.00	0

of 90°, in agreement with the fact that they are very stable. Table 9-1 gives the rise time and bandwidth of second-order systems in terms of the transition frequency for several values of the phase margin and the damping factor. Also tabulated are the overshoot in the step response and the magnitude of the resonance in the frequency response.

9-5 THE PHASE-MARGIN CRITERION

As shown in Fig. 9-9, the phase margin is readily obtained from Bode plots of the loop gain, and it is usually also easy to obtain analytically. It would thus be convenient if it were a measure of stability in systems other than second-order systems. We will argue that this is largely the case if the magnitude of the loop gain decreases monotonically.

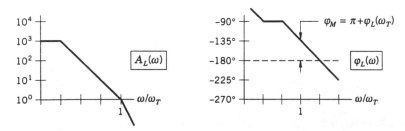

Figure 9-9. Representative Bode plots of the loop gain of an operational amplifier, illustrating the definition of the phase margin.

Let us start by expressing the magnitude of the closed-loop gain in terms of the magnitude and phase of the loop gain:

$$A_C(\omega) = \frac{1}{|1 + [A_L(\omega)]^{-1} \exp[-i\varphi_L(\omega)]|}$$

$$= \frac{A_L(\omega)}{\sqrt{1 + 2A_L(\omega)\cos[\varphi_L(\omega)] + [A_L(\omega)]^2}}$$

At the transition frequency we then have

$$A_C(\omega_T) = \frac{1}{2\cos[\tfrac{1}{2}\varphi_L(\omega_T)]} = \frac{1}{2\sin(\tfrac{1}{2}\varphi_M)}$$

From this expression it is clear that $A_C(\omega_T)$ is larger than 1 if the phase margin is less than 60°; this implies that there is a resonance of magnitude at least $A_C(\omega_T)$ in the vicinity of the transition frequency. The resonance can become arbitrarily large (and the transient response arbitrarily oscillatory) if the phase margin is made too small; for negative phase margins the response is divergent. We also observe that the magnitude of the closed-loop gain at the transition frequency is larger than $1/\sqrt{2}$ if the phase margin is less than 90°, which is always the case in real systems; this implies that the cutoff frequency is at least equal to the transition frequency of the loop gain:

$$\omega_C \geq \omega_T$$

We cannot guarantee the absence of a resonance if the phase margin is greater than 60°, as shown by the fact that second-order systems are slightly resonant for phase margins up to 66° ($\zeta = 1/\sqrt{2}$); nevertheless, an extensive body of experience with systems such as we are considering has established that there is a strong correlation between the phase margin and the degree of stability.

We will refer to the prediction of the closed-loop dynamic properties of a system based only on the transition frequency and phase margin of the loop gain or, alternatively, and more precisely as we will see, on the properties of a second-order system with the same transition frequency and phase margin, as the *phase-margin criterion*.

The empirical support for the phase-margin criterion is examined in Exercises 9-2 through 9-5; the closed-loop response is calculated for third-order loop gains, for loop gains consisting of an integrator in cascade with either a phase rotator or a delay, and for the loop gain of a differentiator like the one discussed in the following section. The results are summarized in Fig. 9-10, which shows the closed-loop step and frequency responses of these systems for a 60° phase margin and a common value $\omega_T = 1/\tau$ of the transition frequency.

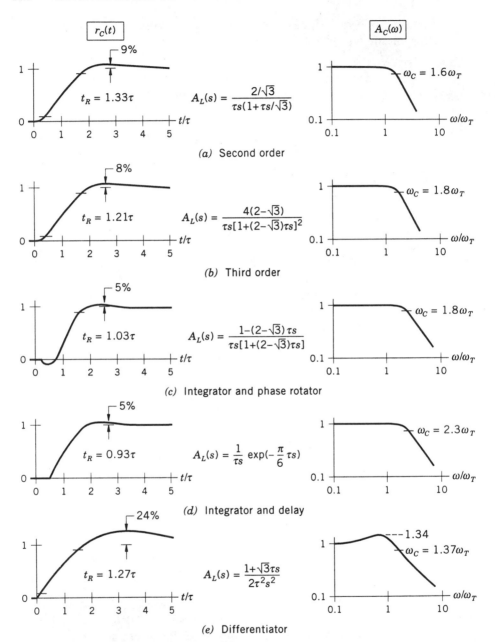

Figure 9-10. Closed-loop step and frequency responses of unity-gain buffers with a variety of loop gains for a 60° phase margin and a transition frequency $\omega_T = 1/\tau$.

From these rather wide-ranging examples we conclude that if we take the transition frequency of the loop gain as an estimate of the cutoff frequency of the closed-loop gain and calculate the rise time of the closed-loop step response using the relation $\omega_T t_R = 2.2$, we can err on the good side by a factor of 2, which is what we set out to do. But we also conclude that our predictions will be in error by less than 50% if we take the values in a second-order system with the same transition frequency and phase margin.

We also observe that the prediction for the overshoot in the closed-loop step response obtained by taking the value in a second-order system (9%) is conservative except in the case of the differentiator, where it is low by almost a factor of 3, although the response is nonetheless definitely stable.

Unusually large overshoots in the step response can be anticipated if, as we should when dealing with an unfamiliar loop gain, we take the trouble to calculate the closed-loop frequency response in the vicinity of the transition frequency. According to Table 9-1, for a 60° phase margin we expect a resonance of 3% or so. Thus the 34% resonance exhibited by the differentiator closed-loop frequency response in Fig. 9-10e tells us that the overshoot in the closed-loop step response is almost surely much larger than predicted by the phase-margin criterion.

9-6   EXAMPLE: UNREALIZABLE IDEAL GAINS—THE DIFFERENTIATOR

Let us consider the differentiator of Fig. 9-11, and let us assume that the operational amplifier has an integrating gain with a time constant τ that is much smaller than the differentiation time constant CR_2. The ideal gain and the loop gain are

$$A_I = \frac{-sCR_2}{1+sCR_1} \quad \text{and} \quad A_L = \frac{A(1+sCR_1)}{1+sC(R_1+R_2)}$$

The Bode plots of the loop gain for $R_1 = 0$ and $99R_1 = R_2$ shown in Fig. 9-12 clearly indicate that stability depends crucially on the phase lead introduced by R_1; the ideal differentiator with $R_1 = 0$ is thus simply unrealizable.

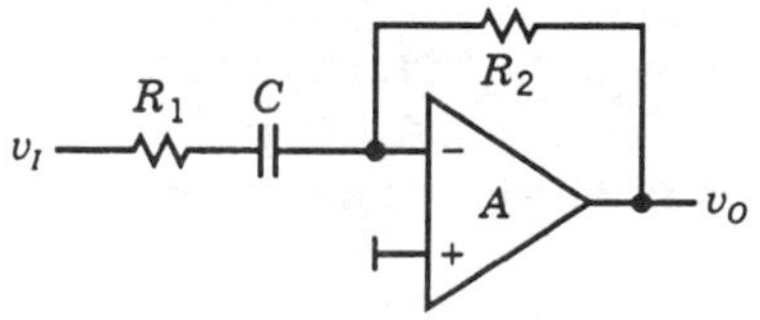

$$A(s) = \frac{1}{\tau s} \qquad \beta_+ = 0$$

$$R_2C = 10^4\tau \qquad \beta_- = \frac{1+sCR_1}{1+sC(R_1+R_2)}$$

Figure 9-11. Operational-amplifier differentiator.

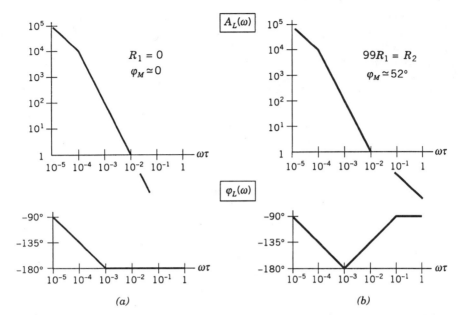

Figure 9-12. Loop gain of the differentiator of Fig. 9-11 (*a*) for $R_1 = 0$ and (*b*) for $99R_1 = R_2$.

As indicated in Section 9-5, the phase-margin criterion is somewhat optimistic in the case of the loop gain of a differentiator. It turns out, however, that circuits with this loop gain often have a lag in the ideal gain, and that the lag tempers the oscillatory behavior in the closed-loop gain. In the present case we have

$$A_C(s) = \frac{-sCR_2}{1+s(CR_1+\tau)+s^2C(R_1+R_2)\tau}$$

For $R_2 = 99R_1$ we can imagine an equivalent circuit consisting of an ideal differentiator $-sCR_2$ followed by a buffer with a second-order loop gain

$$A_L(s) = \frac{1}{s(CR_1+\tau)+s^2C(R_1+R_2)\tau} \approx \frac{1}{100\tau s(1+100\tau s)}$$

The buffer has virtually the same phase margin as the differentiator (52°); as shown in Fig. 9-13, the closed-loop response of the differentiator to a ramp is proportional to the step response of the buffer, and the overshoot in this second-order response (16%) is, by definition, well predicted by the phase-margin criterion.

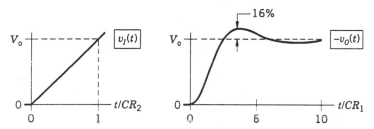

Figure 9-13. Closed-loop ramp response of the differentiator of Fig. 9-11 in the case $99R_1 = R_2$. The lag in the ideal gain results in an overshoot that is well predicted by the phase-margin criterion.

9-7 EXAMPLE: EFFECTS OF OPERATIONAL-AMPLIFIER INPUT IMPEDANCE

Let us consider the inverting amplifier of Fig. 9-14, and let us assume that the operational amplifier has input impedance Z. The ideal gain and the loop gain are

$$A_I = -\frac{R_2}{R_1} \quad \text{and} \quad A_L = \frac{R_1\|Z}{R_2+R_1\|Z}A$$

If Z is a resistor, the loop gain is reduced uniformly; the closed-loop response is therefore slower and less precise but more stable. This result can be used to advantage if we are faced with an amplifier that is oscillating: placing a resistor of sufficiently low value across the inputs of the operational amplifier will suppress the oscillations without modifying the ideal gain, so that measurements can proceed without the perturbing influence of the shifts in operating points that are frequently associated with oscillations.

If Z is a capacitor C—even with the best of layouts, an input capacitance of order 1 picofarad is unavoidable—the loop gain acquires a destabilizing lag, as shown in Fig. 9-15a. If a capacitor C' is placed in parallel with R_2, as shown in Fig. 9-15b, and chosen such that $R_1C = R_2C'$, the lag is transferred

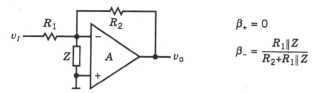

$$\beta_+ = 0$$

$$\beta_- = \frac{R_1\|Z}{R_2+R_1\|Z}$$

Figure 9-14. Inverting amplifier based on an operational amplifier with input impedance Z.

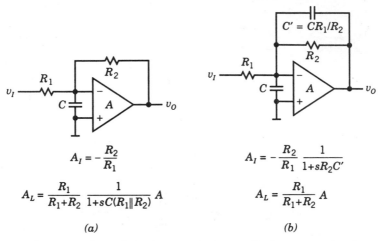

$$A_I = -\frac{R_2}{R_1}$$

$$A_L = \frac{R_1}{R_1+R_2}\frac{1}{1+sC(R_1\|R_2)}A$$

(a)

$$A_I = -\frac{R_2}{R_1}\frac{1}{1+sR_2C'}$$

$$A_L = \frac{R_1}{R_1+R_2}A$$

(b)

Figure 9-15. Compensation of the input capacitance C of an operational amplifier. (a) Capacitor C introduces a destabilizing lag in the loop gain. (b) Capacitor C' transfers the lag to the ideal gain.

from the loop gain to the ideal gain, where it does not affect stability, although it might affect the bandwidth. C' is about 0.5 pF in a well-designed ×10 amplifier; this limits R_2 to about 5 kΩ for a 30 MHz bandwidth.

9-8   EXAMPLE: *RC* INTEGRATOR

The low-frequency behavior of the *RC* integrator shown in Fig. 9-16 is important; we will therefore not assume that the operational amplifier has an integrating gain at all frequencies, although we will assume that it is compensated.

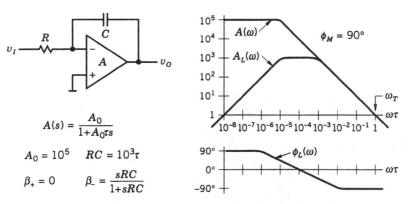

$$A(s) = \frac{A_0}{1+A_0\tau s}$$

$$A_0 = 10^5 \qquad RC = 10^3\tau$$

$$\beta_+ = 0 \qquad \beta_- = \frac{sRC}{1+sRC}$$

Figure 9-16. Operational-amplifier *RC* integrator.

The ideal gain and the loop gain are given by

$$A_I = \frac{-1}{sRC} \quad \text{and} \quad A_L(s) = \frac{sRC}{1+sRC}\frac{A_0}{1+A_0\tau s}$$

The loop gain has a lower transition frequency, which means that an *RC* integrator ceases to be ideal at low frequencies as well as at high frequencies: at $\omega = 0$ the reactance of the feedback capacitor is infinite, and the closed-loop gain is simply the gain of the operational amplifier.

The Bode plots of the loop gain show that an *RC* integrator is stable if the operational amplifier is compensated and, as discussed in Exercise 9-16, if no phase *leads* are introduced at the lower transition frequency.

At high frequencies ($A_0 = \infty$) the closed-loop response is the cascade of an integrator and a lag:

$$A_C(s) = \frac{-1}{s(RC+\tau)}\frac{1}{1+s\dfrac{RC\tau}{RC+\tau}}$$

We note that the time constant τ of the open-loop gain adds to the integration time constant RC, a fact to keep in mind if precision is important.

9-9 EXAMPLE: PROPORTIONAL–INTEGRAL–DERIVATIVE CONTROLLER

We are now going to look into a question that usually arises in the domain of low-frequency hydraulic and electromechanical servomechanisms but that is quite accessible to us with the tools we have acquired. Let us consider a system—perhaps an electric motor whose speed we want to control—with transfer function

$$H(s) = \frac{\text{``1''}}{(1+\tau s)^2}$$

where τ is of order 1 second. The quotes indicate that the system is subject to variations, which is why we will attempt to improve its precision by putting it in a feedback loop. In the process we will see that we can also make it faster if we want to.

We will follow tradition by seeing first what *proportional* control does. As shown in Fig. 9-17, we will use an instrumentation amplifier to multiply the difference between the desired response v_I and the actual response v_O by a constant K and use the resulting signal to drive H. It is clear that what we have is a physically large but otherwise perfectly legitimate operational-

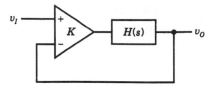

Figure 9-17. Proportional control of a low-frequency system with transfer function $H(s) = 1/(1+\tau s)^2$.

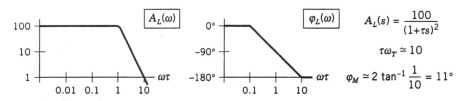

$$A_L(s) = \frac{100}{(1+\tau s)^2}$$

$$\tau \omega_T \approx 10$$

$$\varphi_M \approx 2\, \tan^{-1} \frac{1}{10} = 11°$$

Figure 9-18. Loop gain of the servomechanism of Fig. 9-17 for $K = 100$.

amplifier buffer with loop gain $A_L(s) = KH(s)$, and that we can apply to this buffer what we have learned about operational-amplifier circuits. Because we are trying to overcome the variations in $H(s)$, we want K to be large, let us say 100. The loop gain for this case is shown in Fig. 9-18, and we immediately see that we are in trouble because the phase margin (11°) is much too small.

We know that a system is stable in the measure that its loop gain looks like an integrator in the vicinity of the transition frequency, and we can start toward this objective by introducing a compensating factor

$$G(s) = \frac{1+\tau s}{\tau s}$$

so that the loop gain is an integrator for $\omega \tau \ll 1$:

$$A_L(s) = KG(s)H(s) = K\frac{1}{\tau s(1+\tau s)}$$

This is easily achieved with the operational-amplifier circuit of Fig. 9-19a if we set $RC = \tau$.

Figure 9-19b shows Bode plots of G and GH. For $K = \frac{1}{2}$ the phase margin is 66° and the cutoff frequency is given by $\omega_C \tau = 0.7$, which is slightly better than the cutoff frequency of H, given by $\omega_C \tau = 0.64$. The integrator cuts off at low frequencies, so that precision is determined by the dc gain of the operational amplifier, which can be as large as 10^7; with this exception, we can treat the operational amplifier in the infinite-gain approximation because its bandwidth (typically 1 MHz) is much larger than the servo bandwidth.

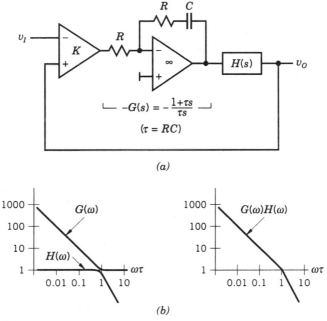

Figure 9-19. (*a*) Proportional-plus-integral control of a system with transfer function $H(s) = 1/(1+\tau s)^2$. (*b*) Bode plots of G and GH.

We can go further: let us introduce an additional factor

$$\frac{1+\tau s}{1+\tau s/10}$$

in $G(s)$ by means of the circuit shown in Fig. 9-20*a*, so that

$$G(s) = \frac{(1+\tau s)^2}{\tau s(1+\tau s/10)}$$

The loop gain now integrates for an additional decade,

$$A_L(s) = KG(s)H(s) = K\frac{1}{\tau s(1+\tau s/10)}$$

Bode plots of G and GH are shown in Fig. 9-20*b*. If we choose $K = \frac{1}{2}$ we will have a larger phase margin (almost 90°) and thus less sensitivity to parameter variations; on the other hand, by choosing $K = 5$ we can increase the bandwidth by a factor of 10.

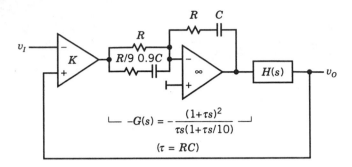

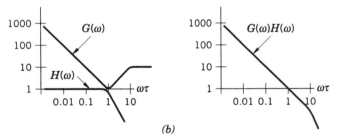

(b)

Figure 9-20. (a) Proportional–integral–derivative control of a system with transfer function $H(s) = 1/(1+\tau s)^2$. (b) Bode plots of G and GH.

In its last version, the transfer function $KG(s)$ can be rewritten as the sum of three terms ("proportional", "integral", and "derivative"):

$$KG(s) = K\left[1 + \frac{1}{\tau s} + \left(\frac{9}{10}\right)\frac{1+\tau s}{1+\tau s/10}\right]$$

An industrial proportional–integral–derivative (PID) controller is a black box with three or more knobs that adjust the absolute and relative weights of terms analogous to these until the response of a servomechanism is satisfactory, that is, until the loop gain looks as much as possible like an integrator.

9-10   EXAMPLE: IDENTIFYING AN OPERATIONAL AMPLIFIER IN TRANSISTOR CIRCUITS

In this example we will see that, with some precautions, we can apply what we have learned about loop gain and dynamics to transistor amplifiers.

The feedback circuit of Fig. 9-21 senses ground at the base of Q1 and drives Q3 so as to keep the base of Q2 at ground. If the current sources are switched off, C remains discharged or, if capacitive feedthrough in the

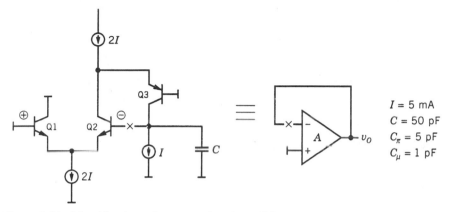

I = 5 mA
C = 50 pF
C_π = 5 pF
C_μ = 1 pF

Figure 9-21. Identification of an operational amplifier in a transistor feedback circuit.

switches is taken into account, at least at some fixed voltage. C thus starts out from a known initial voltage from which it could, for example, be charged up linearly by a separate switched current source in order to convert time to voltage. When the current sources are switched back on, Q2 will be fully cut off if the voltage on C (assumed negative in order to keep Q3 out of saturation) is less than several times $-V_T$; since C is much larger than either C_π or C_μ, it will discharge toward zero at a rate given approximately by $I/C = 100$ mV/ns.

Now we want to see what happens when the transistors return to the active region. The circuit then looks much like an operational amplifier connected as a buffer, and we might be tempted to obtain its loop gain by making a cut at "×" and calculating the negative of the gain from the downstream side of the cut to the upstream side. We must be careful, however: the input impedance of differential amplifier Q1|Q2 is part of the collector circuit of Q3, and should remain associated with Q3 when the cut is made. If we then imagine Q2 as a transistor with infinite input impedance, we will have identified an (equally imaginary) operational amplifier and we can proceed to calculate the loop gain.

The input impedance at the base of Q2 is $2r_\pi$ in parallel with $C_\pi/2$, C_μ, and the Miller capacitance C_μ. The output impedance at the collector of Q3 is simply C_μ (the output resistance is βr_0 and thus negligible). With these elements we can draw the incremental circuit of Fig. 9-22 and obtain the loop gain,

$$A_L = -\frac{v_O}{v_-} = \frac{g_m r_\pi}{1+s(C_\pi+6C_\mu+2C)r_\pi} = \frac{\beta}{1+s\beta(C_\pi+6C_\mu+2C)r_m}$$

This is a highly stable first-order configuration (we have neglected the delay in the transistors because it is of order $C_\pi r_m$). The cutoff frequency is equal

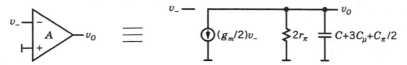

Figure 9-22. Unidirectional equivalent circuit of the operational amplifier identified in Fig. 9-21, once loads on transistors have been accounted for.

to the transition frequency of the loop gain,

$$f_C = f_T = \frac{1}{2\pi(C_\pi + 6C_\mu + 2C)r_m} \simeq 280 \text{ MHz}$$

The transient in the active region is thus very fast ($t_R \simeq 1.2$ ns) and adds virtually nothing to the recovery time.

9-11 EXAMPLE: CIRCUITS WITH SEVERAL OPERATIONAL AMPLIFIERS

The stability of circuits with more than one operational amplifier can be established by studying the loop gain of each operational amplifier. As an example we will consider the tank circuit of Fig. 9-23, which consists of a capacitor C in parallel with an operational-amplifier inductor.

In the infinite-gain approximation we have

$$v_L = \frac{sC_D R_D}{1 + sC_D R_D} v$$

and

$$Ri = v - v_L = \frac{1}{1 + sC_D R_D} v$$

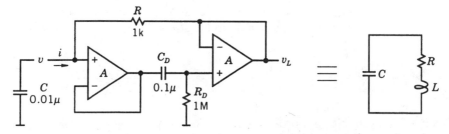

Figure 9-23. Operational-amplifier tank circuit.

The impedance of the circuit to the right of C is therefore

$$\frac{v}{i} = R + sL$$

where

$$L = C_D R_D R$$

The circuit parameters are

$$\omega_0 = \frac{1}{\sqrt{LC}} = \frac{1}{\sqrt{R_D C_D RC}}$$

and, according to Exercise 3-15,

$$Q = \frac{1}{R}\sqrt{\frac{L}{C}} = \sqrt{\frac{R_D C_D}{RC}}$$

For the reasonable values shown in Fig. 9-23 we obtain

$$L = 100 \text{ H} \quad \text{and} \quad Q = 100$$

Real inductors of this high a value require a magnetic core and are thus to some extent nonlinear, a disadvantage not suffered by operational-amplifier inductors in a tank circuit if the loop gain is high enough at the resonant frequency.

We now want to show the circuit is stable (meaning in this case that it will not have high-frequency *parasitic* oscillations) if the operational amplifiers are compensated $[A(s) = 1/\tau s]$ and $RC \gg \tau$. For either amplifier we have

$$\beta_- = 1 \quad \text{and} \quad \beta_+ = \frac{sC_D R_D}{1 + sC_D R_D} \frac{1}{1 + s\tau} \frac{1}{1 + sCR}$$

(we note that β_+ involves the *other* amplifier as a unity-gain buffer). β_+ thus becomes negligible compared with β_- at frequencies well below transition frequency $1/\tau$ of the operational amplifiers, and for ω near $1/\tau$ we have

$$A_L(s) \simeq \frac{1}{\tau s}$$

Both loop gains thus become integrators where it counts, and the circuit is therefore stable.

EXERCISES

9-1 LABORATORY: DYNAMIC RESPONSE OF DIRECT AMPLI-FIERS. Assemble the direct amplifier shown in Fig. 9-24, using a compensated FET-input operational amplifier such as the LF356.

(a) Set $R_1 = 1k\Omega$; $Z_I = \infty$; $R_2 = 0$, $1k\Omega$, $10k\Omega$, $100k\Omega$. Measure the gain and the bandwidth. Verify that $2\pi f_C A_I = 1/\tau$.

(b) Set $R_1 = 1M\Omega$; $Z_I = \infty$; $R_2 = 1M\Omega$. The response should be more oscillatory than it is in the corresponding case in part (a). Explain.

(c) Set $R_1 = 1M\Omega$; $Z_I = 100k\Omega$; $R_2 = 1M\Omega$. The bandwidth should be about five times smaller than it is in part (b). Explain.

Now use an uncompensated amplifier from the same family (LF357).

(d) Set $R_1 = 1k\Omega$; $Z_I = \infty$; $R_2 = 10k\Omega$. Observe that the bandwidth is considerably larger than it is in the corresponding case in part (a).

(e) Set $R_1 = \infty$; $Z_I = \infty$; $R_2 = 0$. The amplifier should oscillate. Explain.

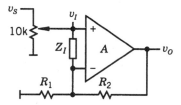

Figure 9-24. Circuit for Exercise 9-1.

9-2 THIRD-ORDER SYSTEMS. The purpose of this exercise is to show that systems with a third-order loop gain

$$A_L(s) = \frac{1}{\tau_1 s(1+\tau_2 s)(1+\tau_3 s)}$$

behave much like second-order systems with the same phase margin. For a given phase margin φ_M, τ_2 and τ_3 are constrained by

$$\varphi_M = \frac{\pi}{2} - \tan^{-1}\omega_T \tau_2 - \tan^{-1}\omega_T \tau_3$$

The furthest departure from second order is obtained for $\tau_2 = \tau_3$; second order is recovered by making τ_2 or τ_3 zero. Consider the case

$$\tau_1 = \frac{2+\sqrt{3}}{4}\tau \quad \text{and} \quad \tau_2 = \tau_3 = (2-\sqrt{3})\tau$$

(a) Check that the transition frequency and the phase margin are given by

$$\varphi_M = 60° \quad \text{and} \quad \omega_T \tau = 1$$

(b) Check that the closed-loop frequency response $A_C(s)$ is

$$A_C(s) = \frac{1}{(1+0.185\tau s)(1+0.748\tau s+0.362\tau^2 s^2)}$$

The closed-loop system is thus the cascade of a damped second-order system and a small lag. According to Exercise 1-12, its step response looks much like the step response of the second-order term except for a delay by the lag time. Verify that the damping factor of the second-order term is $\zeta \approx 0.62$, which corresponds to a 61° phase margin.

(c) Calculate $r_C(t)$ and confirm the following entries:

t/τ	0.5	1.0	1.5	2.0	2.5	3.0	3.5	4.0
$r_C(t)$	0.132	0.493	0.825	1.014	1.075	1.066	1.036	1.011

Observe that the overshoot in the step response is 8%, compared with 9% for a 60° phase margin in a second-order system. Check graphically that the rise time is

$$t_R = 1.21\tau$$

(d) Check that the closed-loop frequency response has a resonance of magnitude 1.02 at $\omega\tau = 0.7$ and that the cutoff frequency is given by

$$\omega_C = 1.78\,\omega_T$$

The rise time and cutoff frequency in a second-order system with a 60° phase margin are $t_R = 1.33\tau$ and $\omega_C = 1.6\omega_T$. It is fair to conclude that the phase-margin criterion works well in third-order systems and that results from second-order systems can be used with confidence.

9-3 LOOP GAIN WITH AN INTEGRATOR AND A PHASE ROTATOR. The main purpose of this exercise is to justify the phase-margin criterion, but you will also be able to verify the assertion, made during the discussion of the direct amplifier in Section 9-3, that it is excess phase shift in the loop gain and not the shape of the magnitude of the loop gain that generates resonances. Consider the loop gain

$$A_L(s) = \frac{1}{\tau s}\frac{1-\tau_R s}{1+\tau_R s}$$

This loop gain has the same magnitude as an integrator $1/\tau s$, but its phase tends to $-270°$ at high frequencies; the closed-loop gain can therefore be oscillatory, and even unstable. The transition frequency and the phase margin are given by

$$\omega_T \tau = 1 \quad \text{and} \quad \varphi_M = \frac{\pi}{2} - 2\tan^{-1}\frac{\tau_R}{\tau}$$

The closed-loop gain is

$$A_C(s) = \frac{1}{1+[A_L(s)]^{-1}} = \frac{1-\tau_R s}{1+(\tau-\tau_R)s+\tau\tau_R s^2} = \frac{1-(\tau_R/\tau_0)\tau_0 s}{1+2\zeta\tau_0 s+\tau_0^2 s^2}$$

This is a sum of second-order responses with parameters

$$\tau_0 = \sqrt{\tau\tau_R} \quad \text{and} \quad 2\zeta = \sqrt{\frac{\tau}{\tau_R}} - \sqrt{\frac{\tau_R}{\tau}}$$

Note that the phase-margin criterion correctly predicts that the response is divergent ($\zeta<0$) if $\tau_R/\tau>1$, that is, if the phase margin is negative.

(a) Choose a 60° phase margin. You should obtain

$$\frac{\tau_R}{\tau} = 2-\sqrt{3} \quad \text{and} \quad \zeta = \frac{1}{\sqrt{2}}$$

(b) Show that the closed-loop step response for $t>0$ is

$$r_C(t) = 1-\left(\cos\omega_2 t+\sqrt{3}\sin\omega_2 t\right)\exp(-\omega_1 t)$$

where

$$\omega_1\tau = \omega_2\tau = \frac{1+\sqrt{3}}{2}$$

(c) Calculate $r_C(t)$ and confirm the following entries:

t/τ	0.25	0.5	1.0	1.5	2.0	2.5	3.0	3.5
$r_C(t)$	−0.082	0.056	0.516	0.861	1.015	1.047	1.033	1.014

Verify graphically that the rise time is

$$t_R = 1.03\tau$$

(d) Check that the closed-loop frequency response has a resonance of magnitude 1.01 at $\omega\tau=0.7$ and that the cutoff frequency is given by

$$\omega_C = 2.21\omega_T$$

(e) Conclude that the phase-margin criterion is safe for this loop gain.

9-4 LOOP GAIN WITH AN INTEGRATOR AND A DELAY. Consider the loop gain

$$A_L(s) = \frac{1}{\tau s} \exp(-\tau_D s)$$

Its transition frequency and phase margin are given by

$$\omega_T \tau = 1 \quad \text{and} \quad \varphi_M = \frac{\pi}{2} - \frac{\tau_D}{\tau}$$

and the corresponding closed-loop response and cutoff frequency are given by

$$A_C(s) = \frac{\dfrac{1}{\tau s} \exp(-\tau_D s)}{1 + \dfrac{1}{\tau s} \exp(-\tau_D s)} \quad \text{and} \quad 1 - 2\omega_C \tau \sin \omega_C \tau_D + \omega_C^2 \tau^2 = 2$$

The closed-loop step response can be obtained by thinking in terms of Laplace transforms and expanding in series:

$$A_C(s) = \sum_{n=1}^{\infty} \frac{(-1)^{n+1}}{(\tau s)^n} \exp(-n\tau_D s)$$

and therefore

$$r_C(t) = \sum_{n=1}^{\infty} (-1)^{n+1} \frac{\left(\dfrac{t}{\tau} - n \dfrac{\tau_D}{\tau} \right)^n}{n!} \, \theta\left(\frac{t}{\tau} - n \frac{\tau_D}{\tau} \right)$$

[Alternatively, you can obtain the closed-loop step response by iteration. First convince yourself that

$$\tau r_C(t) = \int_0^t [\theta(u - \tau_D) - r_C(u - \tau_D)] \, du$$

Now divide time into intervals of width τ_D, and observe that in each interval $\tau r_C(t)$ is its value at the end of the previous interval plus the integral of the difference between $\theta(t)$ and $r_C(t)$ in the previous interval. In the first interval $r_C(t)$ is clearly zero; in the second interval it is a ramp with slope $1/\tau$, and value τ_D/τ at the end of the interval, and so on.]

The phase margin is 60° if $\tau_D/\tau = \pi/6$; this is the case presented in Section 9-5. For ease of calculation, take $\tau_D = \tau/2$; the phase margin in this case is 61.4°, which does not change the results very much.

(a) Calculate $r_C(t)$ and confirm the following entries:

t/τ	0.5	1.0	1.5	2.0	2.5	3.0	3.5
$r_C(t)$	0.000	0.500	0.875	1.021	1.039	1.021	1.006

Verify graphically that the rise time is

$$t_R = 0.95\tau$$

(b) Determine, either graphically or analytically, that $A_C(\omega) \le 1$, that is, that the closed-loop frequency response is not resonant. Verify that the cutoff frequency is given by

$$\omega_C = 2.25\omega_T$$

(c) Conclude that the phase-margin criterion is safe for this loop gain.

9-5   LOOP GAIN OF A DIFFERENTIATOR. Consider the loop gain

$$A_L(s) = \frac{1+\sqrt{3}\,\tau s}{2\tau^2 s^2}$$

It is the high-frequency part of the loop gain of a differentiator like the one discussed in Section 9-6, once the differentiator has been stabilized by placing a small resistor in series with the capacitor, and it shows the phase-margin criterion at its weakest: phase margins near 90° are required to obtain less than 10% overshoot in the closed-loop step response. If you study the *Nyquist diagrams* of Exercise 9-15, you will see that this behavior arises because the loop gain at low frequencies is real and negative.

The transition frequency and the phase margin are given by

$$\omega_T \tau = 1 \quad \text{and} \quad \varphi_M = 60°$$

and the closed-loop gain is

$$A_C(s) = \frac{1+\sqrt{3}\,\tau s}{1+\sqrt{3}\,\tau s + 2\tau^2 s^2}$$

(a) Show that the closed-loop step response for $t>0$ is

$$r_C(t) = 1 - \left(\cos \omega_2 t - \frac{\omega_1}{\omega_2} \sin \omega_2 t \right) \exp(-\omega_1 t)$$

where

$$\omega_1 \tau = \frac{\sqrt{3}}{4} \quad \text{and} \quad \omega_2 \tau = \frac{\sqrt{5}}{4}$$

(b) Calculate $r_C(t)$ and confirm the following entries:

t/τ	0.5	1.0	1.5	2.0	2.5	3.0	3.5
$r_C(t)$	0.398	0.717	0.952	1.109	1.200	1.239	1.240

t/τ	4.0	5.0	6.0	7.0	8.0	9.0	10.0
$r_C(t)$	1.217	1.138	1.061	1.009	0.984	0.979	0.983

Verify graphically that the rise time is

$$t_R = 1.26\tau$$

(c) Verify that $A_C(\omega)$ has a resonance of magnitude 1.34 at $\omega\tau = 0.58$ and that the cutoff frequency is given by

$$\omega_C = 1.37\omega_T$$

(d) Make *stylized* Bode plots of the phase and magnitude of the loop gain on a reasonable scale and estimate the resonance in the closed-loop gain by picking off 10 points from these plots for values of $\omega\tau$ between 0.1 and 1 (you will gain more insight if you plot the reciprocal of the loop gain in the complex plane). You obtain a value of about 1.4, which is close to the value of 1.34 you obtained in part (c). The point here is that even rather rough data allow you to establish the presence of a resonance and thus to check how good the phase-margin criterion is for a given loop gain.

9-6 BENEFICIAL STRAY CAPACITANCE. Stray capacitance is not always detrimental. Consider the buffer of Fig. 9-25, which is driven by a source with a large output resistance R, and assume that there is a small feedback capacitance C_F between the output of the operational amplifier and the positive input, and that there is also a stray capacitance C_S from the positive input to ground. Show that the loop gain is

$$A_L(s) = \frac{1+sRC_S}{1+sR(C_F+C_S)}A(s)$$

On the favorable assumption that $A(s) = 1/\tau s$, the phase margin is almost zero if $C_S = 0$ and $RC_F \gg \tau$. For $R = 10\ \text{M}\Omega$ and $C_F = 0.1\ \text{pF}$, for example,

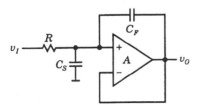

Figure 9-25. Circuit for Exercise 9-6.

you have $RC_F = 1\ \mu s$, which is much larger than τ in most cases. On the other hand, if C_S is larger than C_F, as it usually is, stability is hardly affected by C_F.

Now convince yourself that a direct amplifier driven by a source with a sufficiently large output resistance is unstable if its ideal gain is larger than $1 + C_S/C_F$.

9-7 UNITY-GAIN COMPENSATION OF AN UNCOMPENSATED OPERATIONAL AMPLIFIER

(a) Calculate the ideal gain and the loop gain of the inverting amplifier of Fig. 9-26.

(b) Make Bode plots for $R_I = \infty$. You should obtain a small phase margin, which implies that the step response is oscillatory.

(c) Show that the loop gain is reduced by a factor of 10 if $R_I = R/18$. Check that the phase margin is now about 45°, so that the amplifier is reasonably stable.

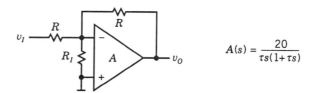

$$A(s) = \frac{20}{\tau s(1 + \tau s)}$$

Figure 9-26. Circuit for Exercise 9-7.

9-8 BANDGAP REFERENCES ARE STABLE.
For the bandgap reference of Fig. 9-27 show that $v_E \simeq 0.962\, v_B$. (*Suggestion:* Calculate the output impedance of Q1|Q2 for $v_B = 0$.) Ignoring the base time constant (the Miller

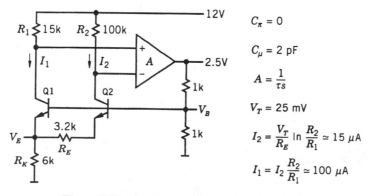

$C_\pi = 0$

$C_\mu = 2$ pF

$A = \dfrac{1}{\tau s}$

$V_T = 25$ mV

$I_2 = \dfrac{V_T}{R_E}\ln\dfrac{R_2}{R_1} \simeq 15\ \mu A$

$I_1 = I_2\dfrac{R_2}{R_1} \simeq 100\ \mu A$

Figure 9-27. Bandgap reference (Exercise 9-8).

effect in Q1 and Q2 is negligible), and assuming that the capacitances from the collectors of Q1 and Q2 to ground are zero, show that

$$\beta = \beta_- - \beta_+ \simeq 1.14 \frac{1-1.1R_KC_\mu s}{1+2.5R_KC_\mu s} - 0.39 \frac{1-21.3R_KC_\mu s}{1+16.7R_KC_\mu s}$$

Calculate β for $0.01 \leq \omega R_K C_\mu \leq 10$ and observe that although β_- and β_+ both change sign at high frequencies, β looks much like a phase lag with a time constant around 20 ns. The phase margin is therefore close to 90° if the operational amplifier has an integrating gain and a bandwidth below 1 MHz $\simeq 1/(2\pi \cdot 160 \text{ ns})$.

9-9 HIGH-SPEED LOW-DRIFT AMPLIFIER. Assume that the operational amplifiers in the $\times 10$ inverting amplifier of Fig. 9-28 have infinite input impedance and zero output impedance. Calculate the β-factors, and show that the phase margins of OA1 and OA2 are 90° and 64° and that the amplifier is therefore stable.

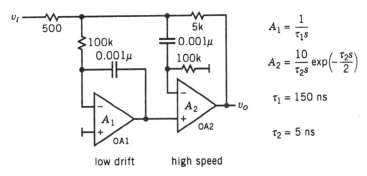

Figure 9-28. High-speed low-drift inverting amplifier (Exercise 9-9).

9-10 SINE-WAVE OSCILLATOR

(a) Consider the circuit of Fig. 9-29. Assuming that the input impedance of differential amplifier Q1|Q2 and the output impedance of Q1 are absorbed in R and C, calculate the loop gain and use Blackman's theorem to show that the impedance at node v is

$$R \frac{s\dfrac{L}{R}}{1+s\dfrac{L}{R}\left(1-\dfrac{g_m R}{2}\right)+s^2 LC}$$

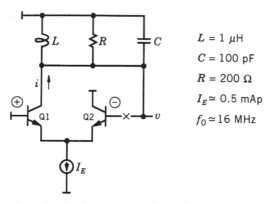

$L = 1\ \mu H$

$C = 100\ pF$

$R = 200\ \Omega$

$I_E \approx 0.5\ mAp$

$f_0 \approx 16\ MHz$

Figure 9-29. Sine wave oscillator (Exercise 9-10).

(b) Setting $I_E = 4V_T/R$ makes $g_m R/2 = 1$, and you get a lossless tank circuit with $Q = \infty$ ($\zeta = 0$) and resonant frequency $\omega_0 = 1/\sqrt{LC}$. Obtain this condition by assuming that v is a sine wave of frequency ω_0 and equating the power dissipated in the tank circuit and the power delivered by Q1.

(c) According to parts (a) and (b), the amplitude of the oscillation is arbitrary, and it diverges or decays if I_E is not exactly equal to $4V_T/R$. This is not quite so. The incremental current into the tank circuit, given in (7-10), is

$$i = \frac{I_E}{2} \tanh \frac{v}{2V_T}$$

Assume that $|v| \leq V_T/2$. You can then use the approximation

$$i \approx \frac{I_E}{2}\left[\frac{v}{2V_T} - \left(\frac{1}{3}\right)\left(\frac{v}{2V_T}\right)^3\right]$$

and obtain for $v = V_0 \sin \omega_0 t$ that

$$i = \frac{I_E V_0}{4V_T}\left[\left(1 - \frac{V_0^2}{16V_T^2}\right)\sin \omega_0 t + \frac{V_0^2}{48V_T^2}\sin 3\omega_0 t\right]$$

If you now equate the power dissipated in the tank circuit and the power delivered by Q1 you should obtain

$$\frac{V_0^2}{16V_T^2} = 1 - \frac{4V_T}{RI_E}$$

The amplitude of the oscillation is controlled by I_E, and it should be clear why: for a given I_E, the nonlinearity in the i–v characteristic makes Q1 deliver too much power at amplitudes below the equilibrium amplitude and too little power at amplitudes above the equilibrium amplitude; the nonlinearity can thus be thought of as a stabilizing feedback.

(d) The calculation in part (c) is not quite self-consistent because the third harmonic in i generates a (small) third harmonic in v. Show that the current in part (c) implies that

$$\frac{v}{V_0} = \sin \omega_0 t + \frac{\left(\dfrac{V_0^2}{V_T^2}\right)\dfrac{RI_E}{4V_T}}{16\sqrt{265}}\sin\left(3\omega_0 t - \tan^{-1}\frac{16}{3}\right)$$

For $V_0 = V_T/2$ the third harmonic is 1000 times smaller than the fundamental, and therefore negligible.

9-11   THE SERIES PAIR. As discussed in Chapter 5, the feedback in a direct amplifier is historically known as *series feedback*. The *series pair* shown in Fig. 9-30 has a rather poor loop gain, but it was nonetheless widely used in nuclear spectroscopy during the 1960s because it requires only one power supply and because it is quite stable—an advantageous consequence of the poor loop gain. Fast operational amplifiers have now made it obsolete.

If you can identify an operational amplifier in this feedback circuit you can calculate the dynamics from the loop gain. Take the input terminals to be the base and emitter of Q1, and the output terminal to be the collector of Q2. The impedance looking into the emitter of Q1 is r_m, and as shown in Fig. 9-31, you can imagine that it shunts the inputs of the operational amplifier; since R_2 is much larger than R_1, you can also imagine that R_2 is connected to ground rather than to the emitter of Q1 when calculating the gain of Q2.

At this time you might want to review the definitions of the Miller resistance and capacitance in (8-2) and the discussion leading to Fig. 8-6. The

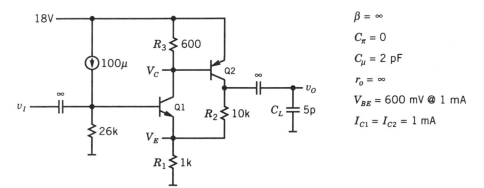

Figure 9-30. Series pair (Exercise 9-11).

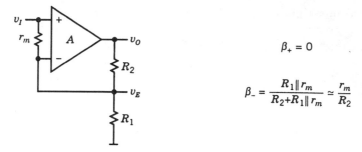

$$\beta_+ = 0$$

$$\beta_- = \frac{R_1 \| r_m}{R_2 + R_1 \| r_m} \approx \frac{r_m}{R_2}$$

Figure 9-31. The series pair viewed as an operational-amplifier circuit (Exercise 9-11).

Miller resistance in Q2 is much smaller than R_3,

$$R_M = r_m\left(1 + \frac{C_L}{C_\mu}\right) = 25\left(1 + \frac{5}{2}\right)\Omega \approx 87\Omega \ll 600\Omega = R_3$$

The Miller effect is thus fully developed, and the Miller capacitance

$$C_M = g_m R_2 C_\mu = 400 C_\mu$$

is much larger than C_μ.

Based on the preceding considerations, the operational-amplifier gain A can be obtained from the incremental circuit of Fig. 9-32,

$$A = \frac{g_m R_3}{1 + sC_\mu g_m R_2 R_3} g_m R_2$$

The β-factors are

$$\beta_+ = 0 \quad \text{and} \quad \beta_- = \frac{R_1 \| r_m}{R_2 + R_1 \| r_m} \approx \frac{r_m}{R_2}$$

and the loop gain is a first-order lag,

$$A_L = \beta A = \frac{g_m R_3}{1 + sC_\mu g_m R_2 R_3}$$

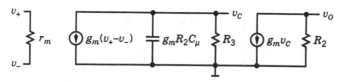

Figure 9-32. Unidirectional equivalent circuit of the operational amplifier of Fig. 9-31 (Exercise 9-11).

You can now obtain that the low-frequency closed-loop gain is

$$A_C(\omega \to 0) = \frac{R_1 + R_2}{R_1} \frac{g_m R_3}{1 + g_m R_3} = 11\left(\frac{24}{25}\right) \approx 10.5$$

and that the cutoff frequency (equal to transition frequency of A_L) is

$$f_C = f_T = \frac{1}{2\pi C_\mu R_2} = 8 \text{ MHz}$$

The gain–bandwidth product is about 80 MHz and thus, as indicated above, well within the range available in operational amplifiers.

The loop gain can be obtained alternatively by making a cut at the base of Q2. In this case you must take into account that the input impedance at the base of Q2 is part of the collector circuit of Q1 and, as discussed in Section 9-10, that it should remain on the the upstream side of the cut; you can then imagine Q2 as a transistor with infinite input impedance. The resulting configuration is essentially a cascode amplifier, and you can obtain the loop gain from the equivalent circuit shown in Fig. 9-33.

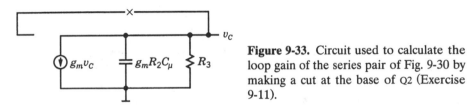

Figure 9-33. Circuit used to calculate the loop gain of the series pair of Fig. 9-30 by making a cut at the base of Q2 (Exercise 9-11).

9-12 EFFECTS OF A FINITE LOW-FREQUENCY OPEN-LOOP GAIN.
Consider a direct amplifier whose ideal gain and open-loop gain are

$$A_I = \frac{R_1 + R_2}{R_1} \quad \text{and} \quad A = \frac{A_0}{1 + A_0 \tau s}$$

Show that the closed-loop gain is

$$A_C = \alpha \frac{A_I}{1 + \alpha A_I \tau s}$$

where

$$\alpha = \frac{A_0}{A_0 + A_I}$$

If the low-frequency loop gain A_0/A_I is large, the effect of assuming that the open-loop gain is an integrator at all frequencies is negligible in the dynamical behavior, and barely perceptible in the closed-loop gain at $\omega = 0$.

9-13   THE TAMING OF UNSTABLE SYSTEMS BY MEANS OF FEEDBACK. Assume that you are faced with an unstable system with transfer function

$$\frac{1}{\tau s - 1}$$

If you put this system in a feedback loop and drive it with a proportional controller with gain K, as shown in Fig. 9-34, the closed-loop gain is stable if $K > 1$:

$$A_C(s) = \frac{K}{K - 1 + \tau s}$$

Now consider an unstable second-order system (an undamped inverted pendulum is a good example) with transfer function

$$\frac{1}{\tau^2 s^2 - 1}$$

Convince yourself that proportional control is not quite enough to tame this system because the resulting closed-loop gain is undamped, but that proportional-plus-derivative control, obtained by inserting a term $1 + \alpha \tau s$ in the loop as shown in Fig. 9-35, will do so if $K > 1$ and α is properly chosen.

For $K = 100$ and $\alpha = \frac{2}{10}$ show that the closed-loop gain is well damped:

$$A_C(s) \simeq \frac{1 + 2\tau s/10}{1 + 2\tau s/10 + \tau^2 s^2/100}$$

Figure 9-34. An unstable first-order system can be stabilized with a proportional controller (Exercise 9-13).

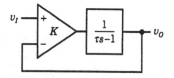

Figure 9-35. An unstable second-order system can be stabilized with a proportional-plus-derivative controller (Exercise 9-13).

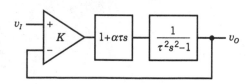

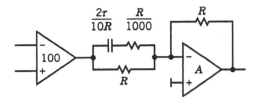

Figure 9-36. Proportional-plus-derivative controller for the unstable system of Fig. 9-35 (Exercise 9-13).

For $A = 2 \cdot 10^7/\tau s$ show that the controller of Fig. 9-36 has a transfer function

$$G(s) \simeq 100 \frac{1+2\tau s/10}{(1+\tau s/10,000)^2}$$

and that it is therefore adequate for this particular case since the time constant of the lags is negligible. For $\tau = 1$ s, the operational-amplifier bandwidth should be 3 MHz; making $R = 100$ kΩ implies $C = 2\tau/10R = 2$ μF. Both requirements are easily satisfied.

9-14 POSITIVE FEEDBACK

(a) For the amplifier with positive feedback of Fig. 9-37, show that

$$\frac{v_O}{v_I} = \frac{-A_0}{\tau s + (1 - A_0)}$$

and that it is therefore stable only if $A_0 < 1$.

(b) Convince yourself that the phase of the loop gain goes from $-180°$ at low frequencies to $-270°$ at high frequencies. For $A_0 > 1$ the loop gain has a well-defined transition frequency and the phase margin is negative; instability is thus correctly predicted by the phase-margin criterion.

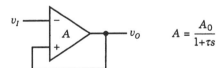

$$A = \frac{A_0}{1+\tau s}$$

Figure 9-37. Circuit for Exercise 9-14.

9-15 THE NYQUIST STABILITY CRITERION. Consider the rational function of a complex variable

$$H(s) = K \frac{(s-z_1) \cdots (s-z_L)}{(s-p_1) \cdots (s-p_M)} = u + iv = \rho \exp(i\varphi)$$

where $s = \sigma + i\omega$, and $z_1 \cdots z_L$ and $p_1 \cdots p_M$ are the zeros and poles of $H(s)$.

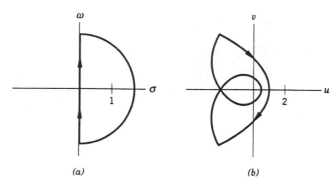

Figure 9-38. (a) Contour C in the $s = (\sigma, \omega)$ plane. (b) Contour C' in the (u, v) image plane for $H(s) = (s-1)^2$. Contour C' encircles the origin twice clockwise if the radius of C is larger than 1 (Exercise 9-15).

Now consider the contour C in the s-plane shown in Fig. 9-38a, which consists of a straight segment along the imaginary axis and a semicircle in the right half-plane. When C is traced clockwise, $H(s)$ maps C into a contour C' in the (u, v) image plane, as shown in Fig. 9-38b for the case $H(s) = (s-1)^2$. C makes a right-hand turn when it leaves the positive imaginary axis to follow the semicircle, and it makes another right-hand turn when it rejoins the imaginary axis; C' also makes right-hand turns at the corresponding image points.

In complex variable theory it is shown that if C encircles Z zeros and P poles of $H(s)$, C' encircles the origin $N = Z - P$ times in the clockwise direction. By letting the radius of C go to infinity, Z and P become the number of zeros and number of poles in the right half of the s-plane. The simple examples of Fig. 9-39 allow you to see why this is so. The phase of $H(s) = s - 1$ in Fig. 9-39a decreases monotonically by 2π as C is traced, so that C' encircles the origin clockwise, whereas the phase of $H(s) = 1/(s-1)$ in Fig. 9-39b increases monotonically and C' encircles the origin counter-clockwise. In contrast, the phases of $H(s) = s + 1$ in Fig. 9-39c and of $H(s) = 1/(s+1)$ in Fig. 9-39d never exceed $\pi/2$ in magnitude, so that C' cannot encircle the origin.

If $H(s)$ has poles at the origin, C can be modified slightly as shown in Fig. 9-40a by making a small semicircular detour around the origin. If the radius of this small circle goes to zero when the radius of the large circle goes to infinity, C will still encircle all the zeros and poles of $H(s)$ in the right half of the s-plane. The case of $H(s) = 1/s$ is illustrated in Fig. 9-40b.

The stability of a system is determined by the zeros of $1 + [A_L(s)]^{-1}$, that is, by the points where the reciprocal of the loop gain is equal to -1. If any of these zeros has a positive real part, that is, if it lies in the right half of the s-plane, the system is unstable.

You can now deduce the *Nyquist stability criterion:* when the radius of C goes to infinity, the image C' of $[A_L(s)]^{-1}$ or *Nyquist diagram* of $A_L(s)$

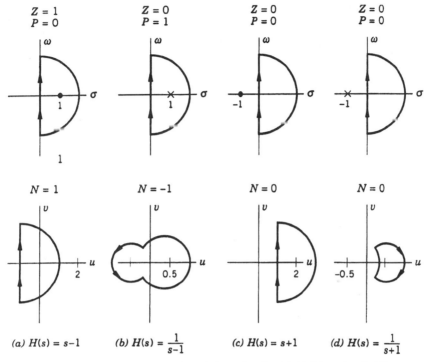

$Z = 1$
$P = 0$

$Z = 0$
$P = 1$

$Z = 0$
$P = 0$

$Z = 0$
$P = 0$

$N = 1$

$N = -1$

$N = 0$

$N = 0$

(a) $H(s) = s-1$ (b) $H(s) = \dfrac{1}{s-1}$ (c) $H(s) = s+1$ (d) $H(s) = \dfrac{1}{s+1}$

Figure 9-39. Simple examples of the rule $N = Z-P$ (Exercise 9-15).

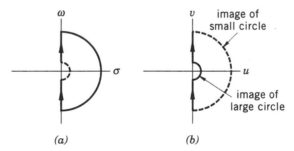

Figure 9-40. Modified contours for the case in which $H(s)$ has poles at the origin (Exercise 9-15). (a) Contour c in the $s = (\sigma, \omega)$ plane. (b) Contour c′ in the (u, v) image plane for $H(s) = 1/s$.

encircles -1 clockwise $N = Z-P$ times, where Z and P are the number of zeros and number of poles of $1+[A_L(s)]^{-1}$ in the right half of the s-plane; a system is therefore stable if and only if $N+P = 0$. Since $1+[A_L(s)]^{-1}$ and $[A_L(s)]^{-1}$ have the same poles and $A_L(s)$ is usually factored, P is usually known, so that Z is determined once N has been obtained.

(a) Figure 9-41 shows Nyquist diagrams (not to scale) of most of the loop gains of Fig. 9-10 (not included is the loop gain with a delay of Fig. 9-10d, to

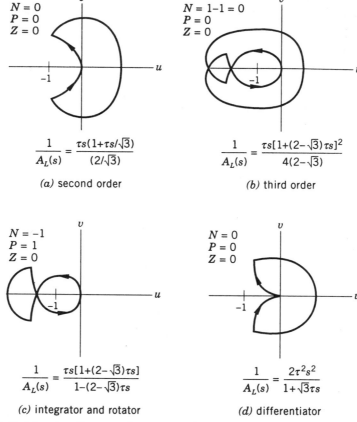

(a) second order

$$\frac{1}{A_L(s)} = \frac{\tau s(1+\tau s/\sqrt{3})}{(2/\sqrt{3})}$$

(b) third order

$$\frac{1}{A_L(s)} = \frac{\tau s[1+(2-\sqrt{3})\tau s]^2}{4(2-\sqrt{3})}$$

(c) integrator and rotator

$$\frac{1}{A_L(s)} = \frac{\tau s[1+(2-\sqrt{3})\tau s]}{1-(2-\sqrt{3})\tau s}$$

(d) differentiator

$$\frac{1}{A_L(s)} = \frac{2\tau^2 s^2}{1+\sqrt{3}\tau s}$$

Figure 9-41. Nyquist diagrams (not to scale) for most of the systems of Fig. 9-10 (Exercise 9-15).

which this discussion does not apply). Verify that the Nyquist stability criterion is satisfied and that it thus validates the phase-margin criterion.

Figure 9-42 shows Nyquist diagrams near the origin of a differentiator loop gain and a second-order loop gain. From Fig. 9-42a you can see why the resonance in the closed-loop gain is poorly predicted by the phase-margin criterion in the case of the differentiator: the phase is 180° at low frequencies, and the distance to −1 can therefore be substantially smaller than it is at the transition frequency. In contrast, the diagram of the second-order loop gain in Fig. 9-42b is essentially a circle centered at −1, and does not come much closer to −1 than it does at the transition frequency.

(b) Use Nyquist diagrams to verify the conclusions of Exercises 9-12 and 9-13.

(c) In most textbooks, N is obtained from the image of $A_L(s)$ rather than its reciprocal (the image of the reciprocal is easier to draw in the examples

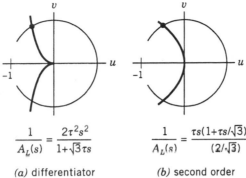

$$\frac{1}{A_L(s)} = \frac{2\tau^2 s^2}{1+\sqrt{3}\tau s}$$

$$\frac{1}{A_L(s)} = \frac{\tau s(1+\tau s/\sqrt{3})}{(2/\sqrt{3})}$$

(a) differentiator   (b) second order

Figure 9-42. Nyquist diagrams of (a) a differentiator loop gain and (b) a second-order loop gain, illustrating why the phase-margin criterion is a poor predictor of overshoot in the case of the former (Exercise 9-15).

considered here). Show that this procedure is also correct and that P is then the number of poles of $A_L(s)$ in the right half of the s-plane and Z is the number of zeros of $1+A_L(s)$ in the right half of the s-plane.

9-16   LOOP GAINS WITH A LOW-FREQUENCY CUTOFF. In the discussion of the phase-margin criterion in Section 9-5 it was assumed that the magnitude of the loop gain was monotonically decreasing. The point of this exercise is to show that the monotonicity requirement can be relaxed and that the phase-margin criterion holds for loop gains that have a low-frequency cutoff as well as the unavoidable high-frequency cutoff.

(a) Start by considering a loop gain with a monotonically increasing magnitude, as indicated schematically in Fig. 9-43. If you follow through the arguments of Section 9-5 and keep in mind that the phase of the loop gain increases at low frequencies, you should be able to convince yourself that one should avoid introducing phase leads at low frequencies and that the phase margin at the low-frequency end should be defined by

$$\varphi_M = \pi - \varphi_L(\omega_T)$$

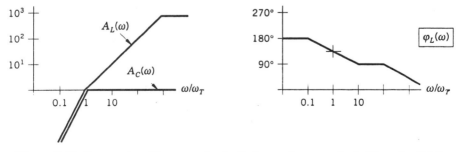

Figure 9-43. Loop gain with a monotonically increasing magnitude (Exercise 9-16).

(b) From another point of view, you have

$$A_C(s) = \frac{1}{1+[A_L(s)]^{-1}} = 1-\frac{1}{1+A_L(s)} \equiv 1-A_C^*(s)$$

You can thus equally well consider the system with loop gain $A_L^*(s) = [A_L(s)]^{-1}$ because $A_L^*(\omega)$ is monotonically decreasing, so that the considerations of Section 9-5 are directly applicable. Notice that $A_L(s)$ and $A_L^*(s)$ have the same transition frequency and, since $\varphi_L^*(s) = -\varphi_L(s)$, the same phase margin. As an example, the reciprocal of the loop gain

$$A_L(s) = \frac{2\tau^2 s^2}{1+\sqrt{3}\,\tau s}$$

is the familiar differentiator loop gain of Exercise 9-5,

$$A_L^*(s) = \frac{1+\sqrt{3}\,\tau s}{2\tau^2 s^2}$$

This approach has the advantage that the overshoot in $r_C^*(t)$ is well correlated with the resonance in $A_C^*(s)$, an empirical fact that does not hold as well for $A_C(s)$ and the undershoot in $r_C(t)$.

(c) Now consider a loop gain that cuts off at both low and high frequencies, like the one shown in Fig. 9-44. The stability of the closed-loop gain can be determined from the Nyquist diagram of $A_L(s)$ (Exercise 9-15). The point to notice is that the region between the transition frequencies, in which $A_L(\omega)$ is larger than 1, is mapped into the interior of a unit circle centered at the origin and can have no part in encircling -1: stability is determined by the behavior of $[A_L(s)]^{-1}$ in the high-frequency and low-frequency regions. Convince yourself that these regions can be studied separately if the upper and lower transition frequencies are well separated.

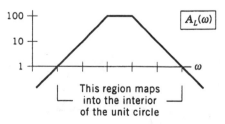

Figure 9-44. Loop gain with well-separated upper and lower transition frequencies (Exercise 9-16).

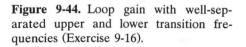

10

DIGITAL CIRCUITS

10-1 INTRODUCTION

Most of the circuits we have seen thus far have dealt with *analog* signals that take on continuous values within a certain range. There is, however, a broad domain of electronics in which signals take on only discrete values and can therefore be represented by integers. The dots and dashes in the Morse code, or the presence or absence of a requirement for a decision, are straightforward examples of such *digital* signals.

Electronic devices with two well-defined states arise much more naturally than devices with ten states; for this reason, digital signals are represented and processed in *binary* form rather than in the familiar *decimal* form; that is, counting is done by twos rather than by tens. It turns out furthermore that a binary representation, while not optimal—counting by threes would in fact be better—results in simpler circuits.

Digital circuits are mandatory for the logic and arithmetic functions inherent in computers, but they are also used for other purposes. Much use is made, for example, of the fact that digital signals are large compared with most analog signals (volts rather than microvolts) and therefore far easier to separate from noise. Analog signals converted into digital form (that is, into a series of integers) lose virtually no information if the sampling resolution and the sampling rate are high enough, and transmission of the *digitized* signals over noisy channels is often perfect. How much can be gained by such procedures is evident in the patent superiority of digital compact discs as a storage medium for audio signals when compared with analog media such as magnetic tapes.

Digital electronics is a dizzying and fast-changing field, but most advances involve speed and versatility rather than fundamental ideas. In this chapter we will present the relatively few concepts that are needed to approach the literature. We will thus look at Boolean algebra and its representation in terms of logic gates, at elementary memory devices in the form of flip-flops, and with an eye on computers, at state machines based on flip-flops.

We will emphasize the fact that digital circuits have delays. On the negative side, delays result in finite processing times and therefore set a limit on speed; but on the positive side, delays are essential for the existence of flip-flops and state machines.

Hardware should not be slighted, and we will therefore look at the basic circuits of the most widely used *logic families*. As their name suggests, logic families are sets of compatible integrated circuits that execute elementary logic and arithmetic functions and are readily interconnected to generate complex functions.

10-2   BINARY ARITHMETIC IN THE 2'S COMPLEMENT REPRESENTATION

Any positive integer m can be represented in the form

$$m = x_n y^n + \cdots + x_1 y^1 + x_0 y^0$$

where y is a positive integer called the *base* and x_i are integers that range from 0 to $y-1$. In *positional* notation, m is written compactly as $x_n \cdots x_0$. In the familiar base-10 or *decimal* representation, for example, we have

$$901 = 9 \cdot 10^2 + 0 \cdot 10^1 + 1 \cdot 10^0$$

In the base-2 or *binary* representation, the string of *bits* $b_n \cdots b_0$, where b_i is either 0 or 1, represents the integer m given by

$$m = b_n 2^n + \cdots + b_1 2^1 + b_0 2^0$$

Where necessary, we will indicate the base by writing it as a subscript in *decimal* representation. Thus, for example, we will write

$$11_{10} = 1011_2$$

If we divide a positive integer m by 2, then divide the quotient by 2, and so on, we obtain

$$m = 2^1 q_0 + 2^0 r_0 = 2^2 q_1 + 2^1 r_1 + 2^0 r_0 = \cdots$$

where q_i are quotients and r_i are remainders. When the quotient becomes 0, we can write the binary representation of m in the form

$$m = 1r_{n-1} \cdots r_0$$

Conversion from decimal to binary representation is thus obtained by writing down in reverse order the remainders of successive divisions by 2 until the quotient is 0. In converting 37, for example, the sequence of quotients and remainders is 18(1), 9(0), 4(1), 2(0), 1(0), 0(1), and we thus have

$$37 = 100101_2$$

In most computers, operations are carried out on 16-bit or 32-bit *words* (storage, which is another matter, is organized in 8-bit *bytes*). For simplicity, we will use 4-bit words in the examples that follow. An N-bit word can be used to represent 2^N nonnegative or *unsigned* integers from 0 to $2^N - 1$. Arithmetic operations, however, require that negative integers also be represented. In the *2's complement* representation, which is by far the most convenient for arithmetic operations, nonnegative integers range from 0 to $2^{N-1} - 1$ and thus always have a 0 in the most significant bit. Negative integers are represented by unsigned integers of the form $2^N - m$; that is, we have the correspondence

$$-m \leftrightarrow 2^N - m$$

Negative integers range from -1 to -2^{N-1}, and their representations therefore always have a 1 in the most significant bit. We note that -2^{N-1} is unpaired because 2^{N-1} is not in the set of positive integers in an N-bit machine. In our 4-bit machine, for example, the representation of -1 is $2^4 - 1 = 1111_2$, and the representation of -8 is $2^4 - 8 = 1000_2$.

Observing that $2^N - m = 2^{N-1} + (2^{N-1} - m)$ and that $-m = -2^{N-1} + (2^{N-1} - m)$, it becomes clear that a negative integer can be recovered from it's 2's complement representation by assigning weight -2^{N-1} rather than 2^{N-1} to the most significant bit, which is always 1 for negative integers. Thus in the case of -1, whose representation is $1111_2 = 1000_2 + 0111_2 = 8 + 7$, we correctly obtain $-1 = -8 + 7$.

Arithmetic in the 2's complement representation is modulo 2^N: operations are carried out as if all integers were unsigned, and integers that differ by 2^N are identified. Thus the sum of the representations of m and $-m$ is 2^N, which is identified with 0. In practice, the identification is automatic: the sum of m and $-m$ gives all zeros plus a carry into the nonexistent bit with weight 2^N; by simply ignoring the carry, we get $m + (-m) = 0$.

The *2's complement* of an unsigned N-bit integer m is $2^N - m$; for a signed N-bit integer we will interpret it as the 2's complement of the unsigned representation. The 2's complement of a positive integer m is thus equal to $-m$; with the exception of -2^{N-1}, whose 2's complement is -2^{N-1}, this result also holds for negative integers. It follows that taking the 2's complement changes the sign of any nonzero integer except -2^{N-1}.

The *1's complement* of an unsigned N-bit integer is obtained by setting all bits that are 1 to 0 and all bits that are 0 to 1. The (unsigned) sum of an integer and its 1's complement is a string of N 1's, and thus equal to $2^N - 1$. It follows that the 2's complement of an integer can be obtained by adding 1 to the 1's complement. For example, the 1's complement of $0001_2 = 1$ is 1110_2; adding 1 to this result then gives the 2's complement $1111_2 = -1$.

In a 2's complement representation, the sum of two integers of opposite sign is always within the set of representable integers. The sum of two integers of the same sign, however, can result in an *overflow*, that is, in an integer that cannot be represented because its magnitude is too large. If the sum is done anyway, the result is an integer of the opposite sign; an overflow condition is thus easily detectable. For example, $7 + 7 = 0111_2 + 0111_2 = 1110_2$ $= -2$, and $(-7) + (-7) = 1001_2 + 1001_2 = 0010_2 = 2$.

Subtraction is carried out by adding the minuend and the negative of the subtrahend; for example, $-7 - 1 = -7 + (-1) = 1001_2 + 1111_2 = 1000_2 = -8$.

Multiplication by 2 of a nonnegative integer less than 2^{N-2} can clearly be obtained by shifting all bits one position to the left, thus throwing away the most significant bit, and writing a 0 into the least significant bit. For example, $3 \cdot 2$ is obtained by shifting 0011_2 to the left, which gives $0110_2 = 6$. The overflow that occurs if the integer is greater than or equal to 2^{N-2} can be detected because the result of the shift is negative. Shifting $0111_2 = 7$ to the left, for example, results in $1110_2 = -2$.

If an unsigned integer of the form $2^N - m$, where $m \le 2^{N-2}$, is multiplied by 2, the result is $2^N + 2^N - 2m$, which can be identified with $2^N - 2m$. It is thus clear that multiplication by 2 of a negative integer can also be obtained by shifting to the left. For example, shifting $1101_2 = -3$ to the left results in $1010_2 = -6$. If $m > 2^{N-2}$ there is an overflow, and since bit $N-2$ is 0 in this case, the result of the shift is positive and the overflow is detectable; shifting $1011_2 = -5$ to the left, for example, results in $0110_2 = 6$.

We conclude that integer multiplication can be reduced to a series of shifts to the left and sums. As we will see, these operations are easily carried out with digital circuits.

Division by 2 of a nonnegative integer is carried out by throwing away the least significant bit (or perhaps storing it elsewhere as the remainder), shifting the remaining bits one position to the right, and writing a 0 into the most significant bit. Shifting $0111_2 = 7$ to the right, for example, results in $0011_2 = 3$.

An unsigned integer of the form $2^N - m$ can be written as $2^N - 2q + r$, where r is 0 or 1. Shifting to the right results in $2^{N-1} - q$; adding 2^{N-1} to this result gives $2^N - q$. It follows that division by 2 of a negative integer can be obtained by shifting to the right and writing a 1 into the most significant bit. The only oddity is that the remainder r is nonnegative and that round-off is therefore toward a more negative integer rather than toward 0; shifting $1111_2 = -1$ to the right thus results in -1 because $-1 = -1 \cdot 2 + 1$.

In general, then, division by 2 of integers is performed by shifting to the right and copying the most significant bit into itself.

10-3 BOOLEAN ALGEBRA, LOGICAL VARIABLES, AND LOGICAL FUNCTIONS

Let us consider an algebra with two elements denoted by 0 and 1. The binary operations in this algebra are *OR* (represented as a sum) and *AND* (represented as a product). In addition, there is a unary *negation* operation *NOT* that is represented by a bar over an element. The results of all possible operations are as follows:

$$0+0 = 0 \qquad 0+1 = 1 \qquad 1+0 = 1 \qquad 1+1 = 1$$
$$0 \cdot 0 = 0 \qquad 0 \cdot 1 = 0 \qquad 1 \cdot 0 = 0 \qquad 1 \cdot 1 = 1$$
$$\bar{0} = 1 \qquad \bar{1} = 0$$

This algebra is the *algebra of propositions*, and it is a particular case of a *Boolean* algebra. In another well-known Boolean algebra, the elements are the subsets of a given set; the operations corresponding to OR and AND are the *union* and *intersection* of two subsets, and the operation corresponding to NOT is the *complement* of a subset.

If we call the elements of the algebra of propositions FALSE and TRUE rather than 0 and 1, it becomes clear after thinking about it for a while that the algebra of propositions governs a world in which propositions are classified as either TRUE or FALSE. Thus if A and B are propositions, A OR B is FALSE if and only if A and B are both FALSE, whereas A AND B is TRUE if and only if A and B are both TRUE; the negation of a TRUE proposition is a FALSE proposition, and the negation of a FALSE proposition is a TRUE proposition.

A *Boolean* or *logical* variable is a variable whose values are the elements 0 or 1 of the algebra of propositions. If A is a logical variable, we clearly have

$$\bar{\bar{A}} = A$$

By assigning values to A, we can verify the following theorems:

$$A+A = A \qquad A+0 = A \qquad A+1 = 1 \qquad A+\bar{A} = 1$$
$$A \cdot A = A \qquad A \cdot 0 = 0 \qquad A \cdot 1 = A \qquad A \cdot \bar{A} = 0$$

With a bit more work, we can also verify that addition and multiplication are commutative and associative; thus if A, B, and C are logical variables, we have

$$A+B = B+A \qquad A+(B+C) = (A+B)+C$$
$$A \cdot B = B \cdot A \qquad A \cdot (B \cdot C) = (A \cdot B) \cdot C$$

We can further verify that multiplication is distributive over addition:

$$A \cdot (B+C) = (A \cdot B)+(A \cdot C)$$

An interesting property of Boolean algebra is that addition is distributive over multiplication:

$$A+(B \cdot C) = (A+B) \cdot (A+C)$$

We have enough machinery that we can prove this theorem. The distributivity of multiplication over addition, commutativity, and associativity imply that

$$(A+B) \cdot (A+C) = (A \cdot A)+(A \cdot B)+(A \cdot C)+(B \cdot C)$$

Setting $A \cdot A = A \cdot 1$ and factoring out A, we obtain

$$(A+B) \cdot (A+C) = A \cdot (1+B+C)+(B \cdot C)$$

But $1+B+C = 1$, and the theorem is proved.

At this point we will make the symbol for multiplication optional, and as in ordinary arithmetic we will usually write AB for $A \cdot B$. To simplify notation further, we will consider that multiplication has priority over addition, so that $A+BC$ should be read as $A+(BC)$.

Central in digital electronics are *DeMorgan's* theorems, both of which are easy to prove:

$$\overline{A+B} = \overline{A}\,\overline{B} \qquad \overline{AB} = \overline{A}+\overline{B}$$

Thus $\overline{A+B}$ is 1 if and only if $A+B$ is 0, which happens if and only if A and B are both 0; but $\overline{A}\,\overline{B}$ is 1 if and only if $\overline{A}$ and $\overline{B}$ are both 1, that is, if and only if A and B are both 0, and the first theorem is proved. Similarly, $\overline{AB}$ is 0 if and only if AB is 1, which happens if and only if A and B are both 1; but $\overline{A}+\overline{B}$ is 0 if and only if $\overline{A}$ and $\overline{B}$ are both 0, that is, if and only if A and B are both 1, and the second theorem is proved.

Using the fact that $A+AB = A(1+B)$ we obtain another useful theorem,

$$A+AB = A$$

Finally, the distributivity of addition over multiplication implies that $A+\overline{A}B = (A+\overline{A})(A+B)$, so we have

$$A+\overline{A}B = A+B$$

A *logical function* Φ is a function that operates on logical variables, using the OR, AND, and NOT operations; the values of a logical function are therefore also logical. Thus if A, B, and C are logical variables,

$$\Phi = A\overline{C}+BC+AB$$

is a logical function. A logical function can be described by a *truth table* that lists all possible arrays of values of the logical variables together with the corresponding values of the function.

A	B	0	$A{\cdot}B$	$A{\cdot}\bar{B}$	A	$\bar{A}{\cdot}B$	B	$A{\oplus}B$	$A{+}B$	$\overline{A{+}B}$	$\overline{A{\oplus}B}$	$\bar{B}$	$A{+}\bar{B}$	$\bar{A}$	$\bar{A}{+}B$	$\overline{A{\cdot}B}$	1
0	0	0	0	0	0	0	0	0	0	1	1	1	1	1	1	1	1
0	1	0	0	0	0	1	1	1	1	0	0	0	0	1	1	1	1
1	0	0	0	1	1	0	0	1	1	0	0	1	1	0	0	1	1
1	1	0	1	0	1	0	1	0	1	0	1	0	1	0	1	0	1

Figure 10-1. Truth tables for logical functions of two variables. $A \oplus B$ is the exclusive-OR operation defined by $A \oplus B = \bar{A}B + A\bar{B}$.

The truth tables for all logical functions of two variables are shown in Fig. 10-1. In these tables, $A \oplus B$ is the *Exclusive-OR (XOR)* operation, defined by

$$A \oplus B = \bar{A}B + A\bar{B}$$

Thus $A \oplus B$ is 1 if and only if A is different from B. The operations $\overline{AB}$, $\overline{A+B}$, and $\overline{A \oplus B}$ are called *NAND, NOR,* and *Exclusive NOR (XNOR)*. Using DeMorgan's theorems we obtain the expected result that $\overline{A \oplus B}$ is 1 if and only if A and B are equal:

$$\overline{A \oplus B} = \overline{\bar{A}B + A\bar{B}} = \left(\overline{\bar{A}B}\right)\left(\overline{A\bar{B}}\right) = (A+\bar{B})(\bar{A}+B) = A\bar{A} + AB + \bar{A}\bar{B} + B\bar{B}$$

and thus

$$\overline{A \oplus B} = AB + \bar{A}\bar{B}$$

Let us consider two 1-bit binary integers A and B. A is greater than than B if and only if A is 1 and B is 0. Looking at the truth tables in Fig. 10-1, we observe that we can detect the condition [A>B] by interpreting A and B as logical variables, executing the logical function $A\bar{B}$, and demanding that the result be 1. If we continue this process, we obtain the following list of correspondences between bit comparisons and logical functions:

$$[A>B] \leftrightarrow A \cdot \bar{B} \qquad [A \le B] \leftrightarrow \bar{A}+B$$
$$[A<B] \leftrightarrow \bar{A} \cdot B \qquad [A \ge B] \leftrightarrow A+\bar{B}$$
$$[A = B] \leftrightarrow \overline{A \oplus B} \qquad [A \ne B] \leftrightarrow A \oplus B$$

10-4 LOGIC GATES

Without getting into the details of hardware for the moment, we can say that digital circuits accept and generate two distinct voltage levels called LOW and HIGH, and that depending on circumstances, we can interpret these *logic levels* as logical 0 and logical 1 or as the integers 0 and 1.

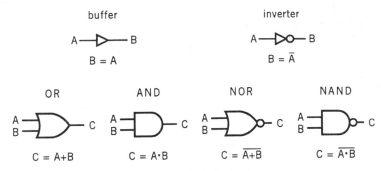

Figure 10-2. Symbols for logic gates.

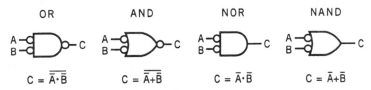

Figure 10-3. Alternative symbols for logic gates, preferably used when LOW inputs initiate an action.

Logic gates are digital circuits that execute the basic logical operations as well as their negations. Symbols and equations for logic gates are shown in Fig. 10-2; in the symbols, negation is indicated by a small circle.

Using DeMorgan's theorems, we obtain the equivalent representations shown in Fig. 10-3. These representations are useful because we often want things to happen when logic levels are LOW, in which case we obtain a better picture of what is intended.

Circuits that execute the XOR and XNOR functions are used so often that they are also classified as logic gates; symbols for XOR and XNOR gates are shown in Fig. 10-4.

Any logical function has a hardware representation in terms of logic gates. As an example, Fig. 10-5 shows circuits that execute the XOR and XNOR functions.

XOR   XNOR

Figure 10-4. Symbols for exclusive-OR (XOR) and exclusive-NOR (XNOR) logic gates.

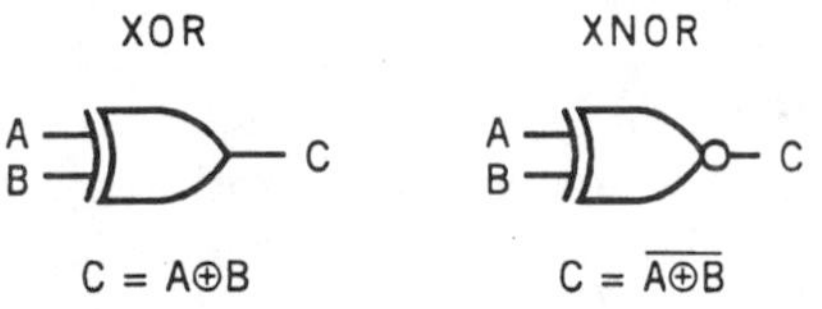

$C = A \oplus B$ $C = \overline{A \oplus B}$

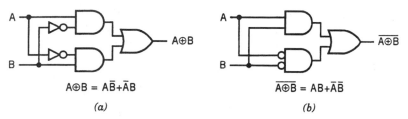

$$A \oplus B = A\overline{B} + \overline{A}B$$

(a)

$$\overline{A \oplus B} = AB + \overline{A}\overline{B}$$

(b)

Figure 10-5. Logic circuits that execute (*a*) the XOR function and (*b*) the XNOR function.

10-5 TRANSISTOR GATES AND LOGIC FAMILIES

Most logic circuits are built using standardized integrated circuits from one of the logic families. In this section we will describe the basic gates of the LS-TTL family developed by Texas Instruments, the CMOS family developed by RCA, and the ECL family developed by Motorola, Inc.

An important requirement on the circuits in a given logic family is that they deliver output voltage levels that are compatible with the voltage levels that they accept as inputs. Another important requirement is that they have a large *fanout*, that is, that they be able to drive many inputs with minimal change in their output voltage levels.

The basic circuit of the LS-TTL family is the NAND gate shown in simplified form in Fig. 10-6a. Actual gates such as the 74LS00 NAND gate are internally more complex, but the structure of their input and output circuits is well represented. The letter L stands for *low power*, and indicates that power consumption is lower than in other versions of the family. The letter S stands for *Schottky*, and indicates that transistors have a Schottky diode from base to collector and thus do not saturate (Q4 is an exception

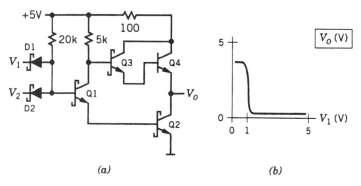

(a)   *(b)*

Figure 10-6. Basic LS-TTL NAND gate. (*a*) Structure. (*b*) Voltage transfer characteristic when one input (V_2) is held HIGH.

because it is kept out of saturation by the Schottky diode in Q3). The letters TTL, which stand for *transistor–transistor logic*, are relics of early versions of the family, in which a multi-emitter transistor was used in place of input Schottky diodes D1 and D2.

Let us assume for a moment that $V_2 = 5V$, so that D2 is reverse biased and can therefore be ignored. If V_1 is zero, V_{B1} is about 300 mV, and Q1 and Q2 are both cut off. If the output load current is within the limits prescribed by family rules, usually 400 μA, the output voltage V_O is 5V minus the voltage drop in Darlington follower Q3|Q4, that is, about 3.6V. V_O remains at this value until V_1 is raised to about 1V—the *threshold* voltage—at which point Q1 and Q2 start to turn on and V_O starts to drop because the base of Q3 is pulled down by Q1.

If V_1 is higher than two diode drops, that is, about 1.4 V, D1 is cut off and Q1 and Q2 are fully on. The base currents in Q1 and Q2 are much more than is required for saturation, so that V_{CE1} and V_{CE2} are both clamped at about 400 mV by their base–collector Schottky diodes; $V_{B3} - V_{E4}$ is thus about one diode drop, and Q3 and Q4 are therefore both cut off.

The resulting voltage transfer characteristic is shown in Fig. 10-6b. If following family rules we define LOW as any voltage under 800 mV, and HIGH as any voltage above 2.4 V, it is clear that a LOW input will give a HIGH output, and vice versa. If V_1 and V_2 are both allowed to take on LOW or HIGH values, V_O will be LOW only if V_1 and V_2 are both HIGH, and we thus have a NAND gate.

A HIGH input to an LS-TTL gate sees a reverse-biased diode; it is thus clear that the fanout in the HIGH state is large in the LS-TTL family. According to family rules, a LOW input to an LS-TTL gate sinks 400 μA in the worst case, and a LOW output must be able to sink 4 mA; the fanout in the LOW state is thus also large, about 10.

Let us now turn to the *complementary MOS* (CMOS) family, so called because it involves *complementary* MOSFETs, that is, n-channel and p-channel enhancement MOSFETs with similar characteristics; for simplicity, we will assume identical characteristics and a threshold voltage $V_{TH} = 1$ V.

We will first consider the inverter shown in Fig. 10-7a. If $V_I = 0$, Q2 is cut off and, assuming that $V_{DD} \geq 5$ V, Q1 is well above threshold and acts like a

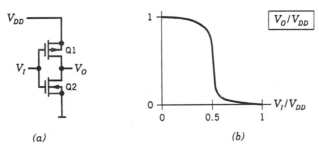

(a)   *(b)*

Figure 10-7. CMOS inverter. (*a*) Structure. (*b*) Voltage transfer characteristic.

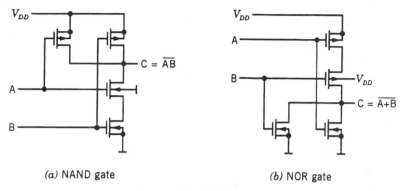

(a) NAND gate   (b) NOR gate

Figure 10-8. CMOS gates.

resistor of order 1 kΩ; the output voltage V_O is thus equal to V_{DD}. V_O remains at this value as long as $V_I \leq 1$ V. Similarly, if V_I is within 1 V of V_{DD}, Q1 is cut off and Q2 is on, and $V_O = 0$. If $V_I = V_{DD}/2$, Q1 and Q2 have the same drain current, and V_O is ideally equal to $V_{DD}/2$. By superposition, the small-signal gain at this point is $2(g_m r_o/2)$, which is of order 100, and V_O changes rapidly as a function of V_I. The voltage transfer characteristic is thus as shown in Fig. 10-7b. At room temperature, taking LOW to be any voltage below $\frac{1}{3}V_{DD}$ and HIGH to be any voltage above $\frac{2}{3}V_{DD}$ is sufficient to guarantee that a LOW input gives a HIGH output, and vice versa.

We can now describe the CMOS gates shown in Fig. 10-8. The output of the NAND gate will be LOW only if both p-channel transistors are off and both n-channel transistors are on; this requires that both inputs be HIGH. In the NOR gate, in contrast, the output will be HIGH only if both p-channel transistors are on and both n-channel transistors are off; this requires that both inputs be LOW. As we might expect, voltage transfer characteristics in gates and inverters are not identical, but the differences are slight and can be ignored.

One of the great advantages of CMOS gates is that they draw negligible current from the power supply when they are in a quiescent state. This advantage can vanish, however, if they change state at a high rate. When an output makes a transition from 0 to V_{DD}, the current that charges the load capacitance C_L is provided by the power supply; on the other hand, when the output makes a transition from V_{DD} to 0, C_L discharges to ground. If this process happens at a frequency f, the power supply must provide a current $fC_L V_{DD}$. For $f = 10$ MHz, $C_L = 10$ pF, and $V_{DD} = 5$ V, the power-supply current is 0.5 mA, which might be far from negligible.

Input currents in CMOS circuits are the gate currents of a few MOSFETs and thus extremely small. In contrast, a CMOS circuit can tolerate a load current of order 1 mA. Fanout in the CMOS family is thus indeed large. Each gate, however, has an input capacitance, so that driving too many gates can result in an unacceptably slow response.

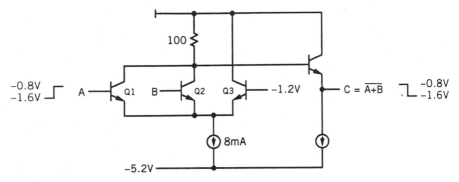

Figure 10-9. ECL NOR gate.

The final logic family we will look at is the *emitter-coupled logic* (ECL) family, whose basic circuit is the NOR gate shown in Fig. 10-9. Let us for a moment assume that the bases of Q1 and Q2 are tied together. The circuit can then be described as a differential amplifier whose single-ended collector output voltage is buffered by an emitter follower. Assuming that $V_{BE} = 0.8$ V in all transistors when they are conducting, the output voltage is -0.8 V when Q1|Q2 is cut off, and -1.6 V when Q3 is cut off and all the current in the 8 mA current source goes through Q1|Q2.

With a bias of -1.2 V at the base of Q3, the current switches between Q1|Q2 and Q3 when the base voltage on Q1|Q2 switches between levels that are about 120 mV above and below -1.2 V. If we define HIGH as a voltage above -1.0 V and LOW as a voltage below -1.4 V, a LOW input at the base of Q1|Q2 will give a HIGH output at C, and vice versa.

If we now untie the bases of Q1 and Q2, it is clear that a HIGH at either A or B will make C LOW and that we are dealing with an ECL NOR gate. In an actual integrated version the logic levels are somewhat different, -0.9 V and -1.8 V, but the circuit structure is essentially the same.

ECL circuits can operate at high frequencies: 100 MHz is standard, and 500 MHz is possible. In addition, they are well suited for driving transmission lines, which are almost mandatory at such frequencies.

10-6 GATE DELAYS

In Section 8-10 we considered the large-signal response of a transistor in the common-emitter configuration. As shown in Fig. 8-25, the step response has a delay and then a transition that can be described in terms of a rise time or a fall time. The transistors that define the dynamic response in logic gates are essentially in the common-emitter or common-source configuration, and like the transistor in Fig. 8-25, they switch back and forth between an off-state and an on-state. The step response of logic gates is thus also characterized by delays and transition times. Transition times are important because they set conditions on the layout of digital circuits; we mentioned in Section 10-5, for example, that ECL circuits must be interconnected with transmission lines in most cases. We will look further into this question in Chapter 11.

Assuming that questions of layout have been properly addressed, we need not consider delay and transition times separately. The only time we really need to know is the time it takes for a transition at the input of a gate to become sufficiently well reflected at the output that it can have an effect on other gates; we are of course thinking of a case in which conditions at the input of the gate are such that the output will change state. This overall time or *gate delay* is equal to the real delay plus a fraction of the transition time.

In the logic families we have described, the gate delay depends on the type of gate, but it does not depend much on the direction of the transition if the capacitive load on a gate is light. The gate delays of the basic gates are about 2 ns in the ECL family, 10 ns in the LS-TTL family, and anywhere from 5 to 100 ns in the CMOS family.

For our purposes it will be sufficient to think in terms of a model in which gate delays are independent of the direction of the transition, and we will therefore describe the dynamics of logic gates in terms of a single time, the gate delay τ. In addition, we will take transition times to be zero so that we can locate events conveniently.

To see how our model works, let us use it in a simple example. If we consider only the static behavior of the circuit shown in Fig. 10-10, we conclude that B is always 0 because $B = A\overline{A}$. But if we take into account the delays in the inverter and in the AND gate, we conclude that B exhibits a pulse of width τ when A goes from 0 to 1, and that the leading edge of the pulse is delayed by τ with respect to the leading edge of A.

Gate delays can thus be profitably used to generate short pulses, but they can also result in unintended pulses that make other circuits malfunction or, more generally, in a *race* condition in which a signal that is meant to block a gate before other inputs change does not arrive in time.

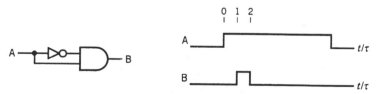

Figure 10-10. Gate delays result in pulses that are not predicted by the static equations of a logic circuit.

10-7 COMBINATIONAL CIRCUITS

A *combinational* circuit is a digital circuit in which the steady-state outputs are logical functions of the steady-state inputs only, and thus do not depend on the internal condition of the circuit. Combinational circuits can therefore always be constructed with gates. Adding or comparing integers, for example, or deciding whether all the conditions for a decision are satisfied, are tasks that can be done with combinational circuits. There is, however, a class of *sequential* tasks that require memory of past actions, for which combinational circuits do not suffice. Thus if we want to detect the *second* occurrence of a given action, we must somehow register the fact that at some point there was a *first* occurrence of this action. We will consider sequential circuits starting in Section 10-9.

Examples of combinational—and sequential—circuits will arise naturally as we go along, but we will not cover all possibilities; the best way to find out what is available at any given time is to leaf through an up-to-date manufacturer's data manual. Since our examples are drawn from industrial sources, it would in fact be valuable to have at hand manuals of LS-TTL and CMOS small-scale integrated circuits to see what complete devices look like.

We have seen that logical functions can be written in more than one way and that some versions are more economical than others; the function $\Phi = A + \overline{A}B$, for example, can be written more succinctly as $\Phi = A + B$. Minimizing the number of gates required to execute a logical function is occasionally worth the effort, but the need for minimization is usually overshadowed by other considerations: speed and logical clarity, for example, almost always have higher priority.

Programmable logic devices or *PLDs* are integrated circuits that can generate arbitrary logical functions of a certain maximum number of variables when the interconnections of their internal gates are suitably programmed. In PLDs it is standardization that has priority over economy, and logical functions are therefore implemented in the well-defined *canonical* forms we will see next.

Let us consider an arbitrary logical function of n variables $\Phi(A_1, \ldots, A_n)$, and let us further consider n-term products of the form

$$\overset{(-)}{A_1} \cdots \overset{(-)}{A_n}$$

in which each variable is represented once, either negated or not; a product of this form is called a *minterm*. For each array of values of the variables $A_1 \cdots A_n$, there is one and only one minterm whose value is 1: it is the minterm in which A_i is not negated if A_i is equal to 1, and is negated if A_i is equal to 0. Since there are 2^n possible arrays of values, there are also 2^n minterms. In view of this correspondence, we can say that Φ is defined on the set of its minterms.

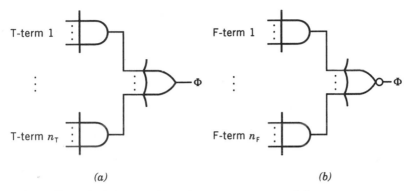

Figure 10-11. Canonical forms of combinational circuits. (*a*) Sum of T-terms. (*b*) Negated sum of F-terms.

Let us now divide the minterms of Φ into *T-terms*, which correspond to arrays of values for which Φ is equal to 1 or TRUE, and *F-terms*, which correspond to arrays of values for which Φ is equal to 0 or FALSE. If the number of T-terms is n_T and the number of F-terms is n_F, we must have

$$n_T + n_F = 2^n$$

By definition, Φ is equal to 1 on its T-terms and equal to 0 on its F-terms; Φ is therefore equal to the sum of its T-terms, and it is also equal to the negated sum of its F-terms. These two canonical forms are shown in Fig. 10-11.

As an example, let us consider $\Phi(A, B) = AB$. The only T-term of Φ is AB, so that $n_T = 1$; the F-terms are $\overline{A}B$, $A\overline{B}$, and $\overline{A}\,\overline{B}$, so that $n_F = 3$, and we correctly obtain $n_T + n_F = 2^2$. The canonical representations of AB are thus as shown in Fig. 10-12. As a check, we have $\overline{A}B + A\overline{B} + \overline{A}\,\overline{B} = \overline{A}(B + \overline{B}) + A\overline{B}$ $= \overline{A} + A\overline{B} = \overline{A} + \overline{B}$, or

$$\overline{A}B + A\overline{B} + \overline{A}\,\overline{B} = \overline{AB}$$

Negating this equality yields $\Phi(A, B) = AB$ as the negated sum of its F-terms.

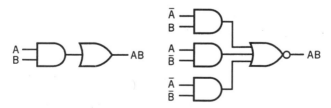

Figure 10-12. Canonical representations of $\Phi(A, B) = AB$.

10-8 ARITHMETIC OPERATIONS WITH LOGIC CIRCUITS

In this section we will show by means of examples how logic circuits are used to execute arithmetic operations on integers. With a slight and well-established license, we will include comparisons in our discussion.

In adding two binary words, at each position we need to know whether there is a carry from the next-lowest position or *carry-in*. We also need to generate the carry for the next-highest position or *carry-out*. At any given position, let A and B be the bits of the words to be added, C the carry-in, and Σ and K the sum and the carry-out. Σ is the low-order bit of the sum $A+B+C$, and it is 1 in only two cases: only one of the addends is 1, or all three are 1. If we interpret A, B, C, and Σ as logical variables, we can write

$$\Sigma = A\overline{B}\overline{C} + \overline{A}B\overline{C} + \overline{A}\overline{B}C + ABC$$

Similarly, K is 1 if any pair of addends is 1, and we obtain

$$K = AB + BC + AC$$

We might initially think that we need to add a term ABC to cover the case in which all three addends are 1, but this is unnecessary because $AB + ABC = AB$ or, equivalently, because $ABC = 1$ implies $AB = BC = AC = 1$.

In terms that parallel the correspondence between T-term and F-term canonical forms, we can also say that Σ is 0 if only one of the addends is 0 or if all three are 0. We then get

$$\overline{\Sigma} = \overline{A}BC + A\overline{B}C + AB\overline{C} + \overline{A}\,\overline{B}\,\overline{C}$$

Analogously, K is 0 if any pair of addends is 0, and we get

$$\overline{K} = \overline{A}\overline{B} + \overline{B}\overline{C} + \overline{A}\overline{C}$$

If we want to add two *N*-bit words we can chain *N* 1-bit adders together as shown in Fig. 10-13. The adder for the lowest-order bit accepts a carry-in C_{in} from a possible lower-order word, and the adder for the highest-order bit

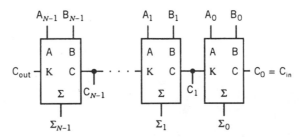

Figure 10-13. Adder for two *N*-bit words, obtained by chaining 1-bit adders. The generation of C_{out} must wait for the decisions in all lower-order bits.

generates a carry-out C_{out} that can feed a possible higher-order word. We observe that the decision in a given bit has to wait for the decisions in all lower-order bits; since each decision takes a finite time, the resulting delay in C_{out} might be considered excessive. This is the reason for the existence of *look-ahead carry generators*, circuits that consider all possible combinations of the inputs and decide with minimal delay whether C_{out} will be generated. We will consider the case $N = 2$, but the argument is valid in general.

We thus want to see if the sum of the 2-bit words A_1A_0 and B_1B_0 will generate a carry. A carry will certainly be generated if A_1 and B_1 are both 1 or, if we interpret A_1 and B_1 as logical variables, if

$$A_1B_1 = 1$$

If this is not the case, either A_1 or B_1 must be 1, and we than have to see what the low-order bits do. A carry will certainly be generated if A_0 and B_0 are both 1, that is, if

$$(A_1+B_1)A_0B_0 = 1$$

or if either of them is 1 and the carry into bit 0 is 1, that is, if

$$(A_1+B_1)(A_0+B_0)C_0 = 1$$

Assuming that bitwise ANDs and ORs of the inputs are generated separately, we can then write

$$C_{out} = A_1B_1+(A_1+B_1)A_0B_0+(A_1+B_1)(A_0+B_0)C_0$$

Let us finally consider a *magnitude comparator* for unsigned integers. We will follow industrial practice and first construct 1-bit magnitude comparators such as the one shown in Fig. 10-14. This is a useful step because magnitude comparators usually have outputs for *equal*, *less than*, and *greater than*.

We know from Section 10-3 that the conditions $[A>B]$, $[A=B]$, and $[A<B]$ can be determined with the logical operations $A\overline{B}$, $\overline{A \oplus B}$, and $\overline{A}B$. We can verify that $A\overline{B}$ and $\overline{A}B$ are correctly generated by the comparator in Fig. 10-14 by observing that

$$A(\overline{AB}) = A(\overline{A}+\overline{B}) = A\overline{A}+A\overline{B} = A\overline{B}$$

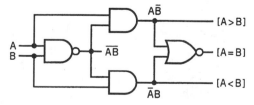

Figure 10-14. 1-bit magnitude comparator.

and similarly for $\overline{AB}$. We will limit ourselves to constructing a circuit that decides whether the 2-bit word A_1A_0 is greater than the 2-bit word B_1B_0. The procedure for *less than* is similar, and both can be used on wider words. We observe first that the condition $[A_1 > B_1] = 1$ is sufficient; if it is not satisfied, we must demand that $[A_1 = B_1]$ and $[A_0 > B_0]$ simultaneously be 1. We thus obtain

$$[A_1A_0 > B_1B_0] = [A_1 > B_1] + [A_1 = B_1][A_0 > B_0]$$

10-9 RS FLIP-FLOPS

A *flip-flop*, as its name suggests, is a memory device with two stable states; by an appropriate choice of inputs, either state can be obtained. Flip-flops are therefore the elements needed to execute sequential logic, which requires memory of past actions. Flip-flops come in many forms; we will only consider the type obtained by cross-coupling logic gates, and we will assume that all gates have the same gate delay τ.

We will begin by considering the *asynchronous* or *RS* NAND flip-flop shown in Fig. 10-15a. If the S (set) and R (reset) inputs are inactive, that is, if $R = S = 1$, the outputs Q and $\overline{Q}$ can be in one of two self-consistent states: either $Q = 0$ and $\overline{Q} = 1$, or $Q = 1$ and $\overline{Q} = 0$. If S is made active, that is, if it is set to 0, Q will become 1 after a time τ (if it was not already 1) and $\overline{Q}$ will become 0 after a time 2τ; after this time, Q will remain at 1 even if S becomes inactive. Similarly, Q can be set to 0 if R is made active for a time longer than 2τ. An analogous description is valid for the NOR flip-flop of Fig. 10-15b; the only difference is that S and R are active when they are equal to 1.

Figure 10-16 shows how a TTL RS flip-flop can be used in a *debouncer* to generate a single transition when a mechanical single-pole double-throw switch changes state. The contacts bounce many times when the switch is thrown toward one pole or the other. The duration of the first closure of the contacts, however, is far larger than any possible gate delay, so that the flip-flop is sure to change state; furthermore, the bounce is never so severe that contact is made with the opposite pole.

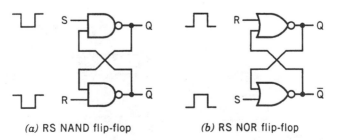

(a) RS NAND flip-flop   (b) RS NOR flip-flop

Figure 10-15. Asynchronous flip-flops.

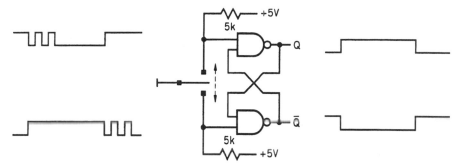

Figure 10-16. Debouncer for a single-pole double-throw switch.

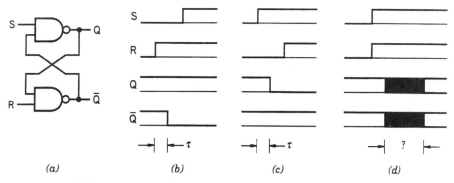

(a)   *(b)*   *(c)*   *(d)*

Figure 10-17. (*a*) RS NAND flip-flop with both inputs initially active. The state of R at the time that S becomes inactive is registered correctly if R becomes inactive (*b*) at least one gate delay earlier or (*c*) at least one gate delay later. (*d*) The decision can take forever in principle if S and R become inactive at the same time.

If both inputs of a NAND RS flip-flop are held active, both outputs are equal to 1 and should perhaps no longer be called Q and $\overline{Q}$. As shown in Fig. 10-17, the flip-flop registers which input first becomes inactive and leaves it in a proper state. If we consider S as a *sampling* signal when it goes to 1, and we then inquire about the state of the flip-flop, we can establish the value of R at the time of sampling. This possibility, as well as others that are similar, is exploited in the clocked flip-flops we will see later; there is a problem, however, and it will be profitable to discuss it now in a simple context.

The *arbitrator* of Fig. 10-17 works well if R and S become inactive at times that are separated by at least one gate delay; this is so because the output that goes to 0 arrives in time to confirm the 0 at the input of the gate that remains at 1. For shorter intervals we must be careful, and we must take a closer look at the structure of RS flip-flops.

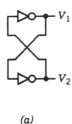

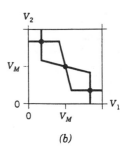

Figure 10-18. (*a*) Cross-coupled inverters; (*b*) static transfer characteristics. The equilibrium point at $V_1 = V_2 = V_M$ is metastable.

(a)   *(b)*

10-10   METASTABILITY IN RS FLIP-FLOPS

When both inputs of an RS flip-flop are inactive, the flip-flop reduces to a pair of cross-coupled inverters. If we superpose the static voltage transfer characteristics of the inverters as shown in Fig. 10-18, we observe that there is a third nondigital equilibrium state in addition to the two stable states we have already described. In this third state, the output voltages V_1 and V_2 both have values that are somewhere between LOW and HIGH; if the inverters are identical their output voltages are equal, and we have $V_1 = V_2 = V_M$.

We have ignored the third state so far because it is *metastable*, meaning that any departure from equilibrium will drive the flip-flop toward one of the two stable states. This happens because the inverters are on the steep part of their voltage transfer characteristic when they are near the metastable state and thus have a large incremental gain; since the feedback is positive, their small-signal response is unstable. To see this point more clearly, let us assume that the inverters have a first-order response and can therefore be represented by the incremental circuit of Fig. 10-19. For small departures from equilibrium we then have

$$v_1 + RCv_1' = -g_m Rv_2$$

$$v_2 + RCv_2' = -g_m Rv_1$$

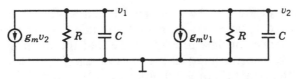

Figure 10-19. Incremental circuit of the cross-coupled inverters of Fig. 10-18 around the metastable state, assuming first-order responses.

Adding and subtracting these equations, we obtain

$$(1+g_m R)(v_1+v_2) +RC(v_1+v_2)' = 0$$
$$(1-g_m R)(v_1-v_2) +RC(v_1-v_2)' = 0$$

Assuming that $g_m R \gg 1$, which is invariably the case, we finally obtain

$$v_1+v_2 \simeq V_+ \exp(-\omega_T t)$$
$$v_1-v_2 \simeq V_- \exp(+\omega_T t)$$

where $\omega_T = g_m/C$ is the gain–bandwidth product of the inverters and V_+ and V_- are constants. It is clear that v_1 and v_2 will grow exponentially if they differ initially in the least degree, and that a large gain–bandwidth product will result in a quick exit from the metastable state. These conclusions hold for higher-order gains.

Returning to the arbitrator of Fig. 10-17, let us imagine that we slowly shorten the interval between the times when R and S become inactive. When the interval becomes shorter than one gate delay, nothing critical happens for some time because even though Q drops slightly, it remains above HIGH. Beyond a certain point, however, Q will drop below HIGH for increasingly longer times, although it will eventually return to a value above HIGH. When the interval is reduced to zero, Q and $\overline{Q}$ are left in the metastable state. If we now make S precede R by an interval shorter than one gate delay, Q will eventually drop below LOW, but it will take longer and longer to do so as the interval is shortened. All these possibilities are illustrated in Fig. 10-20, which shows waveforms of the voltage V_Q at the Q output.

From the preceding discussion it follows that there exists a *metastability window* τ_W of order τ such that if S and R are separated by less than τ_W, the flip-flop will become metastable, that is, that Q will not settle above HIGH

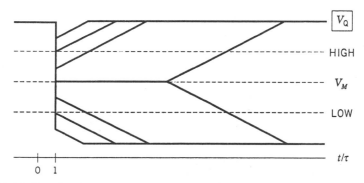

Figure 10-20. Family of responses generated by the RS flip-flop of Fig. 10-17 when R and S are separated by less than one gate delay (schematic).

or below LOW within one gate delay after S becomes inactive. If we assume that R goes from LOW to HIGH and back with an average frequency f, the probability that the flip-flop will become metastable when S becomes inactive is $f\tau_W$. Let us define $\Pi(T)$ as the probability that the flip-flop is still metastable a time T after becoming metastable. The probability $P(T)$ that the flip-flop is still metastable a time $T + \tau$ after S becomes inactive is called the probability of *synchronization failure* or of *arbitration conflict* and is given by

$$P(T) = f\tau_W \Pi(T)$$

The probability of synchronization failure cannot be reduced to zero in an RS flip-flop, but it can be made arbitrarily small by waiting long enough before examining Q because in practice, $\Pi(T)$ decreases exponentially for large T:

$$\Pi(T \gg \tau) \sim \exp(-T/\tau_M)$$

where τ_M is also of order τ. The choice of T, however, is not in general easy and must be researched carefully; failure to do so can result in catastrophes.

Synchronization failure can be avoided *internally* in the clocked systems we will see in the following section by making all signals obey certain timing rules. But even clocked systems need external unsynchronized information; it follows that synchronization failure can affect any system.

10-11   CLOCKED FLIP-FLOPS AND SYNCHRONOUS SYSTEMS

Synchronization failure is avoided in *synchronous* or *clocked* digital systems by gating all flip-flops with an (ideally) square-wave *clock* signal C and demanding that inputs other than the clock be kept steady for a *setup time* t_S before a clock transition that might cause a metastable condition, and that they remain steady for a *hold time* t_H after the transition.

As a first example of a *clocked* flip-flop, let us consider the NOR *latch* shown in Fig. 10-21a. When the clock is 0, G1 and G2 act like inverters, and the output Q of RS flip-flop G3|G4 follows the *data* input D; when the clock goes to 1, the outputs of G1 and G2 are forced to 0, and Q registers the current state of D. The setup time is 2τ when D goes from 1 to 0 because the output of G1 must be kept at 1 for a time 2τ to ensure that G3|G4 is internally locked at $Q = 0$. In contrast, the setup time is 3τ when D goes from 0 to 1 because of the extra delay in G2. The hold time is zero because D loses control over G1 and G2 as soon as the clock goes to 1. Analogously, the NAND latch of Fig. 10-21b registers the current state of D when C goes to 0. The setup time is 3τ when D goes from 1 to 0, and 2τ when D goes from 0 to 1. As in the NOR latch, the hold time is zero.

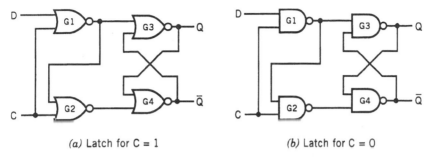

(a) Latch for C = 1   (b) Latch for C = 0

Figure 10-21. Latches based on RS flip-flops.

The most interesting flip-flops we will deal with are *edge-triggered*; that is, they change state only at one of the transitions of the clock signal. We will normally take the transition from 0 to 1 as the *active* transition, and assume that *clock* or *clock transition* refers to the active transition. The state of the output after the clock depends on the state of the output and the inputs just prior to the clock; as we will see, the clock must remain active for a time t_W, the *minimum clock-pulse width*. Outside the interval defined by the setup and hold times, the inputs can change arbitrarily.

Changes of state in edge-triggered flip-flops, if any, occur a *propagation time t_P* after the clock. The propagation time in a family of flip-flops must be larger than the hold time; if this were not so, it would be impossible to construct systems in which the response of a given flip-flop depends on the state of the flip-flop itself or other flip-flops at the time of the clock.

Edge-triggered flip-flops often have asynchronous SET and RESET (or SET and CLEAR) inputs that override all other inputs, including the clock. These inputs can be used to establish a well-defined initial state; this is typically done in *sequencers*, systems that remain idle in an initial state until triggered and return to the initial state after executing a fixed sequence of actions. Examples of sequencers are given in Exercises 10-1 and 10-11.

An edge-triggered flip-flop is described by an equation (or a truth table) that gives the output during clock period $n+1$ as a function of the values of the inputs and the output during clock period n. As an example, the ubiquitous D flip-flop, two versions of which we will see next, is governed only by its single *data* input D; its equation is

$$Q_{n+1} = D_n$$

The *master–slave* edge-triggered D flip-flop of Fig. 10-22 consists of a NOR *master* latch followed by a NAND *slave* latch. When C is equal to 0, the master follows D and we have $X = D$ and $Y = \overline{D}$. When C goes to 1, these values are held by the master and transmitted to the slave; Q thus takes

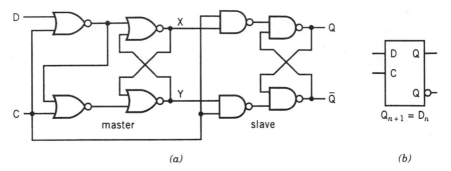

(a)   *(b)*

Figure 10-22. Master–slave edge-triggered D flip-flop. (*a*) Structure. (*b*) Symbol and equation.

on the value that D had just prior to the clock. When C returns to 0, the slave continues to hold the current value of Q while the master accepts new data. The setup and hold times are the same as in the NOR latch of Fig. 10-21*a*. The propagation time is 2τ for the output that goes to 1 and 3τ for the output that goes to 0.

If we connect the $\overline{Q}$ output of a D flip-flop to the D input, we obtain a *toggle* flip-flop that changes state at every (active) clock transition and therefore divides the frequency of the clock signal by 2. Its equation is

$$Q_{n+1} = \overline{Q}_n$$

We note that a toggle flip-flop is the ultimate example of why propagation times must be longer than hold times.

If we now chain together N toggle flip-flops by tying the $\overline{Q}$ output of one flip-flop to the C input of the next, we obtain the divide-by-2^N *ripple counter* shown for $N = 3$ in Fig. 10-23. Although not strictly synchronous, the ripple counter is nonetheless widely used because it is simple and, having no inputs, has no problems with setup or hold times. The drawback it does have, however, is that it can take up to N propagation times to change from one state to the next.

Edge-triggered flip-flops do not always have a master–slave configuration. As an example, we will consider a well-known member of the LS-TTL family, the 74LS74 D flip-flop shown in Fig. 10-24. Flip-flops G1|G2 and G3|G4 are

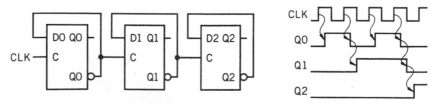

Figure 10-23. Ripple counter.

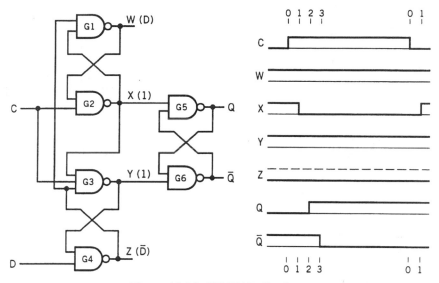

Figure 10-24. 74LS74 D flip-flop.

both arbitrators like the one shown in Fig. 10-17, and they register the value of D when C goes to 1. The values in parentheses indicate the state of the internal nodes when $C = 0$. The waveforms correspond to the case $D = 1$ and $Q = 0$. This example shows that drawing waveforms is often the easiest way to understand how digital circuits work, particularly when feedback is involved.

If $D = 1$, we have $W = 1$ and $Z = 0$ when $C = 0$; X thus becomes 0 one gate delay τ after the 0-to-1 transition of C and confirms $W = 1$ and $Y = 1$. After this time, Z can change without affecting X and Y; this implies that the hold time t_H is zero. X must remain at 0 for at least two gate delays in order to lock output Q of flip-flop G5|G6 at 1; the minimum clock-pulse width t_W is thus 2τ.

If $D = 0$, we have $W = 0$ and $Z = 1$ when $C = 0$; in this case, Y becomes 0 one gate delay τ after the clock transition and confirms $Z = 1$ and therefore also confirms $W = 0$ and $X = 1$. The hold time t_H is τ because D must remain steady until Y becomes 0.

The setup time t_S is 2τ, and the propagation time t_P to the Q output is 2τ if Q goes from 0 to 1 and 3τ if Q goes from 1 to 0.

An interesting advantage of the 74LS74 is that the clock transition times can be arbitrarily long. This is not the case in master–slave flip-flops like the one shown in Fig. 10-22 if the threshold voltage in the master is higher than it is in the slave; if the transition from 1 to 0 is too slow, for example, the master might change before the slave has locked the current value of Q.

The use of D flip-flops in a synchronous system is exemplified in the 4-bit *shift register* shown in Fig. 10-25. The AND–OR gates are simple examples of

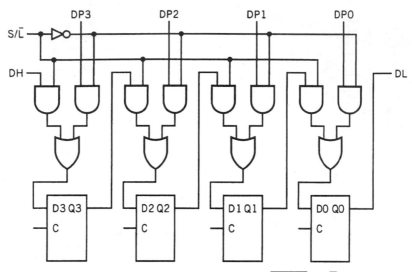

Figure 10-25. 4-bit shift register with SHIFT/$\overline{\text{LOAD}}$ (S/$\overline{\text{L}}$) options.

multiplexers, that is, of circuits that route one of several inputs to their output according to the setting of their *address* or control inputs. In this case, the only control input is S/$\overline{\text{L}}$ (SHIFT/$\overline{\text{LOAD}}$), and it determines whether the D input of a given flip-flop is the Q output of the flip-flop of the next-highest order (DH in the case of D3), or whether it is the corresponding bit of the 4-bit word DP3 $\cdots$ DP0.

If S/$\overline{\text{L}}$ = 1, the 4-bit word in the shift register is shifted one position to the right at the clock transition; DH is copied into Q3, and DL can be copied into the DH input of a lower-order word. Q3 is copied into itself if it is tied to DH; we can thus obtain an *arithmetic shift-right* or *divide-by-2* circuit. If S/$\overline{\text{L}}$ = 0, new *parallel* data are copied into the shift register.

Less common than D flip-flops, but highly useful because of their versatility, are *JK* flip-flops with inputs J and K; their equation is

$$Q_{n+1} = J_n\overline{Q}_n + \overline{K}_n Q_n$$

As shown in Fig. 10-26, a JK flip-flop can be obtained by making the D input of a D flip-flop equal to $J\overline{Q} + \overline{K}Q$. Other designs are considered in Exercises 10-6 and 10-7.

A JK flip-flop changes state if J and K are both equal to 1 and remains in the same state if J and K are both equal to 0. This property is exploited in the 4-bit *synchronous binary counter* shown in Fig. 10-27. If COUNT ENABLE = 0, the J and K inputs to all flip-flops are 0, and the counter remains wherever it is. If COUNT ENABLE = 1, the lowest-order bit Q0 toggles at every clock, and in each succeeding position the AND gates allow the bit to toggle only if all preceding bits are equal to 1; the counter thus cycles through the sequence $0, 1, \ldots, 15, 0$, and so on.

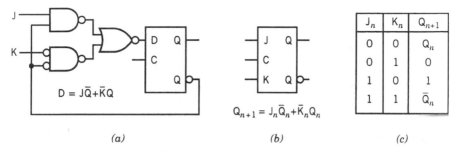

Figure 10-26. JK flip-flops. (*a*) Design based on a D flip-flop. (*b*) Symbol and equation. (*c*) Truth table.

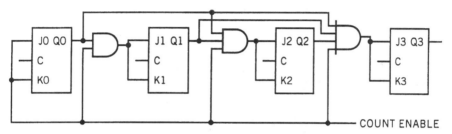

Figure 10-27. 4-bit synchronous binary counter with COUNT ENABLE option.

With more complicated gating, the counter can also count *down*, and like the shift register of Fig. 10-25, it can be set to an arbitrary value. Loading a parallel word and then counting down to zero is useful when generating pulses of variable length because only one gate is needed to detect when the counter arrives at zero; in contrast, counting up to a given value requires a complete magnitude comparator.

We have so far made the implicit assumption that there is no *clock skew*, that is, that the clock transition is fast and that it arrives simultaneously at all flip-flops. Clock skew is never zero, but it must nonetheless be small enough to prevent one flip-flop from firing so far ahead of another that it changes state before the other has made a decision; in the worst of cases, a late clock can make a flip-flop become metastable.

A slow clock rise time can result in a large clock skew if the scatter in the threshold voltages of different flip-flops is large; this problem particularly affects CMOS circuits, whose threshold voltages can lie anywhere between one-third and two-thirds of the power-supply voltage. Clock skew also occurs if there are delays in the clock distribution circuit due either to gates or to transmission lines; this problem is of particular concern in ECL circuits.

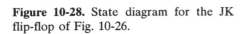
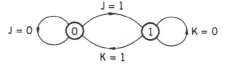

Figure 10-28. State diagram for the JK flip-flop of Fig. 10-26.

10-12 STATE MACHINES

Edge-triggered flip-flops, synchronous counters, and shift registers are simple examples of *state machines*, synchronous systems that consist only of edge-triggered flip-flops and combinational circuits. The inputs to the flip-flops in a state machine are logical functions only of the external inputs, if any, and the outputs of the flip-flops themselves; the external inputs are assumed to be synchronized to the system clock. The state of the machine between clock transitions is completely determined by the state of the flip-flops. In any given state, the combinational logic decides what the next state will be.

One orderly way of describing a state machine is a *state diagram*, in which the states are nodes and the transitions between states are indicated by directed traces labeled with the conditions that enable these transitions. As an example, Fig. 10-28 shows the state diagram of a JK flip-flop. The two states are labeled according to the value of Q. If $Q = 0$, Q will remain at 0 if $J = 0$, and it will go to 1 if $J = 1$. Analogously, if $Q = 1$, Q will remain at 1 if $K = 0$, and it will go to 0 if $K = 1$.

A state machine with N flip-flops has 2^N possible states; state machines thus generally have more states than are needed for a given task. The presence of such *excess* states demands consideration of what action to take if the machine falls into one of these states either at power-up or because of noise. As an example, let us consider the divide-by-6 *Johnson* counter of Fig. 10-29. The counter has three flip-flops and therefore eight possible states; the normal sequence is 000, 100, 110, 111, 011, 001, 000, and so on. In the absence of gating, that is, if G2 is removed and Q1 is connected to D0,

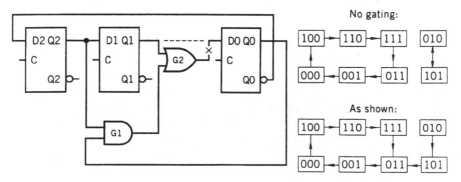

Figure 10-29. Johnson divide-by-6 counter with gating to detect excess states.

states 010 and 101 form a disjoint group; to avoid this possibility, G1 detects state 101 and makes D0 equal to 1, thus simulating state 111 and forcing the counter into state 011.

An outstanding advantage of state machines is that they are insensitive to races in their combinational logic because races can occur only between clock transitions. In the shift register of Fig. 10-25, for example, D3 will exhibit a short pulse if S/$\overline{\text{L}}$ goes from 1 to 0 when DP3 and DH are both 1; but S/$\overline{\text{L}}$ is synchronized to the clock and can thus change only immediately after a clock transition, so that the pulse has no effect on Q3. Another advantage of state machines, also related to the fact that outputs change well after the decision to change is firm, is that sensitivity to inductively or capacitively coupled transients generated by changes of state is greatly reduced; this advantage is particularly valuable when changes of state trigger high-power circuits and there are inputs connected to the outside world by long unshielded lines.

10-13 SYNCHRONIZERS

External inputs to state machines are assumed to be synchronous; asynchronous inputs, which exist except in the rarest of cases, must somehow be converted into synchronous versions. In the standard *synchronizer* shown in Fig. 10-30, asynchronous input ASYN drives the D input of a D flip-flop; a change of state in D is synchronously reflected at output SYN at the first following clock transition. As in RS flip-flops, there exists a chance of synchronization failure, that is, that a change in ASYN will violate setup or hold requirements and that the flip-flop will become metastable.

There are several procedures for reducing the probability P of synchronization failure, of which we will mention only two; both result in a longer response time to ASYN. One procedure is to drive the synchronizer of Fig. 10-30 with a submultiple f/N of the clock frequency f, so that the frequency at which ASYN is sampled—and thus P—is reduced by a factor N. A much more effective procedure is illustrated by the two-stage synchronizer shown in Fig. 10-31. Both ASYN and the intermediate output SYN? are sampled at a frequency f/N, which allows a time $T = N/f$ for SYN? to stabilize before it is copied to SYN. Since P decreases exponentially with T, it can be reduced by orders of magnitude beyond a factor N.

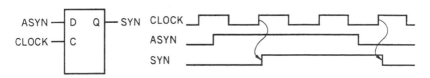

Figure 10-30. Basic synchronizer based on a D flip-flop.

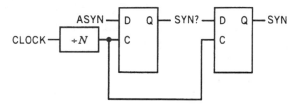

Figure 10-31. Two-stage synchronizer.

The literature quite generally maintains that it is impossible to construct a synchronizer that will not become metastable; this opinion notwithstanding, Fairchild Semiconductor once offered an ECL D flip-flop, the 11C70, that had no requirements on its setup and hold times. The subject of synchronization failure is under intense scrutiny, and there are several manufacturers that claim that the perfect synchronizer in fact exists; the question, however, is not yet settled.

A fundamental rule about synchronization is that it must be done only once for a given asynchronous input because different synchronizers might arrive at different decisions even in the case that they do not become metastable; an otherwise acceptable clock skew, for example, might make one synchronizer recognize a change that another synchronizer ignores.

10-14   CMOS CIRCUITS

The source and drain of a MOSFET are essentially isolated from each other when the MOSFET is off, and they are connected by a small resistance when the MOSFET is on and the drain–source voltage is small. These characteristics are exploited to great advantage in mixed analog and digital CMOS circuits.

One of the basic components of mixed CMOS circuits is the *bilateral switch* or *transmission gate* shown in Fig. 10-32. For proper operation, V_X and

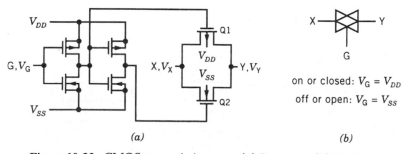

(a)   (b)

Figure 10-32. CMOS transmission gate. (a) Structure. (b) Symbol.

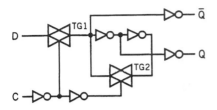

Figure 10-33. CMOS latch based on transmission gates.

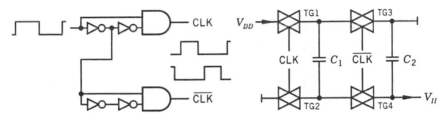

Figure 10-34. Flying-capacitor inverter. V_{II} eventually becomes equal to $-V_{DD}$.

V_Y must both be between V_{SS} and V_{DD}. If control input G is at logical 0, Q1 and Q2 are off and X and Y are isolated from each other. If G is at logical 1, Q1 and Q2 are in principle both on. When V_X and V_Y are near V_{DD}, Q2 cuts off but Q1 has a low resistance; similarly, when V_X and V_Y are near V_{SS}, Q1 cuts off but Q2 has a low resistance. There is thus always a low-resistance path between X and Y when G = 1.

As a first exercise in the use of transmission gates, Fig. 10-33 shows a CMOS latch similar to the ones used in CMOS D and JK master–slave flip-flops. When C = 0, TG1 is on and TG2 is off, and Q follows D. When C goes to 1, TG1 turns off, and TG2 turns on and locks the value of D in Q.

Now let us consider the *flying-capacitor inverter* of Fig. 10-34, often needed when $V_{SS} = 0$. When CLK is 1, TG1 and TG2 are on, and TG3 and TG4 are off; C_2 is therefore isolated from C_1, which charges up to V_{DD}. When CLK is 0, TG1 and TG2 are off, and TG3 and TG4 are on; C_1 shares its charge with C_2, and both are isolated from V_{DD}. Since TG3 connects the positive end of C_1 to ground, the voltage V_{II} at the ungrounded end of C_2 eventually becomes equal to $-V_{DD}$. The positive edges of CLK and $\overline{\text{CLK}}$ are delayed in order to keep TG1|TG2 and TG3|TG4 from turning on simultaneously. For proper operation, TG1|TG2|TG3|TG4 are fed by V_{II} and V_{SS}, and a *level shifter* (not shown) makes CLK and $\overline{\text{CLK}}$ swing between V_{II} and V_{DD}.

Closely related to the flying-capacitor inverter is the *switched-capacitor* low-pass filter shown in Fig. 10-35. As we did above, we will assume that CLK and $\overline{\text{CLK}}$ cannot simultaneously be equal to 1. For a step input $V_I = V_0 \theta(t)$, the output voltage at the end of clock period $n+1$ is given in terms of the output voltage at the end of clock period n by

$$C_1 V_0 + C_2 V_O(nT) = (C_1 + C_2) V_O[(n+1)T]$$

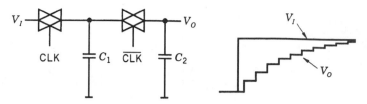

Figure 10-35. Switched-capacitor filter.

where T is the clock period. Assuming that V_O is zero at the beginning of the first clock period, we obtain

$$\frac{V_O(nT)}{V_0} = 1 - \left(\frac{C_2}{C_1 + C_2}\right)^n$$

If we define τ by

$$\frac{C_1 + C_2}{C_2} = \exp\left(\frac{T}{\tau}\right)$$

the step response $r(nT)$ at the end of clock period n can be written as

$$r(nT) = 1 - \exp\left(-\frac{nT}{\tau}\right)$$

If $C_1 \ll C_2$, the step response is similar to the step response of a first-order lag with time constant τ. The advantage of a switched-capacitor filter is that its time constant can be varied by varying the clock frequency. Switched-capacitor filters of any order can be obtained with similar methods.

As shown in Fig. 10-36, flying capacitors can be used to convert differential voltages to single-ended voltages. A square-wave clock alternately connects capacitors C_1 and C_2 to the differential source; while one capacitor is so connected, one end of the other is grounded, so that except for switching transients, $V_I = V_+ - V_-$. A point to be observed in this example is that we are dealing with a sampling system that is subject to aliasing, meaning, as discussed in Chapter 2, that a sine wave with a frequency near the sampling frequency can be confused with a sine wave with a frequency near zero. The sampling theorem requires that the bandwidth of the input signal be less than half the sampling frequency or, in this case, less than the clock frequency, since two samples are taken in one clock period.

As a final example of the potential of mixed analog and digital circuits, let us consider the Intersil ICL7605 *commutating auto-zero* (CAZ) instrumentation amplifier shown schematically in Fig. 10-37, which features an input

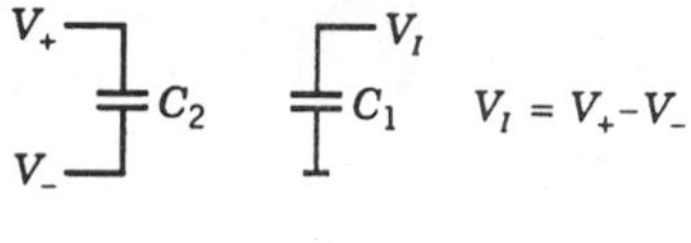

Figure 10-36. Differential-to-single-ended conversion with flying capacitors.

V_+ ─┐

═ C_2 ═ C_1 $V_I = V_+ - V_-$

V_- ─┘   ┌─ V_I

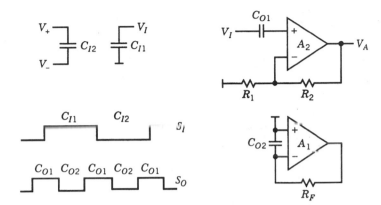

Figure 10-37. Simplified version of the Intersil ICL7605 commutating auto-zero instrumentation amplifier.

offset voltage around 2 μV and a long-term drift in the input offset voltage around 0.5 μV/year.

As in the preceding example, the differential input voltage is converted to a single-ended voltage by switching input capacitors C_{I1} and C_{I2} with clock S_I.

The amplifying section consists essentially of operational amplifiers A_1 and A_2 and of nulling capacitors C_{O1} and C_{O2}. Each amplifier is switched between two alternating configurations by clock S_O. The amplifier in the *auto-zero* configuration has unity closed-loop gain and registers its offset voltage on its nulling capacitor, which is connected across its inputs. The amplifier in the *measuring* configuration effectively has zero voltage offset because the voltage on its nulling capacitor cancels its intrinsic voltage offset; in this configuration, the amplifier output is connected to the outside world, and the closed-loop gain is set by resistors R_1 and R_2.

Clock S_O has twice the frequency of clock S_I, and it is offset so that the switching transients in the input and amplifying sections occur at different times and thus do not mix nonlinearly.

EXERCISES

10-1 LABORATORY: TEST OF A HOMEMADE 74LS74. The circuit shown in Fig. 10-38 will allow you to check out a homemade 74LS74 D flip-flop and observe the waveforms shown in Fig. 10-24. Study the circuit before building it, and convince yourself that the waveforms shown in Fig. 10-38 are correct and that the home-made flip-flop will be clocked in all four possible input–output configurations.

The 74LS132 is a two-input *Schmitt-trigger* NAND gate with a 1.8V threshold for positive-going transitions and a 1V threshold for negative-going

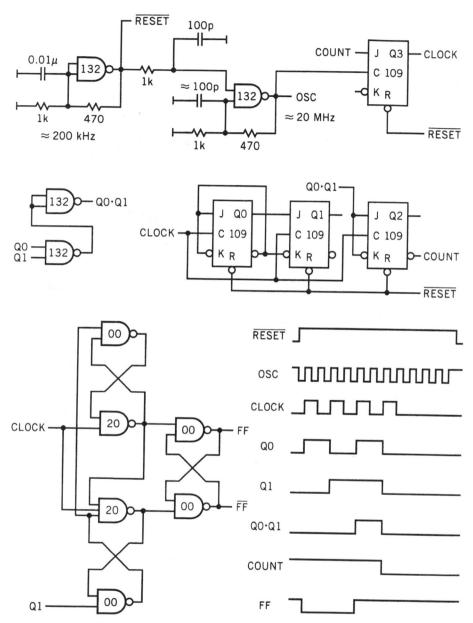

Figure 10-38. Sequencer used in Exercise 10-1 to test a homemade 74LS74.

transitions; with the indicated feedback network, its output is roughly a square wave.

The fact that CLOCK is synchronized to $\overline{\text{RESET}}$ will make oscilloscope triggering and event location easy.

10-2 BOOLEAN ALGEBRA. Prove the following identities:

(a) $\overline{A}+AB = \overline{A}+B$

(b) $A\overline{B}+\overline{A}B+AB = A+B$ (canonical form, sum of T-terms)

(c) $A+B+\overline{A}\overline{B}C = A+B+C$

(d) $A\overline{C}+BC+AB = A\overline{C}+BC$ (used in JK flip-flops in the form $J\overline{Q}+\overline{K}Q+JK = J\overline{Q}+\overline{K}Q$)

(e) $(A\oplus B)\oplus C = A\oplus(B\oplus C) = A\overline{B}\overline{C}+\overline{A}B\overline{C}+\overline{A}\overline{B}C+ABC$ (sum of three 1-bit integers)

(f) $\overline{A}BC+A\overline{B}C+AB\overline{C}+ABC = AB+BC+AC$ (alternative forms for the carry bit in a sum)

(g) $\overline{A}\overline{B}+\overline{B}\overline{C}+\overline{A}\overline{C} = \overline{AB+BC+AC}$ (alternative forms for the negated carry bit in a sum)

10-3 THE IMPLICATION OPERATION. Define the *implication* operation $A \Rightarrow B$ (A *implies* B or *if* A *then* B) by

$$A \Rightarrow B \equiv \overline{A}+B$$

A *tautology* is a logical expression that is always 1. Show that the usual rules of logic hold for implications, that is, that the expressions given in the following theorems are tautologies.

(a) Implication is reflexive; that is, A implies A:

$$A \Rightarrow A$$

(b) Implication is antisymmetric; that is, if A implies B, then $\overline{B}$ implies $\overline{A}$:

$$(A \Rightarrow B) \Rightarrow (\overline{B} \Rightarrow \overline{A})$$

(c) Implication is transitive; that is, if A implies B and B implies C, then A implies C:

$$(A \Rightarrow B)(B \Rightarrow C) \Rightarrow (A \Rightarrow C)$$

(d) If A implies B and B implies A, then A and B are either both true or both false (remember that $\overline{A \oplus B} = AB + \overline{A}\overline{B}$):

$$(A \Rightarrow B)(B \Rightarrow A) \Rightarrow \overline{A \oplus B}$$

10-4 TRUTH TABLE OF A JK FLIP-FLOP. Use the equation for a JK flip-flop to obtain the truth table in Fig. 10-26.

10-5 A DECODER. Using only two-input AND gates fed by the Q or $\overline{Q}$ outputs of the flip-flops, decode the six allowed states of the Johnson counter of Fig. 10-29. You will find the answer in a CMOS data manual.

10-6 JK FLIP-FLOP BASED ON THE 74LS74. Draw waveforms for the circuit shown in Fig. 10-39 and convince yourself that it is in fact a JK flip-flop. Because of the symmetry of the circuit, you need only analyze the case in which $Q = 0$. For $C = 0$, the internal nodes then have the values shown in parentheses, and it is clear that they depend only on J. The analysis then proceeds as for the 74LS74: once X or Y becomes 0 after the clock transition, the inputs cannot affect the output.

Convince yourself that $t_S = 2\tau$, $t_H = 0$, and $t_W = 2\tau$. Also convince yourself that the propagation times t_P are 2τ for positive-going transitions of the outputs and 3τ for negative-going transitions.

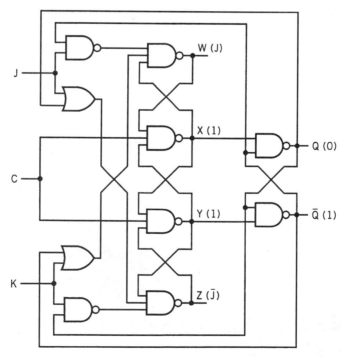

Figure 10-39. JK flip-flop based on the 74LS74 D flip-flop (Exercise 10-6).

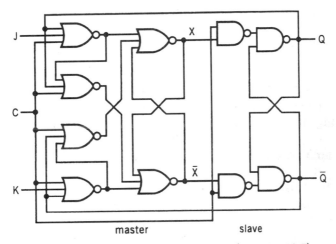

Figure 10-40. JK master–slave flip-flop (Exercise 10-7).

10-7 JK MASTER–SLAVE FLIP-FLOP. The circuit of Fig. 10-40 is an edge-triggered master–slave JK flip-flop. For $C = 0$, show that

$$X = J\overline{Q} + \overline{K}Q + J\overline{K} = J\overline{Q} + \overline{K}Q$$

and that $\overline{X}$ is properly named.

Convince yourself that when C goes to 1, Q becomes equal to X, and J and K lose control. Also convince yourself that the slave retains X and the master can accept new data when C returns to 0.

For $Q = 0$, show that the setup time t_S is 3τ when J goes from 1 to 0 and 4τ when J goes from 0 to 1.

Show also that the hold time t_H is zero and that the propagation times t_P are 2τ for the output that becomes 1 and 3τ for the output that becomes 0.

10-8 AN OLD-FASHIONED JK FLIP-FLOP WITH A MINOR PROBLEM. The now obsolete JK master–slave flip-flop shown in Fig. 10-41 does not comply with a fundamental requirement in modern edge-triggered flip-flops because it does not allow arbitrary changes in the inputs when $C = 1$.

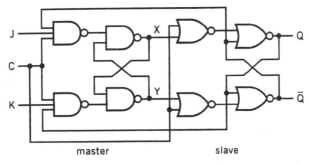

Figure 10-41. Old-fashioned JK flip-flop (Exercise 10-8).

Consider the case in which $C = 1$ and $J = K = X = Q = 0$. If J goes to 1 and then returns to 0 before the active clock transition (from 1 to 0 in this case), X remains at 1, and the flip-flop acts as if $J = 1$ and $K = 0$, and not as if $J = K = 0$.

10-9   A 4-BIT DIGITAL-TO-ANALOG CONVERTER WITH ECL INPUTS. Many commercial digital-to-analog converters have the structure shown in Fig. 10-42. Consider all transistors to be identical, and assume that base losses are negligible. Assuming further that all transistors have the same base–emitter voltage, convince yourself that the $R–2R$ network makes the currents in the current switches equal to powers of 2 times the current in the switch for the least significant bit.

If the logic level that controls the current switch for a given bit is 0, the current from that bit is shunted to ground; if it is 1, the current is fed to the operational-amplifier adder that generates output voltage V_O.

The notation $\times 1$, $\times 2$, etc. indicates the number of parallel transistors in the current sources. All these transistors will have the same base–emitter voltage if the currents in the current switches are proportional to powers of 2. Now convince yourself that the solution you obtained above is self-consistent.

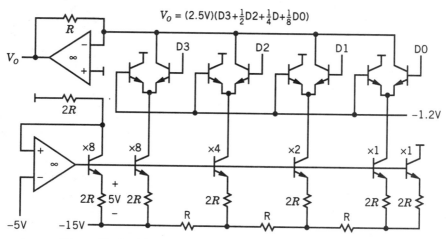

$$V_O = (2.5V)(D3 + \tfrac{1}{2}D2 + \tfrac{1}{4}D1 + \tfrac{1}{8}D0)$$

Figure 10-42. 4-bit digital-to-analog converter (Exercise 10-9).

10-10   ECL LATCH. ECL master–slave flip-flops are built using the latch shown in Fig. 10-43. Two differential amplifiers (A1 and A2) with common collector resistances are fed by a current switch (s) controlled by clock C; depending on the state of C, s activates either A1 or A2.

For $C = 0$, A2 is active and its state, which is transmitted to Q and $\overline{Q}$, is controlled by D. Convince yourself that D loses control when C goes to 1 and A1 becomes active, and that the crossed feedback from the collectors to the bases latches the state of Q and $\overline{Q}$.

By swapping collectors in s, D can be latched when C goes to 0.

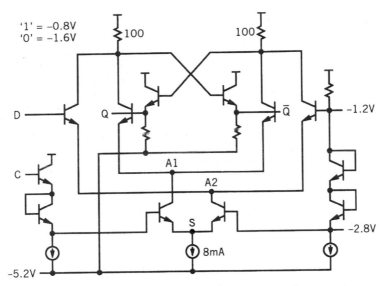

Figure 10-43. ECL latch (Exercise 10-10).

10-11 A 4-BIT SUCCESSIVE-APPROXIMATION ANALOG-TO-DIGITAL CONVERTER. In a successive-approximation analog-to-digital converter (ADC), the bits of a digital-to-analog converter (DAC) are successively activated, starting with the highest-order bit, and the DAC output is compared with the input voltage V_{in}. If the DAC output is lower than V_{in} when a given bit is activated, the bit is latched at 1; if not, it is left at 0. The sequencing is governed by a *successive-approximation register* (SAR).

Convince yourself that the circuit of Fig. 10-44 is a successive-approximation ADC and, assuming that the result of the conversion is 1001_2 and that CLEAR is set to 0 just prior to the conversion, that the waveforms are correct.

10-12 EXCESS STATES IN A DECIMAL COUNTER. In a *binary-coded decimal* (BCD) counter, four flip-flops are used to count by 10. The number of possible states is 16, which implies that as in the divide-by-6 Johnson counter of Fig. 10-29, some gating is needed to keep the counter cycling through the 10-state normal cycle and force it back into the normal cycle if for some reason it falls into one of the excess states.

Convince yourself that the state diagram for the BCD counter of Fig. 10-45 is correct if the states are numbered as in a binary counter. Observe that if Q3 is 0, the counter looks just like a binary counter, so that you need not worry about the transitions from 0 to 8.

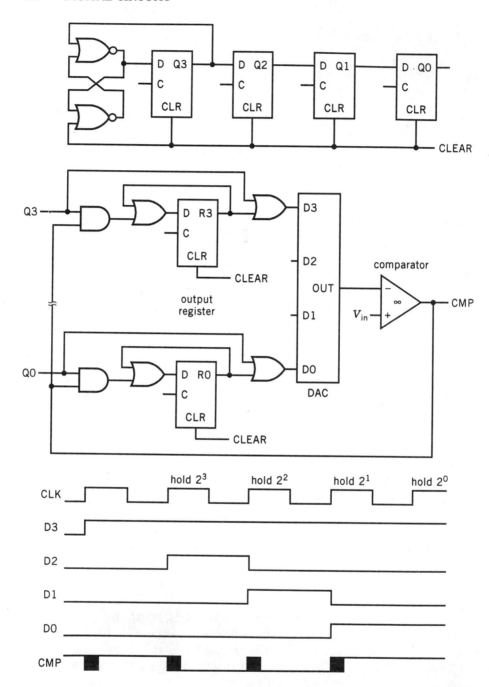

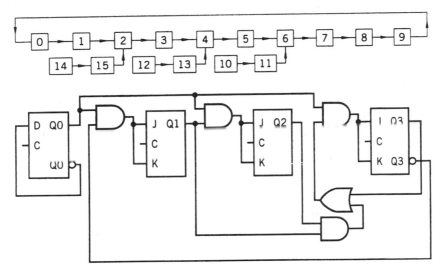

Figure 10-45. Binary-coded decimal counter (Exercise 10-12).

10-13 DIRECTIONAL CONTROL FOR A STEPPER MOTOR. In a certain type of four-winding stepper motor, the windings must be excited by pairs in the sequence 0101, 1001, 1010, 0110 for rotation in one sense. In these binary integers, position indicates the winding number and a 1 indicates that the winding is excited; in the state 0101, for example, windings 0 and 2 are excited. Rotation in the opposite sense is obtained by reversing the order of the sequence.

Convince yourself that the circuit shown in Fig. 10-46 can be used to control the sense of rotation of this type of stepper motor by making R either 0 or 1, and that the state diagrams are correct as shown.

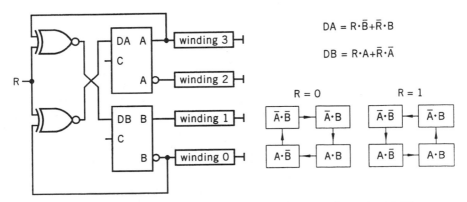

Figure 10-46. Directional control for a stepper motor (Exercise 10-13).

10-14 BENEFICIAL REDUNDANCE. Consider the latch shown in Fig. 10-47, and assume for a moment that NAND gate R is removed and that Y = 1. For C = 0 you then have Q = D. When C goes to 1, there is no problem if D = 1. If D = 0, however, X goes to 1 at the same time that Z goes to 0 if you assume equal gate delays; this state of affairs is marginal. If you now restore gate R, for C = 0 you have

$$Q = D\left(\overline{\overline{DQ}}\right) = D(D+Q) = D+DQ$$

so you still have Q = D. For D = 1 there is again no problem, but for D = 0 the situation is improved when C goes to 1 because Y remains at 0 until after D goes to 1, which can happen only after a propagation time that is always larger than one gate delay.

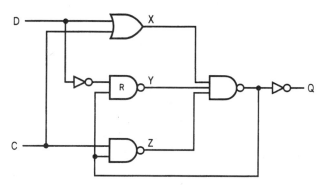

Figure 10-47. Latch protected against races (Exercise 10-14).

10-15 GRAY CODES. Sequences of binary integers from 0 to $2^N - 1$ must sometimes be encoded in an *N*-bit *Gray* code characterized by the fact that sequential codes differ in only one bit. Gray codes are used in shaft encoders, for example, because even minor misalignments at zone boundaries make it impossible to guarantee that bits of different weights will change simultaneously when going from one zone to another. The standard sequence of Gray codes is generated by starting at zero and flipping the lowest-order bit that generates a new code. For 3-bit words, this procedure results in the sequence 000, 001, 011, 010, 110, 111, 101, 100. The point of this exercise is to show that this procedure is correct and that the encoder/decoder pair shown in Fig. 10-48 will take you from a binary code to a Gray code and back.

Begin by accepting that as shown in Fig. 10-48*a*, you get the Gray code for an unsigned binary integer by taking the bitwise XOR of a bit and its left neighbor, making the left neighbor 0 for the highest-order bit. For a 4-bit

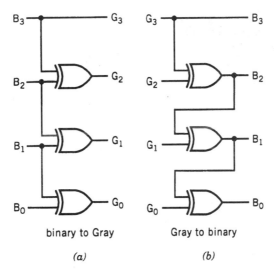

binary to Gray   Gray to binary

(a)   (b)

Figure 10-48. (a) Binary-to-Gray and (b) Gray-to-binary converters (Exercise 10-15).

unsigned integer $B_3 B_2 B_1 B_0$ you then get

$$
\begin{array}{cccc}
B_3 & B_2 & B_1 & B_0 \\
\oplus & \oplus & \oplus & \oplus \\
0 & B_3 & B_2 & B_1
\end{array}
$$

Two different unsigned integers differ in at least one bit. By starting with the highest-order bit, show that their Gray codes also differ.

Starting from the Gray code for an unsigned integer written as above in terms of the binary bit values, show that you can convert back to binary by using the scheme shown in Fig. 10-48b. The highest-order bit is obviously correct. For the next-lowest bit, you get $B_3 \oplus B_2 \oplus B_3 = B_2$ in the case of a 4-bit unsigned integer, and so on.

Any unsigned N-bit integer less than $2^N - 1$ can be written in the form

$$ B_{N-1} B_{N-2} \cdots B_j 0111 \cdots $$

Its Gray code and its successor's Gray code are

$$
\begin{array}{llllllll}
B_{N-1} B_{N-2} \cdots B_j & 0 & 1 & 1 & 1 & \cdots \\
\oplus \quad\quad \oplus & & \oplus & \oplus & \oplus & \oplus & \oplus \\
0 \quad\quad B_{N-1} \cdots B_{j+1} B_j & 0 & 1 & 1 & \cdots
\end{array}
\qquad
\begin{array}{llllllll}
B_{N-1} B_{N-2} \cdots B_j & 1 & 0 & 0 & 0 & \cdots \\
\oplus \quad\quad \oplus & & \oplus & \oplus & \oplus & \oplus & \oplus \\
0 \quad\quad B_{N-1} \cdots B_{j+1} B_j & 1 & 0 & 0 & \cdots
\end{array}
$$

These codes differ only in position $j-1$, and both differ from all previously generated codes; you can thus recover the rule given at the beginning of this exercise.

10-16 TRANSVERSAL FILTERS IN SAMPLED SYSTEMS. If $x(t)$, $h(t)$, and $y(t)$ are the input, impulse response, and output of a linear filter, and $h(t)$ is a *finite impulse response* (FIR), meaning that it is zero for t larger than a finite time T, you have

$$y(t) = \int_0^T h(u)x(t-u)\,du$$

In sampled systems, the corresponding relation for $\{x_k\}$, $\{h_n\}$, and $\{y_l\}$ is

$$y_l = \sum_{n=0}^{N-1} h_n x_{l-n}$$

As shown in Fig. 10-49, this relation is easily implemented in the form of a *transversal filter*. From input to output, the values in a shift register of length N are multiplied by *weights* h_0 through h_{N-1} and then added. If $\{x_k\}$ is presented to the register in increasing order of index, the output of the adder is $\{y_l\}$, also in increasing order of index.

If τ is the period of the clock that drives the shift register, convince yourself that

$$\{h_n\} = \{1,1,1,1\}$$

corresponds to

$$h(t) = \frac{\theta(t)-\theta(t-4\tau)}{\tau}$$

and that

$$\{h_n\} = \{1,1,1,1,-1,-1,-1,-1\}$$

corresponds to

$$h(t) = \frac{\theta(t)-2\theta(t-4\tau)+\theta(t-8\tau)}{\tau}$$

and that the responses shown in Fig. 10-49 are correct.

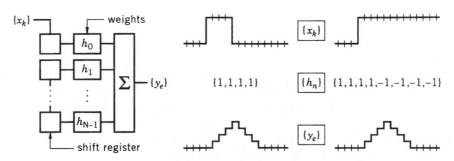

Figure 10-49. Transversal filters (Exercise 10-16).

11

TRANSMISSION LINES

11-1 INTRODUCTION

A signal with a rise time of 1 microsecond can be transmitted with negligible distortion over a distance of 1 meter if minimal precautions are taken. As we will see, this is so because the rise time is much larger than the *propagation time*, that is, the distance divided by the speed of light, 10 nanoseconds at most in this case. In a familiar example, audio signals are transmitted to speakers over a comparable distance, and little thought is given to precisely how this is done: almost any available pair of wires will do the job.

In contrast, a signal with a rise time of order 1 nanosecond can be severely distorted when transmitted over 1 meter. ECL circuits, for example, are characterized by such rise times, and transitions between logic levels can exhibit large overshoots and undershoots at the receiving end if due care is not taken. That such behavior must be avoided can be seen by observing that a 100% overshoot in a LOW-to-HIGH transition will saturate the input transistor of the receiving circuit, whereas a 50% undershoot can be misinterpreted as a return to a LOW level.

ECL circuits generate the type of wideband pulsed signals that motivate many of the developments in this chapter; in addition, they tolerate a certain amount of distortion in the form of overshoots and undershoots and thus offer a setting in which calculations of *imperfect* responses make sense. For this reason, we will choose our examples with ECL circuits preferably though not exclusively in mind. In round numbers, we will consider that logic signals with a 40% overshoot and a 20% undershoot are acceptable in ECL circuits.

Wideband signals can be transmitted with minimal distortion over substantial distances by means of *transmission lines*, two-conductor or three-conduc-

wire over   parallel wire   coaxial   printed circuit   printed circuit
ground   microstrip   stripline

Figure 11-1. Cross sections of commonly used transmission lines.

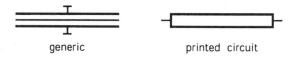

generic   printed circuit

Figure 11-2. Symbols for transmission lines.

tor structures with constant cross section, several examples of which are shown in Fig. 11-1. Symbols for transmission lines are indicated in Fig. 11-2; we will use the symbol labeled *printed circuit* when presenting circuits that are most often implemented in microstrip or stripline versions.

With rare exceptions, transmission lines are at least partially embedded in a dielectric medium other than air or vacuum; the only important consequence, other than some losses in the dielectric at frequencies in the gigahertz range, is that propagation times are increased.

As shown in Fig. 11-3, the *characteristic* or *image* impedance of a two-port is the impedance Z_C such that the input impedance of the two-port is equal to Z_C when the two-port is *terminated* in Z_C, that is, when its load impedance is Z_C; a two-port is *matched* if it is terminated in its characteristic impedance. The $R-2R$ divider of Exercise 3-4 provides a useful illustration of these concepts.

The finite length or *section* of ideal transmission line shown in Fig. 11-4 is a (symmetric) two-port; we will see that it has a characteristic *resistance* R_C of order 100 ohms. We will also see that if the section is matched at its receiving end, signals applied at its sending end are transmitted without distortion and only suffer a delay proportional to its length.

Figure 11-3. A two-port terminated in its characteristic impedance Z_C has input impedance Z_C.

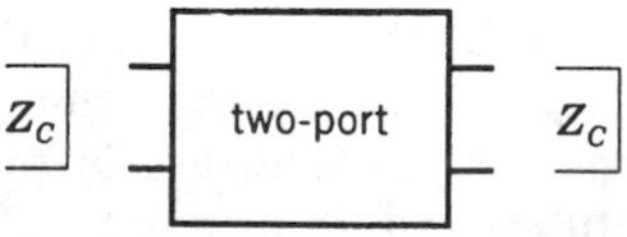

Z_C two-port Z_C

Figure 11-4. A section of ideal transmission line has a real characteristic impedance R_C.

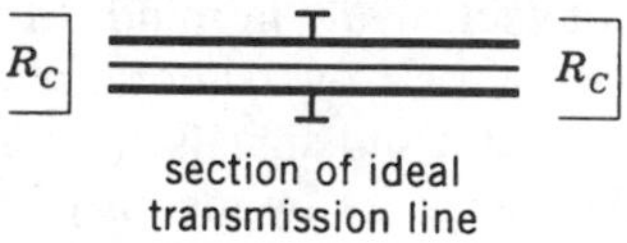

R_C R_C

section of ideal
transmission line

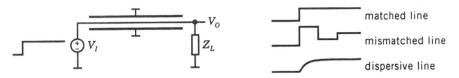

Figure 11-5. Signals are distorted in mismatched or dispersive transmission lines.

Signals in transmission lines are at least partially reflected at an unmatched load; the reflected signal adds to the incident signal, and as shown in Fig. 11-5, the signal at the load is almost invariably distorted.

Perfect matching is not always possible, and occasionally it is not mandatory. Thus stray capacitance at the load makes a line unmatched at high frequencies, whereas short runs in ECL circuits are often not terminated at all. One of our main objectives is to develop machinery that will allow us to determine when the resulting distortion is tolerable.

Real transmission lines inevitably have resistive losses, and these in turn usually imply *dispersion*, meaning that sine waves of different frequencies are transmitted at different speeds and with different attenuations. As shown in Fig. 11-5, this effect also results in distortion in wideband systems.

Although we will emphasize wideband systems, we will also look into the use of transmission lines for narrowband signals, that is, for signals that are almost sinusoidal over many cycles. Such signals can be transmitted with minor distortion over long distances, even if they are considerably attenuated in the process.

Close matching is always possible in narrowband systems, and often essential in transmitting maximum power to initially unmatched loads; we will consider several well-established methods of achieving this purpose.

In narrowband systems we will see that sections of transmission lines can be combined into substructures that are reminiscent of low-frequency circuits in that they have simple and easily visualized functions. In particular, we will analyze substructures that are widely used in microwave stripline circuits, and make use of this context to introduce *scattering parameters*, which provide a description that is both clear and elegant.

Returning briefly to Fig. 11-3, it is clear that the matched load Z_C can be replaced by a copy of the matched two-port, that this procedure can be repeated at will, and as shown in Fig. 11-6, that the characteristic impedance of the resulting network is Z_C; this is, for example, how we generated $R-2R$

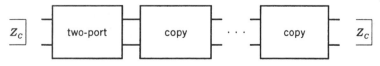

Figure 11-6. The characteristic impedance of a chain of identical two-ports is equal to the characteristic impedance of any member of the chain.

ladder networks in Exercise 3-4. Conversely, the characteristic impedance of a chain of identical two-ports is the characteristic impedance of any member of the chain. We will make use of these facts several times in later sections.

Finally, we note that the analysis of electromagnetic structures that exhibit a plane of symmetry is often simplified by separately considering an *even mode* in which excitations at symmetric points are equal, and an *odd mode* in which excitations at symmetric points are equal in magnitude but opposite in sign. The response to an arbitrary excitation can then be obtained by superposition.

11-2   THE TEM MODE IN IDEAL LINES

Like lumped elements, short transmission lines can behave almost ideally. The analysis of ideal lines will thus be immediately useful, and in addition it will give us a theoretical framework with which we can tackle lossy lines. We will begin by considering ideal lines in vacuum.

The conductors in an ideal line are by definition perfect; this implies that the electromagnetic field is zero in their interior and that the tangential electric field and the normal magnetic field are zero at their surfaces.

For sinusoidal signals, the fields in an ideal line can propagate in an infinite number of configurations or *modes*. All but one of these modes, however, are strongly attenuated for signals with wavelengths longer than a characteristic transverse dimension of the line, generally of order 1 cm and thus corresponding to a cutoff frequency of 30 GHz. In the surviving *transverse electric and magnetic* (TEM) mode, the electric and magnetic fields are both transverse; that is, they are normal to the axis of the line.

With these preliminaries, let us consider the transmission line of arbitrary cross section shown in Fig. 11-7, and let us assume solutions of Maxwell's equations in the form of *forward* and *backward* traveling waves in the plus

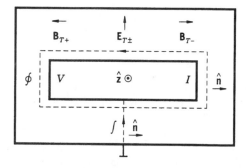

Figure 11-7. Representative cross section of a two-conductor transmission line. ϕ denotes the path used to obtain the charge per unit length and the current, whereas $\int$ denotes the path used to obtain the flux per unit length and the voltage.

and minus z-directions,

$$\mathbf{E}_{\pm} = \mathbf{E}_{T\pm}(x,y)g_{\pm}(z \mp ct)$$
$$\mathbf{B}_{\pm} = \mathbf{B}_{T\pm}(x,y)g_{\pm}(z \mp ct)$$

where $\mathbf{E}_{T\pm}$ and $\mathbf{B}_{T\pm}$ are static transverse fields such that $\mathbf{E}_{T\pm} \cdot \hat{\mathbf{z}} = 0$ and $\mathbf{B}_{T\pm} \cdot \hat{\mathbf{z}} = 0$, and $g_{\pm}(z)$ are arbitrary dimensionless functions. Let us also introduce the *transverse* gradient operator, defined by

$$\nabla_T = \nabla - \hat{\mathbf{z}}\frac{\partial}{\partial z}$$

We then have

$$\nabla \cdot \mathbf{E}_{\pm} = (\nabla_T \cdot \mathbf{E}_{T\pm})g_{\pm} = 0 \quad \text{and} \quad \nabla \cdot \mathbf{B}_{\pm} = (\nabla_T \cdot \mathbf{B}_{T\pm})g_{\pm} = 0$$

Choosing $g_{\pm} = 1$, we obtain that $\mathbf{E}_{T\pm}$ and $\mathbf{B}_{T\pm}$ are solenoidal,

$$\nabla_T \cdot \mathbf{E}_{T\pm} = 0$$
$$\nabla_T \cdot \mathbf{B}_{T\pm} = 0$$

We also have

$$\nabla \times \mathbf{E}_{\pm} = (\nabla_T \times \mathbf{E}_{T\pm})g_{\pm} + (\hat{\mathbf{z}} \times \mathbf{E}_{T\pm})g'_{\pm} = -\frac{\partial \mathbf{B}_{\pm}}{\partial t} = \pm c\mathbf{B}_{T\pm}g'_{\pm}$$

and

$$\nabla \times \mathbf{B}_{\pm} = (\nabla_T \times \mathbf{B}_{T\pm})g_{\pm} + (\hat{\mathbf{z}} \times \mathbf{B}_{T\pm})g'_{\pm} = \frac{1}{c^2}\frac{\partial \mathbf{E}_{\pm}}{\partial t} = \mp \frac{1}{c}\mathbf{E}_{T\pm}g'_{\pm}$$

Choosing $g_{\pm} = 1$, and then choosing $g'_{\pm} = 1$, we obtain that $\mathbf{E}_{T\pm}$ and $\mathbf{B}_{T\pm}$ are also irrotational,

$$\nabla_T \times \mathbf{E}_{T\pm} = 0$$
$$\nabla_T \times \mathbf{B}_{T\pm} = 0$$

and that as in waves in free space, $\mathbf{E}_{T\pm}$ and $c\mathbf{B}_{T\pm}$ are of the same magnitude and orthogonal to each other,

$$\hat{\mathbf{z}} \times \mathbf{E}_{T\pm} = \pm c\mathbf{B}_{T\pm}$$
$$\hat{\mathbf{z}} \times c\mathbf{B}_{T\pm} = \mp \mathbf{E}_{T\pm}$$

Since $\mathbf{E}_{T+}$ and $\mathbf{E}_{T-}$ are irrotational and obey the same boundary conditions, they can be derived from a single transverse scalar potential $\psi(x,y)$. If we use the vector identity

$$\nabla_T \psi \times \hat{\mathbf{z}} = \nabla_T \times (\hat{\mathbf{z}} \psi)$$

we obtain that $\pm \hat{\mathbf{z}} \psi / c$ is the vector potential for $\mathbf{B}_{T\pm}$, and we can write

$$\mathbf{E}_{T\pm} = -\nabla_T \psi$$

$$c\mathbf{B}_{T\pm} = \pm \nabla_T \times (\hat{\mathbf{z}} \psi)$$

Since $\mathbf{E}_{T+}$ and $\mathbf{E}_{T-}$ are also solenoidal, ψ is the solution of Laplace's equation

$$\nabla_T^2 \psi = 0$$

subject to the boundary conditions that it be constant on the surfaces of the conductors.

We note that at least two conductors are required to sustain a TEM mode; in a hollow conducting cylinder of arbitrary cross section, for example, constant ψ on the inner surface implies constant ψ and therefore zero fields everywhere inside the cylinder. We also note that the fields $\mathbf{E}_\pm$ and $\mathbf{B}_\pm$ can be obtained from the potentials

$$\phi_\pm = \psi g_\pm$$

$$c\mathbf{A}_\pm = \pm \hat{\mathbf{z}} \psi g_\pm$$

Once we have the field configurations, we can calculate the voltage $V_\pm$ and the current $I_\pm$ at any transverse plane,

$$V_\pm = V_{T\pm} g_\pm$$

$$I_\pm = I_{T\pm} g_\pm$$

where $V_{T\pm}$ is the static voltage on the inner conductor in Fig. 11-7,

$$V_{T\pm} = -\int \mathbf{E}_{T\pm} \cdot d\mathbf{l} = \int \nabla_T \psi \cdot d\mathbf{l} \tag{11-1}$$

and $I_{T\pm}$ is the static current, obtained by applying Ampère's law to any path that encloses the inner conductor,

$$I_{T\pm} = \varepsilon_0 c^2 \oint \mathbf{B}_{T\pm} \cdot d\mathbf{l} = \pm \varepsilon_0 c \oint \nabla_T \psi \times \hat{\mathbf{z}} \cdot d\mathbf{l} \tag{11-2}$$

The total or *line* voltage and current are then

$$V = V_+ + V_-$$

$$I = I_+ + I_-$$

We will define the inductance per unit length $\mathcal{L}$ and the capacitance per unit length $\mathcal{C}$ by

$$\mathcal{L}I = \mathcal{F}$$
$$\mathcal{C}V = \mathcal{Q} \tag{11-3}$$

where $\mathcal{F}$ is the flux per unit length between the conductors and $\mathcal{Q}$ is the charge per unit length on the inner conductor. Converting surface integrals into line integrals, for forward and backward waves we have

$$\mathcal{L}I_{T\pm} = \mathcal{F}_{T\pm} = \int \hat{\mathbf{z}} \times \mathbf{B}_{T\pm} \cdot d\mathbf{l} = \mp \frac{1}{c} \int \mathbf{E}_{T\pm} \cdot d\mathbf{l} = \pm \frac{1}{c} \int \nabla_T \psi \cdot d\mathbf{l}$$

and

$$\mathcal{C}V_{T\pm} = \mathcal{Q}_{T\pm} = \varepsilon_0 \oint \hat{\mathbf{z}} \times \mathbf{E}_{T\pm} \cdot d\mathbf{l} = \varepsilon_0 \oint \nabla_T \psi \times \hat{\mathbf{z}} \cdot d\mathbf{l}$$

Substituting $I_{T\pm}$ from (11-2) and $V_{T\pm}$ from (11-1), we obtain

$$\mathcal{L} = \frac{\int \nabla_T \psi \cdot d\mathbf{l}}{\varepsilon_0 c^2 \oint \nabla_T \psi \times \hat{\mathbf{z}} \cdot d\mathbf{l}} \qquad \mathcal{C} = \frac{\varepsilon_0 \oint \nabla_T \psi \times \hat{\mathbf{z}} \cdot d\mathbf{l}}{\int \nabla_T \psi \cdot d\mathbf{l}}$$

From these expressions it is clear that $\mathcal{L}\mathcal{C}$ is independent of the shape of the line and that its square root is the reciprocal of the speed of light,

$$\sqrt{\mathcal{L}\mathcal{C}} = \frac{1}{c}$$

We will define the characteristic resistance R_C by

$$R_C = \sqrt{\frac{\mathcal{L}}{\mathcal{C}}}$$

We then have

$$R_C = \frac{1}{\varepsilon_0 c} \frac{\int \nabla_T \psi \cdot d\mathbf{l}}{\oint \nabla_T \psi \times \hat{\mathbf{z}} \cdot d\mathbf{l}}$$

Comparing the expressions for $V_{T\pm}$ and $I_{T\pm}$ in (11-1) and (11-2), we conclude that the forward and backward voltage waves are proportional to the corresponding current waves, and that the constants of proportionality are $\pm R_C$,

$$V_+ = +R_C I_+$$
$$V_- = -R_C I_- \tag{11-4}$$

In ideal lines, the power P transmitted across a transverse plane is given by the VI product because the components of $\mathbf{B}$ and $\mathbf{E}$ normal to the plane are zero, so that the conditions established in Section 4-8 are satisfied. If we now consider an arbitrary superposition of forward and backward waves, we have

$$P = VI = (V_+ + V_-)(I_+ + I_-)$$

Substituting $I_\pm$ in terms of $V_\pm$, we obtain

$$P = \frac{V_+^2 - V_-^2}{R_C}$$

The net power transmitted is thus simply the difference between the power in the forward wave and the power in the backward wave, calculated as if each wave were the only one on the line.

Before proceeding, let us get a feeling for the magnitudes of the parameters we have identified so far. As we will normally do in what follows, we will drop subscripts for *forward* and *static* when the context makes them superfluous. We will first consider the coaxial line shown in Fig. 11-8. If the radius of the center conductor is a and the inner radius of the shield is b, the static transverse fields of a forward wave for $a \le \rho \le b$ are

$$cB_\varphi = E_\rho = \frac{a}{\rho} E_0$$

where E_0 is a constant. The static voltage and current are therefore

$$V = a \ln(b/a) E_0$$

$$I = \frac{2\pi a}{\mu_0 c} E_0 \tag{11-5}$$

whereas the static charge and flux per unit length are

$$\mathcal{Q} = 2\pi a \varepsilon_0 E_0$$

$$\mathcal{F} = \frac{a \ln(b/a)}{c} E_0$$

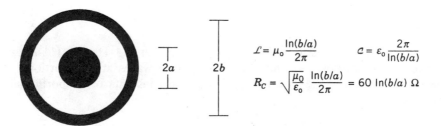

$$\mathcal{L} = \mu_0 \frac{\ln(b/a)}{2\pi} \qquad \mathcal{C} = \varepsilon_0 \frac{2\pi}{\ln(b/a)}$$

$$R_C = \sqrt{\frac{\mu_0}{\varepsilon_0}} \, \frac{\ln(b/a)}{2\pi} = 60 \ln(b/a) \ \Omega$$

Figure 11-8. Parameters of a coaxial line.

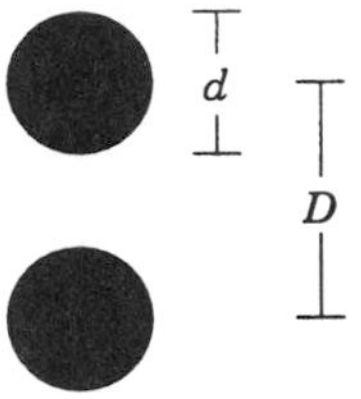

$$\mathcal{L} = \mu_0 \frac{\cosh^{-1}(D/d)}{\pi} \qquad \mathcal{C} = \varepsilon_0 \frac{\pi}{\cosh^{-1}(D/d)}$$

$$R_C = \sqrt{\frac{\mu_0}{\varepsilon_0}} \frac{\cosh^{-1}(D/d)}{\pi} = 120 \cosh^{-1}(D/d) \ \Omega$$

Figure 11-9. Parameters of a parallel-wire line.

From the definitions $\mathcal{F} = \mathcal{L}I$ and $\mathcal{Q} = \mathcal{C}V$ in (11-3), we then obtain

$$\mathcal{L} = \mu_0 \frac{\ln(b/a)}{2\pi}$$

$$\mathcal{C} = \varepsilon_0 \frac{2\pi}{\ln(b/a)}$$

$$R_C = \sqrt{\frac{\mu_0}{\varepsilon_0}} \frac{\ln(b/a)}{2\pi}$$

These expressions are repeated in Fig. 11-8. Observing that $\sqrt{\mu_0/\varepsilon_0} = 120\pi$ ohms, we also have

$$R_C = 60 \ln(b/a) \text{ ohms}$$

We note that although coaxial lines are expensive, they offer the advantage of freedom from pickup, decidedly so if the shield is solid, and up to microwave frequencies if the shield is braided.

Parallel-wire lines are the subject of Exercises 4-8 and 4-9; the expressions for the parameters obtained in these exercises are presented in Fig. 11-9 in terms of the diameter d of the wires and the distance D between the centers of the wires.

With the substitutions $\mathcal{L} \rightarrow 2\mathcal{L}$, $\mathcal{C} \rightarrow \mathcal{C}/2$, $R_C \rightarrow 2R_C$, and $D \rightarrow 2h$, the results for the parallel-wire line of Fig. 11-9 apply to a wire-over-ground line in which d is the diameter of the wire and h is the height of the center of the wire above the ground plane; the expressions for the parameters in this case are presented in Fig. 11-10.

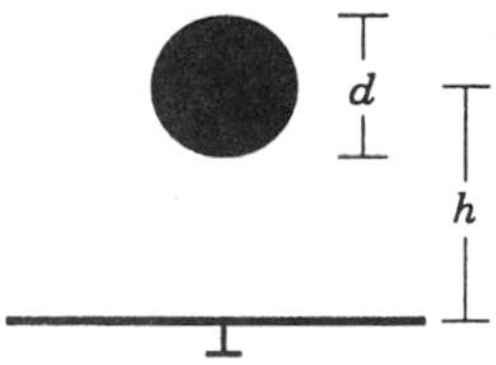

$$\mathcal{L} = \mu_0 \frac{\cosh^{-1}(2h/d)}{2\pi} \qquad \mathcal{C} = \varepsilon_0 \frac{2\pi}{\cosh^{-1}(2h/d)}$$

$$R_C = \sqrt{\frac{\mu_0}{\varepsilon_0}} \frac{\cosh^{-1}(2h/d)}{2\pi} = 60 \cosh^{-1}(2h/d) \ \Omega$$

Figure 11-10. Parameters of a wire-over-ground line.

The results we have obtained so far are essentially unaltered for lines that are fully embedded in an insulating medium of dielectric constant κ. The capacitance per unit length is κ times larger, and the propagation speed $1/\sqrt{\mathcal{L}\mathcal{C}}$ and the characteristic resistance $\sqrt{\mathcal{L}/\mathcal{C}}$ are $\sqrt{\kappa}$ times smaller. In coaxial lines with a polyethylene dielectric, $\sqrt{\kappa}$ is about 1.5; for $R_C = 50\,\Omega$, we have

$$\sqrt{\mathcal{L}\mathcal{C}} \approx 5 \text{ ns/m}$$

$$\mathcal{L} \approx 250 \text{ nH/m}$$

$$\mathcal{C} \approx 100 \text{ pF/m}$$

In the case of lines that are only partially embedded in a dielectric medium, microstrip lines for example, we will accept that they can be described in terms of an *effective* dielectric constant κ_E. For microstrip lines in which the thickness of the strip can be neglected, κ_E and R_C are given roughly by

$$\kappa_E \approx 0.65\kappa + 0.35$$

and

$$R_C \approx \frac{60}{\sqrt{\kappa_E}} \ln\left(\frac{8h}{w} + \frac{w}{4h} \right) \text{ ohms}$$

where w is the width of the strip and h is the height of the strip above the ground plane. These expressions are good to 10% for $0.3 < w/h < 3$ and $\kappa < 10$. As a practical matter, κ is about 5 in the G10 glass epoxy normally used in printed circuit boards; in standard $\frac{1}{16}$ inch boards, R_C is about $100\,\Omega$ for lines 0.025 inches wide, and about $50\,\Omega$ for lines 0.100 inches wide. As a further practical matter, we note that *twisted pairs* of 24AWG insulated wire ($d \approx 0.5$ mm, $\kappa_E \approx 3$), widely used at low frequencies, have $R_C \approx 100\,\Omega$.

Now that we have some examples, we will look into the question of matching transmission lines. Let us consider a section of air coaxial line that is carrying a forward wave, and let us assume that the line can be matched at the receiving end by fitting it with a flat ring of infinitesimal thickness ξ and conductivity σ. Continuity and symmetry then require that the static radial current density in the ring satisfy

$$2\pi\rho\xi J_\rho = I$$

Substituting $I = 2\pi a E_0/\mu_0 c$ from (11-5) and recalling that $E_\rho = (a/\rho)E_0$, we obtain

$$(\sigma\xi\mu_0 c)J_\rho = \sigma E_\rho$$

Ohm's law requires that $\sigma\xi\mu_0 c = 1$, which is compatible with the boundary conditions that B_φ be zero just beyond the ring and that $cB_\varphi = E_\rho$ just before the ring,

$$cB_\varphi = \mu_0 c \xi J_\rho = (\mu_0 c \xi \sigma) E_\rho$$

The line will therefore be matched if

$$\frac{1}{\sigma \xi} = \mu_0 c = 120\pi \text{ ohms}$$

that is, if the ring is made of *spacecloth* with a resistance of 120π *ohms per square* (it should be clear that the resistance between opposite edges of any square is $1/\sigma\xi$). The resulting resistance of the ring is, as we might expect, equal to R_C.[†] For $\sigma = 0.25$ S/cm, which obtains in moderately doped silicon, for example, we have $\xi = 0.1$ mm, a manageable value.

The arguments above can be extended to any line; a parallel-wire line, for example, can be matched by terminating it with an infinite plane of space-cloth. Most matching, however, is done with ordinary resistors of value R_C, so that the geometry is clearly wrong and the boundary conditions are not satisfied by TEM waves. We will nonetheless accept that except when matching requirements are stringent, ordinary resistors give satisfactory performance; we will look into this question further in Section 11-7. In the same spirit we will accept, imperfect boundary conditions at the sending end notwithstanding, that a finite line matched at the receiving end looks like a resistor of value R_C at the sending end. In the language of two-ports, we can now say that the characteristic resistance of the section of line shown in Fig. 11-4 is precisely $R_C = \sqrt{\mathcal{L}/\mathcal{C}}$.

For future purposes, we want coupled first-order differential equations for the line variables V and I. As shown for a microstrip line in Fig. 11-11, the flux Φ and the charge Q in an infinitesimal length Δz are

$$\Phi = \mathcal{F}\Delta z = \mathcal{L}I\Delta z$$

$$Q = \mathcal{Q}\Delta z = \mathcal{C}V\Delta z$$

From Faraday's law and continuity we then get

$$V(z+\Delta z) - V(z) = -\frac{d\Phi}{dt} = -\mathcal{L}\frac{\partial I}{\partial t}\Delta z$$

$$I(z+\Delta z) - I(z) = -\frac{dQ}{dt} = -\mathcal{C}\frac{\partial V}{\partial t}\Delta z$$

[†]There is a minor flaw, however: the electric field is not zero beyond the ring (its tangential component is continuous), so that there must be charges on the external surfaces of the center conductor, the ring, and the shield. A matched termination in spacecloth thus has a (small) parasitic capacitance in parallel.

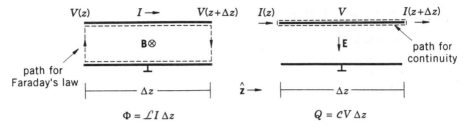

Figure 11-11. Side views of an infinitesimal length of microstrip line, indicating the variables used to obtain the equations of telegraphy.

Taking the limit, we obtain the *equations of telegraphy* for ideal lines,

$$\frac{\partial V}{\partial z} = -\mathcal{L}\frac{\partial I}{\partial t}$$

$$\frac{\partial I}{\partial z} = -\mathcal{C}\frac{\partial V}{\partial t}$$

For sinusoidal or Laplace-transformable excitations we can write

$$\frac{dV}{dz} = -s\mathcal{L}I$$

$$\frac{dI}{dz} = -s\mathcal{C}V$$

(11-6)

From the equations of telegraphy we recover wave equations for V and I,

$$\frac{d^2V}{dz^2} = s^2\mathcal{L}\mathcal{C}V$$

$$\frac{d^2I}{dz^2} = s^2\mathcal{L}\mathcal{C}I$$

(11-7)

In Section 11-4 we will extend the equations of telegraphy to lossy lines, and then discuss solutions.

11-3 *LC* LADDER NETWORKS

To connect with the literature as well as with later developments in this chapter, in this section we will consider transmission lines as limits of lumped circuits. This more general approach covers in less detail what we have already seen, but it has the advantage of opening up new possibilities.

Let us start by considering the two-port *LC* network shown in Fig. 11-12, and let us obtain its characteristic impedance and its transfer function. It is

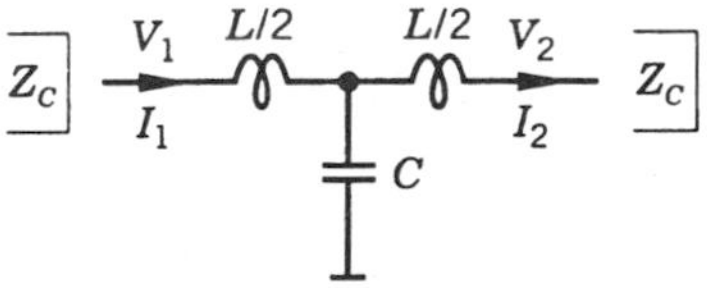

Figure 11-12. Two-port *LC* network.

worth pointing out that we will make good use of the assumption that the input and load impedances are both equal to the characteristic impedance, and that we will do this repeatedly in later sections.

Assuming then that the two-port is matched, we have $I_1 = V_1 Y_C$ and $I_2 = V_2 Y_C$. The impedance looking beyond the input inductor is therefore $Z_C - sL/2$, and its reciprocal must be equal to sC plus the reciprocal of $Z_C + sL/2$. We thus have

$$\frac{1}{Z_C - sL/2} - \frac{1}{Z_C + sL/2} = sC$$

Solving this equation, we obtain

$$Z_C^2 = \frac{L}{C} - \frac{\omega^2 L^2}{4} \tag{11-8}$$

The characteristic impedance is thus real below a cutoff frequency ω_0 given by

$$\omega_0 = \frac{2}{\sqrt{LC}}$$

The voltage on the capacitor is $V_1 - sLI_1/2 = V_2 + sLI_2/2$; substituting $I_1 = V_1 Y_C$ and $I_2 = V_2 Y_C$, we get

$$V_2 = V_1 \frac{1 - sLY_C/2}{1 + sLY_C/2} \qquad I_2 = I_1 \frac{1 - sLY_C/2}{1 + sLY_C/2} \tag{11-9}$$

For $\omega < \omega_0$, Y_C is real, and the magnitude of the transfer function from V_1 to V_2 is 1; sine waves are therefore unattenuated and suffer only a phase shift φ given by

$$\exp(i\varphi) = \frac{V_2}{V_1} \tag{11-10}$$

or after some algebra, by

$$\sin \varphi = -2(\omega/\omega_0)\sqrt{1 - (\omega/\omega_0)^2}$$

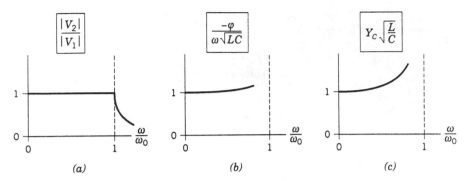

Figure 11-13. Properties of the LC network of Fig. 11-12 when it is matched. (a) Magnitude and (b) phase of the transfer function. (c) Admittance.

For $\omega > \omega_0$, Y_C is imaginary, and sine waves are attenuated,

$$\frac{V_2}{V_1} = \frac{\sqrt{(\omega/\omega_0)^2 - 1} - \omega/\omega_0}{\sqrt{(\omega/\omega_0)^2 - 1} + \omega/\omega_0}$$

Figure 11-13 shows plots of $|V_2/V_1|$, φ, and Y_C. For $\omega < \omega_0/2$ we can say roughly that φ is proportional to ω and that Z_C is constant,

$$\varphi \simeq -\omega\sqrt{LC} \qquad \text{and} \qquad Z_C \simeq \sqrt{\frac{L}{C}}$$

If we now generate the LC ladder network of Fig. 11-14 by chaining together a number of the networks in Fig. 11-12, it is clear that its characteristic impedance is given by (11-8) and that the transfer function for one stage is given by (11-10). For $\omega < \omega_0/2$ we have $Z_C \simeq \sqrt{L/C}$, and if the number of stages is n, the overall phase is approximately proportional to ω times n,

$$\frac{V_{n+1}}{V_1} \simeq \exp(-sn\sqrt{LC})$$

A section of ideal transmission line that is matched at the receiving end has a

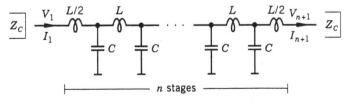

Figure 11-14. LC ladder network.

constant input resistance $R_C = \sqrt{\mathcal{L}/\mathcal{C}}$; for a forward sine wave we can write

$$V(z,t) = V_0 \exp\left[s\left(t - z\sqrt{\mathcal{L}\mathcal{C}}\right)\right] = V_0 \exp(st)\exp\left(-sz\sqrt{\mathcal{L}\mathcal{C}}\right)$$

If the section extends from $z = 0$ to $z = l$, its transfer function is a delay,

$$\frac{V(l)}{V(0)} = \exp\left(-sl\sqrt{\mathcal{L}\mathcal{C}}\right)$$

The magnitude of $V(l)/V(0)$ is equal to 1, and its phase is proportional to ω times l. We conclude that LC ladder networks look like transmission lines if $\omega < \omega_0/2$.

Let us now assume that the number of stages in the ladder is made infinite in such a way that the inductance per unit length $\mathcal{L}$ and the capacitance per unit length $\mathcal{C}$ remain constant, that is, such that

$$L \to \mathcal{L}\Delta z \quad \text{and} \quad C \to \mathcal{C}\Delta z$$

It follows that ω_0 becomes infinite and that

$$Z_C \to \sqrt{\frac{\mathcal{L}}{\mathcal{C}}}$$

From the transfer function for one stage in (11-9) we then obtain

$$\Delta V \equiv V_2 - V_1 = -s(\mathcal{L}\Delta z)Y_C V_1 = -s(\mathcal{L}\Delta z)I_1$$
$$\Delta I \equiv I_2 - I_1 = -s(\mathcal{L}\Delta z)Y_C I_1 = -s(\mathcal{C}\Delta z)V_1$$

We thus recover the equations of telegraphy for ideal lines given in (11-6),

$$\frac{dV}{dz} = -s\mathcal{L}I$$

$$\frac{dI}{dz} = -s\mathcal{C}V$$

We conclude that we can derive the equations of telegraphy whenever we can identify *uniformly distributed* inductances and capacitances, and that at least at low frequencies, transmission lines other than those considered in Section 11-2 are possible; an interesting example is presented in Exercise 11-3.

11-4 LOSSY LINES

In the frequency range from 10 Hz to 1 GHz, losses in transmission lines are due principally to the finite resistivity of the conductors. Losses in the dielectric (if any) are proportional to the frequency, as indicated in Exercise

3-22, and usually become dominant well beyond 1 GHz. We will first consider a coaxial line with a perfectly conducting shield, and then argue that our results can be extended to arbitrary lines.

Let us apply to the center conductor the results we obtained in Section 4-7 for a conducting cylinder that carries a sinusoidal current. According to (4-12), the longitudinal current density at low frequencies is

$$J_z(\rho) = J_z(0)\left(1 + s\frac{\sigma\mu_0\rho^2}{4}\right)$$

If the radius of the center conductor is a, the current is

$$I = 2\pi\int_0^a \rho J_z(\rho)\, d\rho = \pi a^2 J_z(0)\left(1 + s\frac{\sigma\mu_0 a^2}{8}\right)$$

Assuming that $\omega\sigma\mu_0 a^2 \ll 1$ or, equivalently, that the skin depth $\delta = \sqrt{2/\omega\sigma\mu_0}$ is much larger than a, the current density at the surface is given in terms of the current by

$$J_z(a) \simeq \frac{I}{\pi a^2}\left(1 + s\frac{\sigma\mu_0 a^2}{8}\right)$$

The longitudinal electric field at the surface can then be obtained from Ohm's law,

$$E_z(a) = \frac{J_z(a)}{\sigma} = I\left(\frac{1}{\sigma\pi a^2} + s\frac{\mu_0}{8\pi}\right)$$

If we now define the *longitudinal impedance* per unit length $\mathcal{Z}$ by

$$E_z(a) = \mathcal{Z}I$$

we have

$$\mathcal{Z} = \mathcal{R}_L + s\mathcal{L}_I \tag{11-11}$$

where $\mathcal{R}_L = 1/\sigma\pi a^2$ is the dc resistance per unit length and $\mathcal{L}_I = \mu_0/8\pi$ is the *internal inductance* per unit length. We will normally ignore the internal contribution to the inductance since $\mathcal{L}_I = 50$ nH/m is considerably smaller than the *external* inductance per unit length $\mathcal{L}$, which is of order 500 nH/m.

In contrast, at frequencies high enough that $\delta \ll a$, from (4-14) we have

$$J_z(\rho) = J_z(a)\exp\left[(1+i)\frac{\rho - a}{\delta}\right] \tag{11-12}$$

Integrating and extending the upper limit to infinity, we obtain the surface current density K,

$$K = \int_0^\infty J_z(a)\exp\left[-(1+i)\frac{\xi}{\delta}\right]d\xi = \frac{J_z(a)\delta}{1+i} \tag{11-13}$$

In terms of the current $I = 2\pi aK$, we then have

$$E_z(a) = \frac{(1+i)I}{2\pi a\sigma\delta}$$

Assuming that the fields are TEM, the transverse electric field at $\rho = a$ in a forward wave is

$$E_\rho(a) = cB_\varphi(a) = \frac{\mu_0 cI}{2\pi a}$$

It follows that

$$\frac{E_z(a)}{E_\rho(a)} = \frac{1+i}{\mu_0 c\sigma\delta}$$

Even at 10 GHz, $\mu_0 c\sigma\delta$ is larger than 10^4 in copper; the longitudinal electric field is thus but a small perturbation on fields that are basically transverse. We will make use of this observation when we generalize our results.

If we define $\mathscr{R}_H$ as the resistance per unit length at a frequency ω_H chosen high enough that $\delta(\omega_H) \ll a$, the longitudinal impedance per unit length can be written in the form

$$\mathscr{Z} = \mathscr{R}_H\sqrt{\frac{2s}{\omega_H}} \tag{11-14}$$

where

$$\mathscr{R}_H = \frac{1}{2\pi a\sigma\delta(\omega_H)}$$

The cylindrical symmetry of the coaxial line we are considering implies that the current density has no transverse component and that the *longitudinal* magnetic field is zero; the voltage on the center conductor is therefore well defined and constant in transverse planes, and the arguments based on Fig. 11-11 remain valid in the presence of the longitudinal electric field $E_z(a) = \mathscr{Z}I$. We then obtain what will turn out to be the equations of telegraphy for lossy lines,

$$\frac{dV}{dz} = -(s\mathscr{L} + \mathscr{Z})I$$

$$\frac{dI}{dz} = -s\mathscr{C}V \tag{11-15}$$

As we will see below, the voltage in transverse planes is not quite constant in arbitrary transmission lines. We will nonetheless accept for now that the equations of telegraphy that we have derived are valid for transmission lines with no dielectric losses, and that with appropriate values of the parameters the longitudinal impedance is as given in (11-11) or (11-14); we will proceed on this basis, and return to make comments when we have more machinery.

From the equations of telegraphy we obtain second-order differential equations for V and I,

$$\frac{d^2V}{dz^2} = (s\mathcal{L}+\mathcal{Z})s\mathcal{C}V$$

$$\frac{d^2I}{dz^2} = (s\mathcal{L}+\mathcal{Z})s\mathcal{C}I$$

The general solutions of these equations are superpositions of forward and backward waves,

$$V = V_{0+}\exp(-\gamma z) + V_{0-}\exp(+\gamma z)$$

$$I = I_{0+}\exp(-\gamma z) + I_{0-}\exp(+\gamma z)$$

where γ is the *propagation function*,

$$\gamma = \sqrt{(s\mathcal{L}+\mathcal{Z})s\mathcal{C}}$$

For steady-state sine waves we can write

$$\gamma = \alpha + \beta i$$

where α is the *attenuation function* and β is the *phase function*. In lossless lines, we have

$$\alpha = 0 \quad \text{and} \quad \beta = \omega\sqrt{\mathcal{L}\mathcal{C}}$$

The (local) amplitudes of the forward and backward waves are

$$V_{\pm}(z) = V_{0\pm}\exp(\mp\gamma z)$$

$$I_{\pm}(z) = I_{0\pm}\exp(\mp\gamma z)$$

(11-16)

Substituting these amplitudes in the equations of telegraphy in (11-15) and dropping arguments for clarity, we obtain the generalization of (11-4),

$$V_+ = +Z_C I_+$$

$$V_- = -Z_C I_-$$

where Z_C is the *characteristic impedance*,

$$Z_C = \sqrt{\frac{s\mathcal{L}+\mathcal{Z}}{s\mathcal{C}}}$$

Before going on to see the effects of losses, it will be convenient to obtain a few relations that we will often use in the remainder of this chapter. The line voltage and the line current are

$$V = V_+ + V_-$$
$$I = I_+ + I_-$$

Substituting $Z_C I_+ = V_+$ and $Z_C I_- = -V_-$, we obtain

$$V = V_+ + V_-$$
$$Z_C I = V_+ - V_- \tag{11-17}$$

Inverting these equations, we obtain the amplitudes of the forward and backward voltage waves in terms of the line voltage and the line current,

$$V_+ = \tfrac{1}{2}(V + Z_C I)$$
$$V_- = \tfrac{1}{2}(V - Z_C I) \tag{11-18}$$

Returning to the question of losses, we note that the inductive contributions to γ are negligible for frequencies below the crossover point ω_{CO} given by

$$\mathcal{L}\omega_{CO} = \mathcal{R}_L$$

and that rather than slight corrections to the wave equations given in (11-7), we obtain diffusion equations:

$$\frac{d^2 V}{dz^2} = s\mathcal{R}_L \mathcal{C} V$$

$$\frac{d^2 I}{dz^2} = s\mathcal{R}_L \mathcal{C} I$$

In the *diffusion region* we have

$$\gamma \simeq \sqrt{s\mathcal{R}_L \mathcal{C}}$$

The phase and attenuation functions are thus both proportional to the square root of the frequency,

$$\alpha = \beta = \sqrt{\frac{\omega \mathcal{R}_L \mathcal{C}}{2}}$$

Generally speaking, the crossover from the diffusion region to the *wave region* occurs in the neighborhood of 5 kHz. Above crossover we have

$|\mathcal{Z}| \ll \omega\mathcal{L}$, and we can use the approximation

$$\gamma = s\sqrt{\mathcal{L}\mathcal{C}} \sqrt{1 + \frac{\mathcal{Z}}{s\mathcal{L}}} \approx s\sqrt{\mathcal{L}\mathcal{C}}\left(1 + \frac{\mathcal{Z}}{2s\mathcal{L}}\right)$$

The propagation function in the wave region is thus

$$\gamma \approx s\sqrt{\mathcal{L}\mathcal{C}} + \frac{\mathcal{Z}}{2R_C} \tag{11-19}$$

where $R_C = \sqrt{\mathcal{L}/\mathcal{C}}$ is the characteristic resistance in the absence of losses, and the attenuation function is

$$\alpha \approx \frac{\mathcal{R}}{2R_C} \tag{11-20}$$

where $\mathcal{R}$ is the resistive part of $\mathcal{Z}$. As long as the skin effect is negligible we have $\mathcal{R} = \mathcal{R}_L$, and the attenuation is constant,

$$\alpha \approx \frac{\mathcal{R}_L}{2R_C}$$

The attenuation remains constant up to about a decade in frequency above crossover, beyond which point the skin effect becomes dominant, and the attenuation is then again proportional to the square root of the frequency,

$$\alpha \approx \frac{\mathcal{R}_H}{2R_C}\sqrt{\frac{\omega}{\omega_H}}$$

The attenuation as a function of the frequency looks quite generally as shown in Fig. 11-15. An important conclusion we should draw is that constant

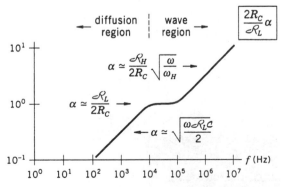

Figure 11-15. Attenuation of a lossy line as a function of the frequency.

attenuation is virtually nonexistent in transmission lines or, equivalently, that dispersive lines are the rule.

Now that we have acquired the appropriate language, let us come back to the question of deriving the equations of telegraphy. In the diffusion region the inductive contributions are negligible, so that regardless of the shape of the transmission line, the voltage and the longitudinal electric fields at the surfaces of the conductors are constant in transverse planes, and the equations of telegraphy as given in (11-15) are correct.

In the wave region, however, inductive contributions are far from negligible, and we must be careful. As shown for a simple case in Exercise 11-4, the fact that power dissipation and power density are uncorrelated in space implies a (small) *transverse* flow of power, which in turn requires a small *longitudinal* magnetic field. The voltage and the flux per unit length in transverse planes are therefore (very slightly) path dependent, and we cannot argue as we did for coaxial lines because we are looking for small effects that are hard to discern if the dominant transverse fields are not perfectly known. These small effects, however, can be isolated by calculating power losses; we will do so only for the case in which the skin effect is well developed because it provides a sufficiently clear picture of the arguments involved.

We will accept that the longitudinal current density depends locally on the normal depth ξ into a conductor as it does in cylinders. From (11-12) we then have

$$J_z(\xi) = J_S \exp\left[-(1+i)\frac{\xi}{\delta}\right]$$

where J_S is the current density at the surface. Integrating this expression, we obtain the generalization of (11-13),

$$J_S = (1+i)\frac{K}{\delta}$$

where K is the local surface current density. It follows from Ohm's law that K and the longitudinal electric field at the surface E_S are locally related by

$$E_S = (1+i)KR_S$$

where

$$R_S = \frac{1}{\sigma\delta}$$

is the *surface resistance*. The (average) power dissipated per unit area can be expressed in terms of K and R_S,

$$\int_0^\infty \frac{J_z J_z^*}{2\sigma}d\xi = \frac{J_S J_S^*}{2\sigma}\int_0^\infty \exp\left(-2\frac{\xi}{\delta}\right)d\xi = J_S J_S^*\frac{\delta}{4\sigma} = \frac{R_S}{2}KK^*$$

Losses can thus be calculated if the magnetic field, and therefore the surface current density, is known. We note incidentally that R_S is consistently named in that it gives the losses according to the usual prescription.

For a forward wave, the power loss per unit length is

$$\frac{dP}{dz} = -\frac{1}{2}\oint R_S K K^* \, dl$$

where the integral is along the perimeters of *all* the conductors and might involve different values of R_S. In the case we are considering, the power transmitted along the line is essentially as it is in ideal lines,

$$P = \text{Re}\left(\frac{1}{2}VI^*\right) = \frac{R_C}{2}II^*$$

We thus have

$$\frac{dP}{dz} = -\frac{\mathscr{R}}{R_C}P$$

where the resistance per unit length $\mathscr{R}$ is given by

$$\mathscr{R} = \frac{\oint R_S K K^* \, dl}{II^*}$$

As a check, we observe that for a coaxial line with a perfectly conducting shield and a center conductor of radius a, we have $K = I/2\pi a$ and therefore $\mathscr{R} = 1/2\pi a \sigma \delta$, which we know is correct. The power decays exponentially,

$$P(z) = P(0)\exp\left(-\frac{\mathscr{R}}{R_C}z\right)$$

Since the voltage is proportional to the square root of the power, we recover the expression for α in (11-20),

$$\alpha = \frac{\mathscr{R}}{2R_C}$$

At this point we will invoke the fact that impedances are necessarily *analytic* functions of s, so that if $\mathscr{R}$ is proportional to $\sqrt{\omega}$, $\mathscr{Z}$ must be proportional to $\sqrt{s}$, and we then recover the expression for $\mathscr{Z}$ in (11-14),

$$\mathscr{Z} = \mathscr{R}_H\sqrt{\frac{2s}{\omega_H}}$$

It should now be reasonably clear that the equations of telegraphy given in (11-15) are valid for arbitrary transmission lines with no dielectric losses.

As an interesting application of what we have seen, we will calculate the step response of a lossy line of length l that is matched at the receiving end and is driven at the sending end by a voltage source with zero internal impedance (Wigington and Nahman, 1957). We will do this for a time scale less than a few microseconds, short enough that the skin effect is important and that the longitudinal impedance $\mathcal{Z}$ and the propagation function γ are therefore as given in (11-14) and (11-19). If we define T_0 by

$$T_0 = \frac{\mathcal{R}_H^2 l^2}{2 R_C^2 \omega_H}$$

we can write

$$\gamma l \simeq sl\sqrt{\mathcal{L}\mathcal{C}} + \sqrt{T_0 s}$$

According to (11-16), for a forward wave we have

$$\frac{V(l)}{V(0)} = \exp\left(-sl\sqrt{\mathcal{L}\mathcal{C}}\right)\exp\left(-\sqrt{T_0 s}\right)$$

Except for a delay by $l\sqrt{\mathcal{L}\mathcal{C}}$, the transfer function from $V(0)$ to $V(l)$ is thus

$$\exp\left(-\sqrt{T_0 s}\right)$$

We will accept that the corresponding step response, shown in Fig. 11-16, is

$$r(l,t) = 1 - \mathrm{erf}\left(\frac{1}{2}\sqrt{\frac{T_0}{t}}\right)$$

where t is the time *after* the delay, and

$$\mathrm{erf}(\eta) = \frac{2}{\sqrt{\pi}}\int_0^{\eta}\exp\left(-\zeta^2\right)d\zeta$$

Since $\mathrm{erf}(0.48) = 0.5$, we can identify T_0 roughly as the time required for $r(t)$

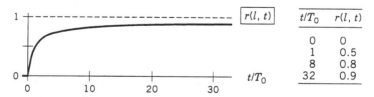

t/T_0	$r(l,t)$
0	0
1	0.5
8	0.8
32	0.9

Figure 11-16. Step response of a lossy line.

to reach 50% of its final value. For $\eta \ll 1$ we have

$$\text{erf}(\eta \ll 1) \simeq \frac{2}{\sqrt{\pi}} \eta$$

and within a few percent for $t \ge 2T_0$,

$$r(l,t) \simeq 1 - \sqrt{\frac{T_0}{\pi t}}$$

The main point to observe is that settling times can be very long and that the time required to reach a given value goes like the *square* of the length of the line. For reference, we note that the time required for $r(t)$ to reach 90% of its final value is $32T_0$. These results are not altered much if the line is terminated in a resistance of value $R_C = \sqrt{\mathcal{L}/\mathcal{C}}$.

The attenuation in decibels for a length l at a frequency ω_H is

$$A_H l = 20 \log[\exp(1)] \sqrt{\frac{\omega_H T_0}{2}} = 8.69 \sqrt{\frac{\omega_H T_0}{2}}$$

where A_H is the attenuation per unit length at ω_H. We can then write

$$T_0 = \frac{2}{\omega_H} \left(\frac{A_H l}{8.69} \right)^2$$

In particular, for $\omega_H/2\pi = 1$ GHz we have

$$T_0 = 4.2(A_H l)^2 \text{ ps}$$

Let us consider the case of RG-58 coaxial cable, which has an attenuation of 0.7 dB/m at 1 GHz. For a 1 meter length we have $T_0 = 2$ ps, so that the step response reaches 90% of its final value in 64 ps; for most purposes, the line can be considered ideal. In contrast, for a 50 meter length we have $T_0 = 5$ ns, and the effects of dispersion are plainly visible: the time required for the step response to reach 90% of its final value is 160 ns.

If the attenuation at ω_H has a contribution from dielectric losses, which grow *linearly* with the frequency, the step response is initially slower than it is if losses are due only to the conductors, but it eventually becomes somewhat faster; the general trend, however, remains as we have described.

11-5   FINITE LINES

The interesting aspects of transmission lines in electronic circuits are most often determined not by losses but by the terminations at the source (sending end) and at the load (receiving end). Although we will assume in the

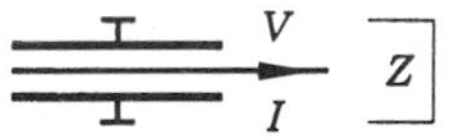

Figure 11-17. Line terminated in an arbitrary impedance Z. The line voltage V and the line current I must satisfy $V = IZ$.

remainder of this chapter that lines are short enough that they can be considered ideal, we will take losses into account for the sake of completeness.

Let us first consider a line terminated in an impedance Z, as shown in Fig. 11-17. We will define the (voltage) *reflection coefficient* at Z as the ratio of the amplitudes of the *reflected* (backward) and *incident* (forward) voltage waves at Z,

$$\rho \equiv \frac{V_-}{V_+}$$

Using the boundary condition $V = ZI$ in (11-18), we obtain

$$\rho = \frac{Z - Z_C}{Z + Z_C}$$

We thus have $\rho = 0$ if the line is matched, $\rho = -1$ if it is short circuited, and $\rho = +1$ if it is open circuited. In terms of the amplitude of the incident voltage wave, the voltage *transmitted* to Z is

$$V = V_+ (1 + \rho)$$

Let us now consider the section of line of length l shown in Fig. 11-18. From (11-16) we have $V_{2\pm} = \exp(\mp \gamma l) V_{1\pm}$, so that

$$\frac{V_{1-}}{V_{1+}} = \rho_2 \exp(-2\gamma l) \tag{11-21}$$

where ρ_2 is the reflection coefficient at $z = l$,

$$\rho_2 = \frac{Z_2 - Z_C}{Z_2 + Z_C}$$

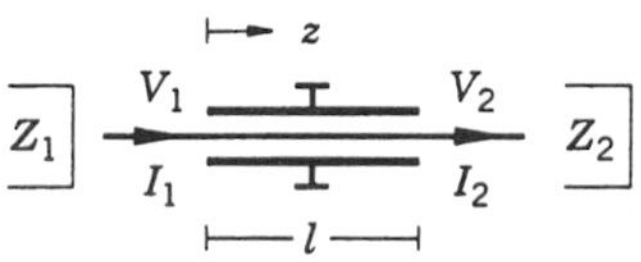

Figure 11-18. The line variables at one end of a section can be obtained in terms of the line parameters and the line variables at the other end, and ditto for the impedances looking in the same direction.

We will make use of this result presently. To continue, let us relate the line variables at $z = 0$ and $z = l$. From (11-18), at $z = 0$ we have

$$2V_{1+} = V_1 + Z_C I_1$$
$$2V_{1-} = V_1 - Z_C I_1$$

Using these relations to express $V_{2\pm} = \exp(\mp \gamma l)V_{1\pm}$ in terms of V_1 and I_1 and then using (11-17), at $z = l$ we get

$$2V_2 = (V_1 + Z_C I_1)\exp(-\gamma l) + (V_1 - Z_C I_1)\exp(+\gamma l)$$
$$2Z_C I_2 = (V_1 + Z_C I_1)\exp(-\gamma l) - (V_1 - Z_C I_1)\exp(+\gamma l)$$

and therefore

$$V_2 = V_1 \cosh \gamma l - Z_C I_1 \sinh \gamma l$$
$$Z_C I_2 = -V_1 \sinh \gamma l + Z_C I_1 \cosh \gamma l \qquad (11\text{-}22)$$

Inverting these equations, we obtain

$$V_1 = V_2 \cosh \gamma l + Z_C I_2 \sinh \gamma l$$
$$Z_C I_1 = V_2 \sinh \gamma l + Z_C I_2 \cosh \gamma l \qquad (11\text{-}23)$$

Using the boundary conditions $V_1 = Z_1 I_1$ and $V_2 = Z_2 I_2$, we can also relate the impedances (admittances) at $z = 0$ and $z = l$,

$$Z_1 Y_C = \frac{Z_2 Y_C + \tanh \gamma l}{1 + Z_2 Y_C \tanh \gamma l} \qquad Z_2 Y_C = \frac{Z_1 Y_C - \tanh \gamma l}{1 - Z_1 Y_C \tanh \gamma l}$$

$$Y_1 Z_C = \frac{Y_2 Z_C + \tanh \gamma l}{1 + Y_2 Z_C \tanh \gamma l} \qquad Y_2 Z_C = \frac{Y_1 Z_C - \tanh \gamma l}{1 - Y_1 Z_C \tanh \gamma l} \qquad (11\text{-}24)$$

Finally, let us consider, as shown in Fig. 11-19, a section of line that is driven at the sending end by a voltage source of impedance Z_S and terminated at the receiving end by a load impedance Z_L. From (11-16) and (11-17) we get

$$V(z) = V_{0+}\exp(-\gamma z) + V_{0-}\exp(\gamma z)$$
$$Z_C I(z) = V_{0+}\exp(-\gamma z) - V_{0-}\exp(\gamma z)$$

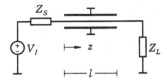

Figure 11-19. Section with arbitrary terminations at the sending and receiving ends.

At $z = 0$ we have

$$V(0) = V_{0+} + V_{0-}$$

$$Z_C I(0) = V_{0+} - V_{0-}$$

From (11-21) we also have

$$V_{0-} = V_{0+} \rho_L \exp(-2\gamma l)$$

where ρ_L is the reflection coefficient at the load,

$$\rho_L = \frac{Z_L - Z_C}{Z_L + Z_C}$$

Using the boundary condition at the source, we obtain

$$V_I = V(0) + Z_S I(0) = (V_{0+} + V_{0-}) + \frac{Z_S}{Z_C}(V_{0+} - V_{0-})$$

$$= V_{0+}\left[\left(1 + \frac{Z_S}{Z_C}\right) + \left(1 - \frac{Z_S}{Z_C}\right)\rho_L \exp(-2\gamma l)\right]$$

and thus

$$\frac{V_{0+}}{V_I} = \frac{Z_C}{Z_S + Z_C} \frac{1}{1 - \rho_S \rho_L \exp(-2\gamma l)}$$

where ρ_S is the reflection coefficient at the source,

$$\rho_S = \frac{Z_S - Z_C}{Z_S + Z_C}$$

Putting everything together, we obtain a key result,

$$\frac{V(z)}{V_I} = \frac{Z_C}{Z_S + Z_C} \frac{\exp(-\gamma z) + \rho_L \exp(-2\gamma l)\exp(+\gamma z)}{1 - \rho_S \rho_L \exp(-2\gamma l)} \tag{11-25}$$

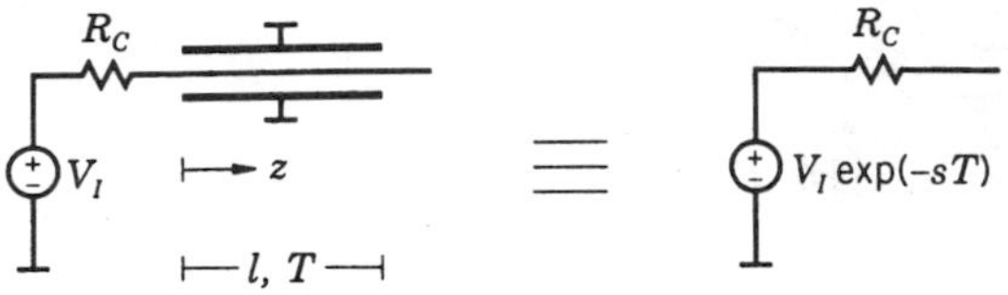

Figure 11-20. Lossless section driven by a matched source. At the receiving end it looks like the matched source delayed by the propagation time of the line.

In particular, and in terms of the propagation time $T = l\sqrt{\mathcal{L}\mathcal{C}}$, the voltages at $z = 0$ and $z = l$ in a lossless section are given by

$$\frac{V(0)}{V_I} = \frac{1}{2}(1-\rho_S)\frac{1+\rho_L \exp(-2sT)}{1-\rho_S \rho_L \exp(-2sT)}$$

$$\frac{V(l)}{V_I} = \frac{1}{2}(1-\rho_S)\frac{(1+\rho_L)\exp(-sT)}{1-\rho_S \rho_L \exp(-2sT)} \qquad (11\text{-}26)$$

Let us consider a simple example. If $Z_S = R_C$ we have $\rho_S = 0$, and thus

$$\frac{V(l)}{V_I} = \frac{1}{2}(1+\rho_L)\exp(-sT) = \frac{Z_L}{Z_L+R_C}\exp(-sT) \qquad (11\text{-}27)$$

We conclude, as indicated in Fig. 11-20, that a lossless section driven by a matched voltage source looks at the receiving end like the matched voltage source except for a delay by the propagation time T. The source impedance is R_C, which is precisely the impedance looking back toward the sending end when the source is nulled. This result is reminiscent of Thévenin's theorem, and rightly so. As shown in Exercise 11-5, Thévenin's theorem holds for circuits with transmission lines; we should expect this result because transmission lines can be considered as limits of ladder networks, for which Thévenin's theorem holds.

11-6 WIDEBAND TRANSIENTS IN LOSSLESS LINES

The steady-state response of a finite line to a sinusoidal input is described by a single forward wave and a single backward wave; the amplitudes of these waves, however, are the result of a (perhaps infinite) number of reflections at the source and load terminations. In this section we will consider step and ramped-step inputs, which generate responses in which we will clearly be able to discern an initial incident wave, a wave reflected at the load, a second incident wave generated by reflection at the source, and so on.

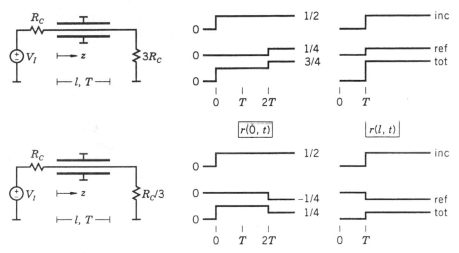

Figure 11-21. Waveforms for lossless lines driven by matched sources in the case of resistive loading.

Let us first consider real impedances $R_S = R_C$ and $R_L \neq R_C$, so that $\rho_S = 0$ and $\rho_L \neq 0$. We already encountered this situation in Section 11-5, and we know that the load sees the source delayed by the propagation time; the step response at the load is therefore

$$r(l,t) = \frac{R_L}{R_L + R_C} \theta(t - T) \qquad (11\text{-}28)$$

Let us see how this works out in detail. As long as $t < 2T$, no information can come back from the load, and the line looks as if it were infinite and thus as if it were a resistance of value R_C. The step response at the source for $t < 2T$ is therefore

$$r(0, t < 2T) = \frac{1}{2} \theta(t)$$

As shown in Fig. 11-21, this is the (first) incident wave that propagates down the line and arrives at the load a time T later. The incident wave generates a (first) reflected wave, given at the load by

$$\frac{1}{2} \rho_L \theta(t - T) = \left(\frac{1}{2}\right) \frac{R_L - R_C}{R_L + R_C} \theta(t - T)$$

The wave transmitted to the load is obtained by adding the incident wave (as if it were unperturbed by the load and continued forever) and the reflected wave,

$$r(l,t) = \frac{1}{2}(1 + \rho_L) \theta(t - T) = \left(\frac{1}{2}\right) \frac{2R_L}{R_L + R_C} \theta(t - T)$$

This is precisely what we obtained in (11-28). The reflected wave arrives back at the source a propagation time later and simply adds to whatever is there, so that

$$r(0,t) = \frac{1}{2}\left[\theta(t) + \rho_L\theta(t-2T)\right] = \frac{1}{2}\left[\theta(t) + \frac{R_L - R_C}{R_L + R_C}\theta(t-2T)\right]$$

The process we have described is automatically and fully contained in (11-26), which in this case becomes

$$\frac{V(0)}{V_I} = \frac{1}{2}\left[1 + \rho_L \exp(-2sT)\right]$$

$$\frac{V(l)}{V_I} = \frac{1}{2}(1 + \rho_L)\exp(-sT)$$

Merely by inspection, we recover the step responses obtained above.

An interesting practical example is offered by the case $R_L = \infty$ shown in Fig. 11-22; in this case $\rho_L = 1$, and the source voltage is reproduced at the load undistorted but with a delay T. This is called *series loading* in ECL circuits, and it has the advantage that the matching resistor dissipates no power when the source is steady. There is a disadvantage, however, in that the line voltage is only half the logic swing for times as long as $2T$, so that this scheme can not be used to drive (high-impedance) gates distributed along the line; when necessary, this can be done with *parallel loading*, which simply means driving a section matched at the receiving end from a source with zero impedance.

Turning things around, if an open section initially charged to a voltage V_0 is suddenly loaded at one end with a resistor of value R_C, the voltage on the resistor is a square pulse of height $V_0/2$ and width $2T$. With LC ladder networks rather than transmission lines, this scheme is used in high-power circuits to apply a voltage pulse of constant amplitude to a resistive load; a single capacitor would discharge exponentially and would therefore need to be much larger than the total capacitance of the ladder network.

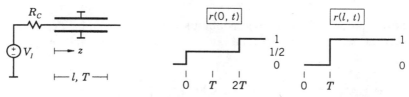

Figure 11-22. Series loading.

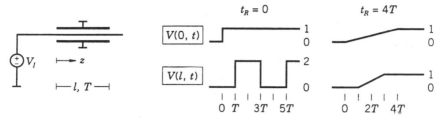

Figure 11-23. Open stub driven by a voltage source. Responses to a step and to a ramped step with slope $1/4T$ are shown.

If the source is not matched, the process we have described at the load is repeated with the second incident wave, equal to the reflection of the reflected wave, and so on; analogous considerations hold at the source. This case is also automatically contained in (11-26).

Let us consider the example of Fig. 11-23, also often seen in ECL circuits, in which a voltage source drives a circuit with infinite input impedance through an open *stub* (a stub is a short length of transmission line). In this case, $\rho_S = -1$ and $\rho_L = +1$, so that (11-26) becomes

$$\frac{V(l)}{V_I} = \frac{2\exp(-sT)}{1+\exp(-2sT)}$$

Thinking in terms of Laplace transforms, we can expand the denominator to obtain

$$\frac{V(l)}{V_I} = 2\exp(-sT)\left[1-\exp(-2sT)+\exp(-4sT)-\exp(-6sT)+\cdots\right]$$

As shown in Fig. 11-23, the step response is a square wave of amplitude twice the amplitude of the step and of period $4T$. This is of course intolerable. ECL signals, however, have a finite rise time t_R of order 1 ns; we will simulate such signals by means of ramped steps with (ramp) slope $1/t_R$. If we choose $T = t_R/4$, we have

$$V_I = V_0 \frac{t\theta(t)-(t-4T)\theta(t-4T)}{4T}$$

Taking Laplace transforms we obtain

$$\frac{V_I}{V_0} = \frac{1-\exp(-4sT)}{4Ts^2}$$

and thus

$$\frac{V(l)}{V_0} = \exp(-sT)\frac{1-\exp(-2sT)}{2Ts^2}$$

Except for a delay by the propagation time T, this is the Laplace transform of a ramped step with amplitude 1 and slope $1/2T$; we conclude that there is no overshoot if $T = t_R/4$ (or less) and, incidentally, that the rise time at the load is half the rise time at the source. For $t_R = 3$ ns, which is typical in the ECL 10K family, we require $T < 0.75$ ns; using $\sqrt{\mathcal{LC}} = 6$ ns/m, we see that runs under 12 cm need not be terminated at all.

In our next example, let us see what happens when, as shown in Fig. 11-24, a line of characteristic resistance R_C is connected to a matched load R_C through a short length of line with a different characteristic resistance R'_C. This is the case, for example, when the shield and center conductor of a coaxial cable are converted into a parallel-wire or wire-above-ground line in order to solder the cable to a printed circuit or to a connector. Setting $\rho_S = \rho_L = \rho$, and using (11-26) once more, we get

$$\frac{V(l)}{V_I} = \frac{1}{2}(1-\rho^2)\frac{\exp(-sT)}{1-\rho^2\exp(-2sT)} \qquad (11\text{-}29)$$

or with a bit of rewriting,

$$\frac{V(l)}{V_I} = \left(\frac{1}{2}\right)\frac{\exp(-sT)}{1+\dfrac{\rho^2}{1-\rho^2}[1-\exp(-2sT)]}$$

Expanding $\exp(-2sT)$ to first order in the denominator, we obtain

$$\frac{V(l)}{V_I} \simeq \left(\frac{1}{2}\right)\frac{\exp(-sT)}{1+\dfrac{2\rho^2}{1-\rho^2}sT}$$

If the rise time of the input signal is slow enough, the response at the load will look as if it had been run through a phase lag with time constant

$$\tau = \frac{2\rho^2}{1-\rho^2}T$$

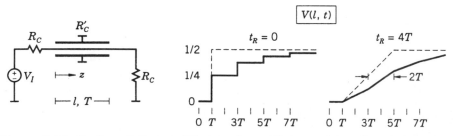

Figure 11-24. Section of characteristic impedance R'_C inserted in a line of characteristic impedance R_C matched at both ends. Responses to a step and to a ramped step with slope $1/4T$ are shown for the case $\rho^2 = \frac{1}{2}$.

A wire of diameter 1 mm, located 1 cm above a ground plane, for example, has $R'_C \simeq 220\ \Omega$; for $R_C = 50\ \Omega$, we have $\rho^2 \simeq 0.4$ and $\tau \simeq 1.3T$. Taking $l = 2$ cm and assuming an air dielectric, we have $T = 67$ ps and $\tau \simeq 90$ ps; a signal with $t_R = 1$ ns will thus barely be affected. Returning to the exact solution in (11-29), we have

$$\frac{V(l)}{V_I} = \frac{1}{2}(1-\rho^2)\big[\exp(-sT) + \rho^2 \exp(-3sT) + \rho^4 \exp(-5sT) + \cdots\big]$$

The step response for $\rho^2 = \frac{1}{2}$ $(\tau = 2T)$ is shown in Fig. 11.24. Also shown is the response to a ramped step with slope $1/4T$; despite the rather short risetime, it agrees well with the approximation made above.

Finally, let us consider an example with a reactive element. In the circuit shown in Fig. 11-25, matching at the load is spoiled at high frequencies by capacitor C. The reflection coefficient at the load is

$$\rho_L = \frac{G_C - Y_L}{G_C + Y_L} = \frac{G_C - (G_C + sC)}{G_C + (G_C + sC)} = \frac{-sCR_C/2}{1 + sCR_C/2}$$

From (11-26) we then obtain

$$\frac{V(0)}{V_I} = \frac{1}{2}\big[1 + \rho_L \exp(-2sT)\big] = \frac{1}{2} - \left(\frac{1}{2}\right)\frac{sCR_C/2}{1 + sCR_C/2}\exp(-2sT)$$

$$\frac{V(l)}{V_I} = \frac{1}{2}(1+\rho_L)\exp(-sT) = \left(\frac{1}{2}\right)\frac{1}{1 + sCR_C/2}\exp(-sT) \quad (11\text{-}30)$$

The step responses are shown in Fig. 11.25. The response at the load, which we could have obtained directly from (11-27), is acceptable except perhaps for the fact that it is distorted by a lag with time constant $CR_C/2$. In contrast, the response at the source exhibits a negative spike with time constant $CR_C/2$, which might represent a problem in very fast systems; the spike will wash out, however, if the rise time of the excitation is long enough.

Capacitor C might represent the input capacitance of an oscilloscope used to observe signals somewhere along a line. For a typical value $C = 15$ pF and

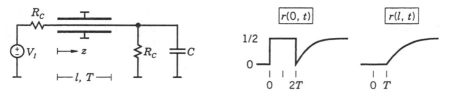

Figure 11-25. Capacitive loading of an otherwise matched line.

a $50\,\Omega$ line, we have $CR_C/2 = 375$ ps, so that there is little distortion for signals with rise times longer than 1 ns. As shown in Exercise 11-6, the response at the source has a *positive* spike with time constant $L/2R_C$ if the load in Fig. 11-25 is an inductor L in series with R_C. The waveform at the input of a line driven by a matched source thus gives us information about the nature and location of imperfections down the line. This is the basis of *time-domain reflectometry* (TDR); with refined techniques, minute imperfections in the line can be detected.

11-7 LOSSLESS NARROWBAND CIRCUITS

By means of examples, in this section we will explore the domain that opens up when signals are almost sinusoidal at a center frequency f, and we will see how sections of particular lengths have easily visualized functions. We will also discuss elementary matching techniques. Mismatches can of course be characterized by reflection coefficients; data sheets and the literature, however, generally use a more compact measure, which we will first define.

The relative phase of the amplitudes of the incident and reflected waves at a load is locked in by the load. Moving away from the load, these amplitudes are successively in and out of phase, so that the magnitude of the line voltage exhibits maxima and minima; a complete cycle takes place in a length $\lambda/2$, where $\lambda = 1/f\sqrt{\mathcal{LC}}$ is the wavelength at f. These results can be obtained from (11-25), which tells us that for a lossless line,

$$|1 + \rho_L| \frac{|V(z)|}{|V(l)|} = \left| 1 + \rho_L \exp\left(-4\pi i \frac{l-z}{\lambda} \right) \right|$$

The maximum and minimum values of $|V(z)|$ are proportional to $1 \pm |\rho_L|$; we will define the *voltage standing-wave ratio* (VSWR) as the ratio of these extrema,

$$\text{VSWR} = \frac{1 + |\rho_L|}{1 - |\rho_L|}$$

Roughly speaking, a VSWR of 1.1 separates good and bad behavior in narrowband circuits. The matching techniques we will consider ideally result in a VSWR of 1.0 at a design frequency f; operation at other frequencies is possible, but good behavior generally requires deviations of only a few percent from f. We will consider only one example of a circuit that tolerates larger deviations because improvements, despite their considerable ingenuity, generally involve conceptually straightforward extensions of what we will cover.

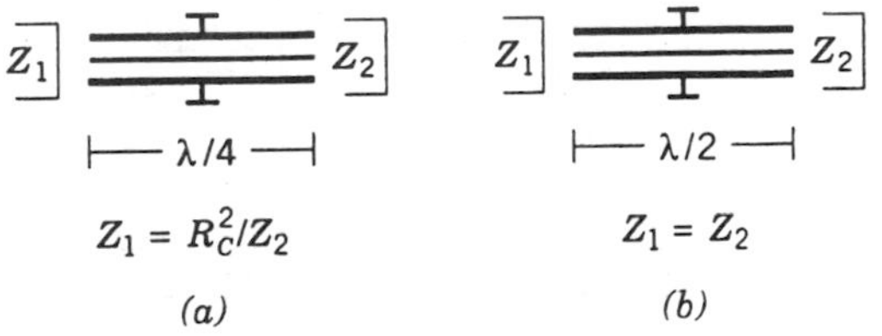

$$Z_1 = R_C^2/Z_2 \qquad\qquad Z_1 = Z_2$$

(a) \hspace{4cm} (b)

Figure 11-26. (a) Quarter-wave and (b) half-wave sections.

Let us start by considering some particular cases of the transformations we obtained in Section 11-5. For lossless lines, (11-24) becomes

$$Z_1 G_C = \frac{Z_2 G_C + i \tan(2\pi l/\lambda)}{1 + i Z_2 G_C \tan(2\pi l/\lambda)}$$

As shown in Fig. 11-26, for a *quarter-wave* section with $l = \lambda/4$ we have

$$Z_1 = \frac{R_C^2}{Z_2} \tag{11-31}$$

whereas for a *half-wave* section with $l = \lambda/2$ we have

$$Z_1 = Z_2$$

A $\lambda/4$ or quarter-wave *transformer* thus effectively converts an impedance into its reciprocal, whereas a $\lambda/2$ or half-wave transformer does nothing. As an example, Fig. 11-27 shows how a $\lambda/4$ transformer with characteristic resistance $R_C = \sqrt{R_{CL} R_{CH}}$ is used to match a line of characteristic resistance R_{CL} to a line of characteristic resistance R_{CH}.

A shorted $\lambda/4$ stub is an open circuit, whereas an open $\lambda/4$ stub is a short circuit; for future use, we also note that the input impedance of a shorted (open) $\lambda/8$ stub is $+iR_C$ ($-iR_C$). These properties are summarized in Fig. 11-28.

As indicated in Fig. 11-29, the input impedance of a shorted $\lambda/4$ stub is not really infinite at f because of the presence of losses, but it is nonetheless

Figure 11-27. A quarter-wave transformer of characteristic resistance $\sqrt{R_{CL} R_{CH}}$ matches lines of characteristic resistances R_{CL} and R_{CH} to each other.

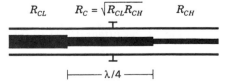

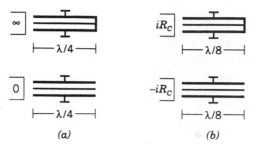

Figure 11-28. Input impedances (*a*) of shorted and open $\lambda/4$ stubs and (*b*) of shorted and open $\lambda/8$ stubs.

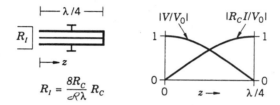

Figure 11-29. Current and voltage in a low-loss shorted $\lambda/4$ stub.

generally hundreds of times larger in magnitude than R_C. To see this point, we will look at what goes on inside the stub. According to (11-22), the amplitudes of the line variables $V(z)$ and $I(z)$ when losses are negligible are

$$V(z) = V_0 \cos \frac{2\pi z}{\lambda}$$

$$R_C I(z) = -iV_0 \sin \frac{2\pi z}{\lambda}$$

Despite the infinite input impedance at $z = 0$, there is an oscillating current distribution in the stub, and there will therefore be losses if the conductors have finite resistivity. Neglecting the dissipation in the short itself, the (average) power P dissipated in the stub is

$$P = \frac{\mathcal{R}}{2} \int_0^{\lambda/4} II^* \, dz = \frac{\mathcal{R}}{2R_C^2} V_0 V_0^* \int_0^{\lambda/4} \sin^2 \frac{2\pi z}{\lambda} \, dz$$

The value of the integral is $\lambda/8$, so we have

$$P = \frac{V_0 V_0^* \, \mathcal{R}\lambda}{16 R_C^2}$$

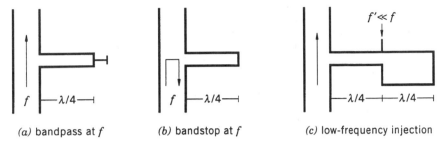

(a) bandpass at f (b) bandstop at f (c) low-frequency injection

Figure 11-30. Open and shorted $\lambda/4$ stubs as filters for transmission lines.

Accepting that the input impedance at f is a resistance R_I because we are at resonance (this point is clarified in Exercise 11-11), we also have

$$P = \frac{V_0 V_0^*}{2R_I}$$

The input resistance at f is therefore

$$R_I = \frac{8R_C}{\mathcal{R}\lambda}R_C \tag{11-32}$$

Taking a conservative value $\mathcal{R} = 5$ Ω/m at $f = 1$ GHz ($\lambda = 30$ cm), for $R_C = 50$ Ω we obtain $R_I \simeq 267R_C$.

The filters illustrated in Fig. 11-30 are widely used in microwave stripline circuits; as suggested earlier, they are often viewed more as lumped elements attached to a main transmission line than as transmission lines in their own right. The bandpass filter (Fig. 11-30a) removes low-frequency signals from the main line, but it is transparent at f. The bandstop filter (Fig. 11-30b) short-circuits the main line at f and thus removes signals at this frequency while leaving low-frequency signals essentially untouched. In the injection filter (Fig. 11-30c), the wide $\lambda/4$ open stub provides the short circuit at f for the narrow $\lambda/4$ bandpass stub and makes a hard short to ground unnecessary; low-frequency signals or bias voltages for semiconductor devices can thus be injected into the main line.

Loads such as antennas require *balanced* inputs, that is, two ground-referenced inputs of equal magnitude that are 180° out of phase. *Unbalanced-to-balanced* transformers, or *baluns* for short, are thus required if the signal has been generated on a single ground-referenced line. In the case of lines on printed circuits, a simple balun is obtained by feeding two lines of characteristic resistance $2R_C$ from a line of characteristic resistance R_C and making one of them longer by $\lambda/2$.

Coaxial lines normally have their shields grounded at some point and cannot drive balanced lines directly in this condition. Figure 11-31 shows a

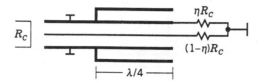

Figure 11-31. Coaxial unbalanced-to-balanced or inverting transformer.

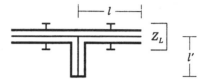

Figure 11-32. Single-stub matching.

structure that can operate as a balun for coaxial lines. A coaxial sleeve is fitted over the shield and shorted to the shield a distance $\lambda/4$ away from the end of the line. The sleeve and the shield form a shorted $\lambda/4$ stub, so that the impedance between the shield and the (grounded) sleeve is infinite. If resistances ηR_C and $(1-\eta)R_C$ are connected from the shield and the center conductor to ground, the line will be matched. Choosing $\eta = \frac{1}{2}$ results in balanced signals that can drive lines with characteristic resistance $R_C/2$. An *inverting* transformer is obtained by choosing $\eta = 1$.

When considerations of losses or of the maximum allowed voltage in a line are important, the preferred procedure is to obtain a match right at the source or at the load. This eliminates standing waves, which increase the line voltage, and therefore the losses, but transmit no power. One method that always works, illustrated in Fig. 11-32, is *single-stub matching*, in which a shorted stub is placed at a distance l from the load. In Exercise 11-12 it is shown that it is always possible to choose l so that the real part of the admittance Y due to the lead at the position of the stub is equal to G_C; for some value of B, we then have

$$Y = G_C + iB$$

The line will be matched if the admittance of the stub is made equal to $-iB$ or, according to (11-24), if its length l' satisfies

$$R_C B = \cot\left(\omega l'\sqrt{\mathcal{L}\mathcal{C}}\right)$$

Figure 11-33 shows how shorted $\lambda/4$ stubs can provide mechanical support for the center conductor of an air coaxial line. Like all the circuits we have seen in this section, this simple support has a large VSWR at frequencies a few percent away from the design frequency.

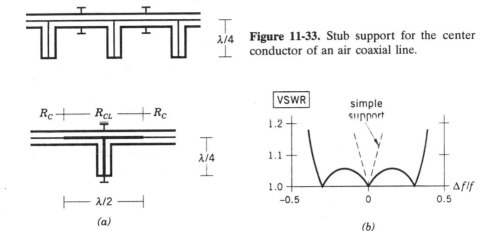

Figure 11-33. Stub support for the center conductor of an air coaxial line.

Figure 11-34. (a) Improved stub support with (b) a low VSWR over a large relative bandwidth.

A low VSWR can be obtained over a broader band using the circuit shown in Fig. 11-34a (Davidson, 1989). The $\lambda/2$ section of lower characteristic impedance R_{CL} inserted in series with line R_C does nothing, and adding a shorted $\lambda/4$ stub of characteristic impedance R_{CL} still keeps line R_C matched at f. In view of this extra degree of freedom, let us inquire whether we can find a match at another frequency $f' = kf$. If so, the input admittance G_C transformed to the position of the stub must equal the sum of the admittance of the stub plus the load admittance G_C transformed to the position of the stub. Using (11-24) with $\gamma' l = (2\pi i k f/c)(\lambda/4) = i k \pi/2$, we obtain

$$\frac{R_{CL}G_C - i\tan(k\pi/2)}{1 - iR_{CL}G_C\tan(k\pi/2)} - \frac{R_{CL}G_C + i\tan(k\pi/2)}{1 + iR_{CL}G_C\tan(k\pi/2)} = \frac{1}{i\tan(k\pi/2)}$$

When the subtraction on the left-hand side is carried out, only the imaginary terms survive, and we get

$$2(1 - R_{CL}^2 G_C^2)\tan^2(k\pi/2) = 1 + R_{CL}^2 G_C^2 \tan^2(k\pi/2)$$

and finally,

$$R_{CL}^2 = \frac{2\tan^2(k\pi/2) - 1}{3\tan^2(k\pi/2)} R_C^2$$

For $0.4 < k < 1.6$, R_{CL} is real, and the line is matched at $f' = (2-k)f$ as well as at $f' = kf$. For $k = 0.7$, for example, we obtain $R_{CL} = 0.76R_C$, which is

Figure 11-35. Parallel-plate transmission line terminated in a slab of resistive material.

reasonable; the resulting VSWR is shown in Fig. 11-34b. From 0.65f to 1.35f, that is, over a 70% relative bandwidth, the VSWR is under 1.06; in contrast, the relative bandwidth for the simple stub support of Fig. 11-33 is only 8%.

To end this section, we will return to the question of perfect matching first discussed in Section 11-2. Let us reconsider the circuit of Fig. 11-24. As shown in Exercise 11-7, the input impedance at $z = 0$ for $\omega T \ll 1$ is R_C in series with a small inductor L if $R'_C > R_C$, and R_C in parallel with a small capacitor C if $R'_C < R_C$. This image is valid out to about 5 GHz for $l = 1$ cm ($T \approx 30$ ps), and in this case we have $L \approx 1$ nH and $C \approx 1$ pF. We argue that in this example we are simulating the effect of an imperfect match, and that equally small corrections obtain at junctions where boundary conditions are not satisfied by TEM waves if the geometry is kept tight. Without going into detailed field calculations, we will justify our argument by means of two further examples.

In the first example, shown in Fig. 11-35, an air parallel-plate line is terminated by inserting a slab of appropriate conductivity and dielectric constant κ between the plates. At low frequencies, boundary conditions are satisfied as they are by a termination in spacecloth: the electric field in the slab is constant and the magnetic field drops linearly to zero. There is, however, a stray capacitance $C = \kappa \varepsilon_0 wl/h$ (the charges on the external surfaces of the slab and the plates remain as they are in spacecloth), and we have the situation depicted in Fig. 11-25. In an air parallel-plate line we have $\mathcal{L} = \mu_0 h/w$ and $\mathcal{C} = \varepsilon_0 w/h$, so that $R_C = (h/w)\sqrt{\mu_0/\varepsilon_0}$; for $l = 1$ cm the time constant in (11-30) is $R_C C/2 = \kappa l/2c \approx 100$ ps, and thus usually negligible.

The second example is shown in Fig. 11-36. Boundary conditions for TEM waves are satisfied if a parallel-wire line is shorted by an infinite conducting plane; in terms of complex amplitudes and at low frequencies, the current I is independent of z and the voltage V is zero at $z = 0$ and grows linearly with z. If the conducting plane is replaced by a wire, the flux Φ and therefore the voltage V grow faster initially than they do at large z. This effect can be simulated by imagining that the line is slightly longer than it really is or, equivalently, by assigning a small inductance to the short and considering that the line is otherwise ideal.

It would seem fair to conclude, as suggested in Section 11-2, that terminating a line with an ordinary resistor results in an almost perfect match if leads are kept short.

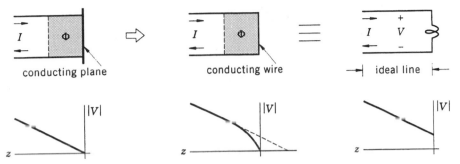

Figure 11-36. A parallel-wire line shorted by a wire rather than an infinite conducting plane looks at the sending end like an ideal line terminated in a small inductor.

11-8 PERIODICALLY LOADED LINES

Transmission lines are often used to distribute signals to uniformly spaced loads; in shared *bus* lines used to exchange information among a large number of interacting systems, *line drivers* and *line receivers* might be uniformly distributed along a line that is matched at both ends. In both cases the lines are *periodically* loaded with lumped impedances. In this section we will consider capacitive loading; inductive loading is the subject of Exercise 11-16.

Let us consider a line that is loaded with capacitances of value C at intervals of length l; one period of this line is shown in Fig. 11-37. Based on what we saw for ladder networks in Section 11-3, we might expect that the low-frequency attenuation is zero but that there is a cutoff frequency. This is in fact the case, although we will see that there is in addition an infinite number of alternating *stopbands* and *passbands* above the lowest cutoff frequency ω_0; below ω_0, however, the loaded line will look quite like an LC ladder network.

As we did for ladder networks, we will calculate the characteristic admittance Y_P for one period by assuming that the circuit in Fig. 11-37 is terminated in Y_P and that its input admittance is therefore also Y_P. The input admittance transformed to $z = 0$ is equal to sC plus the load admittance transformed to $z = 0$; using (11-24), we obtain

$$\frac{Y_P Z_C - \tanh(\gamma l/2)}{1 - Y_P Z_C \tanh(\gamma l/2)} - \frac{Y_P Z_C + \tanh(\gamma l/2)}{1 + Y_P Z_C \tanh(\gamma l/2)} = sCZ_C$$

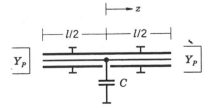

Figure 11-37. One period of a line loaded with capacitances of value C at intervals of length l.

When the subtraction on the left-hand side is carried out, only even powers of $Y_P Z_C$ survive, and we get

$$Y_P^2 Z_C^2 = \frac{1 + (sCZ_C/2)\coth(\gamma l/2)}{1 + (sCZ_C/2)\tanh(\gamma l/2)} \tag{11-33}$$

Transforming the line voltages at $z = -l/2$ and $z = l/2$ to $z = 0$ using (11-22) and (11-23), and recalling that the line currents at $z = -l/2$ and $z = l/2$ are proportional to the line voltages through Y_P, we obtain

$$\frac{V(0)}{V(-l/2)} = \cosh(\gamma l/2) - Y_P Z_C \sinh(\gamma l/2)$$

$$\frac{V(0)}{V(+l/2)} = \cosh(\gamma l/2) + Y_P Z_C \sinh(\gamma l/2)$$

The propagation function γ_P for one period is thus given by

$$\exp(-\gamma_P l) \equiv \frac{V(+l/2)}{V(-l/2)} = \frac{1 - Y_P Z_C \tanh(\gamma l/2)}{1 + Y_P Z_C \tanh(\gamma l/2)} \tag{11-34}$$

As shown in Exercise 11-15, we can also write

$$\cosh(\gamma_P l) = \cosh(\gamma l) + (sCZ_C/2)\sinh(\gamma l)$$

In the remainder of this section we will assume that the line is lossless; in terms of the propagation time $T = l\sqrt{LC}$ we then have $\gamma l = i\omega T$, and Y_P and γ_P are given by

$$Y_P^2 R_C^2 = \frac{1 + (\omega C R_C/2)\cot(\omega T/2)}{1 - (\omega C R_C/2)\tan(\omega T/2)} \tag{11-35}$$

and

$$\cosh(\gamma_P l) = \cos(\omega T) - (\omega C R_C/2)\sin(\omega T) \tag{11-36}$$

As shown in Fig. 11-38, $|\cosh(\gamma_P l)| = 1$ at frequencies given by $\omega T = n\pi$, and $|\cosh(\gamma_P l)| > 1$ in increasingly larger stopbands below these frequencies. In the stopbands γ_P has a real part which, whatever its sign, physically implies attenuation. The stopbands alternate with passbands in which $|\cosh(\gamma_P l)| < 1$, so that $\gamma_P = i\beta_P$ and $\cosh(\gamma_P l) = \cos(\beta_P l)$.

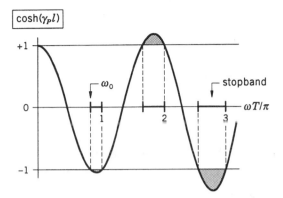

Figure 11-38. Frequency dependence of $\cosh(\gamma_P l)$ for the periodically loaded line of Fig. 11-37.

The lowest cutoff frequency ω_0 is given by the first zero in the denominator of the expression for the characteristic admittance in (11-35),

$$(\omega_0 C R_C/2)\tan(\omega_0 T/2) = 1$$

For $\omega \ll \omega_0$, we observe that the characteristic admittance is constant,

$$Y_P^2 R_C^2 \simeq 1 + C R_C/T$$

Substituting $R_C = \sqrt{\mathcal{L}/\mathcal{C}}$ and $T = l\sqrt{\mathcal{L}\mathcal{C}}$, we obtain

$$Y_P \simeq \sqrt{\frac{\mathcal{C}+C/l}{\mathcal{L}}} \tag{11-37}$$

The loaded line thus looks like a line with capacitance per unit length $\mathcal{C}+C/l$, that is, as if C were distributed over a length l.

Setting $\cosh(\gamma_P l) = \cos(\beta_P l)$ in (11-36) and expanding to second order, we also obtain

$$\beta_P l \simeq \omega T \sqrt{1 + C R_C/T}$$

We thus have the consistent result that the phase function at low frequencies corresponds to a line with capacitance per unit length $\mathcal{C}+C/l$,

$$\beta_P \simeq \omega \sqrt{\mathcal{L}(\mathcal{C}+C/l)} \tag{11-38}$$

The characteristic admittance and the phase function in the lowest passband are graphed in Fig. 11-39 for $R_C C = T$, a case common in ECL circuits. For frequencies less than $\omega_0/2$, we observe that Y_P is approximately constant

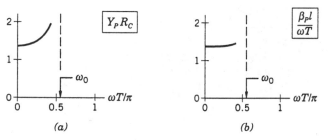

Figure 11-39. (*a*) Characteristic admittance Y_P and (*b*) phase function β_P in the lowest passband of the periodically loaded line of Fig. 11-37.

and that β_P is approximately proportional to ω; these are essentially the results we obtained for *LC* ladder networks in Section 11-3.

We observe that ω_0 is always less than π/T, and that the minutest periodic loading will result in stopbands starting at this frequency; manufacturing defects with a long period, for example, can result in large attenuation at relatively low frequencies. This observation also applies if we decide to reduce dielectric losses in a coaxial line by supporting the center conductor with equally spaced flat rings rather than a continuous dielectric: the spacing between the rings should be appropriately small in order to keep ω_0 above the highest frequency expected on the line.

11-9   COUPLED LINES

Significant coupling can occur between lines—striplines or twisted pairs, for example—that run parallel and are not fully shielded from one other. On the negative side, coupling between lines can be a nuisance, and we then call it *crosstalk*; in a multiconductor cable, for example, we almost invariably have to deal with crosstalk. On the positive side, however, we can profit from the coupling between two lines to obtain in one of them a measure of the forward wave in the other; in this case we will speak of a *directional coupler*. As shown in Fig. 11-40, we will only consider coupled lines that have a plane of symmetry, so that we can find solutions by considering even-mode and odd-mode excitations. Unless otherwise stated, we will also consider that the lines are embedded in a homogeneous dielectric.

Like the isolated lines of Section 11-2, coupled lines can be described in terms of forward and backward TEM waves, and the static transverse fields can be derived from a transverse potential ψ. The significant difference is that except for a scale factor, ψ is unique in isolated lines, whereas it is a linear combination of two independent solutions in coupled lines. The static field configurations and the corresponding characteristic resistances therefore depend on how the lines are excited.

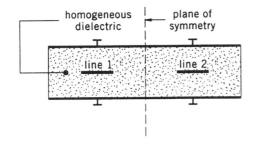

Figure 11-40. Coupled lines with a plane of symmetry.

As shown in Exercise 11-18, the fluxes and charges per unit length in an arbitrary configuration are

$$\mathscr{F}_1 = \mathcal{L}I_1 + \mathcal{L}_M I_2 \qquad\qquad \mathscr{F}_2 = \mathcal{L}I_2 + \mathcal{L}_M I_1$$

$$\mathscr{Q}_1 = \mathcal{C}V_1 - \mathcal{C}_M V_2 \qquad\qquad \mathscr{Q}_2 = \mathcal{C}V_2 - \mathcal{C}_M V_1$$

$$(11\text{-}39)$$

where $\mathcal{L}$ and $\mathcal{C}$ are the *self* inductance and capacitance per unit length, and $\mathcal{L}_M$ and $\mathcal{C}_M$ are the *mutual* inductance and capacitance per unit length. We note that $\mathcal{L}$ and $\mathcal{L}_M$ are obtained in a configuration in which the *current* in one line is zero, whereas $\mathcal{C}$ and $\mathcal{C}_M$ are obtained in a configuration in which the *voltage* in one line is zero. We also note that one line is affected at a given time and at a given position only by what is happening in the other line at the same transverse plane.

We will accept that coupled lines in inhomogeneous media can be described in terms of an effective dielectric constant, and that equations (11-39) hold for such lines.

Let us assume initially that the coupling is *weak*, meaning that $\mathcal{L}_M/\mathcal{L}$ and $\mathcal{C}_M/\mathcal{C}$ are much less than 1; we can then consider that when line 1 is excited it is not affected by line 2, so that we know V_1 and I_1, and we can see what happens in line 2 without solving the coupled equations implicit in (11-39). The magnetic and electric fields are shown schematically in Fig. 11-41 for the case in which line 2 is matched at its sending and receiving ends and line 1

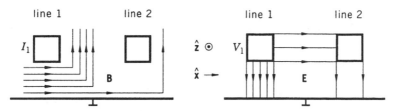

Figure 11-41. Field configurations in weakly coupled lines when line 1 carries a forward wave and line 2 is matched at its sending and receiving ends.

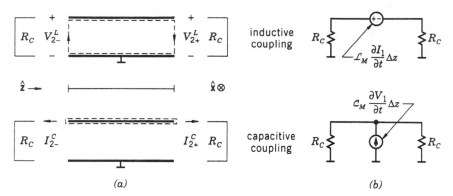

Figure 11-42. Coupled lines. (a) Side views of line 2 in Fig. 11-41 (shown as a microstrip line), indicating the variables used to calculate the densities of induced waves. (b) Circuital representation in terms of a series voltage source and a parallel current source.

carries a forward wave. With these configurations in mind, let us consider the infinitesimal length of line 2 shown, also schematically, in Fig. 11-42a. Ignoring the contribution from line 2 itself because it is of second order in Δz, the flux in line 2 in a length Δz is

$$\Phi_2 = \mathcal{L}_M I_1 \Delta z$$

According to Faraday's law, the voltages at the forward and backward ends of this length must satisfy

$$V_{2+}^L - V_{2-}^L = -\frac{d\Phi_2}{dt} = -\mathcal{L}_M \frac{\partial I_1}{\partial t} \Delta z$$

By symmetry we have $V_{2-}^L = -V_{2+}^l$; the densities of forward and backward induced voltage waves due to inductive coupling are therefore

$$\mathcal{V}_{2\pm}^L = \mp \frac{\mathcal{L}_M}{2} \frac{\partial I_1}{\partial t} = \mp \frac{\mathcal{L}_M}{2R_C} \frac{\partial V_1}{\partial t}$$

On the other hand, and again ignoring the contribution from line 2 itself, the charge on line 2 in a length Δz is

$$Q_2 = -\mathcal{C}_M V_1 \Delta z$$

By continuity, the currents out of this length must satisfy

$$I_{2-}^C + I_{2+}^C = -\frac{dQ_2}{dt} = \mathcal{C}_M \frac{\partial V_1}{\partial t} \Delta z$$

By symmetry, we have $I_{2-}^C = I_{2+}^C$. Each of these currents sees a resistance R_C due to the rest of the line; the densities of forward and backward induced voltage waves due to capacitive coupling are therefore

$$\mathcal{V}_{2\pm}^C = \frac{\mathcal{C}_M R_C}{2} \frac{\partial V_1}{\partial t}$$

Adding the inductive and capacitive contributions, we obtain the total densities of forward and backward induced voltage waves,

$$\mathcal{V}_{2\pm} = \frac{1}{2}\sqrt{\mathcal{L}\mathcal{C}} \left(\mp \frac{\mathcal{L}_M}{\mathcal{L}} + \frac{\mathcal{C}_M}{\mathcal{C}} \right) \frac{\partial V_1}{\partial t}$$

As we will see below, in homogeneous media the inductive coupling and the capacitive coupling are of equal strength; that is, we have $\mathcal{L}_M/\mathcal{L} = \mathcal{C}_M/\mathcal{C}$, and there is no forward induced wave.

An alternative (and more circuital) description of coupled lines, often seen in the literature, is illustrated in Fig. 11-42b. In a length Δz, according to this description, a signal in line 1 gives rise to a voltage source of value

$$\mathcal{L}_M \frac{\partial I_1}{\partial t} \Delta z$$

in series with line 2 and a current source of value

$$\mathcal{C}_M \frac{\partial V_1}{\partial t} \Delta z$$

in parallel with line 2.

Let us now assume, as shown in Fig. 11-43, that the lines are coupled only over a length l with propagation time $T = l\sqrt{\mathcal{L}\mathcal{C}}$. The voltage on line 2 at $z = 0$, the backward-coupled voltage $V_2(0, t)$, is due to the backward induced wave. At any time t, the contribution from position z along the line is generated at time $t - z\sqrt{\mathcal{L}\mathcal{C}}$ by a signal in line 1 that passes through $z = 0$ at

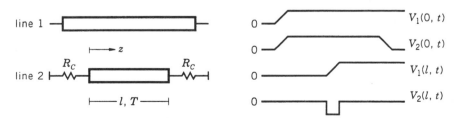

Figure 11-43. Waveforms in lines weakly coupled over a finite length. The amplitudes of $V_2(0, t)$ and $V_2(l, t)$ are not to scale.

time $t - 2z\sqrt{\mathcal{L}\mathcal{C}}$. We thus have

$$V_2(0,t) = \frac{1}{2}\sqrt{\mathcal{L}\mathcal{C}}\left(\frac{\mathcal{L}_M}{\mathcal{L}} + \frac{\mathcal{C}_M}{\mathcal{C}}\right)\int_0^l \frac{\partial V_1(0, t - 2z\sqrt{\mathcal{L}\mathcal{C}})}{\partial t}\, dz$$

Performing the integral and expressing l in terms of T, we obtain

$$V_2(0,t) = \frac{1}{4}\left(\frac{\mathcal{L}_M}{\mathcal{L}} + \frac{\mathcal{C}_M}{\mathcal{C}}\right)[V_1(0,t) - V_1(0, t - 2T)] \qquad (11\text{-}40)$$

For sine waves the backward coupling is maximum for $l = \lambda/4$, and we then have

$$V_2(0) = \frac{1}{2}\left(\frac{\mathcal{L}_M}{\mathcal{L}} + \frac{\mathcal{C}_M}{\mathcal{C}}\right)V_1(0) \qquad (11\text{-}41)$$

The voltage on line 2 at $z = l$, the forward-coupled voltage $V_2(l,t)$, is due to the forward induced wave. At any time t, the contribution from position z is generated at time $t - (l - z)\sqrt{\mathcal{L}\mathcal{C}}$ by a signal in line 1 that passes through z at the same time and arrives at $z = l$ at time t. The contributions from different positions are thus all equal and simply add, and the forward-coupled voltage, which is almost invariably considered to be crosstalk, is proportional to T and to the time derivative of $V_1(l, t)$:

$$V_2(l,t) = -\frac{T}{2}\left(\frac{\mathcal{L}_M}{\mathcal{L}} - \frac{\mathcal{C}_M}{\mathcal{C}}\right)\frac{\partial V_1(l,t)}{\partial t}$$

Figure 11-43 shows waveforms for the case in which the inductive coupling is stronger than the capacitive coupling, as it is in the microstrip lines, and a ramped step is applied to line 1.

As a first exercise in the use of even and odd modes to deal with coupled lines, we will now show that $\mathcal{L}_M/\mathcal{L} = \mathcal{C}_M/\mathcal{C}$ in homogeneous media irrespective of the coupling level. To do so, let us consider the lines shown in Fig. 11-44. In an even-mode excitation we have $V_1 = V_2 = V_E$ and $I_1 = I_2 = I_E$, whereas in an odd-mode excitation we have $V_1 = -V_2 = V_O$ and $I_1 = -I_2 = I_O$. Substituting these conditions in (11-39) and arguing as we did in Section 11-2, we obtain straightforward extensions of the equations of telegraphy given in (11-6),

$$\frac{dV_E}{dz} = -s(\mathcal{L} + \mathcal{L}_M)I_E \qquad\qquad \frac{dV_O}{dz} = -s(\mathcal{L} - \mathcal{L}_M)I_O$$

$$\frac{dI_E}{dz} = -s(\mathcal{C} - \mathcal{C}_M)V_E \qquad\qquad \frac{dI_O}{dz} = -s(\mathcal{C} + \mathcal{C}_M)V_O$$

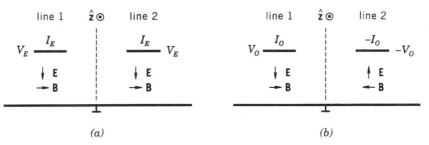

Figure 11-44. (*a*) Even-mode and (*b*) odd-mode excitations in coupled lines; in a homogeneous medium the propagation speed is mode-independent.

Since the propagation speeds in the even and odd modes are equal, we must have

$$(\mathcal{L}+\mathcal{L}_M)(\mathcal{C}-\mathcal{C}_M) = (\mathcal{L}-\mathcal{L}_M)(\mathcal{C}+\mathcal{C}_M)$$

We thus get, as promised above and shown by different means in Exercise 11-18,

$$\frac{\mathcal{L}_M}{\mathcal{L}} = \frac{\mathcal{C}_M}{\mathcal{C}} = k \tag{11-42}$$

where k is the *coupling coefficient* in homogeneous media. The backward-coupled voltage for sine waves in $\lambda/4$ couplers, given in (11-41), can then be expressed as

$$V_2(0) = kV_1(0)$$

Parallel-line directional couplers in homogeneous media are considered in much more detail and for arbitrary coupling in Exercises 11-20 and 11-22. Results there are obtained in terms of the even-mode and odd-mode characteristic resistances R_{CE} and R_{CO} given by

$$R_{CE} = \sqrt{\frac{\mathcal{L}+\mathcal{L}_M}{\mathcal{C}-\mathcal{C}_M}} \qquad R_{CO} = \sqrt{\frac{\mathcal{L}-\mathcal{L}_M}{\mathcal{C}+\mathcal{C}_M}}$$

In terms of k, we can also write

$$R_{CE} = R_C\sqrt{\frac{1+k}{1-k}} \qquad R_{CO} = R_C\sqrt{\frac{1-k}{1+k}}$$

where

$$R_C = \sqrt{R_{CE}R_{CO}} = \sqrt{\frac{\mathcal{L}}{\mathcal{C}}}$$

will turn out to be the characteristic resistance of one line when the other is terminated at both ends in R_C. In the terminology of the following section, R_C *matches* the four-port formed by two finite coupled lines. With some algebra, we can write k in the form obtained in Exercise 11-20,

$$k = \frac{R_{CE} - R_{CO}}{R_{CE} + R_{CO}} \tag{11-43}$$

Finally, it is worth noting, as shown by means of an example in Exercise 11-19, that $1/\sqrt{\mathcal{LC}}$ is *not* equal to the propagation speed in coupled lines.

11-10   SCATTERING PARAMETERS

In Chapter 3 we saw that n-ports can be described by parameters—admittance parameters, for example—that relate voltages and currents. At microwave frequencies, the variables that are convenient to measure are the relative amplitudes of the *incident* voltage wave arriving at a port and the *scattered* voltage waves leaving that port or any other port. A microwave n-port is therefore often described in terms of its *scattering parameters* S_{jk}, which give the amplitudes V_{j-} of the scattered waves in terms of the amplitudes V_{k+} of the incident waves. In matrix form we have

$$[V_{j-}] = [S_{jk}][V_{k+}]$$

As shown in Fig. 11-45a, this description assumes that all ports are connected to lines of characteristic resistance R_C. It follows, as we will see, that the values of the scattering parameters can depend on the value of R_C. Figure 11-45b shows that the *scattering matrix* $[S_{jk}]$ can be obtained one column at a

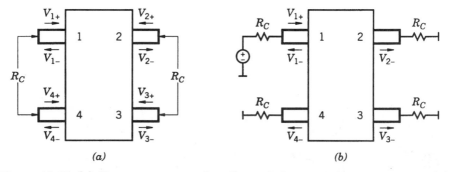

(a)   *(b)*

Figure 11-45. (a) Four-port connected to lines of characteristic resistance R_C. (b) Scattering parameters can be measured by terminating all ports in R_C and exciting one port at a time.

time by terminating all ports in R_C and exciting only one port; in this arrangement, S_{jj} is the reflection coefficient at port j.

We will only consider *reciprocal* n-ports with no dependent sources, for which the scattering matrix is symmetric,

$$S_{kj} = S_{jk}$$

As outlined in Exercise 11-17, the scattering matrix of lossless reciprocal n-ports is *unitary*; that is, its inverse is equal to its complex conjugate, so that if [1] denotes a unit matrix, we have

$$[S_{jl}][S_{lk}^*] = [1]$$

The description of n-ports in terms of a scattering matrix really makes sense only if the driving sources have output impedance R_C so that reflections are absorbed at the source. A point also worth making is that the terms *forward* and *backward* become inadequate for n-ports and that we will therefore speak only of incident waves and of reflected or scattered waves. We will be particularly interested in *matched* n-ports such that the input impedance at any port is R_C when all other ports are terminated in R_C. In matched n-ports we thus have $S_{jj} = 0$.

Not all n-ports can be matched. As an example, let us consider the tandem of $\lambda/4$ sections with characteristic resistances R_{C1} and R_{C2} shown in Fig. 11-46. The input resistances at ports 1 and 2, obtained using (11-31) twice, are

$$R_{I1} = \frac{R_{C1}^2}{R_{C2}^2} R_C \qquad R_{I2} = \frac{R_{C2}^2}{R_{C1}^2} R_C$$

Matching is thus possible only if $R_{C1} = R_{C2}$, in which case we are dealing with a $\lambda/2$ section, which is matched for any value of R_C.

As a warm-up exercise, we will obtain the scattering parameters for the $\lambda/4$ section shown in Fig. 11-47. If the characteristic resistance of the section is R_{CS}, the input resistances at ports 1 and 2 are

$$R_{I1} = R_{I2} = \frac{R_{CS}^2}{R_C}$$

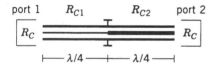

port 1 R_{C1} R_{C2} port 2

R_C ══════════════ R_C

├── $\lambda/4$ ──┼── $\lambda/4$ ──┤

Figure 11-46. Two-port consisting of two $\lambda/4$ sections of characteristic resistances R_{C1} and R_{C2}. Matching is possible only for $R_{C1} = R_{C2}$.

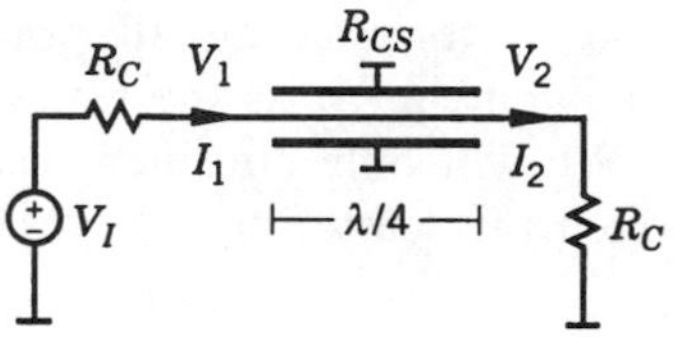

Figure 11-47. Two-port consisting of a $\lambda/4$ section of characteristic resistance R_{CS}. The scattering parameters depend on R_C, and matching is possible only for $R_C = R_{CS}$.

At port 1 we have

$$V_1 = \frac{R_{I1}}{R_{I1}+R_C}V_I = \frac{R_{CS}^2}{R_{CS}^2+R_C^2}V_I \qquad R_C I_1 = \frac{R_C^2}{R_{CS}^2+R_C^2}V_I$$

Using (11-18), we can then obtain the incident and scattered amplitudes at port 1,

$$V_{1+} = \frac{1}{2}(V_1+R_C I_1) = \frac{1}{2}V_I$$

$$V_{1-} = \frac{1}{2}(V_1-R_C I_1) = \left(\frac{1}{2}\right)\frac{R_{CS}^2-R_C^2}{R_{CS}^2+R_C^2}V_I$$

We thus have

$$V_{1-} = \frac{R_{CS}^2-R_C^2}{R_{CS}^2+R_C^2}V_{1+}$$

Using $\gamma l = i\pi/2$ in (11-22), we get

$$V_2 = -iR_{CS}I_1 = -i\frac{R_{CS}R_C}{R_{CS}^2+R_C^2}V_I$$

But $V_2 = V_{2-}$ and $V_I = 2V_{1+}$, so that

$$V_{2-} = -i\frac{2R_{CS}R_C}{R_{CS}^2+R_C^2}V_{1+}$$

By symmetry, the scattering matrix is

$$[S_{jk}] = \frac{1}{R_{CS}^2+R_C^2}\begin{bmatrix} R_{CS}^2-R_C^2 & -2iR_{CS}R_C \\ -2iR_{CS}R_C & R_{CS}^2-R_C^2 \end{bmatrix}$$

A quick check shows that $[S_{jl}][S_{lk}^*] = [1]$. It is clear that the values of the scattering parameters depend on the choice of R_C, and that $S_{jj} = 0$ only if $R_C = R_{CS}$.

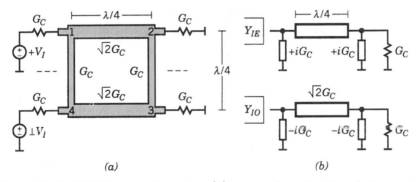

Figure 11-48. (*a*) 3dB two-stub coupler; (*b*) even-mode and odd-mode two-ports.

An important component in microwave stripline circuits is the 3dB two-stub coupler shown in Fig. 11-48*a*. Since it is embedded in a homogeneous medium and has a plane of symmetry, we can analyze it conveniently in terms of even and odd modes. We note that the propagation speed is mode independent, and so therefore are lengths given in terms of the wavelength λ. For an even-mode excitation at ports 1 and 4, the current across the plane of symmetry is zero, and the vertical legs become open $\lambda/8$ stubs with admittance iG_C; for an odd-mode excitation, the voltage at the plane of symmetry is zero, and the vertical legs become shorted $\lambda/8$ stubs with admittance $-iG_C$. We thus need only analyze the simplified even-mode and odd-mode two-ports shown in Fig. 11-48*b*. The input admittance in the even mode is

$$Y_{IE} = iG_C + \frac{\left(\sqrt{2}\,G_C\right)^2}{G_C + iG_C} = G_C$$

The even-mode voltages are thus

$$V_{4E} = V_{1E} = \frac{1}{2}V_I$$

Analogously, the input admittance in the odd mode is

$$Y_{IO} = -iG_C + \frac{\left(\sqrt{2}\,G_C\right)^2}{G_C - iG_C} = G_C$$

and the odd-mode voltages are

$$-V_{4O} = V_{1O} = \frac{1}{2}V_I$$

Since the even-mode and odd-mode input admittances are both equal to G_C, it is clear that port 1 is matched. Superposing even and odd modes, we obtain

$$V_4 = 0$$

This implies that legs 1–4 and 3–4 have infinite impedance as seen from ports 1 and 3 and that leg 2–3 is matched, so that its input admittance is G_C. From conservation of energy it follows that $|V_2| = |V_3| = |V_1|/\sqrt{2}$. In addition, V_2 and V_3 are delayed by 90° and 180° relative to V_1. We thus have

$$V_2 = -\frac{i}{\sqrt{2}} V_1$$

$$V_3 = -\frac{1}{\sqrt{2}} V_1$$

Making use of the fact that all ports are equivalent, we can now obtain the scattering matrix,

$$[S_{jk}] = \frac{1}{\sqrt{2}} \begin{bmatrix} 0 & -i & -1 & 0 \\ -i & 0 & 0 & -1 \\ -1 & 0 & 0 & -i \\ 0 & -1 & -i & 0 \end{bmatrix}$$

An important point to observe is that ports 1 and 4 are *isolated*. Thus, for example, a delicate receiver at port 4 can detect reflected signals returning to ports 2 or 3 without being overloaded by a high-power transmitter at port 1. Also important is that port 1 is matched whatever the load on port 4 as long as ports 2 and 3 are matched.

As an exercise in the use of scattering parameters, let us now load ports 2 and 3 with equal admittances Y. We will assume that port 4 is matched, so that $V_{4+} = 0$. We then have

$$V_{2+} = \frac{G_C - Y}{G_C + Y} V_{2-} = \frac{-i}{\sqrt{2}} \frac{G_C - Y}{G_C + Y} V_{1+}$$

$$V_{3+} = \frac{G_C - Y}{G_C + Y} V_{3-} = \frac{-1}{\sqrt{2}} \frac{G_C - Y}{G_C + Y} V_{1+}$$

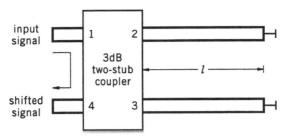

Figure 11-49. Matched phase shifter obtained by shorting the lines connected to ports 2 and 3 of a 3dB two-stub coupler at a distance l from these ports. The phase shift is proportional to l.

We further have

$$V_{1-} = \frac{-i}{\sqrt{2}}V_{2+} + \frac{-1}{\sqrt{2}}V_{3+}$$

$$V_{4-} = \frac{-1}{\sqrt{2}}V_{2+} + \frac{-i}{\sqrt{2}}V_{3+}$$

and we thus get

$$V_{1-} = 0$$

$$V_{4-} = i\frac{G_C - Y}{G_C + Y}V_{1+}$$

If Y is imaginary, we obtain a matched *phase shifter*. In particular, let us place short circuits at a distance l from ports 2 and 3, as shown in Fig. 11-49. We then have

$$Y = -iG_C \cot(\beta l)$$

so that

$$V_{4-} = i\frac{\tan(\beta l) + i}{\tan(\beta l) - i}V_{1+} = -i\frac{1 - i\tan(\beta l)}{1 + i\tan(\beta l)}V_{1+}$$

The phase shift is thus is proportional to l,

$$V_{4-} = -i\exp(-2i\beta l)V_{1+}$$

Good approximations to short circuits can be obtained with *pin* diodes, junction diodes that have an intrinsic region between the p-region and the n-region and look like small resistors at large forward currents. The phase shift can thus be varied electronically by turning on pairs of *pin* diodes at different distances from ports 2 and 3. *Varactor* diodes are junction diodes designed for use as variable capacitors; their junction capacitance varies roughly like the reciprocal square root of the reverse bias. Using varactor diodes in more complicated arrangements, the phase shift can be varied continuously.

A 3dB two-stub coupler can clearly be used as a power divider; as it stands, however, it is inconvenient for balanced loads because its outputs are 90° out of phase. The 3dB *ring coupler* analyzed in Exercise 11-21 can drive balanced loads directly and, quite interestingly, it can also add or subtract signals.

EXERCISES

11-1   LABORATORY.

(a) Take your pick of the circuits discussed in Section 11-6 and verify that the waveforms given there are correct. Use 50 Ω cables with propagation times substantially longer than the rise time of your pulse generator. To observe the effects shown in Fig. 11-24, use short and long sections of 300 Ω television twin-lead (parallel-wire) line.

(b) If you have a long length of RG58 coaxial cable, 50 meters or so, use the setup shown in Fig. 11-50 to observe the step responses at the sending and receiving ends. Using $A_H = 0.7$ dB/m, verify that the discussion around Fig. 11-16 is roughly correct. The signal at the sending end grows slowly until the 50 Ω termination at the receiving end makes itself felt, but it is nonetheless a fair simulation of a step.

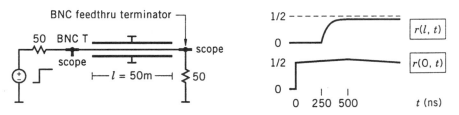

Figure 11-50. Step response of a long coaxial cable (Exercise 11-1).

(c) Use 10 lengths of 50 Ω RG58 coaxial cable, each 20 cm (1 ns) long, and 10 mica capacitors of value $C = 22$ pF, which is close to $\mathcal{C}l$, to assemble the periodically loaded line shown in Fig. 11-51. Use BNC T's to join the cables, and BNC barrels inserted in the T's to hold the capacitors; pressure fits will do.

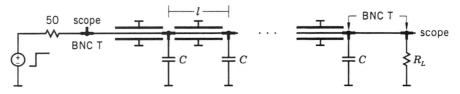

Figure 11-51. Periodically loaded line (Exercise 11-1).

Excite the line with a step from a generator with a $50\,\Omega$ output impedance and a rise time under 3 ns. Observe the reflection at the input when $R_L = R_C = 50\,\Omega$; adjust R_L until the reflection vanishes, that is, until $R_L = R_P$, and measure the propagation time T_P. Now remove the capacitors, set R_L back to $50\,\Omega$, and measure the unloaded propagation time T. According to (11-37) and (11-38), T_P/T and R_C/R_P should be equal and close to $\sqrt{2}$.

11-2 TEM MODE. Assume that the electromagnetic potentials for a forward TEM wave in vacuum are

$$\phi = \psi(x,y)g(z-ct)$$
$$c\mathbf{A} = \hat{\mathbf{z}}\,\psi(x,y)g(z-ct)$$

The transverse potential ψ satisfies Laplace's equation

$$\nabla_T^2 \psi = 0$$

and is constant on the surfaces of conductors. The electromagnetic potentials clearly satisfy the homogeneous wave equation,

$$c^2 \nabla^2 \mathbf{A} = \frac{\partial^2 \mathbf{A}}{\partial t^2}$$

$$c^2 \nabla^2 \phi = \frac{\partial^2 \phi}{\partial t^2}$$

Check that ϕ and $\mathbf{A}$ also satisfy the Lorentz condition

$$c^2 \nabla \cdot \mathbf{A} + \frac{\partial \phi}{\partial t} = 0$$

and that the fields are correctly given by

$$\mathbf{E} = -\nabla\phi - \frac{\partial \mathbf{A}}{\partial t}$$

$$\mathbf{B} = \nabla \times \mathbf{A}$$

11-3 LINE WITH SMALL PROPAGATION SPEED AND LARGE CHARACTERISTIC IMPEDANCE. Consider the delay line with distributed parameters shown in Fig. 11-52. The center conductor is a coil with n turns per unit length rather than a solid wire, and the space between the coil and the shield is filled with a lossless dielectric material. Show that the inductance and capacitance per unit length are

$$\mathcal{L} = \mu_0 \pi a^2 n^2$$

$$\mathcal{C} = \varepsilon_0 \frac{2\pi\kappa}{\ln(b/a)}$$

Take $n = 5000$ m^{-1}, $2a = 3$ mm, $2b = 7.5$ mm, and $\kappa = 2$. You then have $\mathcal{L} \approx 220$ μH/m and $\mathcal{C} \approx 120$ pF/m. The propagation time per unit length is $\sqrt{\mathcal{L}\mathcal{C}} \approx 160$ ns/m, and the characteristic resistance is $R_C = \sqrt{\mathcal{L}/\mathcal{C}} \approx 1400$ Ω; these values are about 30 times higher than they are in RG58 coaxial cable, in which $\sqrt{\mathcal{L}\mathcal{C}} = 5$ ns/m and $R_C = 50$ Ω.

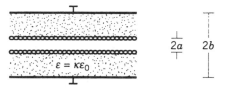

Figure 11-52. Line with small propagation speed and large characteristic impedance (Exercise 11-3).

11-4 LONGITUDINAL MAGNETIC FIELDS IN LOSSY LINES. Consider the asymmetric stripline shown in Fig. 11-53. Neglecting fringing fields and assuming that the skin effect is well developed, show in the case of a forward wave that the power transmitted in the lower region is twice the power transmitted in the upper region, but that the losses in the lower region are four times the losses in the upper region.

Convince yourself that power must be transmitted transversally from the upper region to the lower region and that the magnetic field must therefore have a longitudinal component B_z in order to give the Poynting vector a transverse component S_T. The voltage is thus not strictly constant in transverse planes; variations in the voltage, however, are negligible compared with

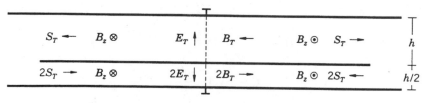

Figure 11-53. Fields in a lossy asymmetric stripline (Exercise 11-4).

the average line voltage. Convince yourself furthermore that B_z must grow linearly with the distance from the plane of symmetry in order to keep the transverse fields in each region uniform as they decay.

As a point of consistency, observe that the transverse surface current densities on the top and bottom surfaces of the center conductor are equal and opposite at equal distances from the plane of symmetry, so that there is no runaway charge accumulation at the edges of the center conductor.

11-5 THÉVENIN'S THEOREM EXTENDED. Consider the circuit of Fig. 11-54. Using (11-25), show that the open-circuit voltage and the short-circuit current are given by

$$V_T = \frac{(1-\rho_S)\exp(-\gamma l)}{1-\rho_S\exp(-2\gamma l)}V_S \qquad Z_C I_T = \frac{(1-\rho_S)\exp(-\gamma l)}{1+\rho_S\exp(-2\gamma l)}V_S$$

The Thévenin impedance is thus

$$Z_T = \frac{1+\rho_S\exp(-2\gamma l)}{1-\rho_S\exp(-2\gamma l)}Z_C$$

Set $\rho_S=(Z_S-Z_C)/(Z_S+Z_C)$ and show that as expected, Z_T is given by (11-24), that is, it is Z_S seen through a section of length l. Load the equivalent circuit with an impedance Z_L and show that the output voltage $V_T Z_L/(Z_T+Z_L)$ is what you get using (11-25).

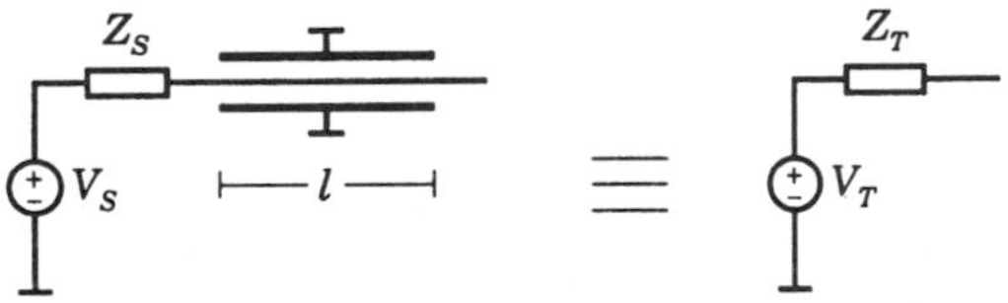

Figure 11-54. Thévenin's theorem is valid in the presence of transmission lines (Exercise 11-5).

11-6 INDUCTIVE MISMATCHES. In the circuit shown in Fig. 11-55, matching at the load is spoiled at high frequencies by inductor L. Use (11-26) to obtain

$$\frac{V(0)}{V_I} = \frac{1}{2}+\left(\frac{1}{2}\right)\frac{sL/2R_C}{1+sL/2R_C}\exp(-2sT)$$

and

$$\frac{V(l)}{V_I} = \frac{1}{2}\exp(-sT)+\left(\frac{1}{2}\right)\frac{sL/2R_C}{1+sL/2R_C}\exp(-sT)$$

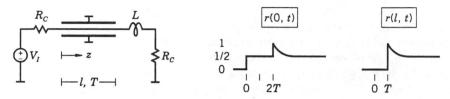

Figure 11-55. Inductive mismatch (Exercise 11-6).

The positive spikes in the step response decay with a time constant $\tau = L/2R_C$. Carbon composition resistors with values below 50 Ω exhibit a series inductance (including the stray inductance of the leads) of order 5 nH; for a 50 Ω line you then have $\tau \approx 50$ ps; distortion is thus negligible for signals with rise times above 1 ns.

11-7 FIRST-ORDER EFFECTS OF MISMATCHES. Using (11-24) and assuming that $\omega T \ll 1$, show that the input impedance at $z = 0$ in Fig. 11-24 is

$$Z_I = R'_C \frac{\dfrac{R_C}{R'_C} + \tanh sT}{1 + \dfrac{R_C}{R'_C}\tanh sT} \approx R_C \frac{1 + \dfrac{R'_C}{R_C}sT}{1 + \dfrac{R_C}{R'_C}sT}$$

If $R'_C > R_C$, Z_I is R_C in series with a small inductance L given by

$$L = \left(\frac{R'_C}{R_C} - \frac{R_C}{R'_C} \right) R_C T$$

whereas if $R'_C < R_C$, Z_I is R_C is parallel with a small capacitor C given by

$$C = \left(\frac{R_C}{R'_C} - \frac{R'_C}{R_C} \right) G_C T$$

For $l = 1$ cm, $\kappa = 1$, and R_C of order R'_C, obtain $L \approx 1$ nH and $C \approx 1$ pF.

To see these effects from a different perspective, use (11-26) to show that the step response at $z = 0$ is

$$r(0,t) = \tfrac{1}{2}\left[(1-\rho)\theta(t) + \rho(1-\rho^2)\theta(t-2T) + \rho^3(1-\rho^2)\theta(t-4T) + \cdots\right]$$

If $R'_C \gg R_C$ you have $\rho \approx -1$, and the initial value of $r(0,t)$ is almost twice the final value; this behavior can be simulated roughly by an inductor L in series with R_C. In contrast, if $R'_C \ll R_C$ you have $\rho \approx 1$, and the initial value of $r(0,t)$ is near zero; again roughly, this behavior can be simulated by a capacitor C in parallel with R_C. Now check that you get similar behavior if you calculate $r(0,t)$ using the approximate expression for Z_I given above.

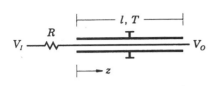
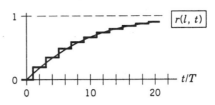

Figure 11-56. Open stub driven through a large resistance by a voltage source (Exercise 11-8)

11-8 OPEN STUBS LOOK LIKE CAPACITORS AT LOW FREQUEN-CIES. Consider the lossless open stub shown in Fig. 11-56. It is driven by a voltage source V_I through a resistance R much larger than R_C. From (11-26), obtain

$$V_O = \frac{(1-\rho_S)\exp(-sT)}{1-\rho_S\exp(-2sT)}V_I$$

Show that the step response at time $2nT$ is

$$r(l,2nT) = 1-\rho_S^n$$

Now define τ by

$$\exp\left(-\frac{2T}{\tau}\right) = \rho_S$$

You then have

$$\tau = -\frac{2T}{\ln\rho_S} \simeq \frac{TR}{R_C} = R\mathcal{C}l$$

In terms of the capacitance $C = \mathcal{C}l$ of the stub, you obtain, as expected, that $r(l,t)$ is close to the step response of a lag with time constant RC,

$$r(l,t) \simeq 1-\exp\left(-\frac{t}{RC}\right)$$

11-9 THE SHORT-LINE LIMIT AT LOW FREQUENCIES. Consider Fig. 11-19, and assume that $Z_S = 0$ and that the load is a resistance R_L. The point of this exercise is to show, infinite reflections and a frequency-dependent propagation function notwithstanding, that a short line at low frequencies behaves as expected; that is, it looks like a resistance and its voltage decreases linearly. Using (11-25), show first that

$$\frac{V(z)}{V(0)} = \frac{R_L\{1+\exp[-2\gamma(l-z)]\}+Z_C\{1-\exp[-2\gamma(l-z)]\}}{R_L[1+\exp(-2\gamma l)]+Z_C[1-\exp(-2\gamma l)]}\exp(-\gamma z)$$

If $|\gamma l| \ll 1$, you obtain

$$\frac{V(z)}{V(0)} \simeq \frac{R_L + Z_C \gamma (l-z)}{R_L + Z_C \gamma l}$$

But $Z_C \gamma = s\mathcal{L} + \mathcal{Z}$, which tends to a constant resistance per unit length $\mathcal{R}_L$ at low frequencies; you thus have

$$\frac{V(z)}{V(0)} \simeq \frac{R_L + \mathcal{R}_L(l-z)}{R_L + \mathcal{R}_L l}$$

This result can of course be obtained much more simply by observing that at low frequencies the equations of telegraphy in (11-15) reduce to

$$\frac{dV}{dz} = -\mathcal{R}_L I \qquad \frac{dI}{dz} = 0$$

11-10 PARALLEL LOADING IN ECL CIRCUITS. Transmission lines in ECL circuits can be terminated by a resistance in series with the source at the sending end (*series loading*) or by a resistance at the receiving end (*parallel loading*). Termination at both ends is not possible because the full logic swing must be transmitted to the load. Series loading is discussed in Section 11-6 and illustrated in Fig. 11-22.

The objective of this exercise is to see the effect of a load capacitance C_L on the parallel-loaded line of Fig. 11-57. From (11-26), obtain

$$V_L = \frac{\dfrac{1}{1 + sR_C C_L/2} \exp(-sT)}{1 - \dfrac{sR_C C_L/2}{1 + sR_C C_L/2} \exp(-2sT)} V_I$$

The first two waves transmitted to the load are given by the transfer function

$$\frac{1}{1 + sR_C C_L/2} \exp(-sT) + \frac{sR_C C_L/2}{(1 + sR_C C_L/2)^2} \exp(-3sT)$$

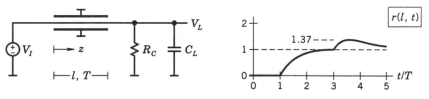

Figure 11-57. Parallel-loaded line (Exercise 11-10).

Ignoring the delays, show that the step responses of the first two waves are of the form

$$r_1(t) = 1 - \exp\left(-\frac{2t}{R_C C_L}\right)$$

$$r_2(t) = \frac{2t}{R_C C_L} \exp\left(-\frac{2t}{R_C C_L}\right)$$

Show also that $r_2(t)$ has a maximum of value $\exp(-1) = 0.37$ at $t = R_C C_L/2$. Conclude that the largest possible overshoot occurs for $T \gg R_C C_L/2$ and that the step response at the load in this case is correctly given in Fig. 11-57. You may take it on faith that overshoots due to further reflections are smaller.

ECL circuits can tolerate a 40% overshoot; parallel loading is thus safe even for zero rise time. You will seldom see overshoots this large, however, because input capacitances are about 5 pF, so that even if the load is 10 gates, time constant $R_C C_L/2$ is 2.5 ns for $R_C = 100 \ \Omega$, and thus less than the typical 3 ns rise time. As you have seen in Section 11-6, long rise times reduce overshoots.

11-11   RESONANT LINES. Consider the shorted quarter-wave section of Fig. 11-58. Taking losses into account and ignoring the reactance of $\mathscr{Z}$ in (11-14) and (11-19), you have

$$\frac{\gamma \lambda_0}{4} = i \frac{\pi}{2} \frac{\omega}{\omega_0} + \frac{\mathscr{R} \lambda_0}{8 R_C}$$

where, assuming that the dielectric is air, $\omega_0 = 2\pi c/\lambda_0$. From (11-24) you have

$$Z G_C = \tanh \frac{\gamma \lambda_0}{4}$$

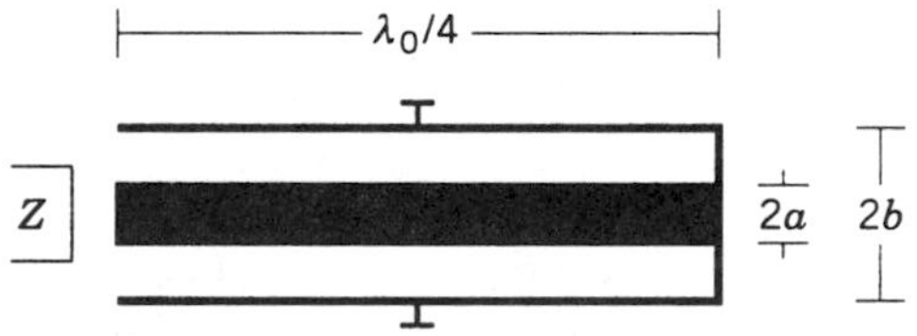

Figure 11-58. Shorted quarter-wave section (Exercise 11-11).

For small deviations from ω_0, now obtain

$$ZG_C \simeq \frac{\dfrac{8R_C}{\mathscr{R}\lambda_0}}{1 + 2iQ\dfrac{\omega - \omega_0}{\omega_0}}$$

where

$$Q = \frac{2\pi R_C}{\mathscr{R}\lambda_0}$$

Observe that $Z(\omega = \omega_0)$ is precisely the resistance R_I in (11-32). In a coaxial line with radii $a = b/5 = 2.5$ mm, you have $R_C \simeq 100$ Ω. For $f_0 = 100$ MHz ($\lambda_0/4 = 75$ cm), you get $\mathscr{R} \simeq 1/2\pi a \sigma\delta \simeq 0.16$ Ω/m if you ignore the losses in the shield, so that $Q \simeq 1300$.

Losses in the shorting plate (which you might wish to calculate) and radiation from the open end result in values of Q that are lower, but nonetheless much larger than those available in lumped resonators.

11-12   SINGLE-STUB MATCHING IS ALWAYS POSSIBLE. If the load on the line in Fig. 11-32 is $Y_L = G_L + iB_L$, the admittance Y at a distance l from the load is given by (11-24),

$$YR_C = \frac{G_L R_C + i[B_L R_C + \tan(\beta l)]}{[1 - B_L R_C \tan(\beta l)] + iG_L R_C \tan(\beta l)}$$

If you demand that the real part of Z be equal to R_C, you obtain

$$1 = \frac{G_L R_C[1 + \tan^2(\beta l)]}{1 - 2B_L R_C \tan(\beta l) + R_C^2(G_L^2 + B_L^2)\tan^2(\beta l)}$$

The discriminant of this quadratic equation is equal to

$$4G_L R_C\left[B_L^2 R_C^2 + (1 - G_L R_C)^2\right]$$

and is thus positive, so that there is always a real solution for $\tan(\beta l)$.

11-13   SINGLE-SECTION MATCHING. Given a source of impedance Z_S, convince yourself first that the power transfer to a load Z_L is maximum when $Z_L = Z_S^*$.

Most matching is done by matching the source and/or the load to a given line impedance. However, it is sometimes possible to achieve a match by inserting a (lossless) section of appropriate length l and characteristic resistance R_C between the source and the load (Somlo, 1960).

Transform the load impedance to the source position and demand that the result be equal to Z_S^*. Show by equating real parts that βl must satisfy

$$\tan \beta l = \frac{(R_L - R_S)R_C}{X_S R_L - X_L R_S}$$

Show further that R_C must satisfy

$$R_C^2 = \frac{R_S Z_I Z_I^* - R_I Z_S Z_S^*}{R_L - R_S}$$

If Z_S and Z_L are both real, you recover a $\lambda/4$ transformer.

11-14 THE ALTERNATED-LINE TRANSFORMER. The alternated-line transformer of Fig. 11-59 has the advantage that it uses sections of the lines to be matched (Bramham, 1961). For small mismatches, $2l$ is approximately $\lambda/6$, which corresponds to 60° in phase; since the signs of the reflections alternate, the amplitude of the reflected wave is the sum of three amplitudes that are equal in magnitude and separated by 120° in phase, and therefore add up to zero.

Let $n = R_{C2}/R_{C1}$; transform to the center junction using (11-24), and show that

$$\frac{n + i \tan(\beta l)}{1 + in \tan(\beta l)} = n \frac{1 - in \tan(\beta l)}{n - i \tan(\beta l)}$$

and that

$$\tan^2(\beta l) = \frac{n}{n^2 + n + 1}$$

For $n = 4$, $2l$ is equal to 0.13λ, about half the length of a $\lambda/4$ transformer.

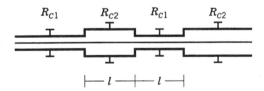

Figure 11-59. Alternated-line transformer (Exercise 11-14).

11-15 PROPAGATION FUNCTION OF A PERIODICALLY LOADED LINE. From the expression for the propagation function given in (11-34), obtain

$$\cosh(\gamma_P l) = \frac{1 + Y_P^2 Z_C^2 \tanh^2(\gamma l/2)}{1 - Y_P^2 Z_C^2 \tanh^2(\gamma l/2)}$$

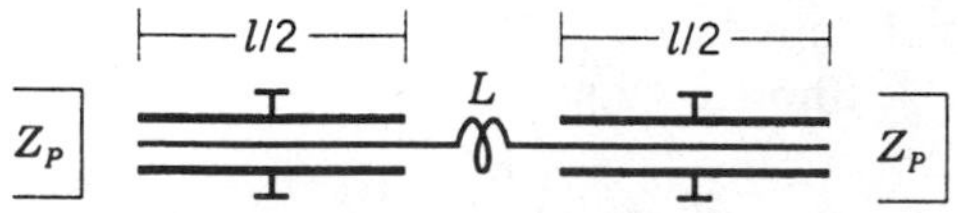

Figure 11-60. One period of a lossless line loaded with inductances of value L at intervals of length l (Exercise 11-16).

Now substitute the expression for $Y_P^2 Z_C^2$ given in (11-33) to obtain

$$\cosh(\gamma_P l) = \cosh(\gamma l) + (sCZ_C/2)\sinh(\gamma l)$$

11-16 INDUCTIVE PERIODIC LOADING. The line shown in Fig. 11-60 is loaded with inductances of value L at intervals of length l. With arguments similar to the ones used in Section 11-8, show that its characteristic impedance and propagation function are given by

$$Z_P^2 Y_C^2 = \frac{1 + (sLY_C/2)\coth(\gamma l/2)}{1 + (sLY_C/2)\tanh(\gamma l/2)}$$

and

$$\cosh(\gamma_P l) = \frac{1 + Z_P^2 Y_C^2 \tanh^2(\gamma l/2)}{1 - Z_P^2 Y_C^2 \tanh^2(\gamma l/2)}$$

As in Exercise 11-15, also obtain

$$\cosh(\gamma_P l) = \cosh(\gamma l) + (sLY_C/2)\sinh(\gamma l)$$

If the line is lossless, you have

$$\gamma l = i\omega l\sqrt{\mathcal{L}\mathcal{C}} = i\omega l\mathcal{L}G_C$$

The lowest cutoff frequency is given by

$$1 = (\omega_C LG_C/2)\tan(\omega_C l\mathcal{L}G_C/2)$$

If $L \gg l\,\mathcal{L}$, which is the case considered below, you get

$$1 \simeq (\omega_C LG_C/2)(\omega_C l\mathcal{L}G_C/2) = \omega_C^2 L\mathcal{C}l/4$$

and thus

$$\omega_C = \frac{2}{\sqrt{L(\mathcal{C}l)}}$$

This is precisely what you get if you look at the line as an infinite ladder network with inductances of value L and capacitances of value $C = \mathcal{C}l$. As you might expect, the low-frequency characteristic resistance R_P is larger than the unloaded characteristic resistance R_C,

$$R_P \simeq R_C \sqrt{1 + \frac{L}{\mathcal{L}l}}$$

A classic case of inductive loading is provided by two-wire telephone lines, which must transmit a *voice band* generally considered to go from 300 Hz to 3 kHz. At these frequencies, the line resistance per unit length is constant. Typical values for the parameters of air telephone lines are

$$\mathcal{L} = 1 \ \mu\text{H/m} \qquad \mathcal{C} = 11 \ \text{pF/m} \qquad \mathcal{R} = 45 \ \text{m}\Omega\text{/m}$$

Neglecting the internal inductance of the line, the propagation function is

$$\gamma = \sqrt{(s\mathcal{L} + \mathcal{R})s\mathcal{C}}$$

The crossover point determined by $\omega\mathcal{L}/\mathcal{R} = 1$ is at about 7 kHz. As shown in Fig. 11-61, the attenuation function is equal to $\sqrt{\omega\mathcal{R}\mathcal{C}/2}$ throughout the entire voice band and therefore results in severe distortion for distances of the order of several kilometers.

If the line is periodically loaded with large inductances, it behaves like an infinite ladder network below the lowest cutoff frequency or, equivalently in this range, like a line with a large inductance per unit length. By loading the line with 200 mH inductances at intervals of about 2 km, the inductance per unit length is raised by a factor of 100. The crossover point is brought down to about 70 Hz, and for the values given, the lowest cutoff frequency is about 5 kHz. Within the voice band, the propagation function of the loaded line is

$$\gamma_{PV} = i\omega\sqrt{\frac{LC}{l}}\sqrt{1 + \frac{\mathcal{R}l}{i\omega L}} \simeq i\omega\sqrt{\frac{LC}{l}} + \frac{\mathcal{R}}{2}\sqrt{\frac{\mathcal{C}l}{L}}$$

As shown in Fig. 11-61, the attenuation function is constant within the voice band and given by

$$\alpha_{PV} \simeq \frac{\mathcal{R}}{2}\sqrt{\frac{\mathcal{C}l}{L}}$$

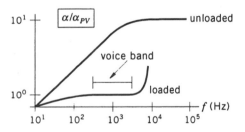

Figure 11-61. Unloaded and loaded attenuation as a function of the frequency for the line of Fig. 11-60 (Exercise 11-16).

From a physical point of view, you should observe that raising the inductance per unit length by a factor of 100 increases the characteristic resistance in the wave region by a factor of 10; for a given transmitted power level, the square of the current decreases by a factor of 10, and so therefore does the attenuation function.

11-17 THE SCATTERING MATRIX OF LOSSLESS RECIPROCAL n-PORTS IS UNITARY. Consider a lossless reciprocal two-port with scattering matrix $[S_{jk}]$. You then have

$$V_{1-} = S_{11}V_{1+} + S_{12}V_{2+}$$
$$V_{2-} = S_{21}V_{1+} + S_{22}V_{2+}$$

Summing squares of magnitudes on both sides yields

$$V_{1-}V_{1-}^* + V_{2-}V_{2-}^*$$
$$= S_{11}S_{11}^*V_{1+}V_{1+}^* + S_{12}S_{12}^*V_{2+}V_{2+}^* + S_{11}S_{12}^*V_{1+}V_{2+}^* + S_{12}S_{11}^*V_{1+}^*V_{2+}$$
$$+ S_{21}S_{21}^*V_{1+}V_{1+}^* + S_{22}S_{22}^*V_{2+}V_{2+}^* + S_{21}S_{22}^*V_{1+}V_{2+}^* + S_{22}S_{21}^*V_{1+}^*V_{2+}$$

Choose $V_{1+} \neq 0$, $V_{2+} = 0$, and then $V_{1+} = 0$, $V_{2+} \neq 0$. Using conservation of energy and reciprocity, obtain

$$S_{11}S_{11}^* + S_{12}S_{21}^* = 1$$
$$S_{21}S_{12}^* + S_{22}S_{22}^* = 1$$

Now choose $V_{2+} = V_{1+}$, and then $V_{2+} = iV_{1+}$. With these choices, show that

$$S_{11}S_{12}^* + S_{12}S_{22}^* = 0$$
$$S_{21}S_{11}^* + S_{22}S_{21}^* = 0$$

because the real and imaginary parts are both zero. This procedure can be extended to nth-order scattering matrices. Outline it for $n = 3$.

11-18 PARAMETERS OF COUPLED LINES IN HOMOGENEOUS MEDIA. Consider the coupled lines shown in Fig. 11-62. The transverse potential ψ is a solution of Laplace's equation that is constant on the surfaces of the conductors. Now consider the static variables in a forward wave. The fields are given by

$$\mathbf{E} = -\nabla_T \psi$$
$$c\mathbf{B} = \nabla_T \psi \times \hat{\mathbf{z}}$$

and the flux per unit length, the voltage, the charge per unit length, and the

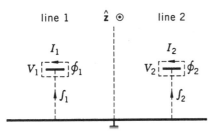

line 1 $\hat{z}$ ⊙   line 2

Figure 11-62. Coupled lines in a homogeneous medium (Exercise 11-18)

current in line 1 and line 2 are given by

$$c\mathcal{F}_1 = V_1 = \int_1 \nabla_T \psi \cdot d\mathbf{l} \qquad c\mathcal{F}_2 = V_2 = \int_2 \nabla_T \psi \cdot d\mathbf{l}$$

$$c\mathcal{Q}_1 = I_1 = \varepsilon_0 c \oint_1 \nabla_T \psi \times \hat{z} \cdot d\mathbf{l} \qquad c\mathcal{Q}_2 = I_2 = \varepsilon_0 c \oint_2 \nabla_T \psi \times \hat{z} \cdot d\mathbf{l}$$

where $\int_1$ and $\int_2$ are along paths from the ground plane to line 1 and line 2 respectively, and ϕ_1 and ϕ_2 are along paths that encircle line 1 and line 2 respectively.

If ψ_1^C is a solution of Laplace's equation for $V_2 = 0$, and ψ_2^C a solution for $V_1 = 0$, the general solution is

$$\psi = \psi_1^C + \psi_2^C$$

The voltages are

$$V_1 = \int_1 \nabla_T \psi_1^C \cdot d\mathbf{l} \qquad V_2 = \int_2 \nabla_T \psi_2^C \cdot d\mathbf{l}$$

and the charges per unit length are

$$\mathcal{Q}_1 = \varepsilon_0 \oint_1 \nabla_T \psi_1^C \times \hat{z} \cdot d\mathbf{l} + \varepsilon_0 \oint_1 \nabla_T \psi_2^C \times \hat{z} \cdot d\mathbf{l}$$

$$\mathcal{Q}_2 = \varepsilon_0 \oint_2 \nabla_T \psi_2^C \times \hat{z} \cdot d\mathbf{l} + \varepsilon_0 \oint_2 \nabla_T \psi_1^C \times \hat{z} \cdot d\mathbf{l}$$

You then have

$$\mathcal{Q}_1 = \mathcal{C} V_1 - \mathcal{C}_M V_2$$

$$\mathcal{Q}_2 = \mathcal{C} V_2 - \mathcal{C}_M V_1$$

where

$$c = \varepsilon_0 \frac{\oint_1 \nabla_T \psi_1^C \times \hat{z} \cdot d\mathbf{l}}{\int_1 \nabla_T \psi_1^C \cdot d\mathbf{l}} \qquad C_M = -\varepsilon_0 \frac{\oint_1 \nabla_T \psi_2^C \times \hat{z} \cdot d\mathbf{l}}{\int_2 \nabla_T \psi_2^C \cdot d\mathbf{l}}$$

Since $c > C_M$, the pair $\{V_1, V_2\}$ uniquely determines the pair $\{Q_1, Q_2\}$ and therefore the pair $\{I_1, I_2\}$.

Now let ψ_1^L be a solution for $I_2 = 0$, and ψ_2^L a solution for $I_1 = 0$. The general solution is

$$\psi = \psi_1^L + \psi_2^L$$

The currents are

$$I_1 = \varepsilon_0 c \oint_1 \nabla_T \psi_1^L \times \hat{z} \cdot d\mathbf{l} \qquad I_2 = \varepsilon_0 c \oint_2 \nabla_T \psi_2^L \times \hat{z} \cdot d\mathbf{l}$$

and the fluxes per unit length are

$$\mathcal{F}_1 = \frac{1}{c} \int_1 \nabla_T \psi_1^L \cdot d\mathbf{l} + \frac{1}{c} \int_1 \nabla_T \psi_2^L \cdot d\mathbf{l}$$

$$\mathcal{F}_2 = \frac{1}{c} \int_2 \nabla_T \psi_2^L \cdot d\mathbf{l} + \frac{1}{c} \int_2 \nabla_T \psi_1^L \cdot d\mathbf{l}$$

You then have

$$\mathcal{F}_1 = \mathcal{L} I_1 + \mathcal{L}_M I_2$$
$$\mathcal{F}_2 = \mathcal{L} I_2 + \mathcal{L}_M I_1$$

where

$$\mathcal{L} = \mu_0 \frac{\int_1 \nabla_T \psi_1^L \cdot d\mathbf{l}}{\oint_1 \nabla_T \psi_1^L \times \hat{z} \cdot d\mathbf{l}} \qquad \mathcal{L}_M = \mu_0 \frac{\int_1 \nabla_T \psi_2^L \cdot d\mathbf{l}}{\oint_2 \nabla_T \psi_2^L \times \hat{z} \cdot d\mathbf{l}}$$

At this point you have justified (11-39). Now reconsider the solution for $V_2 = 0$. The flux in line 2 is zero, so that

$$\mathcal{L}_M I_1 + \mathcal{L} I_2 = 0$$

You therefore have

$$\frac{\mathcal{L}_M}{\mathcal{L}} = -\frac{I_2}{I_1} = -\frac{Q_2}{Q_1} = \frac{C_M}{c}$$

You have now justified (11-42) again.

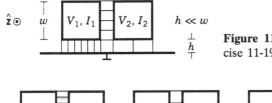

Figure 11-63. Closely coupled lines (Exercise 11-19).

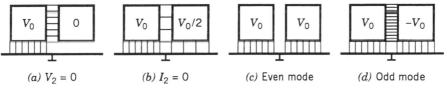

(a) $V_2 = 0$ (b) $I_2 = 0$ (c) Even mode   (d) Odd mode

Figure 11-64. Electric field configurations for the closely coupled lines of Fig. 11-63 (Exercise 11-19).

11-19 A CALCULABLE CASE OF COUPLED LINES. Consider the closely coupled lines shown in Fig. 11-63 (weak coupling implies small effects, and simple *calculable* examples are hard to find). Since $h \ll w$, the fields are important only in the regions between the lines themselves and between the lines and ground. Assume in what follows that the lines are carrying a forward wave and that all variables are static. Figure 11-64 shows the relevant electric-field configurations.

Set $V_1 = V_0$ and $V_2 = 0$, and show that

$$\mathcal{C} = 2\frac{w}{h}\varepsilon_0 \qquad \mathcal{C}_M = \frac{w}{h}\varepsilon_0$$

You thus have $k = \frac{1}{2}$.

Setting $V_1 = V_0$ and $V_2 = V_0/2$ makes $I_2 = 0$. Using $K = B/\mu_0$ and $B = E/c$, obtain

$$I_1 = \frac{3}{2}\frac{V_0}{hc\mu_0}w$$

Now obtain the fluxes,

$$\mathcal{F}_1 = \frac{V_0}{hc}h = \frac{V_0}{c}$$

$$\mathcal{F}_2 = \frac{1}{2}\frac{V_0}{hc}h = \frac{1}{2}\frac{V_0}{c}$$

Conclude that

$$\mathcal{L} = \frac{2}{3}\frac{h}{w}\mu_0 \qquad \mathcal{L}_M = \frac{1}{3}\frac{h}{w}\mu_0$$

Again, you get $k = \frac{1}{2}$. Observe that $\mathcal{L}\mathcal{C} = \frac{4}{3}\varepsilon_0\mu_0$ and *not* $\varepsilon_0\mu_0$.

Now set $V_1 = V_2 = V_0$ to get the even-mode capacitance per unit length $\mathcal{C}_E$ and, using $\mathcal{L}_E \mathcal{C}_E = \mu_0 \varepsilon_0$, the even-mode inductance per unit length $\mathcal{L}_E$,

$$\mathcal{C}_E = \frac{w}{h}\varepsilon_0 \qquad \mathcal{L}_E = \frac{h}{w}\mu_0$$

The even-mode characteristic resistance is therefore

$$R_{CE} = \frac{h}{w}\sqrt{\frac{\mu_0}{\varepsilon_0}}$$

Finally, set $V_1 = -V_2 = V_0$ to obtain the odd-mode parameters,

$$\mathcal{C}_O = 3\frac{w}{h}\varepsilon_0 \qquad \mathcal{L}_O = \frac{1}{3}\frac{h}{w}\mu_0$$

and

$$R_{CO} = \frac{1}{3}\frac{h}{w}\sqrt{\frac{\mu_0}{\varepsilon_0}}$$

You thus have

$$R_C = \sqrt{R_{CE}R_{CO}} = \frac{1}{\sqrt{3}}\frac{h}{w}\sqrt{\frac{\mu_0}{\varepsilon_0}}$$

You can now check that you also have

$$R_C = \sqrt{\frac{\mathcal{L}}{\mathcal{C}}}$$

11-20   PARALLEL-LINE COUPLERS. Coupling between lines is discussed in Section 11-9 assuming that the coupling is weak. In this exercise, the parallel-line coupler of Fig. 11-65 is analyzed for arbitrary coupling in terms of its even-mode and odd-mode characteristic resistances R_{CE} and R_{CO}.

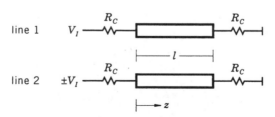

Figure 11-65. Parallel-line coupler (Exercise 11-20).

Assume that the coupler can be matched and that its characteristic resistance is R_C. For an arbitrary length l, let the even-mode and odd-mode input impedances at $z = 0$ be Z_{IE} and Z_{IO}. The even-mode and odd-mode voltages and currents are then given by

$$Z_{IE} I_E(0) = V_E(0) = \frac{Z_{IE}}{R_C + Z_{IE}} V_I \qquad\qquad Z_{IO} I_O(0) = V_O(0) = \frac{Z_{IO}}{R_C + Z_{IO}} V_I$$

Now superpose the even and odd modes. If you demand that line 1 be matched, you must have

$$R_C = \frac{V_E(0) + V_O(0)}{I_E(0) + I_O(0)}$$

Show that this condition implies $R_C^2 = Z_{IE} Z_{IO}$. From (11-24) you have

$$Z_{IE} = R_{CE} \frac{R_C + iR_{CE} \tan(2\pi l/\lambda)}{R_{CE} + iR_C \tan(2\pi l/\lambda)} \qquad\qquad Z_{IO} = R_{CO} \frac{R_C + iR_{CO} \tan(2\pi l/\lambda)}{R_{CO} + iR_C \tan(2\pi l/\lambda)}$$

Convince yourself by inspection that matching requires

$$R_C^2 = R_{CE} R_{CO}$$

Use this condition to show that the even-mode and odd-mode voltages at $z = 0$ are

$$V_E(0) = \frac{R_C + iR_{CE} \tan(2\pi l/\lambda)}{2R_C + i(R_{CE} + R_{CO}) \tan(2\pi l/\lambda)} V_I$$

$$V_O(0) = \frac{R_C + iR_{CO} \tan(2\pi l/\lambda)}{2R_C + i(R_{CE} + R_{CO}) \tan(2\pi l/\lambda)} V_I$$

The (backward) coupling coefficient $k = V_2(0)/V_1(0)$, defined here more generally than in Section 11-9, can then be obtained by superposition:

$$k = \frac{V_E(0) - V_O(0)}{V_E(0) + V_O(0)} = \frac{i(R_{CE} - R_{CO}) \tan(2\pi l/\lambda)}{2R_C + i(R_{CE} + R_{CO}) \tan(2\pi l/\lambda)}$$

Clearly, $|k|$ is maximum for $l = \lambda/4$, as was shown by different means in Section 11-9.

Assume in the rest of the exercise that $l = \lambda/4$. You then have

$$k = \frac{R_{CE} - R_{CO}}{R_{CE} + R_{CO}}$$

This result agrees with (11-43). You also have

$$R_{IE} = \frac{R_{CE}^2}{R_C} \qquad\qquad R_{IO} = \frac{R_{CO}^2}{R_C}$$

$$V_E(0) = \frac{R_{CE}}{R_{CE}+R_{CO}}V_I \qquad\qquad V_O(0) = \frac{R_{CO}}{R_{CE}+R_{CO}}V_I$$

$$R_{CE}I_E(0) = \frac{R_C}{R_{CE}+R_{CO}}V_I \qquad\qquad R_{CO}I_O(0) = \frac{R_C}{R_{CE}+R_{CO}}V_I$$

The forward and backward amplitudes in the even and odd modes are

$$R_{CE}I_{E+} = V_{E+} = \left(\frac{1}{2}\right)\frac{R_{CE}+R_C}{R_{CE}+R_{CO}}V_I \qquad R_{CO}I_{O+} = V_{O+} = \left(\frac{1}{2}\right)\frac{R_{CO}+R_C}{R_{CE}+R_{CO}}V_I$$

$$-R_{CE}I_{E-} = V_{E-} = \left(\frac{1}{2}\right)\frac{R_{CE}-R_C}{R_{CE}+R_{CO}}V_I \qquad -R_{CO}I_{O-} = V_{O-} = \left(\frac{1}{2}\right)\frac{R_{CO}-R_C}{R_{CE}+R_{CO}}V_I$$

and the forward and backward voltage amplitudes in the superposed state are

$$V_{1+} = \left(\frac{1}{2}\right)\frac{R_{CE}+R_{CO}+2R_C}{R_{CE}+R_{CO}}V_I \qquad V_{2+} = \left(\frac{1}{2}\right)\frac{R_{CE}-R_{CO}}{R_{CE}+R_{CO}}V_I$$

$$V_{1-} = \left(\frac{1}{2}\right)\frac{R_{CE}+R_{CO}-2R_C}{R_{CE}+R_{CO}}V_I \qquad V_{2-} = \left(\frac{1}{2}\right)\frac{R_{CE}-R_{CO}}{R_{CE}+R_{CO}}V_I$$

Now show that the current in the backward wave in line 1 is *forward* and that the current in the forward wave in line 2 is *backward*,

$$R_C I_{1+} = +V_{1+} \qquad\qquad R_C I_{2+} = -V_{2+}$$

$$R_C I_{1-} = +V_{1-} \qquad\qquad R_C I_{2-} = -V_{2-}$$

This implies that in either line, the ratio of the *line* voltage and the *line* current is constant and of magnitude R_C. You can then obtain

$$I_1(z)R_C = V_1(z) = V_I\cos(2\pi z/\lambda) - iV_I\frac{2R_C}{R_{CE}+R_{CO}}\sin(2\pi z/\lambda)$$

and

$$-I_2(z)R_C = V_2(z) = V_I\frac{R_{CE}-R_{CO}}{R_{CE}+R_{CO}}\cos(2\pi z/\lambda)$$

With these expressions you can calculate the scattering matrix. If the ports numbered from 1 to 4 are line 1 input, line 1 output, line 2 output, and line 2 input, you have

$$S_{11} = 0 \qquad S_{12} = \frac{-2iR_C}{R_{CE} + R_{CO}} \qquad S_{13} = 0 \qquad S_{14} = \frac{R_{CE} - R_{CO}}{R_{CE} + R_{CO}}$$

The remaining elements of the scattering matrix can be obtained from symmetry and reciprocity.

11-21 THE 3dB RING COUPLER. In power dividers it is sometimes necessary to have the outputs in phase or 180° out of phase. The 3dB ring coupler shown in Fig. 11-66a is an interesting example of a stripline circuit that will do the job. In addition, you will see that this coupler can be used to add and subtract signals. The presence of a plane of symmetry allows you to simplify calculations by using even and odd modes. Figure 11-66b shows the two-ports that result when ports 1 and 4 are fed with even-mode and odd-mode inputs. The input admittances are

$$Y_{IE} = \frac{-iG_C}{\sqrt{2}} + \frac{\left(\dfrac{G_C}{\sqrt{2}}\right)^2}{G_C + \dfrac{iG_C}{\sqrt{2}}} \qquad Y_{IO} = \frac{iG_C}{\sqrt{2}} + \frac{\left(\dfrac{G_C}{\sqrt{2}}\right)^2}{G_C - \dfrac{iG_C}{\sqrt{2}}}$$

or after minor algebra,

$$Y_{IE} = \frac{G_C}{3} - \frac{4iG_C}{3\sqrt{2}} \qquad Y_{IO} = \frac{G_C}{3} + \frac{4iG_C}{3\sqrt{2}}$$

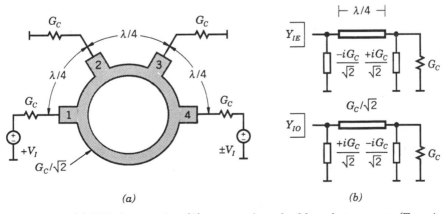

(a)   (b)

Figure 11-66. (a) 3dB ring coupler; (b) even-mode and odd-mode two-ports. (Exercise 11-21.)

It follows that

$$Y_{IE}Y_{IO} = G_C^2$$

The input voltages at port 1 are

$$V_{1E} = \frac{G_C}{G_C + Y_{IE}} V_I \qquad V_{1O} = \frac{G_C}{G_C + Y_{IO}} V_I$$

Superposing even and odd excitations, obtain

$$V_1 = V_I \qquad V_4 = \frac{i}{\sqrt{2}} V_I$$

Port 1 is thus matched. The line currents at the input to leg 1–2 are

$$I_{LE} = \left(Y_{IE} + \frac{iG_C}{\sqrt{2}}\right) V_{1E} = \frac{G_C + Y_{IE}}{4} V_{1E}$$

$$I_{LO} = \left(Y_{IO} - \frac{iG_C}{\sqrt{2}}\right) V_{1O} = \frac{G_C + Y_{IO}}{4} V_{1O}$$

You thus have

$$I_{LE} = I_{LO} = \frac{G_C}{4} V_I$$

Transforming to the output of leg 1–2, you obtain

$$V_{2E} = V_{2O} = \frac{-i}{2\sqrt{2}} V_I$$

Superposing the even and odd modes, you finally get

$$V_2 = -\frac{i}{\sqrt{2}} V_I \qquad V_3 = 0$$

Now feed ports 2 and 3 with even-mode and odd-mode excitations. Convince yourself that this merely swaps the two-ports in Fig. 11-66b relative to the preceding case; the sum signal therefore remains the same, so that port 2 is also matched, but the difference signal changes sign, and you get

$$V_2 = V_I \qquad V_3 = -\frac{i}{\sqrt{2}} V_I \qquad V_1 = -\frac{i}{\sqrt{2}} V_I \qquad V_4 = 0$$

The scattering matrix is thus

$$[S_{jk}] = \frac{1}{\sqrt{2}} \begin{bmatrix} 0 & -i & 0 & i \\ -i & 0 & -i & 0 \\ 0 & -i & 0 & -i \\ i & 0 & -i & 0 \end{bmatrix}$$

Conclude that a signal at port 2 results in signals that are in phase at ports 1 and 3, and that a signal at port 1 results in signals that are 180° out of phase at ports 2 and 4. Conclude also that signals at ports 1 and 3 are added at port 2 and subtracted at port 4.

A microwave *magic T* has the same scattering matrix as a ring coupler. You should be able to understand the excellent descriptions of this fascinating device given in Montgomery (1948) and Pound (1948).

11-22 PULSES IN PARALLEL-LINE COUPLERS. Consider the coupled striplines shown in Fig. 11-67 and assume that line 1 is excited by a pulse of amplitude $2V_0$ and width $\tau \ll T$. The amplitudes of the first incident waves in the even and odd modes are

$$V^I_{1E} = \frac{R_{CE}}{R_C + R_{CE}} V_0 \qquad\qquad V^I_{1O} = \frac{R_{CO}}{R_C + R_{CO}} V_0$$

$$I^I_{1E} = \frac{1}{R_C + R_{CE}} V_0 \qquad\qquad I^I_{1O} = \frac{1}{R_C + R_{CO}} V_0$$

Using the fact that

$$R_C^2 = R_{CE} R_{CO}$$

show that the superposed amplitudes are

$$R_C I^I_1 = V^I_1 = V_0$$

$$-R_C I^I_2 = V^I_2 = \frac{R_{CE} - R_{CO}}{2R_C + R_{CE} + R_{CO}} V_0$$

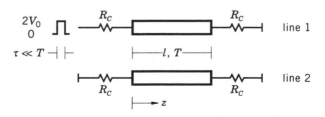

Figure 11-67. Parallel-line coupler driven by a short pulse (Exercise 11-22).

The current in line 2 is backward, which might strike you as an oddity until you realize that it is also backward in the ground plane; the Poynting vector, however, is forward everywhere. The transmission coefficients in the even and odd modes are

$$1+\rho_E = \frac{2R_C}{R_C+R_{CE}} \qquad\qquad 1+\rho_O = \frac{2R_C}{R_C+R_{CO}}$$

Show that the amplitudes transmitted to the loads at the receiving end are

$$V_1^T = \frac{4R_C}{2R_C+R_{CE}+R_{CO}}V_0$$

$$V_2^T = 0$$

By considering even and odd modes, convince yourself that in line 2 the amplitudes V_2^I of the incident wave and V_2^R of the reflected wave must add up to the amplitude V_2^T of the transmitted wave, and therefore that

$$V_2^R = -V_2^I$$

In weakly coupled lines R_{CE} and R_{CO} differ little from R_C, so that

$$V_1^T \simeq V_0$$

$$V_2^I \simeq \frac{k}{2}V_0$$

$$V_2^R \simeq -\frac{k}{2}V_0$$

where

$$k = \frac{R_{CE}-R_{CO}}{R_{CE}+R_{CO}} = \frac{\mathcal{L}_M}{\mathcal{L}} = \frac{\mathcal{C}_M}{\mathcal{C}}$$

is the coupling coefficient. It should be clear by analogy that the reflected pulse in line 2 is almost fully transmitted to the resistance at the sending end. These results agree with (11-40) and Fig. 11-43.

As in the case of the incident wave in line 2, the reflected wave in line 1 will not be transmitted to the resistance at the sending end. This explains why line 1 is matched despite the fact that (multiply) reflected pulses arrive back at the sending end long after the externally applied pulse has gone away.

12

RANDOM PROCESSES AND NOISE

12-1   INTRODUCTION

The free electrons in a resistor at a given temperature—we are thinking in terms of the Drude–Lorentz model—are continually moving about and colliding with the lattice ions. Given the statistical nature of this process, we should expect that at any instant there are more electrons in one half of a resistor than in the other half, and that there is therefore a *thermal noise voltage* $v_R(t)$ across an open-circuited resistor. This is in fact so; as shown in Fig. 12-1, the noise voltage adds to whatever signal voltage $v_S(t)$ is applied in series with a resistor R, and being highly irregular, it can mask a weak signal.

The thermal noise of a resistor is an example of a *random process* in which the value of a variable at any given time is unpredictable, and it is only one of many random processes that generate noise in electronic circuits. We will see that such processes can be described in terms of their statistical measures— that is, in terms of certain averages—and that despite its random nature, we will be able to estimate the degree to which noise is a nuisance.

For the large class of *stationary* random processes, whose properties do not change in time, we will introduce the *autocorrelation function* $R(\tau)$, the average product of values taken at times separated by an interval τ. The Fourier transform of the autocorrelation function, the *power spectrum* $\mathcal{S}(f)$, will then give us a description of the frequency content of a random process, and we will thus be in a position to design filters that maximize the signal relative to the noise. With an eye on the signal-recovery techniques that we will develop in Chapter 13, we will see in several instances how nonlinear devices, which generate products of waveforms, affect the autocorrelation function and the power spectrum.

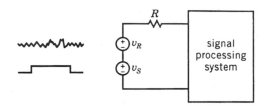

Figure 12-1. The thermal noise generated by a resistor can mask a weak signal.

We will deal preferentially with fundamental noise processes such as thermal noise in resistors and *shot noise* in currents because most other sources of noise can usually be eliminated by a suitable choice of devices. Despite its lack of an equally firm theoretical foundation, we will discuss low-frequency *flicker noise* because it is always present, and unfortunately as we will see, quite indifferent to simple filters, although it too can be reduced to a large extent by device selection.

Before addressing random processes, we will establish the required groundwork of language and notation by presenting a brief review of the elementary theory of probability and random variables. Given our motivation, we will steer a narrow course toward results that we will need later.

12-2   RANDOM VARIABLES

If the outcomes of a probabilistic experiment are assigned values by means of a specific rule, the set of values is a *random variable*. In the throw of a die, for example, one possible assignment is "the value that turns up", in which case the random variable consists of the integers 1 through 6, each with probability $P = \frac{1}{6}$. A given experiment may be described by more than one random variable: again in the throw of a die, the values -1 and $+1$, each with probability $P = \frac{1}{2}$, could be assigned to the outcomes "even" and "odd".

Let us first consider a *continuous* random variable x with values on the real line. In this case, x is described by a nonnegative *probability density (distribution)* $p(x)$ such that $p(x)\,dx$ is the probability that the outcome lies between x and $x+dx$. If we extend x to the entire real line by assigning zero density to regions with impossible values, the probability $P(a < x < b)$ that x lies in an arbitrary interval (a, b) is given by

$$P(a<x<b) = \int_a^b p(x)\,dx$$

and we must have

$$\int_{-\infty}^{+\infty} p(x)\,dx = 1$$

A familiar example of a continuous random variable, defined for an electron in a Drude–Lorentz metal, is "the time t since the last collision", described by a Poisson distribution:

$$p(t) = \frac{1}{\tau_F} \theta(t) \exp\left(-\frac{t}{\tau_F}\right)$$

where τ_F is the mean free time. Another example, also defined for electrons in a Drude–Lorentz metal, is "the velocity u along an axis", described by a normal (or gaussian) distribution:

$$p(u) = \sqrt{\frac{m}{2\pi kT}} \exp\left(-\frac{mu^2}{2kT}\right)$$

where m is the electron mass, k is Boltzmann's constant, and T is the absolute temperature.

We can now extend our discussion to a *discrete* random variable n with values on the integers and a *probability distribution* $P(n)$ such that

$$\sum_{n=-\infty}^{+\infty} P(n) = 1$$

To do so, we will define $p(x)$ as a sum of weighted delta functions,

$$p(x) = \sum_{n=-\infty}^{+\infty} P(n)\delta(x-n)$$

Thus in the "even" or "odd" case given above, we have

$$p(x) = \tfrac{1}{2}\delta(x+1) + \tfrac{1}{2}\delta(x-1)$$

12-2-1   Moments of Random Variables

A random variable x can be at least partially characterized in terms of its nth-order *moments*, defined by

$$\langle x^n \rangle = \int_{-\infty}^{+\infty} x^n p(x)\, dx$$

Of particular importance to us will be the first moment or *mean* μ_x,

$$\mu_x = \langle x \rangle = \int_{-\infty}^{+\infty} x p(x)\, dx$$

(it is simply the average value of x) and the second moment or *mean square*,

$$\langle x^2 \rangle = \int_{-\infty}^{+\infty} x^2 p(x)\, dx$$

TABLE 12-1. Often-Used Distributions and Their Important Moments

Normal or *gaussian*

$$p(x) = \frac{1}{\sqrt{2\pi}\,\sigma} \exp\left[-\frac{(x-\mu)^2}{2\sigma^2}\right] \qquad\qquad \langle x \rangle = \mu \qquad\qquad \sigma_x^2 = \sigma^2$$

Poisson (discrete)

$$P(n) = \frac{\mu^n}{n!}\exp(-\mu) \qquad\qquad \langle n \rangle = \mu \qquad\qquad \sigma_n^2 = \mu$$

Poisson (continuous)

$$p(t) = \frac{1}{\tau}\theta(t)\exp\left(-\frac{t}{\tau}\right) \qquad\qquad \langle t \rangle = \tau \qquad\qquad \langle t^2 \rangle = 2\tau^2$$

We will define the *central moments* by

$$\langle (x-\mu_x)^n \rangle = \int_{-\infty}^{+\infty} (x-\mu_x)^n p(x)\, dx$$

The first central moment is clearly always zero; the second central moment or *variance* σ_x^2 (mean square of the deviation from the mean) is given by

$$\sigma_x^2 = \int_{-\infty}^{+\infty} (x-\mu_x)^2 p(x)\, dx$$

where σ_x is the *standard deviation* (root mean square of the deviation from the mean). Expanding the square binomial in the integral, it becomes clear that

$$\langle x^2 \rangle = \mu_x^2 + \sigma_x^2$$

For reference, the distributions we will use most are given in Table 12-1 together with their important moments.

If x is a random variable with probability density $p(x)$ and $y(x)$ is a differentiable function, then $y(x)$ is also a random variable, and its probability density $p'(y)$ can be obtained from $p(x)$. In the simple case in which $y(x)$ is single-valued, for example, we have

$$p'(y) = p(x)\left|\frac{dx}{dy}\right|$$

We will normally not calculate $p'(y)$ because the moments of y can be obtained directly in terms of $p(x)$:

$$\langle y^n \rangle = \int_{-\infty}^{+\infty} [y(x)]^n p(x)\, dx$$

12-2-2   Joint Probability Densities

If x_1 and x_2 are random variables—on the same experiment or on different experiments, it does not matter—we will define a *joint probability density* $p(x_1, x_2)$ such that $p(x_1, x_2)\, dx_1\, dx_2$ is the probability that the outcomes lie in intervals of length dx_1 and dx_2 at x_1 and x_2. We must then have

$$\int_{-\infty}^{+\infty} \int_{-\infty}^{+\infty} p(x_1, x_2)\, dx_1\, dx_2 = 1$$

The single densities $p(x_1)$ and $p(x_2)$ are implicitly contained in the joint density and can be obtained for one variable by integration over the other; we thus have

$$p(x_1) = \int_{-\infty}^{+\infty} p(x_1, x_2)\, dx_2 \quad \text{and} \quad p(x_2) = \int_{-\infty}^{+\infty} p(x_1, x_2)\, dx_1$$

The joint density can be factored into two terms,

$$p(x_1, x_2) = p(x_1)\, p(x_2 | x_1)$$

where $p(x_2 | x_1)$ is the *conditional probability density* for the occurrence of x_2 given that x_1 has occurred. In the throw of a die, for example, we have

$$\tfrac{1}{6} = P(\text{odd}, 3) = P(\text{odd})\, P(3 | \text{odd}) = \tfrac{1}{2}\left(\tfrac{1}{3}\right)$$

An integration of $p(x_1, x_2)$ over x_2 should give us back $p(x_1)$; it follows that

$$\int_{-\infty}^{+\infty} p(x_2 | x_1)\, dx_2 = 1$$

We thus have, again in the throw of a die,

$$P(1 | \text{odd}) + P(3 | \text{odd}) + P(5 | \text{odd}) = 1$$

The random variables x_1 and x_2 are (*statistically*) *independent* if their joint density is equal to the product of the single densities, that is, if

$$p(x_1, x_2) = p(x_1)\, p(x_2)$$

The noise voltages on different resistors, for example, are independent random variables.

The moments of order (n, m) of a joint density are given by

$$\langle x_1^n x_2^m \rangle = \int_{-\infty}^{+\infty} \int_{-\infty}^{+\infty} x_1^n x_2^m p(x_1, x_2)\, dx_1\, dx_2$$

If x_1 and x_2 are independent, the moments can be factored,

$$\langle x_1^n x_2^m \rangle = \langle x_1^n \rangle \langle x_2^m \rangle$$

More generally, the mean value of a function $g(x_1, x_2)$ is given by

$$\langle g(x_1, x_2) \rangle = \int_{-\infty}^{+\infty} \int_{-\infty}^{+\infty} g(x_1, x_2) p(x_1, x_2) \, dx_1 \, dx_2$$

12-2-3   Sums of Random Variables

Finite sums of random variables are also random variables. For simplicity we will consider sums of two random variables; generalizations are straightforward, and we will therefore use them without further comment. If x and y are random variables, and

$$z = x + y$$

it is clear that *means add*:

$$\langle z \rangle = \langle x \rangle + \langle y \rangle$$

We also have

$$\langle z^2 \rangle = \langle x^2 \rangle + \langle y^2 \rangle + 2\langle xy \rangle$$

If x and y are independent, it is easy to show that

$$\langle z^2 \rangle - \langle z \rangle^2 = \langle x^2 \rangle - \langle x \rangle^2 + \langle y^2 \rangle - \langle y \rangle^2$$

or, succinctly, that *variances add*:

$$\sigma_z^2 = \sigma_x^2 + \sigma_y^2$$

More generally, let x_1, x_2, y_1, and y_2 be random variables such that the *pairs* $\{x_1, x_2\}$ and $\{y_1, y_2\}$ are independent, that is, such that

$$p(x_1, x_2, y_1, y_2) = p(x_1, x_2) p(y_1, y_2)$$

and let

$$z_1 = x_1 + y_1$$
$$z_2 = x_2 + y_2$$

It is then also easy to show—in terminology that will make full sense when we come to random processes—that *fluctuations add*:

$$\langle z_1 z_2 \rangle - \langle z_1 \rangle \langle z_2 \rangle = \langle x_1 x_2 \rangle - \langle x_1 \rangle \langle x_2 \rangle + \langle y_1 y_2 \rangle - \langle y_1 \rangle \langle y_2 \rangle \quad (12\text{-}1)$$

Let us now calculate the probability density $p(z)$ of the sum $z = x + y$ of two independent random variables x and y with probability densities $p'(x)$ and $p''(y)$. For a given z, the probability $P(z)$ that $x + y$ is less than z is given by

$$P(z) = \int_{-\infty}^{+\infty} p'(x) \, dx \int_{-\infty}^{z-x} p''(y) \, dy$$

Observing that $p(z)$ is equal to the derivative of $P(z)$, we obtain

$$p(z) = \int_{-\infty}^{+\infty} p'(u)p''(z-u)\,du$$

In other words, $p(z)$ is given by the convolution of $p'(x)$ and $p''(y)$,

$$p(z) = p'*p''(z)$$

We will define the *characteristic function* $\chi(\nu)$ of a probability density $p(x)$ as the Fourier transform of $p(x)$,

$$\chi(\nu) = \int_{-\infty}^{+\infty} p(x)\exp(-i\nu x)\,dx$$

Recalling that the transform of a convolution is the product of the transforms of the factors of convolution, for the sum $z = x + y$ of two independent random variables we obtain

$$\chi_z(\nu) = \chi_x(\nu)\chi_y(\nu)$$

For a (discrete) Poisson random variable it is easy to show that

$$\chi(\nu) = \exp\{\mu[\exp(-i\nu)-1]\}$$

It follows that a sum of independent Poisson-distributed random variables is Poisson distributed. For a normal random variable we obtain from Exercise 2-17 that

$$\chi(\nu) = \exp\left(-\frac{\sigma^2\nu^2}{2}\right)\exp(-i\mu\nu)$$

It thus also follows that a sum of independent normally distributed random variables is normally distributed. More generally, we will accept the *central-limit* theorem, according to which the sum of N identical independent random variables with a finite second moment becomes normally distributed as N goes to infinity. We note that although the central-limit theorem is valid for Poisson-distributed random variables, convergence far from the mean is slow because the sum remains Poisson distributed for arbitrary N.

12-3 STATIONARY RANDOM PROCESSES

Let us imagine a set containing many resistors of the same value, all at the same temperature. To be realistic, let us also imagine that the resistors are shunted by capacitors, all of the same value; as we will soon see, this will ensure that the mean-square noise voltage is finite. The noise voltage of any particular resistor at any given time is not precisely predictable, but there will nonetheless be certain regularities in the behavior of the *ensemble* $\{v_R(t)\}$ formed by the noise voltages $v_R(t)$ of the resistors.

If we measure all the noise voltages at a given time and plot their relative frequencies, for example, we will almost surely observe what would seem to be a normal distribution. If we then wait for a while and repeat the measurement, we will almost surely conclude that the distribution is *stationary*, meaning that it is invariant in time. We might then be willing to say that observing a voltage that is five standard deviations away from zero is highly improbable, and we would know considerably more than we did before we started. By refining the measurements, we would observe that the voltages at two times separated by an interval τ are not independent if τ is small, but that they do become independent for large values of τ.

More formally, let us consider an ensemble $\{x(t)\}$ of *sample functions* $x(t)$ on the time interval $(-\infty, +\infty)$, and for any $N \geq 1$, let $x_1 \cdots x_N$ be the random variables defined by the values of the sample functions at times $t_1 \cdots t_N$. We will then say that the ensemble constitutes a random process, and that the process is stationary if the N-fold joint probability densities

$$p(x_1, \ldots, x_N)$$

depend on time differences but not on the time itself. We will denote averages using these joint probability densities as *ensemble averages*, and having made the distinction, we will also let $x(t)$ denote the ensemble $\{x(t)\}$ as well as any of its sample functions.

For our purposes, it will be sufficient to consider processes that are *wide-sense stationary*, meaning that the definitions above hold for $N = 2$ and, implicitly, for $N = 1$. We will thus deal only with the *first probability density* $p(x_1)$, which is time independent, and with the *second probability density* $p(x_1, x_2, \tau)$, which depends on the difference $\tau = t_2 - t_1$.

As we know, the first probability density can be obtained from the second probability density,

$$p(x_1) = \int_{-\infty}^{+\infty} p(x_1, x_2, \tau) \, dx_2$$

In addition, the second probability density can be factored,

$$p(x_1, x_2, \tau) = p(x_1) p(x_2 | x_1, \tau)$$

and we clearly must also have

$$\lim_{\tau \to 0} p(x_2 | x_1, \tau) = \delta(x_1 - x_2)$$

Finally, we will say that two wide-sense stationary random processes $x(t)$ and $y(t)$ are *independent* if

$$p(x_1, x_2, y_1, y_2) = p(x_1, x_2) p(y_1, y_2)$$

12-3-1   The Autocorrelation Function

In the notation of the preceding section, we will define the *autocorrelation function* $R_x(\tau)$ of a stationary random process $x(t)$ as the mean value of $x_1 x_2$,

$$R_x(\tau) = \langle x_1 x_2 \rangle$$

We will also define the *cross-correlation* of two random processes $x(t)$ and $y(t)$ by

$$R_{xy}(\tau) = \langle x_1 y_2 \rangle$$

For $\tau = 0$, the autocorrelation function is equal to the mean square of x,

$$R_x(0) = \langle x^2 \rangle$$

Observing that $p(x_2, x_1, -\tau) = p(x_1, x_2, \tau)$ and that $\langle x_2 x_1 \rangle = \langle x_1 x_2 \rangle$, we obtain that the autocorrelation function is symmetric in τ,

$$R_x(-\tau) = R_x(\tau)$$

Observing now that

$$0 \le \langle (x_1 \pm x_2)^2 \rangle = 2 \langle x^2 \rangle \pm 2 \langle x_1 x_2 \rangle$$

we obtain that the absolute value of the autocorrelation function is maximum at $\tau = 0$,

$$|R_x(\tau)| \le R_x(0)$$

12-3-2   Ergodic Random Processes

Returning for a moment to an ensemble of resistors, let us imagine that rather than calculating ensemble averages at a fixed time we observe one resistor for a long time and calculate the *time averages* of the voltage and of the product of voltages taken at times separated by an interval τ. We would then almost surely come to the conclusion that thermal noise is *ergodic*, meaning that time and ensemble averages are equal, and that this is so because any sample function eventually takes on all values accessible to the ensemble, and does so with relative frequencies equal to the ensemble relative frequencies.

More formally, a random process is ergodic if with probability 1, the time averages obtained on a sample function $x(t)$ are equal to the ensemble averages. Unless otherwise indicated, the random processes that we will consider are ergodic as well as stationary, and $\langle x \rangle$ and $R_x(\tau)$ are therefore also given by

$$\langle x \rangle = \lim_{T \to \infty} \frac{1}{2T} \int_{-T}^{+T} x(t)\, dt$$

and

$$R_x(\tau) = \lim_{T \to \infty} \frac{1}{2T} \int_{-T}^{+T} x(t)x(t+\tau)\, dt$$

A simple example of a nonergodic process is given by an ensemble consisting of two sample functions $x_\pm(t)$ that have constant values ± 1 and occur with probabilities $P_\pm = \frac{1}{2}$. The ensemble mean value is zero, and thus differs from the time average of either sample function.

12-3-3   The Power Spectrum

We will define the *power spectrum* $S_x(f)$ of a stationary random process $x(t)$ as the Fourier transform of the autocorrelation function $R_x(\tau)$,

$$S_x(f) = \int_{-\infty}^{+\infty} R_x(\tau) \exp(-2\pi i f\tau)\, d\tau \tag{12-2}$$

The autocorrelation function is then given by the inverse Fourier transform of the power spectrum,

$$R_x(\tau) = \int_{-\infty}^{+\infty} S_x(f) \exp(+2\pi i f\tau)\, df \tag{12-3}$$

Expressions (12-2) and (12-3) are the historically fundamental Wiener–Khintchine relations, originally obtained as a theorem rather than by definition.

We will soon show that we can obtain the average power per unit frequency from the power spectrum; as a preview, we note that the average power $\langle x^2 \rangle$ is equal to the integral of the power spectrum over all frequencies,

$$\langle x^2 \rangle = R_x(0) = \int_{-\infty}^{+\infty} S_x(f)\, df$$

We also note that the power spectrum is symmetric in f because $R_x(-\tau) = R_x(\tau)$,

$$S_x(-f) = S_x(f)$$

In Exercise 12-5 it is shown that the power spectrum is nonnegative. As a corollary, it is shown that the mean square of the integral of a stationary random process $x(t)$, which will be required in Exercises 12-6 and 13-12, can be expressed in terms of the autocorrelation function,

$$\left\langle \left| \int_0^T x(t)\, dt \right|^2 \right\rangle = \int_0^T du \int_0^T dv\, R_x(u-v) = T \int_{-T}^{+T} \left(1 - \frac{|\tau|}{T}\right) R_x(\tau)\, d\tau \tag{12-4}$$

12-3-4   Random Sine Waves

As a first example, we will calculate the autocorrelation function and the power spectrum of a *random sine wave* with sample functions

$$x(t) = x_0 \cos(\omega_0 t + \varphi)$$

where x_0 and $\omega_0 = 2\pi f_0$ are constants, and φ is uniformly distributed in the interval $0 \leq \varphi \leq 2\pi$ in order to make the process both stationary and ergodic. Using either time or ensemble averages, we obtain $\langle x \rangle = 0$. The autocorrelation function is given by

$$R_x(\tau) = x_0^2 \langle \cos(\omega_0 t + \varphi) \cos(\omega_0 t + \omega_0 \tau + \varphi) \rangle$$

Recalling that

$$2 \cos a \cos b = \cos(a+b) + \cos(a-b)$$

we obtain, again using either time or ensemble averages,

$$R_x(\tau) = \frac{x_0^2}{2} \cos \omega_0 \tau$$

We note that a random sine wave illustrates the fairly obvious theorem that periodic sample functions imply a periodic autocorrelation function. From (2-20) we obtain that the power spectrum is concentrated at $f = \pm f_0$,

$$S_x(f) = \frac{x_0^2}{4} [\delta(f+f_0) + \delta(f-f_0)] \tag{12-5}$$

Integrating the power spectrum over all frequencies, we obtain

$$\langle x^2 \rangle = \int_{-\infty}^{+\infty} S_x(f)\, df = \frac{x_0^2}{2}$$

This is of course something we already knew: the average power in a sine wave is one-half the square of the amplitude.

When dealing with nonlinear devices we will need the autocorrelation and power spectrum of $x^2(t)$. The calculations are straightforward, and we will therefore simply quote the results:

$$R_{x^2}(\tau) = \langle x_1^2 x_2^2 \rangle = \frac{x_0^4}{8}(2 + \cos 2\omega_0 \tau)$$

$$S_{x^2}(f) = \frac{x_0^4}{16}[\delta(f+2f_0) + 4\delta(f) + \delta(f-2f_0)] \tag{12-6}$$

12-3-5   The Random Telegraph Wave

As a more physical example of a random process (it describes a form of *burst noise* in transistors), let us consider the *random telegraph wave*. As shown in Fig. 12-2, the sample functions of this process alternate between the values 0

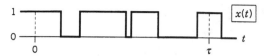

Figure 12-2. A sample function of the random telegraph wave alternates between the values 0 and 1 at times governed by a Poisson distribution.

and 1 at times governed by a Poisson distribution with average frequency λ. The first probability distribution is clearly given by

$$P(0) = P(1) = \tfrac{1}{2}$$

and we therefore have

$$\langle x \rangle = \tfrac{1}{2}$$

Observing that $x_1 x_2$ differs from zero only if x_1 and x_2 are both equal to 1, we have

$$\langle x_1 x_2 \rangle = P(1,1) = P(1)P(1|1)$$

Observing further that x_1 and x_2 are equal only if the number n of transitions in the interval τ is even, and using $P(n)$ as given in Table 12-1, we obtain

$$P(1|1) = \sum_{n=0}^{\infty} \frac{(\lambda\tau)^{2n}}{(2n)!} \exp(-\lambda\tau) = \cosh(\lambda\tau)\exp(-\lambda\tau) = \frac{1+\exp(-2\lambda\tau)}{2}$$

The autocorrelation function of the random telegraph wave is thus

$$R_x(\tau) = \frac{1+\exp(-2\lambda|\tau|)}{4}$$

From (2-14) and (2-20) we obtain that the power spectrum is

$$S_x(f) = \frac{1}{4}\delta(f) + \frac{1}{4}\frac{\lambda}{\lambda^2+\pi^2 f^2}$$

These results are typical of noise processes: as shown in Fig. 12-3, $R_x(\tau)$ is maximum and equal to $\langle x^2 \rangle$ at $\tau = 0$, but becomes equal to $\langle x \rangle^2$ at large values of $|\tau|$ because $x(0)$ and $x(\tau)$ become independent or *uncorrelated*. Correspondingly, $S_x(f)$ has an impulse $\langle x \rangle^2 \delta(f)$ at $f = 0$, and a bump that is flat around $f = 0$ and falls off to zero at large values of $|f|$.

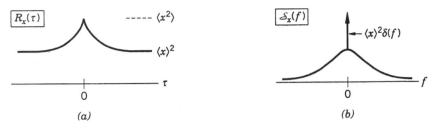

Figure 12-3. (*a*) Autocorrelation and (*b*) power spectrum of the random telegraph wave.

12-3-6 Gaussian Processes

Many important noise processes are *gaussian* because they are actually sums of large numbers of individual processes. Thus thermal noise is always gaussian in practice, and shot noise becomes gaussian when it is intense enough; in addition, signal processes are often gaussian. We will look into these processes because useful statements can be made about them that do not hold in general.

The second probability density of a gaussian process $x(t)$ is a two-dimensional normal distribution,

$$p(x_1, x_2) = \frac{1}{2\pi\sigma^2\sqrt{1-\rho^2}} \exp\left[-\frac{x_1^2 - 2\rho x_1 x_2 + x_2^2}{2\sigma^2(1-\rho^2)}\right] \qquad (12\text{-}7)$$

If we complete the square in the argument of the exponential, we can write

$$p(x_1, x_2) = p(x_1)p(x_2|x_1)$$

where $p(x_1)$ is a normal distribution with mean zero and variance σ^2,

$$p(x_1) = \frac{1}{\sqrt{2\pi}\,\sigma} \exp\left(-\frac{x_1^2}{2\sigma^2}\right)$$

and $p(x_2|x_1)$ is a normal distribution with mean ρx_1 and variance $\sigma^2(1-\rho^2)$,

$$p(x_2|x_1) = \frac{1}{\sqrt{2\pi}\,\sigma\sqrt{1-\rho^2}} \exp\left[-\frac{(x_2-\rho x_1)^2}{2\sigma^2(1-\rho^2)}\right]$$

The autocorrelation function can now be obtained in terms of ρ,

$$R_x(\tau) = \int_{-\infty}^{+\infty} x_1 p(x_1)\,dx_1 \int_{-\infty}^{+\infty} x_2 p(x_2|x_1)\,dx_2 = \int_{-\infty}^{+\infty} \rho x_1^2 p(x_1)\,dx_1 = \rho\sigma^2$$

We can thus identify ρ as the normalized autocorrelation function,

$$\rho = \frac{R_x(\tau)}{\sigma^2}$$

As an example of the statements that can be made by knowing that a process is gaussian, we will obtain the autocorrelation $R_{x^2}(\tau) = \langle x_1^2 x_2^2 \rangle$ of $x^2(t)$, which we will need when dealing with nonlinear devices. We have

$$R_{x^2}(\tau) = \int_{-\infty}^{+\infty} x_1^2 p(x_1)\, dx_1 \int_{-\infty}^{+\infty} x_2^2 p(x_2|x_1)\, dx_2$$

$$= \int_{-\infty}^{+\infty} \left[\rho^2 x_1^2 + \sigma^2(1-\rho^2) \right] x_1^2 p(x_1)\, dx_1$$

Using the easily verified fact that $\langle x_1^4 \rangle = 3\sigma^4$, we obtain

$$R_{x^2}(\tau) = R_x^2(0) + 2R_x^2(\tau)$$

Correspondingly, the power spectrum of $x^2(t)$ is given by

$$S_{x^2}(f) = R_x^2(0)\,\delta(f) + 2S_x * S_x(f) \tag{12-8}$$

12-3-7 Linear Filters: Interpretation of the Power Spectrum

To grasp the significance of the power spectrum of a random process $x(t)$, we need to know how the power spectrum is affected by a linear filter with impulse response $h(t)$ and frequency response $H(f)$.[†] We will make the assumption that $\langle x^2 \rangle$ is finite, and we will then accept that the response of the filter to a sample function, given by

$$y(t) = \int_{-\infty}^{+\infty} h(u) x(t-u)\, du$$

exists with probability 1 if the filter is *stable*, that is, if the abscissa of convergence of the Laplace transform of $h(t)$ is negative or, equivalently, if

$$\int_{-\infty}^{+\infty} |h(u)|\, du < \infty$$

If $x(t)$ is gaussian, it is painful but not hard to prove that for arbitrary values of u_1 and u_2, $h(u_1)x(t-u_1)+h(u_2)x(t-u_2)$ is also gaussian. It follows that $y(t)$ is gaussian if $x(t)$ is gaussian.

[†]In this and the following chapter, in which Fourier rather than Laplace transforms are prominent, we will write frequency responses as functions of the real variables f or $\omega = 2\pi f$.

Given that a filter is stable, we will also accept that the order in which integration and averaging are carried out can be exchanged, so that

$$\langle y \rangle = \left\langle \int_{-\infty}^{+\infty} h(u)x(t-u)\, du \right\rangle = \int_{-\infty}^{+\infty} h(u)\langle x \rangle\, du = \langle x \rangle \int_{-\infty}^{+\infty} h(u)\, du$$

We will further accept that such exchanges are allowed in the calculation of the autocorrelation of $y(t)$. On this basis, in Exercise 12-3 it is shown that

$$R_y(\tau) = R_x * R_h(\tau) \qquad\qquad (12\text{-}9)$$

where $R_h(\tau)$ is the autocorrelation of the filter impulse response,

$$R_h(\tau) = \int_{-\infty}^{+\infty} h(u)h(u+\tau)\, du$$

It is also shown that the Fourier transform of $R_h(\tau)$ is given by

$$\mathcal{S}_h(f) = |H(f)|^2 \qquad\qquad (12\text{-}10)$$

It follows that

$$\mathcal{S}_y(f) = \mathcal{S}_x(f)|H(f)|^2 \qquad\qquad (12\text{-}11)$$

We can now address the question of what the power spectrum means. Let us choose $H(f)$ so that its response is unity in narrow bands of width Δf centered at frequencies $\pm f_0$, and zero outside these bands. We note that this convenient *brick-wall* filter is not causal because $h(t)$ is not zero for negative t, but we will use it notwithstanding because it can be approximated to any required degree by causal *Butterworth* filters (Exercise 5-5). If Δf is small enough and contains no impulses, the average power at the output of the filter is

$$\left[\mathcal{S}_x(-f_0) + \mathcal{S}_x(f_0) \right] \Delta f = 2\mathcal{S}_x(f_0)\,\Delta f$$

The power spectrum is defined for negative as well as positive frequencies, but a measurement will not make this distinction. We will therefore identify the *power density* $x^2(f)$, defined for arbitrary $f>0$ by

$$x^2(f) = 2\mathcal{S}_x(f)$$

as the average power per unit frequency in random process $x(t)$. We note that $x^2(f)$ does *not* have an impulse $\langle x \rangle^2 \delta(f)$ at $f=0$, and that it therefore gives the power in the *fluctuations* of $x(t)$.

The average power at the output of a linear filter (including contributions from impulses, if any) is given by

$$\langle y^2 \rangle = \int_{-\infty}^{+\infty} \mathcal{S}_x(f)|H(f)|^2\, df$$

If $\mathcal{S}_x(f)$ has a constant value $\mathcal{P}_0$ within the passband of the filter, we may write

$$\langle y^2 \rangle = 2\mathcal{P}_0 B$$

where B is the *power bandwidth* of the filter,

$$B = \tfrac{1}{2}\int_{-\infty}^{+\infty} |H(f)|^2 \, df = \tfrac{1}{2}\int_{-\infty}^{+\infty} |h(t)|^2 \, dt \tag{12-12}$$

12-3-8   Sums and Products of Independent Random Processes

Noise generally adds to signals of interest, and nonlinear devices generate products of random processes. We will therefore determine the power spectra of the sum and product of two independent random processes $x(t)$ and $y(t)$ on the assumption that the resulting processes are ergodic. Let us first consider the sum

$$z(t) = x(t) + y(t)$$

For times t_1 and t_2 separated by an interval τ, from (12-1) we have

$$\langle z_1 z_2 \rangle - \langle z \rangle^2 = \langle x_1 x_2 \rangle - \langle x \rangle^2 + \langle y_1 y_2 \rangle - \langle y \rangle^2$$

or

$$R_z(\tau) - \langle z \rangle^2 = R_x(\tau) - \langle x \rangle^2 + R_y(\tau) - \langle y \rangle^2$$

If as will often be the case for us, either

$$\langle x \rangle = 0 \quad \text{or} \quad \langle y \rangle = 0$$

we simply have

$$R_z(\tau) = R_x(\tau) + R_y(\tau)$$

and therefore

$$\mathcal{S}_z(f) = \mathcal{S}_x(f) + \mathcal{S}_y(f)$$

Now let us consider the product

$$w(t) = x(t) y(t)$$

We then have

$$\langle w_1 w_2 \rangle = \langle x_1 x_2 y_1 y_2 \rangle = \langle x_1 x_2 \rangle \langle y_1 y_2 \rangle$$

or

$$R_w(\tau) = R_x(\tau) R_y(\tau)$$

and therefore

$$\mathcal{S}_w(f) = \mathcal{S}_x * \mathcal{S}_y(f)$$

In particular, let $x(t)$ be a random sine wave of unit amplitude,

$$x(t) = \cos(\omega_0 t + \varphi)$$

From (12-5) we have

$$\mathcal{S}_x(f) = \tfrac{1}{4}[\delta(f+f_0) + \delta(f-f_0)]$$

and therefore

$$\mathcal{S}_w(f) = \tfrac{1}{4}[\mathcal{S}_y(f+f_0) + \mathcal{S}_y(f-f_0)] \qquad (12\text{-}13)$$

Finally, let $x(t)$ and $y(t)$ be unit-amplitude random sine waves with different frequencies ω_1 and ω_2,

$$x(t) = \cos(\omega_1 t + \varphi_1) \quad \text{and} \quad y(t) = \cos(\omega_2 t + \varphi_2)$$

and let $f_+ = f_1 + f_2$ and $f_- = |f_1 - f_2|$ be the sum and difference frequencies; we then have

$$\mathcal{S}_w(f) = \tfrac{1}{16}[\delta(f+f_+) + \delta(f+f_-) + \delta(f-f_-) + \delta(f-f_+)] \quad (12\text{-}14)$$

We observe that $w(t)$ is stationary but not ergodic if $\omega_1 = \omega_2$ and that its *time* average $\tfrac{1}{2}\cos(\varphi_1 - \varphi_2)$ is a measure of the phase between sample sine waves from $x(t)$ and $y(t)$. This observation, and similar ones on square waves and other periodic functions, are put to use in *phase detectors*.

12-3-9 Poisson Processes: Carson's Theorem

We will now consider the important class of random processes called *Poisson* processes. Shot noise, as we will soon see, belongs to this class. A sample function of a Poisson process is of the form

$$x(t) = \sum_{n=-\infty}^{+\infty} a_n h(t-t_n)$$

where the amplitudes a_n are *independent* samples from a random variable a, $h(t)$ is a pulse that is zero for $t < 0$, and the times t_n are Poisson distributed with average frequency λ.

To get a more physical picture, we might imagine that we are using some form of detector to observe the gamma rays produced by the radioactive decay of a substance. The timing of events is governed by a Poisson distribution, and the detector output has a fairly constant shape from event to event, but its amplitude depends on the energy deposited in the detector by a

Figure 12-4. The autocorrelation of a Poisson process can be obtained by considering the effect at times $t = 0$ and $t = \tau$ of events that occur in a small interval dt centered at $t = -T$ and then using the fact that events in disjoint intervals are independent.

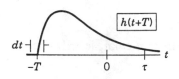

gamma ray. To complete the picture, we will assume, as is usually the case, that the detector response is linear in energy.

With reference to Fig. 12-4, we will obtain the statistical measures of $x(t)$ by calculating the effect at times $t = 0$ and $t = \tau$ of events that occur in a small interval dt centered at $t = -T$. The contributions to $x(0)$ and $x(\tau)$ from N events with amplitudes $a_1 \cdots a_N$ are

$$dx_T(0) = (a_1 + \cdots + a_N)h(T)$$
$$dx_T(\tau) = (a_1 + \cdots + a_N)h(T+\tau)$$

Taking averages for fixed N and then averaging over N and setting $\langle N \rangle = \lambda \, dt$, we obtain

$$\langle dx_T(0) \rangle = \langle a \rangle \langle N \rangle h(T) = \langle a \rangle h(T) \lambda \, dt$$
$$\langle dx_T(\tau) \rangle = \langle a \rangle \langle N \rangle h(T+\tau) = \langle a \rangle h(T+\tau) \lambda \, dt$$

Since means add, we can now integrate either of these expressions to obtain Campbell's *theorem of the mean*,

$$\mu_x = \lambda \langle a \rangle \int_{-\infty}^{+\infty} h(t) \, dt \qquad (12\text{-}15)$$

To obtain the autocorrelation function, let us consider the product

$$dx_T(0) \, dx_T(\tau) = (a_1 + \cdots + a_N)^2 h(T) h(T+\tau)$$

Taking the average for fixed N and then averaging over N as we did above, we obtain

$$\langle dx_T(0) \, dx_T(\tau) \rangle = (\langle N \rangle \langle a^2 \rangle + \langle N(N-1) \rangle \langle a \rangle^2) h(T) h(T+\tau)$$

Substituting $\langle N \rangle = \lambda \, dt$ and $\langle N^2 \rangle - \langle N \rangle = \langle N \rangle^2 = (\lambda \, dt)^2$, we get

$$\langle dx_T(0) \, dx_T(\tau) \rangle - \langle dx_T(0) \rangle \langle dx_T(\tau) \rangle = \langle a^2 \rangle h(T) h(T+\tau) \lambda \, dt$$

Since events in disjoint intervals are independent, we can use (12-1) to integrate the fluctuations. We then obtain Carson's theorem,

$$R_x(\tau) - \mu_x^2 = \lambda \langle a^2 \rangle \int_{-\infty}^{+\infty} h(t) h(t+\tau) \, dt$$
$$\mathcal{S}_x(f) - \mu_x^2 \delta(f) = \lambda \langle a^2 \rangle |H(f)|^2 \qquad (12\text{-}16)$$

Evaluating the autocorrelation function at $\tau = 0$, we obtain Campbell's *theorem of the mean square*, which we will write compactly in the form

$$\sigma_x^2 = \lambda \langle a^2 \rangle \int_{-\infty}^{+\infty} |h(t)|^2 \, dt \qquad (12\text{-}17)$$

An interesting property of Poisson processes, often used to prove Carson's theorem, is discussed in Exercise 12-2: Once an interval and a fixed number of events in the interval are specified, the times at which the events occur are uniformly distributed in the interval. For this reason, Poisson processes are said to be *purely random*.

12-3-10   Impulse Processes

Impulse processes are idealized Poisson processes in which the pulse $h(t)$ is a delta function, that is, an impulse of unit area:

$$h(t) = \delta(t)$$

From Campbell's theorem of the mean (12-15) and Carson's theorem (12-16), we then have

$$\mu_x = \lambda \langle a \rangle$$

$$R_x(\tau) - \mu_x^2 = \lambda \langle a^2 \rangle \delta(\tau)$$

$$S_x(f) - \mu_x^2 \delta(f) = \lambda \langle a^2 \rangle$$

Campbell's theorem of the mean square (12-17) fails for impulse processes, but we will not worry about this failure because impulse processes always drive filters that make the mean square finite.

Let us now assume that the amplitudes have values $\pm a_0$ with equal probabilities $P = \frac{1}{2}$. We then have a random process with mean zero and a constant or *white* power spectrum,

$$R_x(\tau) = \lambda a_0^2 \delta(\tau)$$

$$S_x(f) = \lambda a_0^2$$

Conversely, we can imagine that *any* power spectrum with a constant magnitude $\mathcal{P}_0$ arises from an impulse process with an (arbitrary) average frequency λ and equally probable amplitudes $\pm \sqrt{\mathcal{P}_0/\lambda}$. This point of view will be used profitably in Exercise 13-8. It is also clear that any random process with a power spectrum of the form

$$S_x(f) = \mathcal{P}_0 |H(f)|^2$$

may be considered to be generated by the passage of an impulse process with mean zero through a filter with frequency response $H(f)$.

12-4   FUNDAMENTAL NOISE PROCESSES

Thermal noise and shot noise are fundamental in the sense that they depend on essential device parameters and cannot be reduced by device selection. The thermal noise generated by a resistor, for example, depends on the resistance and on the temperature but not at all on the composition of the resistor. Flicker noise is not fundamental in this sense and arises for many reasons that we will not look into, but it is ubiquitous and will therefore be discussed in order to obtain credible noise models of semiconductor devices.

12-4-1   Shot Noise

Shot noise is associated with currents—but only certain currents, as we will see—and arises because charge is quantized in units of e. Rather than attempt an explanation by analogy, we will proceed directly to calculate shot noise in temperature-limited thermionic diodes because the physics involved is elementary and will not obscure our understanding of the conditions required for shot noise to manifest itself fully.

Let us thus consider the parallel-plate vacuum diode shown in Fig. 12-5a, and let us assume that the cathode is heated to the point that it emits electrons at an average rate λ. If the anode voltage V is made high enough, all the electrons emitted are quickly swept to the anode; the average diode current therefore depends only on the cathode temperature, and remains constant even if V is increased further. In this *temperature-limited regime*, the cathode-to-anode *transit time* τ_A is so short that *space-charge* effects are negligible, meaning that electrons in the region between the plates are few and that the electric field they generate is negligible. The electric field is then equal to V/l, where l is the separation of the plates. In addition, V is much larger than the thermal voltage kT/e; we may therefore assume that electrons are emitted with zero initial velocity.

Let us now consider one electron after it is emitted from the cathode. Since it is subjected to a constant acceleration eV/ml, its velocity $u(t)$

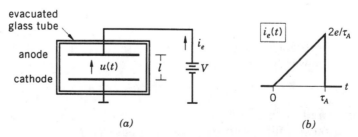

Figure 12-5. Parallel-plate vacuum diode. (a) Structure. (b) Current pulse due to the transit of one electron in the temperature-limited regime.

increases linearly in time. Observing that the final velocity is twice the average velocity l/τ_A, we obtain

$$u(t) = \frac{2l}{\tau_A}\frac{t}{\tau_A}$$

To obtain the current $i_e(t)$ induced in the external circuit, we will equate the power $(eV/l)u(t)$ absorbed from the electric field by the electron and the power $Vi_e(t)$ delivered by the source of V. We thus obtain

$$i_e(t) = \frac{e}{l}u(t)$$

Substituting $u(t)$ in this expression, we get the triangular pulse shown in Fig. 12-5b,

$$i_e(t) = eh(t)$$

where

$$h(t) = \frac{2t}{\tau_A^2}\left[\theta(t) - \theta(t - \tau_A)\right]$$

Since

$$\int_{-\infty}^{+\infty} h(t)\, dt = 1$$

it is clear that the transit of the electron makes a charge e circulate in the external circuit.

If we accept the experimental fact that emission times are Poisson distributed, it follows that the total current $i(t)$ is a Poisson process with pulses of shape $h(t)$, constant amplitude e, and average frequency λ; the power spectrum of $i(t)$ is therefore given by Carson's theorem, and we have

$$\mathcal{S}_i(f) = (\lambda e)^2 \delta(f) + \lambda e^2 |H(\omega)|^2$$

or in terms of the average current $I = \lambda e$,

$$\mathcal{S}_i(f) = I^2 \delta(f) + eI |H(\omega)|^2$$

We note without proof that

$$|H(\omega)|^2 = 4\frac{2 + \omega^2\tau_A^2 - 2\cos(\omega\tau_A) - 2\omega\tau_A\sin(\omega\tau_A)}{\omega^4\tau_A^4}$$

and that $|H(\omega)|^2$ is equal to 1 at $\omega\tau_A = 0$, drops to 0.95 at $\omega\tau_A = 1$, and falls off to zero at higher frequencies. For $\omega\tau_A < 1$ (160 MHz for $\tau_A = 1$ ns), we can safely consider that $|H(\omega)| = 1$. We then obtain the *Schottky formula* for the power density $i_S^2(f)$ of shot noise,

$$i_S^2(f) = 2eI$$

The importance of the Schottky formula for us is not so much that it gives the noise in temperature-limited thermionic diodes, although such diodes are in fact used as calibrated noise sources, but that it gives the noise associated with any current that can be described as a Poisson process with short pulses of total charge e.

The Schottky formula is also important because despite essential differences in physical mechanisms, other noise processes—notably the forward current in junction diodes—exhibit a shot-noise power spectrum at low frequencies.

Full shot noise is *not* associated with arbitrary currents. As an example, in Exercise 12-7 it is shown that the noise associated with the drift current in a powered resistor is many orders of magnitude less than the value predicted by the Schottky formula.

Finally, we note that the instantaneous current has contributions from many overlapping pulses if $\lambda\tau_A$ is much larger than 1, let us say 100; the central-limit theorem then implies that shot noise is gaussian if the average current is sufficiently large. To see when this might be so, let us take $\tau_A = 1$ ns; the noise will then reasonably gaussian for currents above 10 nA, that is, above $6.25 \cdot 10^{10}$ electrons per second.

12-4-2   Thermal Noise

We will first calculate the power spectrum $\mathcal{S}_R(f)$ of the thermal noise voltage generated by a resistor of value R using arguments from statistical mechanics and show that $\mathcal{S}_R(f)$ depends only on R and on the absolute temperature T (a more detailed but model-dependent calculation is offered in the following section). Let us thus imagine, as shown in Fig. 12-6a, that the noise voltage

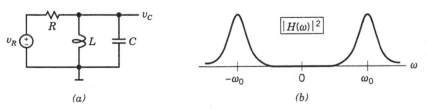

(a)   (b)

Figure 12-6. (*a*) Tank circuit used to calculate the power spectrum of thermal noise at $\omega = \pm\omega_0 = \pm 1/\sqrt{LC}$; (*b*) square of the magnitude $|H(\omega)|$ of its frequency response.

v_R is filtered in a tank circuit. The mean-square voltage on capacitor C is then given by

$$\langle v_C^2 \rangle = \int_{-\infty}^{+\infty} |H(\omega)|^2 S_R(f) \frac{d\omega}{2\pi}$$

where

$$H(\omega) = \frac{i\omega \dfrac{L}{R}}{1 + i\omega \dfrac{L}{R} - \omega^2 LC}$$

If we make inductor L infinitesimal while holding $\omega_0 = 1/\sqrt{LC}$ constant, we can write $|H(\omega)|^2$ as a sum of two terms with sharp peaks at $\mp \omega_0$, as shown in Fig. 12-6*b*:

$$|H(\omega)|^2 = \frac{1}{1 + [2RC(\omega + \omega_0)]^2} + \frac{1}{1 + [2RC(\omega - \omega_0)]^2}$$

We may now assume that $S_R(f)$ is constant over the passbands of $H(\omega)$, and we obtain

$$\langle v_C^2 \rangle = 2 S_R(f_0) \int_{-\infty}^{+\infty} \frac{1}{1 + [2RC(\omega - \omega_0)]^2} \frac{d\omega}{2\pi}$$

The integral is equal to $1/4RC$; since f_0 is arbitrary, we get

$$S_R(f) = 2RC \langle v_C^2 \rangle$$

But v_C is one of the degrees of freedom in the tank circuit, and equipartition of energy requires that its associated mean-square energy $(C/2)\langle v_C^2 \rangle$ be equal to $kT/2$. We thus have

$$\langle v_C^2 \rangle = \frac{kT}{C}$$

Substituting $\langle v_C^2 \rangle$ in the expression for $S_R(f)$ yields

$$S_R(f) = 2kTR$$

and we obtain the *Nyquist formula* for the power density of thermal noise,

$$v_R^2(f) = 4kTR$$

For a 1 kΩ resistor, for example, we have

$$v_R = \sqrt{(4kT/e)eR} = \sqrt{(100 \text{ mV})(1.6 \cdot 10^{-19} \text{ C})(1 \text{ k}\Omega)} = 4 \text{ nV}/\sqrt{\text{Hz}}$$

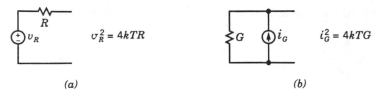

Figure 12-7. The thermal noise of a resistor can be represented either (*a*) as a series voltage source or (*b*) as a parallel current source.

The rms voltage noise in a power bandwidth $B = 1$ MHz is thus

$$\sqrt{\langle v^2 \rangle} = v_R \sqrt{B} = 4 \ \mu\text{V}$$

We may draw the rough conclusion that signal levels above 1 mV are not much affected by thermal noise.

The relative frequencies of the values of v_C are given by Boltzmann factors; the first probability density of v_C is therefore a normal distribution,

$$p(v_C) = \sqrt{\frac{C}{2\pi kT}} \exp\left(-\frac{Cv_C^2}{2kT} \right)$$

We will accept that the second probability density of v_C is a two-dimensional normal distribution like the one given in (12-7), and that v_C is therefore a gaussian process. More generally, we will accept that thermal noise is gaussian at the output of any linear filter with a high-frequency cutoff.

If we replace v_R and R in Fig. 12-6 with their Norton equivalents i_G and G, as shown in Fig. 12-7, and observe that if a voltage and a current are related by an impedance Z their power densities are related by $|Z|^2$, we obtain that the power density $i_G^2(f)$ of the noise current generated by a conductance G is given by an alternative form of the Nyquist formula,

$$i_G^2(f) = 4kTG$$

12-4-3   Thermal Noise in the Drude–Lorentz Model

As a preparation for the exercises and, more generally, for discussions in terms of models in the literature, we will now use the Drude–Lorentz model to obtain the power spectrum of thermal noise. Let us consider, as shown in Fig. 12-8*a*, a cylindrical resistor with perfectly conducting end plates of area A and separation l, and let us assume that the resistor is composed of a material of conductivity σ and permittivity ε, and that the density of free electrons in the material is n.

To obtain the current $i_e(t)$ generated in the external circuit by the motion of one electron, we will imagine that we apply a voltage V to the resistor. If

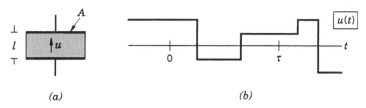

Figure 12-8. (*a*) Cross section of a cylindrical resistor with perfectly conducting end plates of area *A* and separation *l*. (*b*) Sample function of the axial velocity $u(t)$ of an electron.

the dielectric relaxation time $\tau_D = \varepsilon/\sigma$ is much smaller than the mean free time τ_F, charge neutrality is quickly reestablished as the electron moves, so that the axial electric field on the electron is essentially V/l. We may therefore obtain $i_e(t)$ as we did in the case of a thermionic diode in Section 12-4-1 by equating the power $Vi_e(t)$ absorbed by the source of V and the power delivered by the electron, which is $(eV/l)u(t)$ if the (axial) velocity of the electron is $u(t)$. We then have

$$i_e(t) = \frac{e}{l}u(t) \tag{12-18}$$

This result is independent of the applied voltage V and therefore valid if the resistor is short circuited, which is the situation we want to consider.

A sample function of the velocity is shown in Fig. 12-8*b*. If there are no collisions between $t = 0$ and $t = \tau$, which occurs with probability $\exp(-\tau/\tau_F)$, we have $u(\tau) = u(0)$. On the other hand, $u(\tau)$ and $u(0)$ are independent if there is at least one collision, and contribute nothing to $\langle u(0)u(\tau)\rangle$ because $\langle u \rangle$ is zero. We thus obtain

$$\langle u(0)u(\tau)\rangle = \langle u^2 \rangle \exp\left(-\frac{\tau}{\tau_F}\right)$$

and therefore

$$\langle i_e(0)i_e(\tau)\rangle = \frac{e^2}{l^2}\langle u^2 \rangle \exp\left(-\frac{\tau}{\tau_F}\right)$$

Since free electrons act independently we may use (12-1) to add fluctuations, which in this case simply means multiplying by the number of free electrons in the resistor, equal to nAl. The autocorrelation of the noise current generated by the resistor—of conductance G—is thus

$$R_G(\tau) = nAl\frac{e^2}{l^2}\langle u^2 \rangle \exp\left(-\frac{|\tau|}{\tau_F}\right) \tag{12-19}$$

But from equipartition of energy we have $\langle u^2 \rangle = kT/m$, and from (4-8) we have

$$G = \sigma \frac{A}{l} = \frac{ne^2 \tau_F}{m} \frac{A}{l}$$

We thus finally get

$$R_G(\tau) = \frac{kTG}{\tau_F} \exp\left(-\frac{|\tau|}{\tau_F}\right)$$

From (2-14) we then obtain

$$\mathcal{S}_G(f) = \frac{2kTG}{1 + \omega^2 \tau_F^2}$$

We may now neglect τ_F because it is typically 10^{-13} s, and we thus recover the Nyquist formula.

12-4-4 Flicker Noise

The distinguishing characteristic of flicker noise is that its power density is inversely proportional to f raised to a power that is close to 1. We will not miss essential points if we limit our discussion to $1/f$ noise, which has a power density of the form

$$x^2(f) = \frac{x_0^2}{f}$$

where x_0^2 is a constant. We note, however, that in the literature the term "$1/f$ noise" often refers to flicker noise in general. We also note that certain forms of fundamental noise can have a $1/f$ power density. The case of dielectric noise in capacitors is discussed in Exercise 13-4.

Flicker noise is a nonequilibrium phenomenon and thus arises only in powered devices. An unpowered resistor will therefore be free of flicker noise. Transistors, on the other hand, which are interesting only when they are powered, always exhibit some flicker noise.

Flicker noise retains its $1/f$ dependence at the lowest frequencies that have been explored, of order 10^{-6} Hz. As a consequence, we might worry about the possibility of a low-frequency divergence if we do not have a high-pass filter somewhere. We need not worry, however, because there is an effective cutoff at a frequency $1/T$, where T is of the order of the time during which a device is observed: components with frequencies less than the reciprocal of this time are bucked out in the initial setup of the device and do not change in the course of the observation.

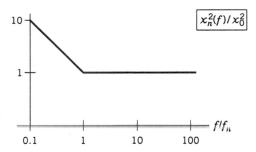

Figure 12-9. Flicker noise is almost always accompanied by fundamental forms of noise that have a constant power density at low frequencies. The contributions from flicker noise and fundamental noise are equal at the $1/f$ corner frequency f_n.

The problem with flicker noise is that unlike thermal or shot noise it is hard to suppress with low-pass filters. Let us consider what happens if we apply a signal contaminated by flicker noise to a brick-wall low-pass filter with bandwidth B and observe the output for a time T. For anything interesting to happen, clearly we must have $BT \gg 1$. The mean-square noise $\langle x^2 \rangle$ at the output of the filter is

$$\langle x^2 \rangle = \int_{1/T}^{B} \frac{x_0^2}{f} df = x_0^2 \ln BT$$

The logarithmic dependence of $\langle x^2 \rangle$ tells us that it is pointless to decrease B and that we will have to find other methods to get rid of flicker noise.

Flicker noise is almost invariably accompanied by fundamental forms of noise, thermal noise for example, that have a constant power density x_0^2 at low frequencies. As shown in Fig. 12-9, low-frequency power densities thus generally have the form

$$x_n^2(f) = x_0^2 \left(1 + \frac{f_n}{f} \right)$$

where f_n is the $1/f$ *corner frequency*, that is, the frequency at which the contributions from $1/f$ and fundamental noise are equal. The location of the corner frequency is highly device dependent: it can range from 1 Hz in bipolar transistors to tens of megahertz in microwave devices.

Flicker noise is decidedly a nuisance, but there is one good thing in its favor: historically speaking, it has diminished steadily as a result of improved manufacturing techniques.

EXERCISES

12-1 LABORATORY: THERMAL NOSE. The setup in Fig. 12-10 will allow you to get a rough measure of the thermal noise generated by resistor $R = 100$ kΩ. Look at Exercise 9-6 if it is not clear to you that buffer A1 stabilizes amplifier A2 and thus reflects prudent design. Thermal noise was

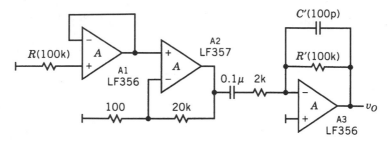

Figure 12-10. Circuit for a measurement of thermal noise (Exercise 12-1).

first studied systematically by Johnson (1928); you will profit by looking up his paper.

The noise voltage density due to R is

$$\sigma_R = \sqrt{4kTR} = \sqrt{4eV_T R} = 40 \text{ nV}/\sqrt{\text{Hz}}$$

To see whether this value is measurable, you have to check whether it is above the *amplifier noise* generated by A1 and A2 (A3 does not count because A2 has a large gain). You will learn more about amplifier noise in Chapter 13, but accept for now that an operational amplifier has a voltage noise source in series with, say, the positive input, and a current noise source from each input to ground. The input noise current density of an LF356 is 10 fA/$\sqrt{\text{Hz}}$. The amplifier input noise current density, due only to A1, is thus

$$i_A = 10 \text{ fA}/\sqrt{\text{Hz}}$$

The equivalent voltage density is negligible,

$$Ri_A = 1 \text{ nV}/\sqrt{\text{Hz}}$$

The input noise voltage density of an LF356/357 is 12 nV/$\sqrt{\text{Hz}}$, and the amplifier input noise voltage density, due to A1 and A2, is $\sqrt{2}$ times larger,

$$\sigma_A = 17 \text{ nV}/\sqrt{\text{Hz}}$$

You can live with this number because σ_R and σ_A add quadratically.

The time constants in the third-stage filter are 200 μs (0.8 kHz) and 10 μs (16 kHz), so that the power bandwidth B is determined mainly by R' and C' (the low-frequency cutoff is of course meant to get rid of flicker noise). From (12-12) you have

$$B \simeq \frac{1}{4R'C'} = 25 \text{ kHz}$$

The bandwidths of an LF356 and an LF357 at closed-loop gains of 50 and 200 are both 80 kHz, so you are reasonably safe. Given the relatively low value of R, the requirements on stray capacitance are not stringent. Since the midband gain is 10^4, the rms output voltage is easily observed on an oscilloscope,

$$\sqrt{\langle v_O^2 \rangle} \simeq 10^4 \sqrt{\sigma_R^2 B} \simeq 60 \text{ mV}$$

To ensure that things are coming from where they should, ground the positive input of A1 and check that the rms output voltage drops by at least a factor of 2. If you have been careful in your layout, you should be able to make R as high as 1 MΩ without affecting the power bandwidth and thus check that σ_R^2 is proportional to R.

12-2 POISSON PROCESSES ARE PURELY RANDOM. Consider a Poisson process with average frequency λ. The probability of N events in an interval T is

$$P(N,T) = \frac{(\lambda T)^N}{N!} \exp(-\lambda T)$$

Now consider N nonoverlapping subintervals $\tau_1 \cdots \tau_N$ inside T. Since Poisson processes in nonoverlapping intervals are independent, the probability of (exactly) one event in each subinterval and of zero events in the remainder of T is

$$P(1, \tau_1; \cdots; 1, \tau_N) = (\lambda \tau_1)(\lambda \tau_2) \cdots (\lambda \tau_N) \exp(-\lambda T)$$

The conditional probability of one event in each subinterval given that there are N events in T is therefore

$$P(1, \tau_1; \cdots; 1, \tau_N | N, T) = N! (\tau_1/T)(\tau_2/T) \cdots (\tau_N/T)$$

But this is precisely what you get if you calculate the probability of one event in each subinterval on the assumption that the times at which events occur are independent random variables with a uniform distribution $p(t) = 1/T$. To see this, observe that you have N choices for which event falls in τ_1, then $N-1$ choices for which event falls in τ_2, and so on.

To connect with the literature, consider a Poisson process $x(t)$ with pulses of shape $h(t)$, and consider further an interval T, much longer than the characteristic time of $h(t)$, in which N pulses occur at times $t_1 \cdots t_N$. For t not too close to the beginning of the interval, you have

$$x_N(t) = \sum_{n=1}^{N} h(t - t_n)$$

Neglecting edge effects again, and using the fact that event times are uniformly distributed in T, for the *ensemble* average of x_N you obtain

$$\langle x_N \rangle = \int_{-T/2}^{+T/2} \frac{dt_1}{T} \cdots \int_{-T/2}^{+T/2} \frac{dt_N}{T} \sum_{n=1}^{N} h(t - t_n) = \frac{N}{T} \int_0^\infty h(t) \, dt$$

(The result for the *time* average is the same, as you can quickly verify.) Averaging over N you then get the simplest version of Campbell's theorem of the mean,

$$\langle x \rangle = \langle\langle x_N \rangle\rangle = \lambda \int_0^\infty h(t)\, dt$$

With similar arguments you can prove Carson's theorem.

12-3   LINEAR FILTERS. Let $x(t)$, $h(t)$, and $y(t)$ be the input, impulse response, and output of a linear filter. Verify that the steps below lead to (12-9) and (12-10), and therefore to (12-11). To begin with, you have

$$R_y(\tau) = \left\langle \int_{-\infty}^{+\infty} h(u)x(t-u)\, du \int_{-\infty}^{+\infty} h(v)x(t+\tau-v)\, dv \right\rangle$$

$$= \int_{-\infty}^{+\infty} h(u)\, du \int_{-\infty}^{+\infty} h(v) R_x(\tau+u-v)\, dv$$

Substitute $w = v - u$ to obtain

$$R_y(\tau) = \int_{-\infty}^{+\infty} h(u)\, du \int_{-\infty}^{+\infty} h(w+u) R_x(\tau-w)\, dw$$

$$= \int_{-\infty}^{+\infty} R_x(\tau-w)\, dw \int_{-\infty}^{+\infty} h(u)h(w+u)\, du$$

or

$$R_y(\tau) = R_x * R_h(\tau)$$

Now calculate the Fourier transform of $R_h(\tau)$:

$$\mathcal{S}_h(\omega) = \int_{-\infty}^{+\infty} \exp(-i\omega\tau)\, d\tau \int_{-\infty}^{+\infty} h(u)h(u+\tau)\, du$$

$$= \int_{-\infty}^{+\infty} h(u)\, du \int_{-\infty}^{+\infty} h(u+\tau) \exp(i\omega\tau)\, d\tau$$

Substitute $v = \tau + u$ to obtain

$$\mathcal{S}_h(\omega) = \int_{-\infty}^{+\infty} h(u) \exp(i\omega u)\, du \int_{-\infty}^{+\infty} h(v) \exp(-i\omega v)\, dv$$

or

$$\mathcal{S}_h(\omega) = |H(\omega)|^2$$

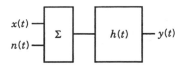

$x(t)$
$n(t)$
Σ $h(t)$ $y(t)$

Figure 12-11. Block diagram for Exercise 12-4.

12-4 NOISE CAN BE PUT TO GOOD USE. In the linear system shown in Fig. 12-11, a small amount $n(t)$ of noise with mean zero is added to the input $x(t)$. The cross-correlation of the noise and the output $y(t)$ is

$$R_{ny}(\tau) = \lim_{T \to \infty} \frac{1}{2T} \int_{-T}^{+T} n(t) \, dt \int_{-\infty}^{+\infty} h(u)[x(t+\tau-u)+n(t+\tau-u)] \, du$$

Exchanging the order of integration and observing that $R_{nx}(\tau) = 0$ because the input and the noise are independent random processes and $\langle n \rangle = 0$, you get

$$R_{ny}(\tau) = \int_{-\infty}^{+\infty} h(u) R_n(\tau-u) \, du$$

If the noise bandwidth is large enough you have

$$R_n(\tau-u) \sim \delta(\tau-u)$$

so that

$$R_{ny}(\tau) \sim h(\tau)$$

You can thus obtain $h(\tau)$ without ever interrupting system operation. You will find an example in Lee (1960), p. 341 ff.

12-5 POWER SPECTRA ARE NONNEGATIVE. First observe that

$$I(T) = \left\langle \left| \int_0^T x(t) \exp(-i\omega t) \, dt \right|^2 \right\rangle \geq 0$$

and that

$$I(T) = \left\langle \int_0^T x(u) \exp(+i\omega u) \, du \int_0^T x(v) \exp(-i\omega v) \, dv \right\rangle$$

$$= \int_0^T du \int_0^T dv \, R_x(v-u) \exp[-i\omega(v-u)]$$

Using the substitution $v-u = \tau$, obtain

$$I(T) = \int_0^T du \int_{-u}^{T-u} d\tau \, R_x(\tau) \exp(-i\omega\tau)$$

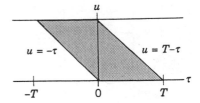

Figure 12-12. Domain of integration for Exercise 12-5.

The domain of integration is shown in Fig. 12-12. Exchanging the order of integration, you get

$$I(T) = \int_{-T}^{0} R_x(\tau) \exp(-i\omega\tau)\, d\tau \int_{-\tau}^{T} du + \int_{0}^{T} R_x(\tau) \exp(-i\omega\tau)\, d\tau \int_{0}^{T-\tau} du$$

Integrating over u, you obtain a generalization of (12-4),

$$I(T) = T \int_{-T}^{+T} \left(1 - \frac{|\tau|}{T} \right) R_x(\tau) \exp(-i\omega\tau)\, d\tau$$

For simplicity, consider only the case in which $R_x(\tau)$ is absolutely integrable. Now look up Lebesgue's theorem on dominated convergence in any book on integration theory, and use it show that $\mathcal{S}_x(\omega) \geq 0$ by considering the family of functions

$$R_T(\tau) = \frac{|\tau|}{T} R_x(\tau)\, \theta(T - |\tau|) \exp(-i\omega\tau)$$

12-6 BROWNIAN MOTION AND ITS CIRCUIT ANALOGS. (Einstein, 1905, 1906; Langevin, 1908). In his 1905 paper on Brownian motion, Einstein predicted that the mean-square one-dimensional displacement $\langle x^2(t) \rangle$ of a small particle in a viscous medium is given by

$$\langle x^2(t) \rangle = 2\mu kTt$$

where μ is the mobility of the particle (in m/N-s) and t is the time since the first observation. In this exercise you will obtain Einstein's result by using a method first proposed by Langevin.

If the particle has a velocity u, it suffers an *average* drag force

$$\langle F \rangle = -\frac{u}{\mu}$$

Following Langevin, you can write

$$mu' = -\frac{u}{\mu} + F(t)$$

where $F(t)$ is now the (rapidly) fluctuating part of the force and m is the mass of the particle. In terms of

$$a(t) = \frac{F(t)}{m}$$

and the correlation time

$$\tau_C = m\mu$$

you have

$$\tau_C u' + u = \tau_C a(t)$$

For reference, note that $\tau_C \simeq 100$ ns in water for a sphere 1 μm in diameter. Now make the assumption that

$$\langle a(t)a(t+\tau)\rangle = K\delta(\tau)$$

where K is a constant to be determined. From the convolution theorem you have

$$u(t) = \exp\left(-\frac{t}{\tau_C}\right)\int_{-\infty}^{t} \exp\left(\frac{\xi}{\tau_C}\right)a(\xi)\,d\xi$$

You also have

$$\langle u(t)u(t+\tau)\rangle = \exp\left(-\frac{2t+\tau}{\tau_C}\right)\int_{-\infty}^{t} d\xi \int_{-\infty}^{t+\tau} d\eta\, \exp\left(\frac{\xi+\eta}{\tau_C}\right)\langle a(\xi)a(\eta)\rangle$$

The integral over η only goes out to t because $\langle a(\xi)a(\eta)\rangle = K\delta(\xi-\eta)$, and you get

$$R_u(\tau) = \frac{K\tau_C}{2}\exp\left(-\frac{|\tau|}{\tau_C}\right)$$

But from equipartition you know that

$$R_u(0) = \langle u^2\rangle = \frac{kT}{m}$$

So finally you get

$$R_u(\tau) = \frac{kT}{m}\exp\left(-\frac{|\tau|}{\tau_C}\right)$$

Using (2-14) you can now obtain the power spectrum of the velocity,

$$\mathcal{S}_u(f) = \frac{kT}{m}\frac{2\tau_C}{1+\omega^2\tau_C^2}$$

Substituting $\tau_C = m\mu$ you get an expression that is, as you will see below, an analog of the power spectrum of bandlimited thermal noise,

$$\mathcal{S}_u(f) = \frac{2\mu kT}{1+\omega^2\tau_C^2}$$

Now consider the displacement of the particle,

$$x(t) = \int_0^t u(\xi)\,d\xi$$

From (12-4) you have

$$\langle x^2(t)\rangle = t\int_{-t}^{+t}\left(1-\frac{|\tau|}{t}\right)R_u(\tau)\,d\tau$$

Substituting $R_u(\tau)$ as obtained above and $\tau_C = m\mu$, you get

$$\langle x^2(t)\rangle = (2\mu kTt)\left\{1-\frac{\tau_C}{t}\left[1-\exp\left(-\frac{t}{\tau_C}\right)\right]\right\}$$

For $t \gg \tau_C$ you obtain Einstein's result,

$$\langle x^2(t)\rangle = 2\mu kTt$$

For $t \ll \tau_C$, however, the inertia of the particle still plays a role, so that memory of the initial velocity is not lost, and you have

$$\langle x^2(t)\rangle = \frac{kT}{m}t^2 = \langle u^2\rangle t^2$$

Brownian motion in viscous media has direct analogs in electric circuits. Consider, for example, the circuit of Fig. 12-13, in which $v_R(t)$ is the noise voltage associated with resistor R. The current in inductor L is given by

$$\tau_C i' + i = \tau_C\frac{v_R}{L}$$

where $\tau_C = LG$. If you assume that $\langle v_R(t)v_R(t+\tau)\rangle = K\delta(\tau)$ and you make the appropriate substitutions in the arguments above, you obtain that the

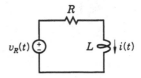

Figure 12-13. *RL* circuit for Exercise 12-6.

power spectrum of $i(t)$ is, as you should expect, equal to the power spectrum $2RkT$ of $v_R(t)$ divided by $R^2 + \omega^2 L^2$,

$$\mathcal{S}_i(f) = \frac{2GkT}{1 + \omega^2 \tau_C^2}$$

Completing the analogy, you obtain that the mean square of the charge $q(t)$ through the inductor increases linearly with time, a result also predicted by Einstein (1906):

$$\langle q^2(t) \rangle = 2GkTt$$

For a highly readable account of Brownian motion, see MacDonald (1962).

12-7   THE CURRENT IN IDEAL RESISTORS IS VIRTUALLY NOISE-LESS. In this exercise you will see that the noise due to current flow in a resistor is normally much less than the thermal noise, and also much less than would be predicted by the Schottky formula.

Consider the resistor of Fig. 12-8a and assume that it can be described by the Drude–Lorentz model. If a constant voltage is applied to the resistor, the free electrons acquire an average axial drift velocity u_D. If the time elapsed since its last collision is t, an electron has an (excess axial) velocity

$$u(t) = \frac{t}{\tau_F} u_D$$

where τ_F is the mean free time between collisions. According to (12-18), the current induced in the external circuit is

$$i_e(t) = \frac{e}{l} u(t)$$

A sample function of the velocity of an electron is shown in Fig. 12-14. To obtain the autocorrelation $\langle u(0)u(\tau) \rangle$ of the velocity, you need to know the density $p(t'' \,|\, t')$ of the conditional probability that the last collision before $t = \tau$ occurs at $t = \tau - t''$ given that the last collision before $t = 0$ occurs at $t = -t'$. If N is the number of collisions between $t = 0$ and $t = \tau$, you have either $N = 0$ or $N > 0$, and you can write

$$p(t'' \,|\, t') = p(t'' \,|\, t'; N = 0) + p(t'' \,|\, t'; N > 0)$$

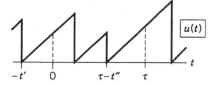

Figure 12-14. Sample function of the excess axial velocity of an electron in a powered resistor (Exercise 12-7).

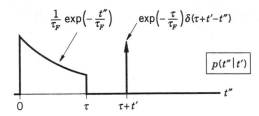

Figure 12-15. Density $p(t''\mid t')$ of the conditional probability that the last collision of an electron before $t = \tau$ occurs at $t = \tau - t''$ given that the last collision before $t = 0$ occurs at $t = -t'$ (Exercise 12-7).

If $N = 0$, you have $t'' = \tau + t'$ with probability $\exp(-\tau/\tau_F)$ or, expressed as a density,

$$p(t'' \mid t'; N = 0) = \exp\left(-\frac{\tau}{\tau_F}\right)\delta(\tau + t' - t'')$$

If $N > 0$, t'' becomes independent of t', but it is restricted to the interval between 0 and τ; you therefore have

$$p(t'' \mid t'; N > 0) = \frac{1}{\tau_F}\exp\left(-\frac{t''}{\tau_F}\right)[\theta(t'') - \theta(t'' - \tau)]$$

A graph of $p(t'' \mid t')$ is shown in Fig. 12-15. As a check, observe that $p(t'' \mid t')$ approaches $\delta(t'' - t')$ as $\tau \to 0$, that it approaches $p(t'')$ as $\tau \to \infty$, and that its integral over t'' is equal to 1.

The autocorrelation of the velocity is thus

$$\langle u(0)u(\tau)\rangle$$

$$= \frac{u_D^2}{\tau_F^2}\int_0^\infty dt' \frac{t'}{\tau_F}\exp\left(-\frac{t'}{\tau_F}\right)\left[(t'+\tau)\exp\left(-\frac{\tau}{\tau_F}\right) + \int_0^\tau dt'' \frac{t''}{\tau_F}\exp\left(-\frac{t''}{\tau_F}\right)\right]$$

Observing that

$$\int \xi \exp(-\xi)\, d\xi = -(\xi+1)\exp(-\xi) \quad \text{and that} \quad \int_0^\infty \xi^2 \exp(-\xi)\, d\xi = 2$$

you get

$$\langle u(0)u(\tau)\rangle - u_D^2 = u_D^2 \exp\left(-\frac{|\tau|}{\tau_F}\right)$$

You can now use (12-1) to add fluctuations and obtain the autocorrelation $R_D(\tau)$ of the drift current. If l and A are the length and cross-sectional area of the resistor, n the density of free electrons, and $I = enAu_D$ the average drift current, you have

$$R_D(\tau) - I^2 = nAl\frac{e^2}{l^2}u_D^2 \exp\left(-\frac{|\tau|}{\tau_F}\right)$$

Comparing with the expression for the autocorrelation of the thermal noise current in (12-19), it is clear that the power density due to current flow is $u_D^2/\langle u^2 \rangle$ times the thermal power density; except for some cases of velocity saturation in field-effect transistors with submicron lengths, this factor is orders of magnitude smaller than 1.

Ignoring the impulse at $f = 0$, the power spectrum of the current is

$$S_D(f) = \frac{nAe^2u_D^2}{l}\frac{2\tau_F}{1+\omega^2\tau_F^2}$$

Observing again that $I = enAu_D$, you can write the low-frequency power spectrum as

$$S_D(0) = eI\frac{2u_D\tau_F}{l}$$

The mean free time is of order 1 ps, so that even if the drift velocity saturates and is of order 10^7 cm/s, l must be well below 1 μm for the current to have a shot-noise spectrum.

13

SIGNAL RECOVERY

13-1   INTRODUCTION

In this chapter we will look into the question of extracting weak signals from noise. The subject is huge, and we will be able to cover only a few basic techniques. Even then, we will have to progress through fairly long preliminaries in disparate domains before we are ready to analyze significant experiments from the literature.

We will begin by examining the degree to which transistor amplifiers increase the noise in signals from resistive sources. To do so, we will first need to obtain noise models of bipolar transistors and JFETs. (MOSFETs are like JFETs except for a generally much higher level of flicker noise, so that we need not treat them separately.)

Nonlinear devices are essential in some of the signal-recovery techniques we will develop, and we will therefore study how they affect signals and noise. In doing so, we will learn the basics about *modulation* (shifting of signals in frequency) and about *detection* (recovery of the original signals in the case of *lock-in* detection). We will proceed in two stages: first discuss signal manipulation in the absence of noise, and then see how the *signal-to-noise ratio* is affected.

As an interesting topic in its own right, but also as a preparation for the experiments on correlations between photons and on optical beats presented in Exercises 13-12 and 13-13, we will discuss *photomultipliers* both as nonlinear devices for the detection of optical signals and as current amplifiers, and obtain their noise characteristics.

We will introduce the subject of signal recovery by studying the case in which the signal is a (fast) pulse of known shape that is masked by additive white noise, and we will see that there exist *matched* filters that are *optimal* in the sense that they maximize the signal-to-noise ratio. We will also see that we can extend our results to cover arbitrary additive noise spectra.

Of fundamental importance, we will see in the case of a slowly varying signal that flicker noise can be obviated—and less stubborn forms of noise strongly suppressed—if the signal is shifted in frequency *before* it gets contaminated and is then amplified and recovered in a lock-in detector.

We will also see that lock-in detectors are optimal because they can be viewed as matched filters, and for analogous reasons, that (slow) repetitive signals can be optimally recovered by time slicing in *multichannel analyzers*.

Optimal filters are sometimes noncausal, and sometimes simply inconvenient to implement. In several instances we will therefore consider nonoptimal filters, and conclude that we can do rather well with them.

Finally, we will discuss x-ray spectrometers based on lithium-drifted silicon detectors because they offer a compact setting in which noise calculations and signal-recovery techniques can be applied to a complete system.

13-2   NOISE MODELS OF SEMICONDUCTOR DEVICES

The noise models we will discuss are the best possible: except for unavoidable concessions to flicker noise, only fundamental noise sources are taken into account. These models are nonetheless valid for carefully selected devices, and provide bench marks that tell us when to stop looking for a better device. The presentation that follows is brief and qualitative; detailed discussions are offered in Exercises 13-9 and 13-10.

13-2-1   Active Bipolar Transistors

The noise generators of an active bipolar transistor are given in Fig. 13-1. We will consider that these generators are external to a *noiseless* transistor described by the Shockley model and thus by the large-signal and small-signal models discussed in Section 6-3. The principal consequence of this assumption is that noise due to recombination in depletion regions is negligible.

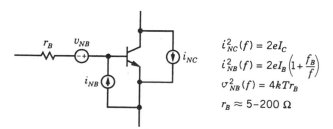

$$i^2_{NC}(f) = 2eI_C$$

$$i^2_{NB}(f) = 2eI_B\left(1+\frac{f_B}{f}\right)$$

$$\sigma^2_{NB}(f) = 4kTr_B$$

$$r_B \approx 5\text{--}200 \ \Omega$$

Figure 13-1. Noise generators of an active bipolar transistor.

The *base series resistance* r_B is a rough representation of the distributed resistance in the neutral regions of the base; it is adequate, however, for noise calculations below the transition frequency f_T. Typically, r_B is around 100 Ω, although values as low as 5 Ω are possible.

Minority carriers injected into the base arrive at the collector–base junction at random times and are swept into the collector by a strong electric field. This is essentially what happens in temperature-limited thermionic diodes, and i_{NC}^2 is therefore given by the Schottky formula. This result is obtained from a strikingly different point of view in Exercise 13-9.

At frequencies low enough that we can ignore junction capacitances, the noise in the recombination component of the base current is given by the Schottky formula because recombination events occur at random times and the base terminal supplies one majority carrier for each event. As shown in Exercise 13-9, the noise in the diffusion component of the base current, that is, the component that crosses the base–emitter junction, is also given by the Schottky formula at low frequencies, but we must emphasize that this result *cannot* be obtained from simple arguments.

We will accept that a reasonable (actually slightly pessimistic) noise model of an active bipolar transistor is obtained by assuming that i_{NB}^2 as well as i_{NC}^2 is given by the Schottky formula at high frequencies, as indicated in Fig. 13-1, and letting junction capacitances and the external circuit account for the frequency dependence of the noise. We will also assume that the $1/f$ corner frequency f_B in i_{NB}^2 is constant for a given transistor.

13-2-2   Junction Field-Effect Transistors in Saturation

The noise generators of a junction FET in the saturation region are given in Fig. 13-2. As we did for bipolar transistors, we will consider that these generators are external to a noiseless device, described in this case by the large-signal and small-signal models discussed in Section 6-4-1.

The channel in a junction FET—clearly so in the ohmic region—is a chain of differential resistors of varying magnitude. The transconductance g_m and the channel conductance at zero drain–source voltage are equal, and we

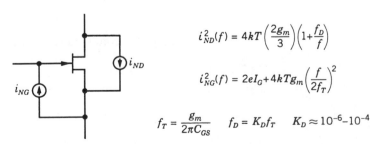

$$i_{ND}^2(f) = 4kT\left(\frac{2g_m}{3}\right)\left(1+\frac{f_D}{f}\right)$$

$$i_{NG}^2(f) = 2eI_G + 4kTg_m\left(\frac{f}{2f_T}\right)^2$$

$$f_T = \frac{g_m}{2\pi C_{GS}} \qquad f_D = K_D f_T \qquad K_D \approx 10^{-6}-10^{-4}$$

Figure 13-2. Noise generators of a saturated junction field-effect transistor.

might therefore expect that some factor times g_m is the conductance that should be inserted in the Nyquist formula to give the thermal contribution to i_{ND}^2. In Exercise 13-10 it is shown that this factor is $\frac{2}{3}$.

The thermal noise in the differential resistors modifies the voltage along the channel; if the gate is shorted to the source, a noise current is injected into the gate through the gate–source capacitance C_{GS}. Ignoring the $1/f$ term in i_{ND}^2, a dimensional argument suggests that the thermal contribution to i_{NG}^2 should be proportional to $(i_{ND}^2/g_m^2)\omega^2 C_{GS}^2$, and therefore to $i_{ND}^2(f^2/f_T^2)$ if we neglect the gate–drain capacitance C_{GD}. In Exercise 13-10 it is shown that this argument is correct.

As in all junction diodes, the reverse current in the gate–channel diode of a junction FET is due overwhelmingly to generation of electron–hole pairs in the depletion region. If we assume that generation of a carrier of one sign is quickly followed by generation of a carrier of the opposite sign, and that a short pulse of total charge e thus circulates in the external circuit for each generation event, we conclude that the power density of the noise in the reverse current is given by the Schottky formula because events occur at random times. This explains the term $2eI_G$ in i_{NG}^2. In a more refined analysis that takes into account that there may be a delay between the generation of the carriers in a single event, the Schottky formula is multiplied at high frequencies by a factor that is somewhat less than 1. We will ignore this refinement, however, because it comes into play well above the frequency— generally several hundred kilohertz—at which the thermal part of i_{NG}^2 becomes dominant.

As pointed out by Radeka (1984), the $1/f$ corner frequency f_D in JFETs of a given technology is proportional to g_m/C_{GS} or, equivalently at constant drain current, to the reciprocal square root of the device width. This is one of the reasons why low-noise JFETs are often wide. As we saw in Section 7-8, another good reason for this choice is that it increases the amplification factor μ at constant drain current and therefore the maximum gain obtainable in a single stage. The price is a lower transition frequency.

13-3   LINEAR AMPLIFIERS

The input stage of a low-noise amplifier is invariably in the common-emitter or common-source configuration if we accept that these terms cover cascode and differential amplifiers. Just as invariably, the conductance of the collector (drain) load is chosen much smaller than the transconductance g_m in order to make its noise current negligible compared with the collector (drain) noise current. As a result, the input stage has a large gain, and it therefore determines the noise behavior of the entire amplifier because noise from succeeding stages, when referred to the input, is reduced by the overall gain of intervening stages. In addition, as shown in Exercise 13-1, feedback increases the noise only slightly if the closed-loop gain is above 10 or so. It follows that if we want to understand a low-noise amplifier we need only consider its input stage in an open-loop configuration.

13-3-1 Figures of Merit

Noise in linear amplifiers is often quantified in the literature by the two figures of merit we will now introduce, the *noise figure* and the *noise temperature*. Noting that it applies far beyond the domain of amplifiers, we will also introduce the *signal-to-noise ratio* (SNR).

Let us consider, as shown in Fig. 13-3, a noisy amplifier driven by a voltage source v_S and by the noise voltage v_R generated by the output resistance R_S of the source. As we will see below in the case of bipolar transistors, the noise sources in the amplifier may all be represented by a single voltage source v_A in series with v_S and v_R, and the amplifier proper is then noiseless. We note, however, that v_A may depend on R_S as well as on the frequency.

Since v_R and v_A are independent and v_R has mean zero, from Section 12-3-8 we obtain that the power density of the total noise $v_N = v_R + v_A$ is given by

$$\sigma_N^2 = \sigma_R^2 + \sigma_A^2 = \sigma_R^2 \left(1 + \frac{\sigma_A^2}{\sigma_R^2} \right)$$

We will define the *noise figure F* of the amplifier by

$$F = 1 + \frac{\sigma_A^2}{\sigma_R^2}$$

For a noiseless amplifier we thus have $F = 1$. Recalling that σ_R^2 is given by the Nyquist formula, we may write

$$\sigma_N^2 = 4kT_S R_S F$$

where T_S is the absolute temperature of source resistance R_S. We note that a quoted value of F is in principle meaningless if T_S is not quoted at the same time; in the literature, however, it is often tacitly assumed that $T_S = 293$ K.

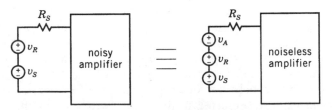

Figure 13-3. The noise generators of an amplifier driven by a voltage source v_S and by the noise voltage v_R of the source output resistance R_S can be represented by a single voltage source v_A in series with v_S and v_R.

We also note that F is expressed in (power) decibels in the literature; the quoted value is thus

$$F \text{ (dB)} = 10 \log F$$

A source with a nonzero output resistance has an irreducible thermal noise level, and the noise figure indicates how much this level is increased by an amplifier. This may be significant or not: if we are measuring the voltage in a thermocouple of negligible resistance, for example, the fact that the noise figure is huge is irrelevant because we only require that the total noise be less than the minimum variation we want to detect. The noise figure is often the only specification available, however, and if it is given as a function of the source resistance and the frequency, we can use it to determine σ_A.

Closely related to the noise figure is the *noise temperature* T_A defined by

$$\sigma_N^2 = 4k(T_S + T_A)R_S$$

The noise temperature thus tells us how much hotter the source resistance has to be in order to simulate the noise generated by an amplifier. We note that the noise temperature has little to do with the actual temperature of an amplifier and that it can range from a few kelvin to thousands of kelvin. We also note that the noise temperature is given in terms of the noise figure by

$$T_A = (F-1)T_S$$

As suggested by the example of a thermocouple mentioned above, we often want a measure of how clearly a signal stands out above the noise—or does not. Let us consider the common case in which a noise process $n(t)$ with mean zero and power spectrum $\mathcal{S}_n(f)$ is added to a signal process $s(t)$ with power spectrum $\mathcal{S}_s(f)$. Assuming that $s(t)$ and $n(t)$ are independent, the power spectrum of the sum $s(t)+n(t)$, again according to Section 12-3-8, is $\mathcal{S}_s(f)+\mathcal{S}_n(f)$, and it thus makes sense to speak separately of the average power $S^2 \equiv \langle s^2 \rangle$ in the signal and the average power $N^2 \equiv \langle n^2 \rangle$ in the noise, and it also makes sense to define the *signal-to-noise power ratio* by

$$\frac{S^2}{N^2} = \frac{\langle s^2 \rangle}{\langle n^2 \rangle} = \frac{\int_{-\infty}^{\infty} \mathcal{S}_s(f)\, df}{\int_{-\infty}^{\infty} \mathcal{S}_n(f)\, df}$$

In common usage, which we will follow on occasion, "signal-to-noise ratio" can denote the ratio S/N of the (rms) *signal* S to the (rms) *noise* N as well as the power ratio S^2/N^2, so that caution must be exercised with quoted numerical values.

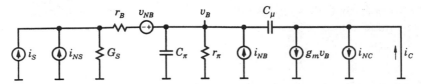

Figure 13-4. Noise model of a bipolar transistor in the common-emitter configuration.

13-3-2   Noise Figure of Bipolar Transistors

The noise model of a bipolar transistor in the common-emitter configuration is shown in Fig. 13-4. Current source i_{NS} is the noise generator associated with the output conductance G_S of the source, and v_{NB}, i_{NB}, and i_{NC} are the noise generators given in Fig. 13-1. The collector may be grounded as shown because a (noiseless) finite collector impedance does not affect the noise figure.

Although this model looks formidable, we can simplify it considerably with a bit of foreknowledge. It will turn out that low noise figures require $r_B \ll R_S$ and $r_\pi = \sqrt{\beta} R_S$. We may therefore open-circuit r_π and, while keeping its noise generator v_{NB}, short-circuit r_B. We may then also transform v_{NB} into a current source $G_S v_{NB}$ and represent i_{NC} with a source $i'_{NC} = r_m[G_S + i\omega(C_\pi + C_\mu)]i_{NC}$ in the base circuit if we neglect the minute current that i'_{NC} injects into the collector circuit through C_μ. We then obtain the circuit shown in Fig. 13-5.

Assuming that the transistor and the source are at the same temperature T, the power density i_N^2 of the total noise current is given by

$$i_N^2 = 4kTG_S + \left(\frac{2}{\beta}\right)eI_C\left(1+\frac{f_B}{f}\right) + 2eI_C r_m^2\left[G_S^2 + \omega^2(C_\pi + C_\mu)^2\right] + 4kTr_B G_S^2$$

Substituting $eI_C = kTg_m$, and defining

$$f_0 = \frac{1}{2\pi R_S(C_\pi + C_\mu)}$$

we can express the noise figure $F = i_N^2/4kTG_S$ as

$$F = 1 + \frac{g_m R_S}{2\beta}\left(1+\frac{f_B}{f}\right) + \frac{1}{2g_m R_S}\left(1+\frac{f^2}{f_0^2}\right) + \frac{r_B}{R_S}$$

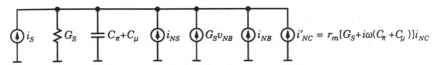

Figure 13-5. Simplified version of Fig. 13-4.

The base contribution becomes large at large collector currents, whereas the collector contribution becomes large at low collector currents. Minimizing F with respect to g_m in the midband range $f_B \ll f \ll f_0$, we obtain

$$g_m R_S = \sqrt{\beta}$$

and thus

$$F = 1 + \frac{1}{\sqrt{\beta}} \left(1 + \frac{f_R}{2f} + \frac{f^2}{2f_0^2} \right) + \frac{r_R}{R_S}$$

In the midband range we have

$$F = 1 + \frac{1}{\sqrt{\beta}} + \frac{r_B}{R_S}$$

The best possible noise figure F_0 is obtained for $r_B \ll R_S$,

$$F_0 = 1 + \frac{1}{\sqrt{\beta}}$$

Values of β in the hundreds are easily available, and F_0 can thus be quite close to 1. For $\beta = 400$, for example, we have

$$F_0 = 1.05$$

Since r_B ranges from 5 to 200 Ω in small-signal transistors, midband noise figures comparable to F_0 are generally obtained for source resistances above 1 kΩ, and the corresponding values of the collector current are reasonable. For $R_S = 10$ kΩ and $\beta = 400$, for example, we have $g_m = \sqrt{\beta}/R_S = 2$ mS, so that

$$I_C = \frac{kT}{e} g_m = 50 \ \mu A$$

Good noise figures can be obtained for larger source resistances, but only up to a point because β decreases at low collector currents. As shown in Exercise 13-2, JFETs can have better noise figures than bipolar transistors for source resistances above 100 kΩ, but because of their relatively low transconductance they generally have worse noise figures for source resistances under 10 kΩ. As an interesting and perhaps unexpected oddity, we note that high-frequency bipolar transistors, which by necessity have small values of r_B, C_π, and C_μ, are prime candidates for low-noise service.

A given amplifier is not in general tailored to a given source resistance. The noise figure of the amplifier can be optimized, however, by inserting a low-loss *matching transformer* of turns ratio n between the source and the amplifier, as shown in Fig. 13-6. If the short-circuit current and the output conductance of the source are i_S and G_S, the amplifier sees a source with short-circuit current ni_S and output conductance $n^2 G_S$. For simplicity, let us

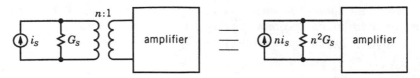

Figure 13-6. An ideal transformer can be used to choose the source resistance seen by an amplifier so as to optimize the noise figure.

assume that i_S is a random sine wave of amplitude i_0 and that the amplifier has a brick-wall response over a band of width B centered at the frequency of the sine wave. The signal-to-noise power ratio

$$\frac{S^2}{N^2} = \frac{\frac{1}{2}(ni_0)^2}{4kT_S(n^2G_S)B} = \frac{\frac{1}{2}i_0^2}{4kT_SG_SB}$$

is not affected by the transformer, and the turns ratio can thus be chosen to optimize the source conductance seen by the amplifier.

13-3-3   Low-Frequency Noise Models

At low frequencies we can ignore the capacitances in transistors. In the case of bipolar transistors biased for low noise, we have seen that we must have $r_\pi \gg R_S$, and we may therefore assume that r_π is infinite. If we represent the collector current source in Fig. 13-1 with a voltage source in series with the base and substitute $I_B = I_C/\beta$ and $eI_C = kTg_m$, we obtain the low-frequency model shown in Fig. 13-7a. Analogous arguments based on Fig. 13-2 lead to the JFET model shown in Fig. 13-7b.

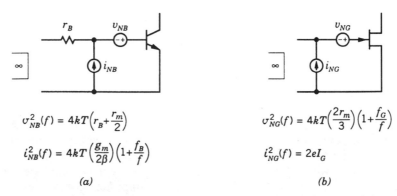

$$v_{NB}^2(f) = 4kT\left(r_B + \frac{r_m}{2}\right)$$

$$i_{NB}^2(f) = 4kT\left(\frac{g_m}{2\beta}\right)\left(1+\frac{f_B}{f}\right)$$

$$v_{NG}^2(f) = 4kT\left(\frac{2r_m}{3}\right)\left(1+\frac{f_G}{f}\right)$$

$$i_{NG}^2(f) = 2eI_G$$

(a)   (b)

Figure 13-7. Low-frequency noise models of (a) bipolar transistors and (b) JFETs.

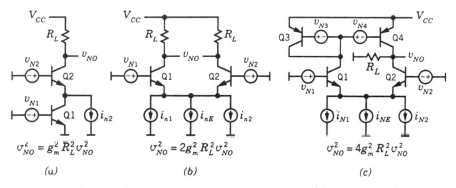

$$\sigma^2_{NO} = g^2_m R^2_L \sigma^2_{NO}$$

(a)

$$\sigma^2_{NO} = 2g^2_m R^2_L \sigma^2_{NO}$$

(b)

$$\sigma^2_{NO} = 4g^2_m R^2_L \sigma^2_{NO}$$

(c)

Figure 13-8. Noise performance of standard input stages. (*a*) Cascode. (*b*) Differential amplifier with resistive collector loads. (*c*) Differential amplifier with a current-mirror load.

To put these models to work, we will compare the low-frequency noise behavior of the standard bipolar input stages shown in Fig. 13-8. For simplicity we will assume that $r_B \ll r_m$ and that the output resistances of signal sources are small, not more than a few times r_m, so that inputs may be grounded as shown. The signal gain (*differential* gain in Fig. 13-8*b*) is $g_m R_L$ in all cases, and we may therefore compare the output noise densities σ^2_{NO}. We will also assume that $g_m R_L$ is large, so that noise due to load resistors is negligible.

In the cascode amplifier of Fig. 13-8*a*, voltage source v_{N2} has no effect on σ^2_{NO} because the emitter load on output transistor Q2 is the large output resistance of input transistor Q1. Current source i_{N2} has a significant effect only at frequencies below f_B/β, and then only if the noise generated in the source resistance by Q1 is utterly negligible. We may therefore conclude that except perhaps at very low frequencies, cascode and common-emitter amplifiers have the same noise behavior, and that the advantages of cascode amplifiers in other respects are not compromised.

In the differential amplifier of Fig. 13-8*b*, we observe that the noise i_{NE} in the emitter bias-current source generates only a common-mode output voltage if the transistors and the load resistors are matched, and that i_{NE} may therefore be ignored. The currents injected into the emitter circuit by sources i_{N1} and i_{N2} also generate common-mode output voltages, so that the emitter terminals of these sources may be grounded; this is normally done in noise models of operational amplifiers. We also observe that differential amplifiers driven by low-resistance sources are twice as noisy as common-emitter amplifiers.

Finally, we observe that the differential amplifier with a current-mirror load of Fig. 13-8*c*, so favored as an input stage in operational amplifiers, is not optimal because load transistors Q3 and Q4 contribute as much to σ^2_{NO} as input transistors Q1 and Q2.

13-4 NONLINEAR DEVICES

We will discuss nonlinear devices at a systems level in order to get quickly to basic techniques of signal manipulation. Real circuits are not implemented quite as we will indicate, but our descriptions will be functionally correct notwithstanding. We will deal mainly with two ideal devices: the (full-wave) *square-law* device and the *multiplier* or *balanced mixer*. For completeness, and because it behaves much like a square-law device under circumstances that we will point out, we will also define the (half-wave) *linear-law* device.

The output y of a square-law device is proportional to the square of the input x; since the constant of proportionality can always be absorbed by the gain of another device, we may set this constant to 1. We will therefore consider that a square-law device is defined by

$$y = x^2$$

To see that small-signal square-law devices are readily obtained, let us consider a semiconductor diode with a forward bias current I_0; if v and i are the incremental voltage and current, for $v \ll V_T = kT/e$ we have

$$I_0 + i = I_0 \exp\left(\frac{v}{V_T}\right) \simeq I_0\left(1 + \frac{v}{V_T} + \frac{v^2}{2V_T^2}\right)$$

If v is a narrowband signal, which will usually be the case for us, the term linear in v can easily be filtered out, and we get

$$\frac{i}{I_0} \simeq \frac{v^2}{2V_T^2}$$

The output of a linear-law device is given by

$$y(x < 0) = 0 \quad \text{and} \quad y(x \geq 0) = x$$

For large signals, a half-wave rectifier consisting of Schottky diode and a resistor in a divider configuration is a good approximation to a linear-law device up to microwave frequencies. At low frequencies, on the other hand, the operational-amplifier half-wave rectifier discussed in Exercise 5-19 is close to ideal even for signals in the millivolt range.

A balanced mixer has two inputs x_1 and x_2, and its output y is the product of x_1 and x_2:

$$y = x_1 x_2$$

Low-frequency examples are the small-signal Gilbert multiplier presented in Section 7-12 and its large-signal extension discussed in Exercise 7-15. High-frequency balanced mixers are usually obtained by subtracting the outputs of square-law devices (called *unbalanced mixers* in this role) that square the sum and difference of the inputs:

$$y = \tfrac{1}{4}[(x_1+x_2)^2-(x_1-x_2)^2] = x_1x_2$$

13-4-1 Noiseless Modulation and Detection

Let us consider a unit amplitude random carrier sine wave of frequency f_c, and a signal process $s(t)$ whose power spectrum is confined to a narrow baseband of width $2f_s \ll 2f_c$ around $f=0$. If we multiply sample functions from these two processes in a balanced mixer (called a *modulator* in this role), the result $x(t)$ is an *amplitude-modulated* sine wave,

$$x(t) = s(t)\cos(\omega_c t + \varphi_c)$$

Assuming now that $s(t)$ and the carrier sine wave are independent, from (12-13) we have

$$\mathcal{S}_x(f) = \tfrac{1}{4}[\mathcal{S}_s(f+f_c)+\mathcal{S}_s(f-f_c)]$$

As shown in Fig. 13-9, the spectrum of $s(t)$ is shifted in frequency to bands centered at $\pm f_c$ but it conserves its shape.

In another form of amplitude modulation, used for example in the AM broadcast band from 550 kHz to 1.6 MHz, a term proportional to the carrier sine wave is added to $x(t)$, so that $\mathcal{S}_x(f)$ exhibits impulses at $\pm f_c$ even in the absence of a signal. Simply multiplying a random sine wave by $s(t)$, as we have done, is therefore called *suppressed-carrier* modulation.

We will now see that *detectors* consisting of a nonlinear device followed by a low-pass filter can be used to recover $s(t)$ or to put $x(t)$ into other useful forms. As shown in Fig. 13-10, the outputs of the nonlinear device and the low-pass filter are denoted by $y(t)$ and $z(t)$. For simplicity, in what follows we will not exhibit the phases of random sine waves.

To begin with, let us consider a *homodyne* or *synchronous* detector, in which the nonlinear device is a balanced mixer whose second input is a copy

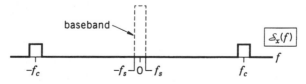

Figure 13-9. Power spectrum of a random sine wave modulated by a low-frequency baseband signal.

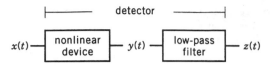

Figure 13-10. Structure of a detector.

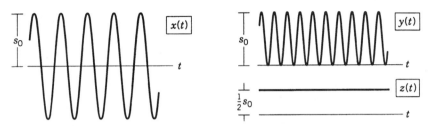

Figure 13-11. Synchronous detection in the case that the baseband signal $s(t)$ has a constant value s_0. Waveforms are labeled according to Fig. 13-10.

of the carrier sine wave. The output of the balanced mixer is

$$y(t) = s(t)\cos^2(\omega_c t) = \tfrac{1}{2}s(t)[1+\cos(2\omega_c t)]$$

The low-pass filter removes frequencies of order $2f_c$, so that except for an irrelevant factor of $\tfrac{1}{2}$, we recover $s(t)$:

$$z(t) = \tfrac{1}{2}s(t)$$

Waveforms for the case in which $s(t)$ has a constant value s_0 are shown in Fig. 13-11; it should be clear that this picture will hardly change if $s(t)$, while not strictly constant, varies little over many cycles of the carrier sine wave.

Let us now consider a *square-law detector*, in which, as the name implies, the nonlinear device is square law. In this case we have

$$y(t) = s^2(t)\cos^2(\omega_c t) = \tfrac{1}{2}s^2(t)[1+\cos(2\omega_c t)]$$

The low-pass filter removes frequencies of order $2f_c$, and we thus obtain that the output of a square-law detector is proportional to the instantaneous power in $s(t)$:

$$z(t) = \tfrac{1}{2}s^2(t)$$

The sign of $s(t)$ is lost in square-law—and linear-law—detection; since a change in the sign of $s(t)$ shifts the phase of $x(t)$ by 180°, we can now see why, in contrast, synchronous detection is said to be *phase sensitive*.

Finally, let us assume that the nonlinear device is a balanced mixer in which $x(t)$ is multiplied by a *local-oscillator* signal, that is, by a random sine wave of unit amplitude and frequency f_l different from f_c. We then have

$$y(t) = s(t)\cos(\omega_c t)\cos(\omega_l t) = \tfrac{1}{2}s(t)[\cos(\omega_- t) + \cos(\omega_+ t)]$$

where $f_+ = f_c + f_l$ and $f_- = |f_c - f_l|$. The low-pass filter removes frequencies of order f_+, and we get

$$z(t) = \tfrac{1}{2}s(t)\cos(\omega_- t)$$

Except again for a factor of $\tfrac{1}{2}$, and assuming that the *intermediate frequency* f_- is much larger than f_s, we have simply shifted the carrier frequency from f_c to f_- in an *intermediate-frequency* (i-f) *converter*. This procedure is used, for example, to shift the carrier frequencies of stations in the AM broadcast band down to 455 kHz, so that most of the signal processing chain need not be modified when changing stations.

13-4-2 Signal-to-Noise Ratio After Detection

Having seen what the ideal behavior of detectors is like, we will now consider the interesting situation in which the modulated signal at the input to the detector is buried in noise. We will assume that the noise $n(t)$ is either gaussian white noise with a constant power spectrum $\mathcal{P}_n$ or, if the total waveform

$$x(t) = s(t)\cos(\omega_c t) + n(t)$$

has been run through a brick-wall filter, that it is *bandlimited* gaussian noise with a constant power spectrum $\mathcal{P}_n$ in bands of width B centered at $\pm f_c$. For simplicity we will further assume that the baseband signal $s(t)$ is either a random sine wave or a constant. We note that we will freely use the results for random sine waves in Section 12-3-4 and for sums and products of independent random processes in Section 12-3-8, and that for economy we will let *signal-to-noise ratio* refer to the *power* ratio.

Let us first consider an intermediate-frequency converter with a total input waveform

$$x(t) = s_0\cos(\omega_c t) + n(t)$$

If the local-oscillator frequency is f_l, the output of the balanced mixer is

$$y(t) = s_0\cos(\omega_c t)\cos(\omega_l t) + n(t)\cos(\omega_l t)$$

We will assume that $n(t)$ is bandlimited and that $\tfrac{1}{2}B < f_- = |f_c - f_l|$. The power spectra $\mathcal{S}_x(f)$ and $\mathcal{S}_z(f)$ at the input and output of the detector are

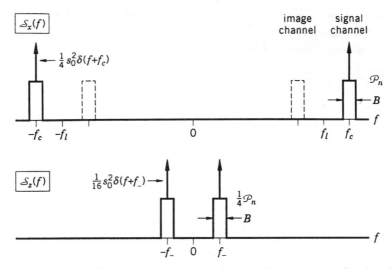

Figure 13-12. Intermediate-frequency conversion in the presence of noise. Power spectra are labeled according to Fig. 13-10.

then as shown in Fig. 13-12. The input and output signal-to-noise ratios are equal,

$$\left(\frac{S^2}{N^2}\right)_O = \left(\frac{S^2}{N^2}\right)_I = \frac{\frac{1}{2}s_0^2}{2\mathscr{P}_n B}$$

We conclude that shifting the carrier frequency of a signal does not affect the signal-to-noise ratio if the input noise is properly bandlimited. The noise that does get through comes from *signal channels* centered at $\pm f_c$; if the band-limiting is done *after* the detector, noise from *image channels* centered at frequencies symmetric around $\pm f_I$ is also shifted to frequencies around $\pm f_-$, so that the magnitude of the in-band power spectrum of the noise becomes $\frac{1}{2}\mathscr{P}_n$ rather than $\frac{1}{4}\mathscr{P}_n$, and the output signal-to-noise ratio is halved.

Now let us consider a synchronous detector with a total input waveform

$$x(t) = s_0 \cos(\omega_c t) + n(t)$$

The output of the balanced mixer is

$$y(t) = s_0 \cos^2(\omega_c t) + n(t) \cos(\omega_c t)$$

We will take $n(t)$ to be white noise. The input and output spectra are then as shown in Fig. 13-13. If we apply a low-pass filter with a (small) power bandwidth B' to the output, we obtain a narrowband synchronous detector or *lock-in* detector, and we have

$$\left(\frac{S^2}{N^2}\right)_O = \frac{\frac{1}{4}s_0^2}{\mathscr{P}_n B'} \tag{13-1}$$

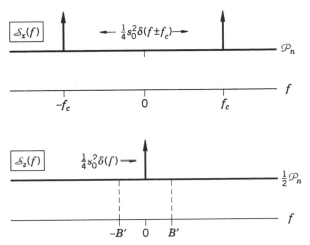

Figure 13-13. Synchronous detection in the presence of noise. Power spectra are labeled according to Fig. 13-10.

The signal-to-noise ratio of a lock-in detector can thus be made as large as is compatible with the time available for a measurement and, when the baseband signal is not constant, with the width of the baseband spectrum. We observe that we get the same signal-to-noise ratio if we run $x(t)$ through a brick-wall filter with passbands of width B' centered at $\pm f_c$. The advantages of lock-in detection are clear, however: the design of low-pass filters with extremely small values of B' is no challenge, and the stability of f_c is not critical.

Finally, let us consider a square-law detector. We will assume that $s(t)$ is a random sine wave of amplitude s_0 and frequency f_m, so that the waveform at the detector input is

$$x(t) = s_0 \cos(\omega_m t)\cos(\omega_c t) + n(t)$$

We will also assume that $n(t)$ is bandlimited and that $f_m \ll \frac{1}{2}B < f_c$; using (12-14) we then obtain the spectrum shown in Fig. 13-14,

$$\mathcal{S}_x(f) = \tfrac{1}{16}s_0^2[\delta(f+f_+)+\delta(f+f_-)+\delta(f-f_-)+\delta(f+f_+)]+\mathcal{S}_n(f)$$

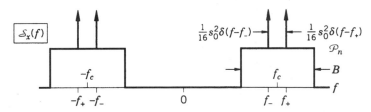

Figure 13-14. Square-law detection in the presence of noise: input spectrum in the case that the baseband signal is a random sine wave of amplitude s_0 and frequency f_m.

where $f_+ = f_c + f_m$ and $f_- = f_c - f_m$. The output of the square-law device is

$$y(t) = [s_0 \cos(\omega_m t) \cos(\omega_c t) + n(t)]^2$$

We are assuming that the noise swamps the signal; we may therefore drop the cross-term when we expand the square, and we then get

$$y(t) \simeq s_0^2 \cos^2(\omega_m t) \cos^2(\omega_c t) + n^2(t)$$

Since the noise is gaussian, from (12-8) we have

$$\mathcal{S}_{n^2}(f) = R_n^2(0)\,\delta(f) + 2\mathcal{S}_n \ast \mathcal{S}_n(f) = 4\mathcal{P}_n^2 B^2 \delta(f) + 2\mathcal{S}_n \ast \mathcal{S}_n(f)$$

Using (12-6) to obtain the power spectra of $\cos^2(\omega_m t)$ and $\cos^2(\omega_c t)$, and then convolving the results and dropping the high-frequency terms, we obtain that the power spectrum at the detector output can be expressed as

$$\mathcal{S}_z(f) = \mathcal{S}_{ss}(f) + \mathcal{S}_{nn}(f)$$

where the *signal–signal* and *noise–noise* contributions $\mathcal{S}_{ss}(f)$ and $\mathcal{S}_{nn}(f)$, shown in Fig. 13-15, are given by

$$\mathcal{S}_{ss}(f) = \tfrac{1}{64}s_0^4 \delta(f + 2f_m) + \tfrac{1}{16}s_0^4 \delta(f) + \tfrac{1}{64}s_0^4 \delta(f - 2f_m)$$

and

$$\mathcal{S}_{nn}(|f| < B) = 4\mathcal{P}_n^2 B^2 \delta(f) + 4\mathcal{P}_n^2 B\left(1 - \frac{|f|}{B}\right)$$

The input signal-to-noise ratio is

$$\left(\frac{S^2}{N^2}\right)_I = \frac{\tfrac{1}{4}s_0^2}{2\mathcal{P}_n B}$$

On the assumption that the dc terms are blocked in the process of eliminating flicker noise, we obtain that the output signal-to-noise ratio is proportional to the square of the input signal-to-noise ratio,

$$\left(\frac{S^2}{N^2}\right)_O = \frac{\tfrac{1}{32}s_0^4}{4\mathcal{P}_n^2 B^2} = \frac{1}{2}\left(\frac{S^2}{N^2}\right)_I^2$$

A poor signal-to-noise ratio thus gets even worse. Let us now assume that we apply the detector output to a brick-wall filter with passbands of width $B' \ll B$ centered at $\pm 2f_m$ (or equivalently, as we have seen, to a lock-in

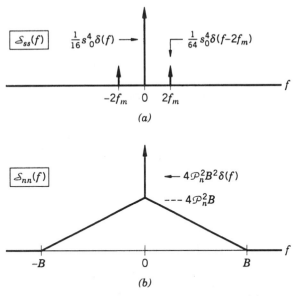

Figure 13-15. Square-law detection in the presence of noise: (a) signal–signal and (b) noise–noise contributions to the output spectrum.

detector with power bandwidth B'). The output signal-to-noise ratio then becomes

$$\left(\frac{S^2}{N^2}\right)_O = \frac{\frac{1}{32}s_0^4}{8\mathcal{P}_n^2 BB'}$$

It is clear that B and B' should both be as small as possible. On the other hand, as shown in Exercises 13-7 and 13-13, if the signal itself is, like the noise, a bandlimited gaussian process, the output signal-to-noise ratio becomes proportional to B/B', and B should then be made as *large* as possible. For reference, we note that the output signal-to-noise ratio of linear-law detectors is also proportional to the square of the input signal-to-noise ratio.

13-4-3 Heterodyne Receivers

Microwave amplifiers are expensive, and it is thus often the case that the first element in a microwave receiver is a linear-law or square-law detector. It is also the case that microwave detectors have $1/f$ corner frequencies in the tens of megahertz. The flicker noise can be obviated, however, by using a *heterodyne* receiver like the one shown in Fig. 13-16: the carrier frequency is shifted to an intermediate frequency above the $1/f$ corner, and the resulting

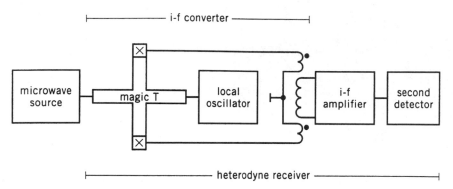

Figure 13-16. Heterodyne receiver.

wave is (inexpensively) amplified and then detected in what is known in the literature as the *second* detector.

We have depicted the i-f converter as a waveguide circuit in order to prepare for Exercises 13-7 and 13-13, but except for unimportant details, we can think of it as if it were a stripline or even a low-frequency circuit. The balanced mixer in the i-f converter is based on a *magic T*, which is the waveguide equivalent of the 3dB ring coupler discussed in Exercise 11-21; the crosses on the upper and lower arms represent square-law detectors. The source and local-oscillator waves add at one detector and subtract at the other, and the outputs of the detectors are subtracted in a transformer whose secondary feeds the i-f amplifier.

The local oscillator in microwave circuits usually generates enough noise in the signal and image channels to swamp a weak signal; a balanced mixer, even an imperfect one, can reduce *local-oscillator noise* to tolerable proportions (Pound, 1948). If we denote the local-oscillator noise by $l(t)$ and the incoming wave by $x(t)$, the input to the i-f amplifier is ideally of the form

$$\tfrac{1}{4}\{[\cos(\omega_l t)+l(t)+x(t)]^2 - [\cos(\omega_l t)+l(t)-x(t)]^2\}$$

$$= \cos(\omega_l t)x(t)+l(t)x(t)$$

The cross-term $\cos(\omega_l t)l(t)$, which contains contributions from the signal and image channels that might be much larger than the cross-term of interest $\cos(\omega_l t)x(t)$, is eliminated by the balanced mixer; the remaining undesirable cross-term $l(t)x(t)$, which contains contributions from frequencies around the local-oscillator frequency, is insignificant because the power of $l(t)$ in the passbands of the i-f amplifier is negligible compared with the power in the local-oscillator wave $\cos(\omega_l t)$.

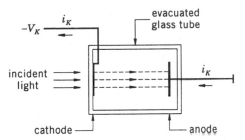

Figure 13-17. Structure of a vacuum photodiode (schematic).

13-4-4 Photodetectors

The front-end detectors in the classic experiments analyzed in Exercises 13-12 and 13-13 convert an incident light flux into an electron current by means of the photoelectric effect. We will therefore briefly describe these square-law *photodetectors* and obtain their noise characteristics.

The prototype of the photodetectors we will consider is the *vacuum photodiode* shown in Fig. 13-17. The cathode is a semitransparent layer of appropriately chosen conducting material deposited on the interior surface of one of the end caps of an evacuated glass tube. Photoelectrons ejected from the cathode by the incident light are swept to the anode by the anode–cathode voltage V_K and generate the photocurrent i_K. If V_K is made large enough the transit time is negligible.

Let us assume that the cathode is uniformly illuminated, and for simplicity of notation, that the light is linearly polarized. If the power density of the electric field $E(t)$ of the light is confined to a band whose width is much smaller than the average frequency, we will accept that the average value I_K of the photocurrent is proportional to the mean square of $E(t)$,

$$I_K \sim \langle E^2 \rangle$$

[From a quantum-mechanical perspective we can think of the incident light as a beam of photons with different frequencies. We can then speak of the (average) density $n(f)$ of incident photons per unit time and per unit frequency, which is of the form $(1/f)\mathscr{S}_E(f)$, where $\mathscr{S}_E(f)$ is the power spectrum of $E(t)$. We can also speak of the *efficiency* $\alpha(f)$ of the photocathode, that is, of the probability that an incident photon of frequency f will generate a photoelectron. We then have

$$I_K = e\int_0^\infty \alpha(f)n(f)\,df \sim \int_0^\infty \alpha(f)\left(\frac{1}{f}\right)\mathscr{S}_E(f)\,df$$

It follows that I_K is proportional to $\langle E^2 \rangle$ if f and $\alpha(f)$ can be taken to be constant, which is why we required that the bandwidth of $E(t)$ be narrow.

For reference, we note that efficiencies above 10% can be obtained for wavelengths between 300 and 550 nm.]

In Exercise 13-11 it is shown that if the incident light is *coherent*, meaning that $E(t)$ is independent of position on the cathode, the noise in i_K is shot noise plus an extra contribution that is important only when I_K is large and the bandwidth of $E(t)$ is small. The extra contribution washes out if, as is usually the case, the light has a large spread in solid angle and is therefore coherent only over small areas of the cathode. We may thus normally assume that the power density of i_K is given by the Schottky formula,

$$ i_K^2 = 2eI_K $$

If I_K is measured on a resistor R, and we want the current noise from R to be at most equal to the shot noise in i_K, we must have

$$ R = \frac{2kT}{eI_K} $$

For $I_K = 1$ nA, R must be at least 50 MΩ, and we run into limitations on speed. Photodiodes are therefore used only for large signals in the microampere range when speed is important.

At low currents and high speeds we must turn to *photomultipliers*. As indicated schematically in Fig. 13-18, a photomultiplier is a vacuum photodiode in which the current emitted from the cathode is increased by orders of magnitude in a low-noise *electron multiplier* consisting of a series of conducting *dynodes* placed at increasingly higher voltages relative to the cathode. Although it is not true in practice, for our purposes we may imagine the dynodes as grids.

An electron incident on a dynode generates several *secondary* electrons which are then directed to the next-higher dynode; the electrons emitted by the last dynode are collected by the anode. The average gain of a dynode is

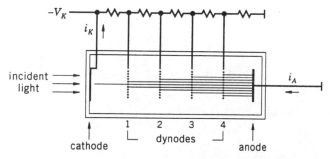

Figure 13-18. Structure of a photomultiplier (schematic).

typically between 3 and 6. With 12-stage multipliers, cathode-to-anode gains of order 10^8 can be obtained. Photomultipliers are fast: cathode-to-anode transit times under 35 ns and anode rise times under 2 ns are common.

If the signal at the cathode is the average current I_K, and we imagine that the shot noise is bandlimited by a low-pass filter with power bandwidth B, the signal-to-noise power ratio at the cathode is in principle given by

$$\frac{S^2}{N^2} = \frac{I_K^2}{2eI_K B} = \frac{I_K}{2eB}$$

As shown in Exercise 13-3, the signal-to-noise power ratio at the anode is smaller than it is at the cathode:

$$\frac{S^2}{N^2} = \frac{I_K}{2eBF}$$

where $F \geq 1$ is the noise figure (defined in this case relative to shot noise rather than thermal noise). In the simple case in which all dynodes have the same gain μ, for example, F is given by

$$F = \frac{\mu}{\mu - 1} \tag{13-2}$$

For $\mu = 4$ we have $F = 1.33$, which for photocurrents in the nanoampere range and large bandwidths is far better than what is obtained with conventional amplifiers.

Finally, we observe that photomultipliers can detect single photons. If the multiplier gain is A, the average voltage generated on a capacitor C by the anode charge due to one photoelectron is eA/C. Since C is necessarily in parallel with a resistor, its mean-square thermal noise voltage is kT/C, so that the signal-to-noise power ratio is $(eA/C)^2(C/kT)$, equal to 10^6 for reasonable values $C = 10$ pF and $A = 10^6$, and thus comfortably large.

13-5 OPTIMAL SIGNAL RECOVERY

The few techniques of signal recovery that we will discuss are elementary, but they are adequate for a surprisingly large fraction of the problems that arise in practice. In all cases some form of synchronization is needed, but we will assume that it is either obviously available or easily obtained, and not worry further about this question. Again in all cases, we will assume that the signal is masked by a noise process $n(t)$ that is independent of the signal, has mean zero, and *adds* to the signal.

13-5-1 Matched Filters

Certain experiments generate pulses of known shape whose *amplitudes* are the signals of interest; this is true, for example, of the x-ray spectrometers that we will discuss in Section 13-6. For such signals there exist *optimal* filters that maximize the signal-to-noise ratio. If the noise is white, which is the case we will consider first, optimal filters are said to be *matched* according to a definition we will give below.

Let us thus consider signals of the form $ax(t)$, where $x(t)$ is a pulse of known shape and a is an amplitude to be determined, and let us assume that $n(t)$ is white noise with a constant power spectrum $\mathscr{P}_n$ and that the total input waveform $x(t)+n(t)$ is processed by a filter with frequency response $H(f)$ and impulse response $h(t)$. Let us further assume that the pulse occurs at $t=0$, so that $x(t)$ is zero for $t<0$, and that the filter output is measured at a *peaking* time t_P (the reason for this name will become clear shortly). The value of a will not affect our arguments, and we will therefore set it to 1. If $y(t)$ is the response of the filter to $x(t)$, we have

$$y(t) = h*x(t) = \int_{-\infty}^{+\infty} h(u)x(t-u)\, du$$

Using the Parseval–Plancherel theorem, the output signal-to-noise ratio at $t=t_P$ can be expressed as

$$\frac{S^2}{N^2} = \frac{\left| \int_{-\infty}^{+\infty} h(t)x(t_P-t)\, dt \right|^2}{\mathscr{P}_n \int_{-\infty}^{+\infty} |H(f)|^2\, df} = \frac{\left| \int_{-\infty}^{+\infty} h(t)x(t_P-t)\, dt \right|^2}{\mathscr{P}_n \int_{-\infty}^{+\infty} |h(t)|^2\, dt}$$

But from the Schwarz inequality obtained in Exercise 2-10 we have

$$\frac{\left| \int_{-\infty}^{+\infty} h(t)x(t_P-t)\, dt \right|^2}{\int_{-\infty}^{+\infty} |h(t)|^2\, dt} \leq \int_{-\infty}^{+\infty} |x(t_P-t)|^2\, dt$$

Equality holds—and the signal-to-noise ratio is maximum—if $h(t)$ is any constant λ times $x(t_P-t)$,

$$h(t) = \lambda x(t_P-t) \tag{13-3}$$

In words, the optimal impulse response $h(t)$ in the presence of white noise is proportional to $x(t)$ played backward starting at the peaking time t_P and is, by definition, *matched* to $x(t)$.

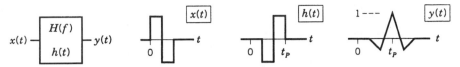

Figure 13-19. Representative waveforms in a matched filter.

We observe that $h(t)$ is causal only if $x(t)$ is zero beyond a certain time, and then only if t_P is greater than or equal to that time. Whether or not these two conditions hold, we have

$$y(t) = \int_{-\infty}^{+\infty} x(u)h(t-u)\, du = \lambda \int_{-\infty}^{+\infty} x(u)x(t_P-t+u)\, du \qquad (13\text{-}4)$$

The output signal $y(t)$ is thus proportional to the autocorrelation of $x(t)$ evaluated at t_P-t, and it is therefore symmetric around t_P and maximum at t_P (hence *peaking* time), where its value is proportional to the integrated square of $x(t)$,

$$y(t_P) = \lambda \int_{-\infty}^{+\infty} |x(t)|^2\, dt$$

We also have

$$\frac{S^2}{N^2} = \frac{\int_{-\infty}^{+\infty} |x(t)|^2\, dt}{\mathcal{P}_n} \qquad (13\text{-}5)$$

Waveforms for $x(t)$, $h(t)$, and $y(t)$ are shown schematically in Fig. 13-19. We will always choose λ to have the dimension of reciprocal time, and a value such that

$$y(t_P) = 1$$

Even if $x(t)$ is not zero beyond a certain time it must eventually become negligible, and we can thus obtain an essentially causal matched filter by making t_P sufficiently large. This tells us, at least in principle, that the maximum signal-to-noise ratio can be attained experimentally; in the discussion of x-ray spectrometers we will see that it can be approached closely, let us say within 20%, with easily implemented causal filters.

For t_P large enough that $x(t)$ is negligible for $t>t_P$, the response to $x(t)+n(t)$ of any filter with an impulse response $h(t)$ that is negligible for $t<0$ and zero for $t>t_P$ can be expressed in terms of a *weighting function* $x'(t) = h(t_P-t)$ as

$$\int_0^{t_P} [x(t)+n(t)]x'(t)\, dt$$

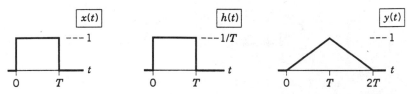

Figure 13-20. Waveforms in a matched filter in the case that the input $x(t)$ is a square pulse of unit height and width T and the peaking time is equal to T.

We can then say that the weighting function that is zero for $t<0$ and maximizes the signal-to-noise ratio at the peaking time is $x(t)$ itself (the weighting function thus *matches* the signal), and we see that we can implement a matched filter in the form of a balanced mixer followed by an integrator if we generate a copy of $x(t)$ in synchronism with $x(t)$ and multiply it in the balanced mixer by the total waveform $x(t)+n(t)$. This is reminiscent of a lock-in detector, and rightly so, as we will see in Section 13-5-3.

As an example from which we can immediately profit, let us consider the case illustrated in Fig. 13-20, in which $x(t)$ is a square pulse of unit height and width T,

$$x(t) = \theta(t) - \theta(t-T)$$

We will take the peaking time to be equal to T; setting $\lambda = 1/T$ in (13-3) we then have

$$h(t) = \frac{\theta(t) - \theta(t-T)}{T}$$

and from (13-5) we obtain

$$\frac{S^2}{N^2} = \frac{1}{\mathscr{P}_n} \int_{-\infty}^{+\infty} |x(t)|^2 \, dt = \frac{T}{\mathscr{P}_n}$$

For $t = T$ the response of the matched filter to $x(t)$ is $1/T$ times the integral from 0 to T of $x(t)$, so that

$$S = y(T) = \frac{1}{T} \int_0^T x(t) \, dt = 1$$

But S does not change if $x(t) = 1$ for $t<0$ because $h(t) = 0$ for $t>T$. On the other hand, thinking of the noise $n(t)$ as an impulse process, for the same reason it is clear that $n(t)$ can be set to zero for $t<0$. Recalling that $R_n(\tau) = \mathscr{P}_n \delta(\tau)$ we then obtain

$$N^2 = \left\langle \left| \frac{1}{T} \int_0^T n(t) \, dt \right|^2 \right\rangle = \frac{1}{T^2} \int_0^T du \int_0^T dv \, R_n(u-v) = \frac{\mathscr{P}_n}{T}$$

Since $S^2 = 1$, we recover the signal-to-noise ratio obtained above. It follows that if we are allowed a finite time interval to extract a presumably constant signal from white noise, the best we can do is integrate the total waveform over the whole interval. This conclusion is obtained from scratch in Exercise 13-6.

13-5-2 Whitening Filters

To generalize the results of the preceding section, let us now assume that the power spectrum $\mathcal{S}_n(f)$ of the noise can be expressed as

$$\mathcal{S}_n(f) = \frac{\mathcal{P}_n}{|W(f)|^2} \tag{13-6}$$

where $\mathcal{P}_n$ is a constant and $W(f)$ is the frequency response of a causal filter. An arbitrary filter can be replaced with no loss of generality by the cascade of a *whitening* filter with frequency response $W(f)$ and a filter with frequency response $H(f)$. As indicated in Fig. 13-21, at the output of the whitening filter the signal becomes $w*x(t)$ and the power spectrum of the noise has a constant magnitude $\mathcal{P}_n$. The overall filter is therefore optimal if $h(t)$ is matched to $w*x(t)$,

$$h(t) = \lambda w * x(t_p - t)$$

To take an example that will be useful when we come to x-ray spectrometers, let us consider the case in which

$$x(t) = \theta(t)$$

and

$$\mathcal{S}_n(f) = \frac{1 + \omega^2 \tau_C^2}{\omega^2 \tau_C^2} \mathcal{P}_n \tag{13-7}$$

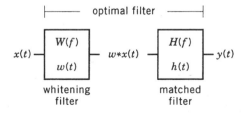

whitening
filter

matched
filter

Figure 13-21. A whitening filter W placed before a filter H converts a noise power spectrum of the form $\mathcal{P}_n/|W(f)|^2$ into a white spectrum of constant magnitude $\mathcal{P}_n$. But then H can be chosen to match $w*x(t)$.

We then have

$$W(f) = \frac{i\omega\tau_C}{1+i\omega\tau_C}$$

The signal at the output of the whitening filter is

$$w*x(t) = \theta(t)\exp\left(-\frac{t}{\tau_C}\right)$$

Substituting $w*x(t)$ in place of $x(t)$ in (13-3) and setting $\lambda = 2/\tau_C$, we have

$$h(t) = \frac{2}{\tau_C}\theta(t_P-t)\exp\left(-\frac{t_P-t}{\tau_C}\right)$$

and from (13-4) we obtain

$$y(t) = \exp\left(-\frac{|t-t_P|}{\tau_C}\right) \tag{13-8}$$

The matched filter is noncausal, and we will therefore use its signal-to-noise ratio only as a bench mark.

Returning to our discussion, let us take the Fourier transform of the overall response

$$y(t) = h*w*x(t) = \lambda\left[w*x(t_P-t)\right] * \left[w*x(t)\right]$$

Using the fact that the Fourier transform of $w*x(-t)$ is $W^*(f)X^*(f)$, we obtain

$$\mathcal{F}(y) = \lambda|W(f)|^2|X(f)|^2\exp(-i\omega\tau_P)$$

We can thus write $y(t)$ as an inverse Fourier transform:

$$y(t) = \lambda\int_{-\infty}^{+\infty}|W(f)|^2|X(f)|^2\exp\left[i\omega(t-t_P)\right]df$$

If the power spectrum of the noise cannot be expressed in terms of the frequency response of a causal filter as in (13-6), but is given by

$$\mathcal{S}_n(f) = \frac{\mathcal{P}_n}{G(f)}$$

where $G(f)$ is nonnegative and symmetric in f (this is the case, for example, if flicker noise contributes to the power spectrum), we can take the frequency response of the whitening filter to be the positive square root of $G(f)$,

$$W(f) = \sqrt{G(f)}$$

The overall response is then given by

$$y(t) = \lambda \int_{-\infty}^{+\infty} |X(f)|^2 G(f) \exp[i\omega(t-t_P)] \, df$$

In the example considered above we have

$$|X(f)|^2 = \frac{1}{\omega^2}$$

and

$$G(f) = \frac{\omega^2 \tau_C^2}{1 + \omega^2 \tau_C^2}$$

Taking $\lambda = 2/\tau_C$ and using (2-14), we recover $y(t)$ as given in (13-8),

$$y(t) = \frac{1}{2\pi} \int_{-\infty}^{+\infty} \frac{2\tau_C}{1 + \omega^2 \tau_C^2} \exp[i\omega(t-t_P)] \, d\omega = \exp\left(-\frac{|t-t_P|}{\tau_C}\right)$$

[As shown in Exercise 2-20, the Fourier transform of $\theta(t)$ is not quite $(i\omega)^{-1}$, but we can take it to be so because of the factor ω^2 in the numerator of $G(f)$. Alternatively, we can obtain $y(t)$ for $x(t) = \theta(t)\exp(-\sigma t)$ and take the limit as $\sigma \to 0$.]

13-5-3 Lock-in Detectors as Matched Filters

The techniques of (suppressed-carrier) amplitude modulation and lock-in detection discussed in Sections 13-4-1 and 13-4-2 can be instrumented with any periodic carrier wave $c(t)$ of frequency f_c, and in fact $c(t)$ is often a square wave. We will therefore reconsider lock-in detectors more generally before proceeding.

In the notation of Fig. 13-10, the modulated carrier for a signal $s(t)$ and a carrier $c(t)$ is

$$x(t) = s(t)c(t)$$

and the input to the low-pass filter is

$$y(t) = s(t)c^2(t)$$

But $c^2(t)$ is equal to its average value $\langle c^2 \rangle$ plus sinusoidal terms at $2f_c$ and its harmonics, so that if the maximum significant signal frequency f_s and the power bandwidth B' of the low-pass filter satisfy

$$f_s \ll B' \ll f_c$$

the low-pass filter eliminates the sinusoidal terms but does not affect $s(t)$, and its output is therefore

$$z(t) = \langle c^2 \rangle s(t)$$

For $c(t) = \cos \omega_c t$ we have $\langle c^2 \rangle = \frac{1}{2}$, and as we saw in Section 13-4-1,

$$z(t) = \tfrac{1}{2} s(t)$$

If $x(t)$ is contaminated by (independent additive) white noise $n(t)$ with a constant power spectrum $\mathcal{P}_n$ and autocorrelation $\mathcal{P}_n \delta(\tau)$, we have

$$x(t) = s(t)c(t) + n(t)$$

and thus

$$y(t) = s(t)c^2(t) + n(t)c(t)$$

The autocorrelation of $n(t)c(t)$ is

$$R_n(\tau)R_c(\tau) = \mathcal{P}_n \delta(\tau)R_c(\tau) = R_c(0)\mathcal{P}_n \delta(\tau) = \langle c^2 \rangle \mathcal{P}_n \delta(\tau)$$

so that the (constant) power spectrum $\mathcal{P}'_n$ at the input to the low-pass filter is given by

$$\mathcal{P}'_n = \langle c^2 \rangle \mathcal{P}_n$$

Assuming now that $s(t)$ has a constant value s_0, the signal S and the mean-square noise N^2 at the output of the low-pass filter are

$$S = s_0 \langle c^2 \rangle \quad \text{and} \quad N^2 = 2\mathcal{P}_n B' \langle c^2 \rangle$$

Again for $c(t) = \cos \omega_c t$, and in agreement with (13-1) and Fig. 13-13, we have

$$\mathcal{P}'_n = \tfrac{1}{2} \mathcal{P}_n$$

and

$$\frac{S^2}{N^2} = \frac{\frac{1}{4} s_0^2}{\mathcal{P}_n B'}$$

Subject to the conditions on B' given above, we will now argue that a lock-in detector is essentially a matched filter applied to $s(t)c(t)$ and that in the presence of white noise it is an excellent if not strictly optimal detector for $s(t)$. We will first assume that $s(t)$ has a constant value s_0 and that we are given a finite time T to measure s_0 in the presence of white noise $n(t)$. Under these conditions we can consider a pulse equal to $s_0c(t)$ for $0 \le t \le T$ and zero elsewhere, and design a matched filter according to Section 13-5-1. Setting the peaking time equal to T, an optimal measurement of s_0 is obtained by multiplying $s_0c(t)+n(t)$ by a matched weighting function $c'(t) = c(t)$ and integrating the product $[s_0c(t)+n(t)]c(t)$ from $t=0$ to $t=T$ or, equivalently, by averaging it over a time T in a low-pass filter with impulse response

$$h_T(t) = \frac{\theta(t) - \theta(t-T)}{T}$$

and power bandwidth

$$B' = \frac{1}{2} \int_{-\infty}^{+\infty} |h_T(t)|^2 \, dt = \frac{1}{2T}$$

It should be clear the except for being quite specific about the low-pass filter, we have just described a lock-in detector. We have

$$S = \frac{s_0}{T} \int_0^T |c(t)|^2 \, dt$$

and, again using the fact that $R_n(\tau) = \mathscr{P}_n \delta(\tau)$,

$$N^2 = \left\langle \left| \frac{1}{T} \int_0^T n(t) c(t) \, dt \right|^2 \right\rangle$$

$$= \frac{1}{T^2} \int_0^T du \int_0^T dv \, R_n(u-v) c(u) c(v) = \frac{\mathscr{P}_n}{T^2} \int_0^T |c(t)|^2 \, dt$$

Since $B' \ll f_c$ we have $f_c T \gg 1$, so that

$$S = s_0 \langle c^2 \rangle$$

and

$$N^2 = \frac{\mathscr{P}_n}{T} \langle c^2 \rangle = 2 \mathscr{P}_n B' \langle c^2 \rangle$$

The matched (optimum) value of S^2/N^2 is thus

$$\frac{S^2}{N^2} = \frac{s_0^2}{2\mathcal{P}_n B'} \langle c^2 \rangle$$

But these are precisely the results we obtain using any low-pass filter with power bandwidth B' as long as $s(t)$ has been held steady at a value s_0 for a time sufficiently longer than $1/B'$ that S has settled to its final value $s_0 \langle c^2 \rangle$, and we conclude that lock-in detectors are essentially matched filters. Since $f_s \ll B' = 1/2T$, $s(t)$ changes negligibly in a time T, and we conclude more generally that a lock-in detector gives an excellent measurement of $s(t)$.

If the weighting function $c'(t)$ differs from $c(t)$, meaning in physical terms that the balanced mixers in the detector and the modulator are driven by different waveforms, for a constant signal s_0 we have

$$S = s_0 \langle cc' \rangle \quad \text{and} \quad N^2 = 2\mathcal{P}_n B' \langle c'^2 \rangle$$

and thus

$$\frac{S^2}{N^2} = \frac{s_0^2}{2\mathcal{P}_n B'} \frac{\langle cc' \rangle^2}{\langle c'^2 \rangle}$$

It follows that S^2/N^2 is equal to the matched value times a factor $\eta^2 \le 1$ given by

$$\eta^2 = \frac{\langle cc' \rangle^2}{\langle c^2 \rangle \langle c'^2 \rangle}$$

Let us take $c(t)$ to be a sine wave and $c'(t)$ to be a square wave with amplitudes ± 1, which is often what we have in the lock-in amplifiers described in the following section. Assuming of course that $c(t)$ and $c'(t)$ have the same zero crossings, we obtain

$$\eta^2 = \frac{(2/\pi)^2}{(1/2)(1)} = \frac{8}{\pi^2} \simeq 0.81$$

From this example we learn that signal-to-noise ratios quite close to the matched value can be obtained with unmatched filters. As a minor point, we note without proof that in this particular case the matched value can be achieved by placing before the detector a filter with passbands of width B centered at $\pm f_c$ and choosing B such that

$$2B' \ll B \ll 2f_c$$

because it turns out that doing so is equivalent to applying the filter to $c'(t)$, which clearly converts $c'(t)$ into a sine wave matched to $c(t)$.

13-5-4 Lock-in Amplifiers

In the second class of experiments that we will consider, the signal is minute but its power spectrum is limited to a narrow baseband around $f = 0$. What we mean by *minute* depends on circumstances, but to be somewhat more specific, in the case of a voltage we will take it to mean of order 1 μV. A signal of this magnitude must unavoidably be amplified, and flicker noise will just as unavoidably add to the signal, so that clean recovery with a direct-coupled amplifier followed by a low-pass filter is unlikely. Even in the absence of flicker noise, other low-frequency noise sources are important at this level; these include thermal drift in the amplifier, and thermocouple effects if junctions of conductors are not kept at constant temperatures.

As indicated schematically in Fig. 13-22a, these problems are obviated if the functional equivalent of a *noiseless* balanced modulator is used to shift the spectrum of the signal to a frequency f_c beyond the $1/f$ corner frequency

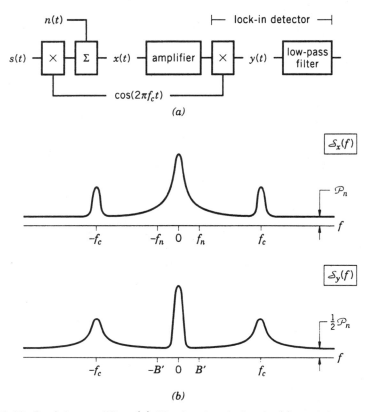

(a)

(b)

Figure 13-22. Lock-in amplifier. (*a*) The baseband signal $s(t)$ modulates a random sine wave of frequency f_c before contamination with flicker noise. (*b*) The signal and flicker-noise spectra exchange positions at the input to the low-pass filter, and the flicker noise falls well outside the bandwidth of the low-pass filter.

f_n. Once this is accomplished—it is not always feasible—the signal can be amplified and then recovered in a lock-in detector with a low-pass filter with power bandwidth $B' \ll f_c$. Systems designed to execute these tasks are called *lock-in amplifiers*. As shown in Fig. 13-22b, the flicker-noise and signal spectra exchange positions at the input to the low-pass filter. Since the residual noise can be imagined as part of a white spectrum, it is clear according to the discussion in Section 13-5-3 that we have an optimal solution. This conclusion also holds for nonsinusoidal carrier waves; the power spectra, however, are more complicated.

One form of noiseless modulation consists in periodically varying a parameter on which the signal depends; if the signal is the charge on a parallel-plate capacitor, for example, a sinusoidal voltage proportional to the charge can be generated by varying the plate separation. Another possibility, much used when the signal is a light beam incident on a noisy photodetector, is to interrupt the beam periodically; this is done, for example, in the experiment discussed in Exercise 13-13. A third possibility, used when measuring imbalances in bridge circuits such as the Wheatstone bridge of Fig. 3-8, is quite simply to feed the bridge with a sinusoidal voltage.

13-5-5 Multichannel Analyzers

In the final class of experiments that we will consider, the signal is in the form of repeated pulses whose shape, assumed to be constant, must be determined; this is the case, for example, if an experiment that generates such a pulse is triggered at a constant rate or if a feature such as a resonance is explored by varying the frequency of the excitation in a sawtooth pattern.

Figure 13-23 illustrates in principle how the signal pulse shape is recovered in *multichannel analyzers*. Each pulse is divided into a number M of narrow time slices or *channels*, so that the signal is essentially constant in any given channel, and in each channel the values of the signal-plus-noise waveform are extracted to form a *sample pulse train*; each of these sample pulse trains is then integrated. In the presence of white noise, this procedure is clearly equivalent to applying a matched filter to each sample pulse train and is thus optimal. The signal grows linearly in time, whereas the rms noise grows like the square root of the time.

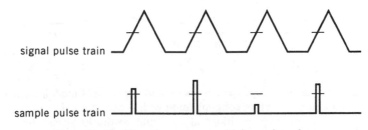

signal pulse train

sample pulse train

Figure 13-23. Waveforms in a multichannel analyzer.

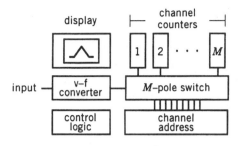

Figure 13-24. Essential components of a multichannel analyzer.

As shown in Fig. 13-24, modern multichannel analyzers are digital instruments except for a voltage-to-frequency converter that generates pulses at a rate proportional to the amplitude of the input waveform; to accommodate inputs of either sign, which are to be expected when noise is dominant, the converter can be biased so that its output frequency is centered in the operating range when the input is zero. Integration of the sample pulses is performed in *channel counters* that are connected to the converter during the appropriate time slice by means of a digital switch.

We note that multichannel analyzers are also normally used as *pulse-height analyzers* to display the relative frequencies of the amplitudes or areas of a number of pulses. All that is needed is an analog-to-digital converter that gives the relevant magnitude in digital form; the rest is already available.

13-6   SEMICONDUCTOR DETECTORS

As an interesting example of signal recovery techniques, we will now consider x-ray spectrometers based on lithium-drifted silicon detectors or, as they are usually denoted, $Si(Li)$ detectors. We will focus on x-rays with energies in the range from 100 eV to 1 keV, which we might obtain, for example, by shining an electron beam on samples containing light elements such as boron or carbon. As we will see, electronic noise is always significant in this range, and at the lowest energies, dominant.

The essential structure of a planar Si(Li) detector is shown in Fig. 13-25. A region of intrinsic silicon several millimeters thick and typically 1 cm^2 in area is sandwiched between thin p^+ and n^+ layers. Ohmic contacts to metallic terminals are made off to one side on the p^+ and n^+ layers. The intrinsic silicon region is obtained by drifting lithium (a donor) into p-silicon; once this process is complete, the detector is kept at liquid-nitrogen temperature (77 K) in order to immobilize the lithium ions.

When in operation, the detector is subjected to a reverse voltage in the kilovolt range; the electric field in the intrinsic region is constant, and high enough that charge carriers drift across the detector in about 100 ns. As a result of operation at liquid-nitrogen temperature, the *leakage* current I_D can be below 10^{-15} A.

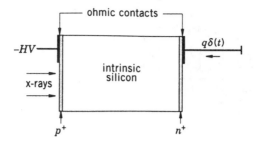

Figure 13-25. Cross section of a Si(Li) detector.

X-rays with energies below 30 keV interact with silicon mainly through the photoelectric effect. The result is essentially as if an electron with the energy of the x-ray were ejected and then stopped in a distance much smaller than the size of the detector. In the stopping process, electron–hole pairs are created at an average cost $\varepsilon = 3.8$ eV per pair. Holes and electrons then drift to the detector electrodes and generate a current pulse about 100 ns wide; for most purposes we will replace this pulse with a delta function times the charge q in the pulse.

The number n of electron–hole pairs generated in an event is a statistical process; given the large numbers involved, we can assume that it is normally distributed. Since the (additive) noise is also normally distributed, for a mixture of x-rays from different elements we get a *line* spectrum like the one shown in Fig. 13-26.

The average number of electron–hole pairs generated by an x-ray of energy E is

$$\langle n \rangle = \frac{E}{\varepsilon}$$

Pair generation is less random than a Poisson process because the total energy dissipated must be equal to E. This constraint is accounted for by

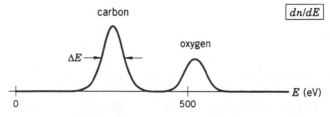

Figure 13-26. Spectrum of a mixture of x-rays from carbon and oxygen. The linewidth ΔE has contributions from pair-generation statistics and from electronic noise.

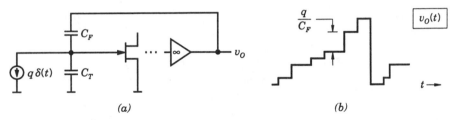

Figure 13-27. (*a*) A charge preamplifier is always the first element in the signal-processing chain of a Si(Li) detector. (*b*) A current pulse of charge q generates a step of amplitude q/C_F in the output v_O.

expressing the variance of n as $\langle n \rangle$ times the *Fano factor* $F \approx 0.12$,

$$\sigma_n^2 = \langle n \rangle F = \frac{E}{\varepsilon} F$$

Converting to energy units, the *intrinsic linewidth* ΔE_I (full width at half maximum) is

$$\Delta E_I = \sqrt{8 \ln 2} \sqrt{\varepsilon E F} = 2.35 \sqrt{\varepsilon E F}$$

For $E = 1$ keV we thus have

$$\Delta E_I \approx 50 \text{ eV}$$

The first element in the signal processing chain of a Si(Li) detector is invariably a JFET-input *charge preamplifier* like the one shown schematically in Fig. 13-27a. (A more detailed view of a high-gain amplifier designed specifically for use in charge preamplifiers is offered in Exercise 7-9.) The feedback capacitance C_F is at least 10 times smaller than the sum C_T of the detector capacitance (typically 1 pF) and the JFET gate–source capacitance (typically 10 pF). The input stage is also invariably a cascode amplifier, so that the Miller effect is negligible.

Mobilities increase as the temperature decreases; the transconductance of a JFET thus also increases initially, but it reaches a maximum at a temperature around 150 K because the charge-carrier density starts to decrease. The input JFET is usually operated near the temperature of maximum transconductance, and its gate current I_G, like the detector leakage current I_D, can be below 10^{-15} A.

A pulse of charge q results in a step of amplitude q/C_F in the output v_O, so that as shown in Fig. 13-27b, v_O looks like a (random) staircase. When v_O exceeds a certain value it is reset to a lower value and a new staircase is generated. For our purposes we will imagine that v_O is amplified by a constant-gain direct-coupled amplifier that raises the magnitudes of the steps to levels suitable for the processing electronics.

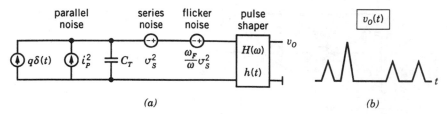

Figure 13-28. (*a*) Noise model of Si(Li) detector. (*b*) The pulse shaper converts the staircase waveform of Fig. 13-27*b* into a pulse train.

According to Exercise 13-1, we can calculate the noise referred to the input in an open-loop configuration and neglect C_F because it is much smaller than C_T. In the JFET noise model of Fig. 13-2 we will replace the drain current generator with an equivalent voltage generator in series with the gate and neglect the frequency-dependent piece of the gate current generator; we will thus in effect use the JFET model of Fig. 13-7*b*. Taking into account the shot noise generated by the detector leakage current I_D, we obtain the noise model shown in Fig. 13-28*a*, in which the *series* and *parallel* noise power densities v_S^2 and i_P^2 are given by

$$v_S^2 = \tfrac{8}{3}kTr_m \quad \text{and} \quad i_P^2 = 2e(I_G + I_D)$$

The pulse *shaper* used to improve the signal-to-noise ratio is shown as following the preamplifier immediately, as it does in traditional systems. We will take the shaper to be a time-invariant filter with frequency response $H(\omega)$ and impulse response $h(t)$. Time-variant filters, which have been used with success, are considered briefly in Exercise 13-8. As shown in Fig. 13-28*b*, the shaper converts the steps in the staircase waveform of Fig. 13-27*b* into pulses of constant shape whose amplitude in each case is proportional to the charge generated by an x-ray. What we learned about matched filters in Sections 13-5-1 and 13-5-2 is thus relevant, but we will bring it to bear only gradually.

The power density of the total noise voltage referred to the input is

$$v_N^2(f) = v_S^2 + \frac{\omega_F}{\omega} v_S^2 + \frac{1}{\omega^2 C_T^2} i_P^2$$

In terms of the *series–parallel* time constant

$$\tau_C = \frac{C_T v_S}{i_P}$$

we have

$$\sigma_N^2(f) = \left(1 + \frac{\omega_F}{\omega} + \frac{1}{\omega^2 \tau_C^2}\right)\sigma_S^2$$

We note that in the absence of flicker noise, the noise power density is of the form given in (13-7). We will write the (dimensionless) step response $r(t)$ of the shaper as a function of the dimensionless variable t/τ, where τ is the scaling time,

$$r(t) = r_0\left(\frac{t}{\tau}\right)$$

By varying τ we can then generate a family of responses with the same functional form. The impulse response $h(t)$ is given by

$$h(t) = \frac{1}{\tau}h_0\left(\frac{t}{\tau}\right)$$

where

$$h_0\left(\frac{t}{\tau}\right) = r_0'\left(\frac{t}{\tau}\right)$$

Correspondingly, the (dimensionless) frequency response is given by

$$H(\omega) = H_0(\omega\tau)$$

We will further assume that $r(t)$ is measured at a *peaking* time t_P, that it is maximum at t_P, and that

$$r(t_P) = 1$$

A one-pair impulse $e\delta(t)$ generates a voltage e/C_T; the mean-square number of (equivalent) *noise electrons* is therefore

$$N^2 = \left(\frac{1}{\tau}K_S^2 + \omega_F K_F^2 + \frac{\tau}{\tau_C^2}K_P^2\right)\frac{C_T^2\sigma_S^2}{2e^2} \qquad (13\text{-}9)$$

where

$$K_S^2 = \frac{1}{2\pi} \int_{-\infty}^{+\infty} |H_0(\eta)|^2 \, d\eta = \int_{-\infty}^{+\infty} |h_0(\eta)|^2 \, d\eta$$

$$K_F^2 = \frac{1}{2\pi} \int_{-\infty}^{+\infty} \frac{|H_0(\eta)|^2}{|\eta|} \, d\eta \qquad (13\text{-}10)$$

$$K_P^2 = \frac{1}{2\pi} \int_{-\infty}^{+\infty} \frac{|H_0(\eta)|^2}{|\eta|^2} \, d\eta = \int_{-\infty}^{+\infty} |r_0(\eta)|^2 \, d\eta$$

We note that K_S, K_F, and K_P are all of order 1, and that from the Schwarz inequality (Exercise 2-10) we must have $K_F^2 \leq K_S K_P$. Empirically, it turns out that

$$K_F^2 \simeq 0.7 K_S K_P$$

We also note that N^2 achieves its minimum value N_0^2 at

$$\tau_0 = \frac{K_S}{K_P} \tau_C$$

and that

$$N_0^2 = \left(K_S K_P + \tfrac{1}{2} \omega_F \tau_C K_F^2 \right) \frac{C_T \sigma_S i_P}{e^2} \qquad (13\text{-}11)$$

The *noise linewidth* ΔE_N is given by

$$\Delta E_N = 2.35 \varepsilon \sqrt{N}$$

and the *total linewidth* ΔE is obtained by adding ΔE_I and ΔE_N quadratically:

$$\Delta E = \sqrt{(\Delta E_I)^2 + (\Delta E_N)^2}$$

Assuming that $t_P \gg \tau$, let us first consider the (noncausal) family of step responses

$$r(t) = \exp\left(-\frac{|t - t_P|}{\tau} \right)$$

We then have

$$K_S^2 = K_P^2 = 2 \int_0^\infty \exp(-2\eta) \, d\eta = 1$$

and therefore

$$\tau_0 = \tau_C$$

This is of course compatible with the fact that the optimal step response in the absence of flicker noise, given in (13-8), is

$$r(t) = \exp\left(-\frac{|t - t_p|}{\tau_C}\right)$$

Taking $C_T = 5$ pF, $I_G + I_D = 1$ fA, $g_m = 5$ mS ($r_m = 200$ Ω), and $T = 150$ K, we obtain

$$\sigma_S = \sqrt{\tfrac{8}{3}kTr_m} = 1.05 \text{ nV}/\sqrt{\text{Hz}}$$

$$i_p = \sqrt{2e(I_G + I_D)} = 0.018 \text{ fA}/\sqrt{\text{Hz}}$$

and thus

$$\tau_C = \frac{C_T \sigma_S}{i_p} = 294 \ \mu\text{s}$$

We will assume that $\omega_F/2\pi = 10^4$ Hz. This is a high value, but one nonetheless found in practice and attributed to the dielectric noise discussed in Exercise 13-4. We then obtain $\tfrac{1}{2}\omega_F \tau_C K_F^2 = 6.5$, so that the flicker noise terms in (13-9) and (13-11) are important. The parallel noise is insignificant for τ less than 1 ms. The noise linewidth ΔE_N is shown as a function of τ in Fig. 13-29. For τ above 50 μs the flicker noise is dominant, and ΔE_N remains essentially constant at a value of approximately 50 eV; that is, it exhibits a $1/f$ *plateau*. We note that the noise linewidth is comparable to the intrinsic linewidth, which is 50 eV or less for energies below 1 keV.

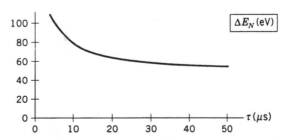

Figure 13-29. Noise linewidth of a Si(Li) detector as a function of the shaper scaling time τ.

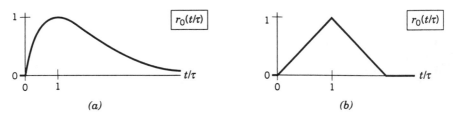

Figure 13-30. Causal pulse shapers. (*a*) *CR|RC*. (*b*) Triangular.

In drifting to the detector electrodes, charge carriers can be captured in *traps* from which they may later be emitted or not; thus either the collected charge is less than it should be or, in good detectors, the rise time becomes longer. In addition, at larger energies there are variations in the rise time that depend on where pairs are generated. For these reasons, the step response of shapers must have a flat top, so that the optimal response would not be used even in the absence of flicker noise because it has a cusp at $t = t_P$. Let us therefore consider two simple examples of (nonoptimal) causal shapers to see how well we do.

The first example is the *CR|RC* shaper, that is, the cascade of a lead and a lag with equal time constants, which has the step response shown in Fig. 13-30*a*,

$$r_0\left(\frac{t}{\tau}\right) = \left(\frac{t}{\tau}\right)\theta\left(\frac{t}{\tau}\right)\exp\left(1 - \frac{t}{\tau}\right)$$

We then have $t_P = \tau$, and

$$h_0\left(\frac{t}{\tau}\right) = \frac{1}{\tau}\left(1 - \frac{t}{\tau}\right)\theta\left(\frac{t}{\tau}\right)\exp\left(1 - \frac{t}{\tau}\right)$$

Using these expressions in (13-10), we obtain

$$K_S^2 = K_P^2 = \tfrac{1}{4}\exp 2 = 1.85$$

and thus

$$\tau_0 = \tau_C$$

$$\sqrt{K_S K_P} = 1.36$$

According to (13-11), and recalling that $K_F^2 \approx 0.7\, K_S K_P$, the noise linewidth is 1.36 times broader than it is with an optimal shaper. This might not seem like much, but as we will see below, it can have a significant effect in the case of closely spaced lines.

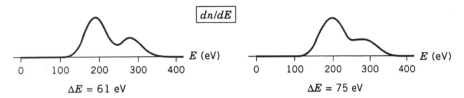

Figure 13-31. Spectra with closely spaced lines.

The second example is the *triangular* shaper, which has the step response shown in Fig. 13-30*b*. In this case we have

$$K_P^2 = \tfrac{2}{3} \qquad K_S^2 = 2 \tag{13-12}$$

and thus

$$\tau_0 = \sqrt{3}\,\tau_C$$

$$\sqrt{K_S K_P} = 1.08$$

Despite its excellent characteristics, the triangular shaper is seldom seen in analog form in traditional systems; as shown in Exercise 10-16, however, it can easily be implemented as a transversal filter in sampled systems like the one we will consider below. To avoid problems with variations in the detector rise time, triangular shapers are invariably replaced by *trapezoidal* shapers with a short flat top in the step response and only slightly inferior characteristics.

To illustrate the importance of good resolution, Fig. 13-31 shows spectra with a $2:1$ mixture of lines from boron (185 eV) and carbon (277 eV). At these low energies the intrinsic linewidth $2.35\sqrt{\varepsilon EF}$ is about 25 eV, so that the total linewidth is dominated by the noise linewidth. The difference between using a triangular shaper with $\Delta E = 61$ eV and a $CR|RC$ shaper with $\Delta E = 75$ eV is patent.

Traditional systems run into trouble at high counting rates because shaped pulses that come too close together distort one another. In principle this problem is solved by using a *pile-up detector* that looks at the preamplifier output and rejects events that will result in distorsion. The solution is far from perfect, however, among other things in the case of low-energy x-rays because the pile-up detector must be slow enough that it does not trigger on noise and therefore misses many events.

The architecture of a computer-based spectrometer with excellent performance at high counting rates, designed by Mott and Friel (1994) and indicative of future trends, is shown in Fig. 13-32. The preamplifier output is sampled and converted into a digital data stream by a free-running 16-bit analog-to-digital converter with a 5 MHz conversion frequency. A trapezoidal

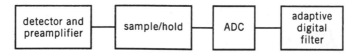

Figure 13-32. Architecture of the computer-based spectrometer of Mott and Friel.

transversal filter then extracts the data; the resulting linewidth is about 60 eV at 200 eV. Having overcome the hurdle of fast and accurate conversion, this system is superior in all respects to traditional systems. Among other invaluable advantages we note (1) that closely spaced events can be detected at some leisure with pattern-recognition techniques; (2) that the width of the trapezoidal shaper can be adapted on an event-by-event basis to the time available for a measurement, so that the increase in linewidth at high counting rates is small; (3) that events of a given energy are converted at random positions in the ADC, so that the effect of uneven channel widths (differential nonlinearity) is virtually eliminated.

EXERCISES

13-1 FEEDBACK WILL ONLY WORSEN THE SIGNAL-TO-NOISE RATIO. Consider the amplifiers shown in Fig. 13-33. Z can be a resistor or, in the case of a Si(Li) detector, a capacitor. In the feedback version v_N and i_N are the noise contributions of the operational amplifier, and i_Z represents the thermal noise, if any, due to Z. The signal gains are clearly equal. Now convince yourself that the amplifiers are also equivalent as far as noise is concerned if

$$v_N' = v_N\left(1+A_I^{-1}\right) \quad \text{and} \quad i_Z' = i_Z\sqrt{1+A_I^{-1}}$$

If A_I is high enough, say 10 or more, the benefits of feedback are obtained for a small price.

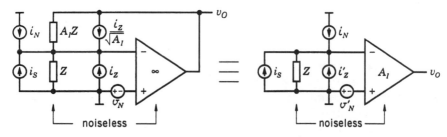

Figure 13-33. Noise models for Exercise 13-1.

13-2 NOISE FIGURE OF JFETs. The noise model of a JFET in the common-source configuration, neglecting C_{GD}, is shown in Fig. 13-34. The drain noise source i_{ND} in Fig. 13-2 is represented by source i'_{ND}, and i_{NS} is the thermal noise due to source resistance R_S. The power density i_N^2 of the total noise current is given by

$$i_N^2 = 4kTG_S + 2eI_G + 4kTg_m \left(\frac{f}{2f_T}\right)^2 + \left(\frac{2}{3}\right)4kTg_m \frac{G_S^2 + 4\pi^2 f^2 C_{GS}^2}{g_m^2}\left(1 + \frac{f_D}{f}\right)$$

Observing that $g_m/C_{GS} = 2\pi f_T$ and that $\frac{1}{4} + \frac{2}{3} \simeq 1$, and recalling that the noise figure F is given by

$$i_N^2 = 4kTG_S F$$

you obtain

$$F \simeq 1 + \frac{eI_G}{2kT}R_S + g_m R_S \left(\frac{f^2}{f_T^2} + \frac{2}{3}\frac{f_D f}{f_T^2}\right) + \frac{2}{3}\frac{1}{g_m R_S}\left(1 + \frac{f_D}{f}\right)$$

If you now define $f_0 = 1/2\pi R_S C_{GS} = f_T/g_m R_S$ and set $f_D = K_D f_T$, you have

$$F \simeq 1 + \frac{eI_G}{2kT}R_S + \frac{1}{g_m R_S}\left(\frac{2}{3} + \frac{f^2}{f_0^2}\right) + \frac{2}{3}K_D\left(\frac{f_0}{f} + \frac{f}{f_0}\right)$$

The lesson is clear: for any given source resistance, you should make I_G as small as possible and g_m as large as possible. To obtain good high-frequency noise performance you should in addition choose C_{GS} as small as possible. If low-frequency performance is what you are interested in, however, within a given technology you should choose the device width—and thus C_{GS}—as large as possible in order to minimize the $1/f$ corner frequency; at constant drain current, you will then also obtain a larger value of g_m but, of course, a larger value of I_G, which might be a problem for large source resistances.

Taking reasonable values $I_G = 10$ pA, $g_m = 3$ mS, $C_{GS} = 10$ pF, and $f_D = 100$ Hz, you have $2kT/eI_G = 5$ GΩ and $K_D = 2 \cdot 10^{-6}$, and for

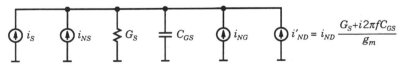

Figure 13-34. Noise model of a JFET in the common-source configuration (Exercise 13-2).

$R_S = 100$ kΩ you have $g_m R_S = 300$ and $f_0 \approx 170$ kHz, and you obtain an excellent midband noise figure,

$$F \approx 1.002$$

In contrast, for $R_S = 1$ kΩ you have $g_m R_S = 3$, and the midband noise figure becomes

$$F \approx 1.2$$

In this case you will do better with bipolar transistors at low as well as at midband frequencies because f_D is seldom under 100 Hz, whereas f_B in Fig. 13-1 can be as low as a few hertz.

13-3 ELECTRON-MULTIPLIER NOISE (Shockley and Pierce, 1938). Assume that secondary emission in the dynodes of an electron multiplier is a Poisson process. Let n_k be the number of electrons emitted by stage k and m_{ki} $(1 \le i \le n_k)$ the number of electrons generated by each electron incident on stage $k+1$. You then have

$$n_{k+1} = \sum_{i=1}^{n_k} m_{ki}$$

The random variables m_{ki} are independent and have mean and variance equal to the average gain μ_{k+1} of stage $k+1$. Now let σ_k^2 be the variance of n_k, and let $k = 0$ denote the primary source, a photocathode for example. As in Section 12-3-9, average first with n_k fixed and then average over n_k to obtain

$$\sigma_{k+1}^2 = \mu_{k+1}\langle n_k \rangle + \mu_{k+1}^2 \sigma_k^2$$

Assume that the second and higher dynodes have the same gain μ. If you start off with one primary electron and the multiplier has K stages, you obtain

$\langle n_0 \rangle = 1$ $\qquad\qquad \sigma_0^2 = 0$

$\langle n_1 \rangle = \mu_1$ $\qquad\qquad \sigma_1^2 = \mu_1$

$\langle n_2 \rangle = \mu_1 \mu$ $\qquad\qquad \sigma_2^2 = \mu_1 \mu + \mu_1 \mu^2$

$\qquad\vdots$ $\qquad\qquad\qquad\vdots$

$\langle n_K \rangle = \mu_1 \mu^{K-1}$ $\qquad \sigma_K^2 = \mu_1 \mu^{K-1} + \mu_1 \mu^K + \cdots + \mu_1 \mu^{2(K-1)}$

You therefore have

$$\langle n_K^2 \rangle = \mu_1^2 \mu^{2(K-1)}\left(1 + \frac{1+\mu^{-1}+\cdots+\mu^{-(K-1)}}{\mu_1}\right) = \mu_1^2 \mu^{2(K-1)}\left[1 + \frac{1-\mu^{-K}}{\mu_1(1-\mu^{-1})}\right]$$

Even for $K = 5$ and $\mu = 3$, which are low values, μ^{-K} is negligible compared with 1, so you get

$$\langle n_K^2 \rangle = \langle n_K \rangle^2 F$$

where F is the noise figure of the multiplier,

$$F = 1 + \frac{\mu}{\mu_1(\mu - 1)}$$

For typical values $\mu_1 = 6$ and $\mu = 4$ you have $F = 1.22$, which is good enough for most purposes. In contrast, in devices with a GaAs first dynode, such as an RCA 8850 photomultiplier, you have $\mu_1 \simeq 40$ and $F \simeq 1.03$, so that $\sigma_K / \langle n_K \rangle = \sqrt{F-1} \simeq 0.18$; as you can easily verify, this implies that the probability distribution of n_K for one primary electron or *one-electron peak* is reasonably well separated from the distributions for two or more primary electrons.

Now assume that the primary source emits electrons randomly with an average frequency λ. If the multiplier response is instantaneous you have an impulse process with amplitudes n_K, and you can use Carson's theorem to obtain that the power spectrum of the anode current is

$$\mathcal{S}(f) = \lambda e^2 A^2 F + \lambda^2 e^2 A^2 \delta(f)$$

where $A = \mu_1 \mu^{K-1}$ is the (average) multiplier gain. In terms of the primary current $I = \lambda e$, the signal-to-noise power ratio after a low-pass filter with power bandwidth B is therefore, as stated in Section 13-4-4,

$$\frac{S^2}{N^2} = \frac{I}{2eBF}$$

13-4 DIELECTRIC NOISE CAN HAVE A $1/f$ POWER DENSITY. As indicated in Exercise 3-22, the dielectric losses in a capacitor C are proportional to the frequency and are taken into account by placing in parallel with C a conductance G_C given by

$$G_C = \omega C D$$

where D, the loss factor, is usually between 10^{-4} and 10^{-2}.

Consider the common situation, shown in Fig. 13-35, in which C is in parallel with a resistor R. Note that G_C itself is not represented because it is negligible compared with ωC. The power densities of the noise currents due to G_C and R are

$$i_C^2 = 4kT\omega CD \quad \text{and} \quad i_R^2 = \frac{4kT}{R}$$

The dielectric noise is insignicant as long as $\omega \ll 1/RCD$ and can be ignored

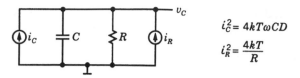

Figure 13-35. Noise model of a resistor in parallel with a lossy capacitor (Exercise 13-4).

in most circuits. If R is infinite, however, as it is effectively if C is the input capacitance of a Si(Li) detector, you must be careful. Observe first that the noise voltage v_C has a $1/f$ power density:

$$v_C^2 = \frac{4kTD}{2\pi C}\frac{1}{f}$$

To see that v_C can be significant, assume that it drives a JFET common-source amplifier, and consider the (corner) frequency f_C at which v_C^2 equals the thermal contribution to the power density v_{NG}^2 given in Fig. 13-7b. Setting

$$\frac{4kTD}{2\pi Cf_C} = \frac{2}{3}\frac{4kT}{g_m}$$

you obtain

$$f_C = \frac{3g_m D}{4\pi C}$$

For values that are typical in Si(Li) detectors, $g_m = 5$ mS, $C = 5$ pF, and $D = 10^{-4}$, you have $f_C = 24$ kHz, which can be more than an order of magnitude higher than the corner frequency f_D of the JFET flicker noise.

13-5   DETECTORS IN THE FREQUENCY DOMAIN. Using power spectra as in Fig. 13-9, and brick-wall filters with power bandwidth f_c, go through the arguments of Section 13-4-1 in the frequency domain. In the case of the square-law detector assume that the signal is a gaussian process, and in the case of the i-f converter take $f_I = \frac{1}{2}f_c$.

13-6   INTEGRATING ADCs ARE OPTIMAL. Consider the time-variant filter of Fig. 13-36. A constant signal of amplitude 1 masked by additive white noise $n(t)$ is applied through a linear gate to a time-invariant causal filter with impulse response $h(t)$. The gate is turned on from $t = 0$ to $t = T$ and the filter output is measured at $t = T$. The task is to show that an integrator is the best choice for $h(t)$. This is essentially the problem of the square pulse in Fig. 13-20, except that the noise is zero for $t < 0$. You have

$$S = \int_0^T h(T-t)\, dt$$

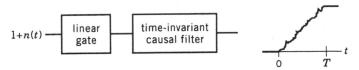

Figure 13-36. Time-variant filter. The gate is turned on from $t = 0$ to $t = T$ (Exercise 13-6).

and

$$N^2 = \left\langle \left| \int_0^T n(t)h(T-t)\, dt \right|^2 \right\rangle$$

Interchanging the order of averaging and integration in N^2 and recalling that $R_n(\tau) = \mathcal{P}_n \delta(\tau)$, obtain

$$\frac{S^2}{N^2} = \frac{\left| \int_0^T h(t)\, dt \right|^2}{\mathcal{P}_n \int_0^T |h(t)|^2\, dt}$$

Applying the Schwarz inequality to the functions 1 and $h(t)$, now show that S^2/N^2 achieves its maximum value $T/\mathcal{P}_n$ if the time-invariant filter is an integrator during the time that the gate is on,

$$h(t) = \frac{\theta(t) - \theta(t-T)}{T}$$

Integrating ADCs use a gated RC integrator to obtain a filtered sample of the (approximately) constant input and are therefore optimal. An incidental advantage of this technique is that power-line hum is in principle completely suppressed if T is chosen equal to an integral number of power-line cycles.

13-7 THE DICKE RADIOMETER (Dicke, 1946). Consider the heterodyne receiver shown in Fig. 13-37, and assume that it is connected to a source of resistance R at a temperature T. Assume further that the gain of the intermediate-frequency amplifier is constant over passbands of width B. The in-band power spectrum due to R at the input to the square-law detector then has a constant magnitude

$$\mathcal{P}_0 = 2kTRA$$

The factor A takes into account the loss in the mixer and the gain in the i-f amplifier. According to Fig. 13-15 and to the discussion in Section 13-4-2, the contribution from R to the dc term at the output of the detector (which simply adds to the contribution from the receiver noise, as you can quickly verify) is

$$S = 2(2kTRA)B$$

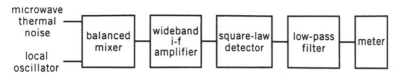

Figure 13-37. Heterodyne receiver looking at microwave thermal noise (Exercise 13-7).

and can thus be considered a measurement of T. If the receiver has a noise figure F and the filter at the output of the detector has a bandwidth $B' \ll B$, the noise, also according to Fig. 13-15, is

$$N = 2(2kTRA)F\sqrt{2B'B}$$

If the minimum change in S that can be detected is equal to N, you can measure T with a sensitivity ΔT given by

$$\Delta T = TF\sqrt{2B'/B}$$

It is clear that the sensitivity is maximized by making the bandwidth B of the i-f amplifier as large as possible and the bandwidth B' of the low-pass filter as small as possible. For $T = 300$ K, $F = 10$, $B = 10$ MHz, and $B' = 0.1$ Hz you get $\Delta T = 0.4$ K, which was Dicke's objective.

The preceding calculation ignores flicker noise and variations in amplifier parameters, grouped together by Dicke as *gain-variation noise*. In practice, this noise reduces the sensitivity considerably. To get around these problems, Dicke devised a chopper that alternately sensed the source whose temperature was to be measured and a source at a known temperature not too different from the unknown temperature; in doing so, he invented the lock-in amplifier. Get the paper and read about the details. You should be able to understand what the calculations mean even though they are carried out using a presumably unfamiliar approach.

13-8 PULSE SHAPERS IN THE TIME DOMAIN (Goulding, 1972; Wilson, 1950). In Section 12-3-10 you saw that white noise can be represented as an impulse process. This fact opens up a fertile time-domain approach to the analysis of how pulse shaping affects noise in nuclear spectrometers. Comparisons between different shapers are easy to make, and the effects of changes in parameters are readily visualized. In addition, it becomes possible to study time-variant shapers whose parameters change in synchronism with the signal; this is by definition impossible in the frequency domain.

In this exercise you will see briefly how the time-domain approach is applied to the pulse shapers in x-ray spectrometers based on Si(Li) detectors. In the noise model shown in Fig. 13-38, noise generators i_σ and v_Δ are

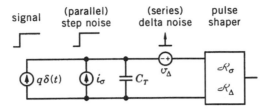

Figure 13-38. Noise model of a Si(Li) detector in the absence of flicker noise (Exercise 13-8),

actually the parallel and series noise generators of Fig. 13-28; the reason for the change in nomenclature will become apparent immediately. To simplify the analysis, $1/f$ noise is assumed to be negligible.

The *step-noise* power density i_σ^2 can be imagined as generated by an impulse process with frequency λ_σ and amplitudes $\pm i_\sigma/\sqrt{2\lambda_\sigma}$; the choice of λ_σ is arbitrary because the noise has mean zero. The impulses are integrated in capacitor C_T to produce voltage steps of amplitudes $\pm i_\sigma/C_T\sqrt{2\lambda_\sigma}$. Analogously, the *delta-noise* power density v_Δ^2 can be imagined as generated by an impulse process with frequency λ_Δ and amplitudes $\pm v_\Delta/\sqrt{2\lambda_\Delta}$.

Assume that an x-ray at $t=0$ generates a charge q on C_T and a signal $(q/C_T)s(t)$ at the output of the shaper, and assume further that $s(t_P)=1$, where t_P is the *peaking* time, that is, the time at which the signal is measured. Now define the (dimensionless) *step weighting function* $\mathcal{R}_\sigma(u)$ as the response of the shaper at time t_P to a unit step function occurring at time t_P-u, and define the *delta weighting function* $\mathcal{R}_\Delta(u)$ as the corresponding response to a delta function. Clearly, you have

$$\mathcal{R}_\Delta(u) = \mathcal{R}_\sigma'(u)$$

Following the steps that led to Carson's theorem in Section 12-3-9, you can now obtain that the mean-square number of noise electrons is

$$N^2 = \frac{i_\sigma^2}{2e^2}N_\sigma^2 + \frac{C_T^2 v_\Delta^2}{2e^2}N_\Delta^2$$

where N_σ^2 and N_Δ^2 are the *noise indices* for step and delta noise,

$$N_\sigma^2 = \int_0^\infty |\mathcal{R}_\sigma(u)|^2 \, du \qquad N_\Delta^2 = \int_0^\infty |\mathcal{R}_\Delta(u)|^2 \, du$$

To connect with what you already know, consider a time-invariant shaper with the trapezoidal response $s(t)$ shown in Fig. 13-39. The weighting functions are shown in Fig. 13-40. In this case, $\mathcal{R}_\sigma(u)$ is equal to the step

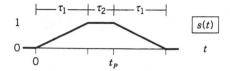

Figure 13-39. Response $s(t)$ of a time-invariant trapezoidal shaper (Exercise 13-8).

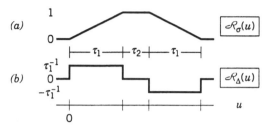

Figure 13-40. (*a*) Step and (*b*) delta weighting functions for the trapezoidal shaper of Fig. 13-39 (Exercise 13-8).

response $r(u) = s(u)$, and $\mathcal{R}_\Delta(u)$ is equal to the impulse response $h(u)$. You then have

$$N_\sigma^2 = \int_0^\infty |\mathcal{R}_\sigma(u)|^2 \, du = \int_0^\infty |r(u)|^2 \, du = \tfrac{2}{3}\tau_1 + \tau_2$$

$$N_\Delta^2 = \int_0^\infty |\mathcal{R}_\Delta(u)|^2 \, du = \int_0^\infty |h(u)|^2 \, du = 2\tau_1^{-1}$$

If you now express $r(u)$ and $h(u)$ in terms of a constant τ that defines the time scale, as was done in Section 13-6, it is clear that for time-invariant shapers you have

$$N_\sigma^2 = \tau K_P^2 \qquad N_\Delta^2 = \tau^{-1} K_S^2$$

where K_P^2 and K_S^2 are the integrals given in (13-10). Setting $\tau_2 = 0$ and $\tau_1 = \tau$, you recover the results for a triangular shaper given in (13-12).

To get an idea of why time-domain analysis is convenient, you might consider the effect of making the trapezoid asymmetric while keeping its flat-top and total durations constant. It should quickly become clear that the step noise remains invariant but that the delta noise increases, with no advantage in terms of counting rate or insensitivity to detector rise time.

To see now that time-variant shapers can be dealt with, consider the tandem of a square shaper and a *gated integrator* shown in Fig. 13-41. The square shaper is time invariant, and its step response is

$$r(t) = \theta(t) - \theta(t - \tau_1)$$

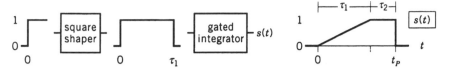

Figure 13-41. Time-variant shaper formed by a time-invariant square shaper followed by a gated integrator (Exercise 13-8).

The gated integrator is dormant until triggered by the arrival of a signal at $t = 0$, and it integrates the output of the shaper on a time scale τ_1,

$$s(t) = \tau_1^{-1}\theta(t)\int_0^t r(u)\, du$$

The integrator is reset to zero immediately after the measurement, and you may assume that this is done instantaneously.

To obtain $\mathcal{R}_\sigma(u)$, which differs from $s(u)$ in this case, slide $r(t)$ backward through the integrator starting at $t = t_P$ and continuing until its effect is no longer felt, which occurs at $t = t_P - 2\tau_1 - \tau_2$. The result is equal to what you obtained above for the trapezoidal shaper, and the noise indices are therefore also equal.

The gated integrator offers an advantage in counting rate, however, because the duration of its pulse is $\tau_1 + \tau_2$ rather than $2\tau_1 + \tau_2$, and it is almost always the case that τ_2 is made just long enough to make $s(t_P)$ insensitive to variations in detector rise time, so that it is much smaller than τ_1.

13-9   THE NOISE MODEL OF BIPOLAR TRANSISTORS (Buckingham and Faulkner, 1974; Buckingham, 1983). This exercise will lead you through the main arguments used by Buckingham and Faulkner to obtain the noise generators of a bipolar transistor.

To begin with, charge carriers give rise to *events*, whose nature depends on the process being considered. Each event in turn generates a current pulse. Since events occur at random times, the power spectrum can be obtained from Carson's theorem if the pulse shape is constant; if the pulse shape depends on position, the power spectrum can be obtained by a volume integration once Carson's theorem has been applied locally. This method is used in part (a) to recover the familiar noise spectrum of a resistor and, in the process, to justify a simplifying assumption. Using this assumption, in part (b) it is shown that in agreement with the argument advanced in Section 13-2-1, the recombination component of the base current of a transistor exhibits the power spectrum of shot noise at low frequencies. Finally, the power spectrum of the collector current is obtained in part (c).

(a) Consider the short-circuited resistor of length l and cross-sectional area A shown in Fig. 13-42. Take an event to be the axial displacement between

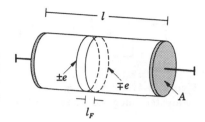

Figure 13-42. Resistor of length l and cross-sectional area A (Exercise 13-9). The axial displacement between collisions of a carrier gives rise to charge sheets that are separated by the free path l_F and have total charges $\pm e$.

collisions of a carrier, and assume that rather than a point dipole, it gives rise to *charge sheets* that are separated by the free path l_F and have total charges $\pm e$. As you will see immediately, this assumption simplifies calculations but does not affect the physics. Assume further that the mean free time between collisions τ_F is much smaller than the dielectric relaxation time $\tau_D = \varepsilon / \sigma$.

The voltage $v_E(t)$ between the charge sheets is initially equal to $e l_F / \varepsilon A$ and, as discussed in Section 4-5, it decays exponentially with a time constant τ_D because $\tau_F \ll \tau_D$. You thus have

$$v_E(t) = \frac{e l_F}{\varepsilon A} \theta(t) \exp\left(-\frac{t}{\tau_D}\right)$$

Since $l_F \ll l$, the short-circuit current due to the event is $i_E = G v_E(t)$, where G is the conductance of the resistor. Substituting $G = \sigma A / l$, you obtain

$$i_E(t) = \frac{e l_F}{l \tau_D} \theta(t) \exp\left(-\frac{t}{\tau_D}\right)$$

Observe now that $i_E(t)$ can be written as a convolution,

$$i_E(t) = \left[\frac{e u t_F}{l} \delta(t)\right] * \left[\frac{1}{\tau_D} \theta(t) \exp\left(-\frac{t}{\tau_D}\right)\right]$$

where l_F has been expressed as the product of the (axial) thermal velocity u and the free time t_F. Since $t_F \ll \tau_D$, the first term of the convolution can be replaced by a current pulse of amplitude eu/l and width t_F; this is precisely the pulse that was used in Section 12-4-3 to calculate thermal noise in the context of the Drude–Lorentz model. As you can now see, that pulse should be convolved with the impulse response of a lag with time constant τ_D if $\tau_F \ll \tau_D$. Since the only effect of τ_D is that it reduces the noise bandwidth, there is no harm in setting it to zero. For the purposes of this exercise you may thus assume that

$$i_E(t) = \frac{e l_F}{l} \delta(t)$$

The average number of events per unit time λ is equal to the collision rate in the entire resistor; if n is the density of charge carriers, you have

$$\lambda = \frac{n l A}{\tau_F}$$

With these preliminaries, you can now use Carson's theorem to obtain the power spectrum of the current,

$$\mathcal{S}_G(f) = \frac{e^2 \langle l_F^2 \rangle}{l^2} \frac{nlA}{\tau_F}$$

Since the velocity and the free time are independent, you have

$$\langle l_F^2 \rangle - \langle u^2 \rangle \langle t_F^2 \rangle - \frac{kT}{m} 2\tau_F^2$$

But from (4-8) you also have

$$G = \sigma \frac{A}{l} = \frac{ne^2 \tau_F}{m} \frac{A}{l}$$

and you finally recover the Nyquist formula,

$$i_G^2(f) = 4kTG$$

(b) Consider now the neutral region in the base of an active *pnp* transistor. The hole density at the emitter end is held constant by the forward bias voltage on the emitter–base junction, whereas the hole density at the collector end is zero because the collector–base junction is reverse biased. Take an event to be the generation of a hole. Such an event does not affect charge neutrality, but since it is a fluctuation in the hole density, it will be followed by a relaxation process, in this case a flow of holes away from the location of the event until the fluctuation is smoothed out.

As in part (a), assume that the initial fluctuation can be represented by a sheet rather than a point, and assume further that the hole lifetime is infinite. The time-dependent diffusion equation for the hole density per unit length p in the fluctuation, given in (6-15), then reduces to

$$\frac{\partial p}{\partial t} = D \frac{\partial^2 p}{\partial x^2}$$

Since the hole density is held constant at the emitter end ($x = 0$) and at the collector end ($x = w$), the boundary conditions are $p(0, t) = p(w, t) = 0$. If the event occurs at $x = \xi$, continuity imposes the further boundary condition

$$i(\xi+, t) - i(\xi-, t) = e\delta(t)$$

To visualize this condition you can imagine that a short pulse of total charge e is injected *laterally* into the base. Taking Fourier transforms on both sides of the diffusion equation and setting $s = i\omega$ yields

$$sp = D \frac{d^2 p}{dx^2}$$

The boundary conditions for the transformed equation are $p(0, s) = p(w, s) = 0$ and

$$i(\xi+, s) - i(\xi-, s) = e$$

You can verify that a solution that is continuous at $x = \xi$ and satisfies the boundary conditions at $x = 0$ and $x = w$ is

$$p(x<\xi, s) = \frac{\sinh\left[(w-\xi)\sqrt{s/D}\right]\sinh\left(x\sqrt{s/D}\right)}{\sinh\left(w\sqrt{s/D}\right)\sqrt{Ds}}$$

$$p(x>\xi, s) = \frac{\sinh\left(\xi\sqrt{s/D}\right)\sinh\left[(w-x)\sqrt{s/D}\right]}{\sinh\left(w\sqrt{s/D}\right)\sqrt{Ds}}$$

The current

$$i = -eD\frac{dp}{dx}$$

for this solution is

$$i(x<\xi, s) = -e\frac{\sinh\left[(w-\xi)\sqrt{s/D}\right]\cosh\left(x\sqrt{s/D}\right)}{\sinh\left(w\sqrt{s/D}\right)}$$

$$i(x>\xi, s) = +e\frac{\sinh\left(\xi\sqrt{s/D}\right)\cosh\left[(w-x)\sqrt{s/D}\right]}{\sinh\left(w\sqrt{s/D}\right)}$$

You can now check that the current satisfies the boundary condition at $x = \xi$ and that you are therefore dealing with the right solution. The low-frequency limits of $p(x, s)$ and $i(x, s)$, shown together with snapshots of $p(x, t)$ and $i(x, t)$ in Fig. 13-43, are thus

$$p(x<\xi, s=0) = \left(1 - \frac{\xi}{w}\right)\frac{x}{D} \qquad p(x>\xi, s=0) = \frac{\xi}{D}\left(1 - \frac{x}{w}\right)$$

$$i(x<\xi, s=0) = -e\left(1 - \frac{\xi}{w}\right) \qquad i(x>\xi, s=0) = e\frac{\xi}{w}$$

Recalling that the base charging time is given by $\tau_B = w^2/2D$, it is clear that *low-frequency* means $\omega\tau_B \ll 1$, as you might have expected.

With the appropriate changes of sign, the preceding discussion is clearly also valid for recombination events. Charge neutrality in the base is maintained by the flow of electrons from the base lead and must compensate for the inflow of holes at $x = 0$ and $x = w$. For each recombination event, the base lead must therefore supply a current pulse whose low-frequency amplitude spectrum is

$$i(w, s=0) - i(0, s=0) = -e$$

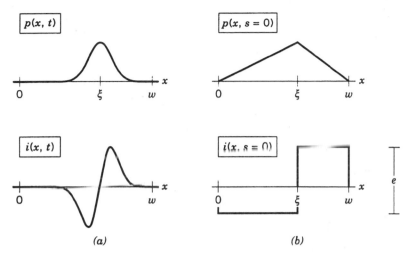

Figure 13-43. (a) Snapshots and (b) low-frequency amplitude spectra of the hole density p and the hole current (i) in the fluctuation due to the generation of a hole (Exercise 13-9).

or equivalently, whose integrated charge is $-e$. Since recombination events occur at random times, Carson's theorem tells you that the recombination component I_{BR} of the base current exhibits a shot-noise spectrum at low frequencies,

$$\mathcal{S}_{BR}(0) = eI_{BR}$$

(c) Consider again the neutral region in the base of an active *pnp* transistor as described in part (b), but now take an event to be the axial displacement between collisions of a hole. As you did in the case of the resistor in part (a), assume that the event gives rise to charge sheets that are separated by the free path l_F and have total charges $\pm e$. Charge neutrality is quickly reestablished by the majority carriers, electrons in this case; the ensuing currents contribute to the noise due to the resistance of the neutral region and can be ignored in this analysis. You thus need only consider relaxation by diffusion.

Assuming that $l_F \ll w$, the hole density in the fluctuation is (approximately) l_F times the derivative with respect to ξ of the density that you obtained for a recombination event in part (b),

$$p(x<\xi,s) = -\frac{l_F}{D}\frac{\cosh\left[(w-\xi)\sqrt{s/D}\,\right]\sinh\left(x\sqrt{s/D}\,\right)}{\sinh\left(w\sqrt{s/D}\,\right)}$$

$$p(x>\xi,s) = +\frac{l_F}{D}\frac{\cosh\left(\xi\sqrt{s/D}\,\right)\sinh\left[(w-x)\sqrt{s/D}\,\right]}{\sinh\left(w\sqrt{s/D}\,\right)}$$

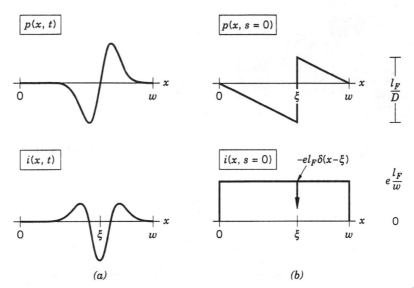

Figure 13-44. (*a*) Snapshots and (*b*) low-frequency amplitude spectra of the hole density p and the hole current i in the fluctuation due to the axial displacement between collisions of a hole (Exercise 13-9).

The current is then

$$i(x<\xi,s) = el_F\sqrt{s/D}\ \frac{\cosh\left[(w-\xi)\sqrt{s/D}\right]\cosh\left(x\sqrt{s/D}\right)}{\sinh\left(w\sqrt{s/D}\right)}$$

$$i(x>\xi,s) = el_F\sqrt{s/D}\ \frac{\cosh\left(\xi\sqrt{s/D}\right)\cosh\left[(w-x)\sqrt{s/D}\right]}{\sinh\left(w\sqrt{s/D}\right)}$$

Observe incidentally that the current between the charge sheets, which you should expect to be important at small times, is represented by

$$i(\xi,s) = -el_F\,\delta(x-\xi)$$

The low-frequency limits of $p(x,s)$ and $i(x,s)$, shown together with snapshots of $p(x,t)$ and $i(x,t)$ in Fig. 13-44, are thus

$$p(x<\xi,s=0) = -\frac{l_F}{D}\frac{x}{w} \qquad p(x>\xi,s=0) = \frac{l_F}{D}\left(1-\frac{x}{w}\right)$$

$$i(x<\xi,s=0) = e\frac{l_F}{w} \qquad\qquad i(x>\xi,s=0) = e\frac{l_F}{w}$$

Recalling that the hole density per unit length $p_B(x)$ has a triangular shape you can now obtain the power spectrum of the collector current,

$$\mathcal{S}_C(f) = e^2\frac{\langle l_F^2\rangle}{w^2}\frac{p_B(0)w}{2\tau_F} = e\frac{kT}{e}\frac{e\tau_F}{m}\frac{ep_B(0)}{w}$$

From the discussion leading up to (4-8) in Section 4-4, it follows that $e\tau_F/m$ is the hole mobility and therefore that $(kT/e)(e\tau_F/m)$ is equal to the diffusion constant D. Observing further that $I_C = ep_B(0)D/w$, you finally obtain

$$S_C(0) = eI_C$$

Analogous arguments hold for the hole current at the emitter junction and, in fact, for the currents in any short-base junction diode. It follows that the diffusion component of the base current exhibits a shot-noise spectrum at low frequencies.

You can now show that the collector current exhibits a shot-noise spectrum at *all* frequencies. First show that

$$|i(w, i\omega)|^2 = \frac{e^2 l_F^2 \omega}{D} \left| \frac{\cosh(\xi\sqrt{i\omega/D})}{\sinh(w\sqrt{i\omega/D})} \right|^2$$

$$= \frac{e^2 l_F^2 \omega}{D} \frac{\cosh(\xi\sqrt{2\omega/D}) + \cos(\xi\sqrt{2\omega/D})}{\cosh(w\sqrt{2\omega/D}) - \cos(w\sqrt{2\omega/D})}$$

After some algebra, and making use of the integrals

$$\int_0^w \left(1 - \frac{\xi}{w}\right) \cosh(\xi\sqrt{2\omega/D}) \, d\xi = \frac{D}{2\omega w} \left[\cosh(w\sqrt{2\omega/D}) - 1\right]$$

$$\int_0^w \left(1 - \frac{\xi}{w}\right) \cos(\xi\sqrt{2\omega/D}) \, d\xi = \frac{D}{2\omega w} \left[1 - \cos(w\sqrt{2\omega/D})\right]$$

you should finally obtain

$$S_C(f) = eI_C \frac{2\omega w}{D} \int_0^w \left(1 - \frac{\xi}{w}\right) \frac{\cosh(\xi\sqrt{2\omega/D}) + \cos(\xi\sqrt{2\omega/D})}{\cosh(w\sqrt{2\omega/D}) - \cos(w\sqrt{2\omega/D})} \, d\xi = eI_C$$

This result is *not* true for the hole current at the emitter junction, which supplies the hole charge stored in the base and thus has some excess noise at high frequencies; it follows that the diffusion component of the base current, which supplies an equal and opposite electron charge, also has excess noise at high frequencies.

13-10 THE NOISE MODEL OF JFETs. Consider the n-channel JFET shown in Fig. 13-45a, and assume that it is in the saturation region. According to (6-50), the (average) drain current I_D is

$$I_D = \sigma W x_C \frac{d\psi_{CS}}{dy}$$

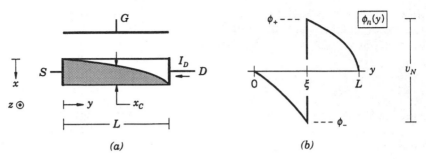

Figure 13-45. (*a*) Cross section of a saturated JFET showing the channel width $x_C(y)$ as a function of the position y along the channel. (*b*) Potential $\phi_N(y)$ generated by a series noise voltage v_N at $y = \xi$ (Exercise 13-10).

For the purposes of this exercise, the channel width x_C obtained in Exercise 6-15 can be expressed in a more convenient form using (6-56) and (6-52),

$$x_C(y) = x_M \frac{-V_P + V_{GS}}{-2V_P} \sqrt{1 - \frac{y}{L}} = \frac{g_m L}{\sigma W} \sqrt{1 - \frac{y}{L}}$$

The channel can be considered to be a chain of infinitesimal resistors, each of which generates thermal noise in series with the channel. Assume then that there is a series noise voltage source v_N at $y = \xi$. As shown in Fig. 13-45*b*, v_N generates an (infinitesimal) noise potential ϕ_N that adds to ψ_{CS}; it also generates a noise current i_N that, like I_D, is independent of y. At any point along the channel the total drain current $I_D + i_N$ must satisfy

$$I_D + i_N = \sigma W \left(x_C + \frac{dx_C}{d\psi_{CS}} \phi_N \right) \left(\frac{d\psi_{CS}}{dy} + \frac{d\phi_N}{dy} \right)$$

To first order in ϕ_N you then have

$$i_N = \sigma W \frac{d(x_C \phi_N)}{dy}$$

Integrating i_N from the source ($y = 0$) to ξ and from ξ to the drain ($y = L$), and using the boundary conditions $\phi_N(0) = \phi_N(L) = 0$, you obtain

$$i_N \xi = \sigma W x_C(\xi) \phi_-$$

$$i_N(L - \xi) = -\sigma W x_C(\xi) \phi_+$$

where ϕ_- and ϕ_+ are the values of ϕ_N just to the left and just to the right of ξ. Now use the fact that $\phi_+ - \phi_- = v_N$ to solve for the noise current,

$$i_N = -\frac{\sigma W x_C(\xi)}{L} v_N$$

Integrating i_N once more from $y = 0$ to $y < \xi$ and from $y > \xi$ to L, you obtain the noise potential,

$$\phi_N(y<\xi) = -\left(\frac{y}{L}\right)\frac{x_C(\xi)}{x_C(y)}v_N \qquad \phi_N(y>\xi) = \left(1-\frac{y}{L}\right)\frac{x_C(\xi)}{x_C(y)}v_N$$

The voltage noise generated by a length $d\xi$ of the channel is

$$dv_N^2 = \frac{4kT}{\sigma W x_C(\xi)}d\xi$$

The corresponding contribution to the drain thermal noise is

$$di_{NDT}^2 = \frac{\sigma^2 W^2 x_C^2(\xi)}{L^2}dv_N^2 = 4kT\frac{\sigma W x_C(\xi)}{L^2}d\xi = 4kTg_m\sqrt{1-\frac{\xi}{L}}\frac{d\xi}{L}$$

Integration from 0 to L then yields Van der Ziel's result for the thermal piece of the drain noise generator given in Fig. 13-2,

$$i_{NDT}^2 = 4kT\left(\frac{2g_m}{3}\right)$$

Using (6-49) and (6-57), you can now obtain that the variation in the channel electron charge due to v_N is

$$dq_N = -eN_D W \int_0^L \frac{dx_C}{d\psi_{CS}}\phi_N \, dy$$

$$= \frac{eN_D W x_M}{-2V_P}\int_0^L \phi_N(y)\,dy = \left(\frac{3}{2}\right)C_{GS}\left[\frac{1}{L}\int_0^L \phi_N(y)\,dy\right]$$

Observing that

$$\frac{1}{L}\int_0^L \phi_N(y)\,dy = v_N\sqrt{1-\frac{\xi}{L}}\left(\int_0^\xi \frac{-\dfrac{y}{L}}{\sqrt{1-\dfrac{y}{L}}}\frac{dy}{L} + \int_\xi^L \sqrt{1-\frac{y}{L}}\frac{dy}{L}\right)$$

$$= v_N\sqrt{1-\frac{\xi}{L}}\left(2\sqrt{1-\frac{\xi}{L}} - \frac{4}{3}\right)$$

you obtain

$$dq_N = C_{GS}\left[3\left(1-\frac{\xi}{L}\right) - 2\sqrt{1-\frac{\xi}{L}}\right]v_N$$

The voltage noise generated by a length $d\xi$ of the channel can be rewritten as

$$dv_N^2 = \frac{4kT}{\sigma W x_C(\xi)}d\xi = \frac{4kT}{g_m}\frac{1}{\sqrt{1-\dfrac{\xi}{L}}}\frac{d\xi}{L}$$

The (short-circuit) thermal current noise injected into the gate terminal is then given by

$$i_{NGT}^2 = \omega^2 C_{GS}^2 \frac{4kT}{g_m}\int_0^1 \frac{[3(1-\eta)-2\sqrt{1-\eta}]^2}{\sqrt{1-\eta}}d\eta = \left(\frac{4}{15}\right)\omega^2 C_{GS}^2 \frac{4kT}{g_m}$$

Assuming that C_{GS} is the dominant contributor to the transition frequency f_T, you finally obtain the thermal piece of the gate noise generator given in Fig. 13-2,

$$i_{NGT}^2 \simeq 4kTg_m\left(\frac{f}{2f_T}\right)^2$$

13-11 THE POWER SPECTRUM OF PHOTOCURRENTS IS NOT ALWAYS JUST SHOT NOISE. Light incident on photoelectric surfaces is generally uncorrelated except at points that are close according to a criterion that depends on the wavelength and on the spread in solid angle of the light. As a result, the power spectrum of the photocurrent is essentially white and given by the Schottky formula. As you will now see, this is not the case if experimental conditions are chosen appropriately.

Consider the idealized situation in which linearly polarized plane-wave light is normally incident on a plane photoelectric surface and assume, as shown in Fig. 13-46, that the power spectrum of the electric field E consists of lines of gaussian shape centered at $\pm f_0$. You then have

$$\mathcal{S}_E(f) = \tfrac{1}{2}\langle E^2\rangle[G(f+f_0)+G(f-f_0)]$$

where

$$G(f) = \frac{1}{\sqrt{2\pi}\,\sigma}\exp\left(-\frac{f^2}{2\sigma^2}\right)$$

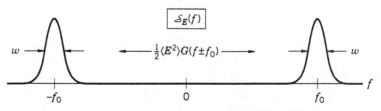

Figure 13-46. Power spectrum of the electric field E of a light wave, assumed to consist of lines of gaussian shape centered at $\pm f_0$ (Exercise 13-11).

The linewidth w (full width at half maximum) is given in terms of σ by

$$w = \sqrt{8 \ln 2}\, \sigma = 2.35\sigma$$

You of course also have

$$\int_{-\infty}^{+\infty} G(f)\, df = 1$$

Let P be the average of E^2 over a few cycles, and accept as an experimental fact that the probability of emission of a photoelectron in a short interval is proportional to P. As you will see below, you will need the power spectrum of P to obtain the power spectrum of the photocurrent I.

Now assume that the electric field is a gaussian process. The power spectrum of E^2 is then given by (12-8),

$$\mathcal{S}_{E^2}(f) = R_E^2(0)\,\delta(f) + 2\mathcal{S}_E * \mathcal{S}_E(f)$$

But P can be considered to be obtained by running E^2 through a low-pass filter with a cutoff frequency much larger than w but much smaller than f_0, so that only the spectrum around $f = 0$ survives. In terms of $\langle P \rangle = \langle E^2 \rangle$, the power spectrum of P is thus as shown in Fig. 13-47,

$$\mathcal{S}_P(f) = \langle P \rangle^2 \delta(f) + \langle P \rangle^2 G * G(f)$$

Observe that $G * G(f)$ has a gaussian shape, that its width is $\sqrt{2}\, w$, and that

$$G * G(0) = \int_{-\infty}^{+\infty} G^2(f)\, df = \tau_0$$

where

$$\tau_0 = \frac{\sqrt{8 \ln 2}}{\sqrt{4\pi}\, w}$$

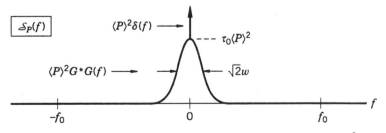

Figure 13-47. Power spectrum of P, the average over a few cycles of E^2 as given in Fig. 13-46 (Exercise 13-11).

The autocorrelation of P is given by

$$R_P(\tau) = \langle P \rangle^2 \left[1 + R_G^2(\tau) \right]$$

where, according to Exercise 2-17,

$$R_G(\tau) = \int_{-\infty}^{+\infty} G(f) \exp(i\omega\tau)\, df = \exp\left(-\frac{\sigma^2 \tau^2}{2} \right)$$

The probability of electron emission in a short interval Δ is $\alpha P \Delta$, where α is a constant, and the number n of electrons emitted in Δ is Poisson distributed. For $t_1 \neq t_2$ you thus have

$$\langle I_1 I_2 \rangle = \frac{e^2}{\Delta^2} \langle n_1 n_2 \rangle = e^2 \alpha^2 \langle P_1 P_2 \rangle$$

For $t_1 = t_2$, however, you have

$$\langle I^2 \rangle = \frac{e^2}{\Delta^2} \langle n^2 \rangle = \frac{e^2}{\Delta^2} \left(\alpha^2 \langle P^2 \rangle \Delta^2 + \alpha \langle P \rangle \Delta \right) = e^2 \alpha^2 \langle P^2 \rangle + \frac{1}{\Delta} e^2 \alpha \langle P \rangle$$

Observing that $e\alpha \langle P \rangle = \langle I \rangle$ and that

$$\lim_{\Delta \to 0} = \int_{-\Delta/2}^{+\Delta/2} \langle I^2 \rangle\, d\tau = e^2 \alpha \langle P \rangle = e \langle I \rangle$$

you obtain that the autocorrelation of the current is

$$R_I(\tau) = \langle I \rangle^2 \left[1 + R_G^2(\tau) \right] + e \langle I \rangle \delta(\tau)$$

The power spectrum of the current is therefore

$$\mathcal{S}_I(f) = \langle I \rangle^2 \delta(f) + \langle I \rangle^2 G*G(f) + e \langle I \rangle$$

It should come as no surprise that you get shot noise in the current, but as shown in Fig. 13-48, it is also clear that the correlation in the electric field at the photoelectric surface is reflected in the current. For $w = 10^{12}$ Hz and $\langle I \rangle = 1$ nA, both of which are reasonable values, you have $\tau_0 = 0.66$ ps and

$$\frac{\tau_0 \langle I \rangle}{e} \simeq 4 \cdot 10^{-3}$$

Recovering such a small signal is well within the capabilities of, say, lock-in amplifiers, and you should thus expect that the correlation in the current can

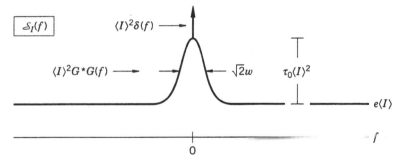

Figure 13-48. Power spectrum of the photocurrent I due to the electric field of Fig. 13-46 (Exercise 13-11).

be measured. This has in fact been done in the experiment by Brown and Twiss described in Exercise 13-12 and, from a somewhat different perspective, in the experiment by Forrester, Gudmundsen, and Johnson described in Exercise 13-13. Lasers have now made these experiments trivial, but that is not what they were in their time.

13-12   CORRELATION BETWEEN PHOTONS IN COHERENT BEAMS OF LIGHT [Brown and Twiss, 1956a,b (briefly 1956a and 1956b); Purcell, 1956]. Consider the setup shown in Fig. 13-49. An unpolarized plane-wave light beam is normally incident on the plane cathodes of two identical photomultipliers that are at the same position in the direction of the beam. The anode currents I_1 and I_2 are applied to bandpass filters with a power bandwidth $B = 24$ MHz (3–27 MHz). The filtered currents J_1 and J_2 are then multiplied together, and the product $J_1 J_2$ is integrated for a time $T = 5400$ s. In simplified form, this is the setup used by Brown and Twiss to show that photocurrents emitted by coherently illuminated photoelectric surfaces are correlated or, equivalently, that photons in coherent beams of light are correlated.

Before proceeding do Exercise 13-11, and assume that the power spectrum of the electric field is as given there.

Consider first what happens in one photomultiplier, and for simplicity, assume that multiplication is noiseless. For an unpolarized beam the anode

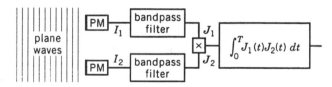

Figure 13-49. Simplified layout of the Brown–Twiss experiment (Exercise 13-12).

current I can be expressed as

$$I = I_x + I_y$$

where x and y denote orthogonal linear polarizations of the electric field and

$$\langle I_x \rangle = \langle I_y \rangle = \tfrac{1}{2} \langle I \rangle$$

At times t_1 and t_2 separated by an interval τ you have

$$\langle (I_{x1} + I_{y1})(I_{x2} + I_{y2}) \rangle = \langle I_{x1} I_{x2} \rangle + \langle I_{y1} I_{y2} \rangle + \langle I_{x1} I_{y2} \rangle + \langle I_{y1} I_{x2} \rangle$$

In the notation of Exercise 13-11, the autocorrelation of the anode current for an unpolarized beam is therefore

$$R_I(\tau) = \langle I \rangle^2 \left[1 + \tfrac{1}{2} R_G^2(\tau) \right] + e \langle I \rangle \delta(\tau)$$

The response $H(f)$ of the bandpass filter is significant only for $0 < |f| \ll w$; it follows that the power spectrum after the filter is

$$S_J(f) = \left(\tfrac{1}{2} \tau_0 \langle I \rangle^2 + e \langle I \rangle \right) |H(f)|^2$$

The autocorrelation after the filter is thus

$$R_J(\tau) = \left(\tfrac{1}{2} \tau_0 \langle I \rangle^2 + e \langle I \rangle \right) R_h(\tau)$$

where $R_h(\tau)$ is the autocorrelation of the impulse response $h(t)$ of the filter,

$$R_h(\tau) = \int_{-\infty}^{+\infty} h(t) h(t+\tau) \, dt$$

Now come back to the two coherently illuminated detectors shown in Fig. 13-49. The average anode currents are equal,

$$\langle I_1 \rangle = \langle I_2 \rangle = \langle I \rangle$$

and the autocorrelations of J_1 and J_2 are equal to $R_J(\tau)$ as given above. Denoting the cross-correlation of J_1 and J_2 by $R_{12}(\tau)$ and observing that $R_{12}(\tau) = R_{21}(\tau)$ by symmetry, for the autocorrelation $R_+(\tau)$ of the sum current $J_+ = J_1 + J_2$ you obtain

$$R_+(\tau) = 2 R_J(\tau) + 2 R_{12}(\tau) = 2 \left(\tfrac{1}{2} \tau_0 \langle I \rangle^2 + e \langle I \rangle \right) R_h(\tau) + 2 R_{12}(\tau)$$

But the two detectors can be considered to be a single detector with twice the average anode current, so you also have

$$R_+(\tau) = \left(2 \tau_0 \langle I \rangle^2 + 2 e \langle I \rangle \right) R_h(\tau)$$

You can thus conclude that

$$R_{12}(\tau) = \tfrac{1}{2}\tau_0 \langle I \rangle^2 R_h(\tau)$$

The signal S at the output of the integrator is

$$S = \left\langle \int_0^T J_1(t) J_2(t)\, dt \right\rangle = \int_0^T \langle J_1(t) J_2(t) \rangle\, dt = TR_{12}(0) = \tfrac{1}{2}\tau_0 \langle I \rangle^2 TR_h(0)$$

From the Parseval–Plancherel theorem you have

$$R_h(0) = \int_{-\infty}^{+\infty} h^2(t)\, dt = \int_{-\infty}^{+\infty} |H(f)|^2\, df$$

Assume that the bandpass filter has a brick-wall response over a bandwidth B. The integral is then equal to $2B$, and you get

$$S = \tau_0 \langle I \rangle^2 BT$$

It will turn out that $\tau_0 \langle I \rangle / e \ll 1$; you may therefore assume that shot noise is dominant in I_1 and I_2. Since the bandpass filter removes dc components, the noise N is given by

$$N^2 = \left\langle \left[\int_0^T J_1(t) J_2(t)\, dt \right]^2 \right\rangle = \int_0^T du \int_0^T dv\, \langle J_1(u) J_2(u) J_1(v) J_2(v) \rangle$$

The shot-noise contributions to J_1 and J_2 are independent, so that

$$N^2 = \int_0^T du \int_0^T dv\, \langle J_1(u) J_1(v) \rangle \langle J_2(u) J_2(v) \rangle = e^2 \langle I \rangle^2 \int_0^T du \int_0^T dv\, R_h^2(u-v)$$

Using (12-4) and observing that $BT \gg 1$, you get

$$N^2 = e^2 \langle I \rangle^2 T \int_{-\infty}^{+\infty} R_h^2(\tau)\, d\tau$$

Observing further that $R_h(\tau)$ and $|H(f)|^2$ are a Fourier transform pair, you can use the Parseval–Plancherel theorem again to obtain

$$N^2 = e^2 \langle I \rangle^2 T \int_{-\infty}^{+\infty} |H(f)|^4\, df$$

For a brick-wall filter the integral is equal to $2B$, so you finally get

$$N^2 = 2e^2 \langle I \rangle^2 BT$$

The signal-to-noise ratio is thus

$$\frac{S}{N} = \frac{\tau_0 \langle I \rangle}{e} \sqrt{\frac{BT}{2}}$$

To compare with the results given in Eqs. (1) and (2) of 1956a, observe that if the number of photons per second per hertz incident on a photocathode is described by a one-sided density $n(f)$, you can write

$$n(f) = \frac{1}{\alpha} \frac{\langle I \rangle}{e} G(f - f_0)$$

where α is the efficiency of the photocathode, assumed constant. You then have

$$\frac{\int_0^\infty \alpha^2 n^2(f)\, df}{\int_0^\infty \alpha n(f)\, df} = \frac{\langle I \rangle}{e} \frac{\int_{-\infty}^{+\infty} G^2(f)\, df}{\int_{-\infty}^{+\infty} G(f)\, df} = \frac{\tau_0 \langle I \rangle}{e}$$

To complete the comparison, you should take into account that the signal is multiplied by a factor $\eta \simeq 0.5$ that describes the lack of complete coherence in the light incident on the photocathodes, and you should include the noise figure of the multipliers, given in (13-2), on the assumption that all dynodes have the same gain $\mu = 4$. In the notation used here, you then obtain

$$\frac{S}{N} = \eta \frac{\mu - 1}{\mu} \frac{\tau_0 \langle I \rangle}{e} \sqrt{\frac{BT}{2}}$$

This is essentially the result given in 1956a. From Table 1 in 1956b you have $\langle I \rangle / e \simeq 5 \cdot 10^9 /\text{s}$ and $w \simeq 10^{13}$ Hz, so that $\tau_0 \simeq 7 \cdot 10^{-14}$ s and $S/N \simeq 30$, which is amply sufficient to obtain a clean result.

13-13   DETECTION OF BEATS AT OPTICAL FREQUENCIES [Forrester, Gudmundsen, and Johnson, 1955 (briefly FGJ); Forrester, 1956]. Before proceeding do Exercise 13-11: you will need the notation as well as the procedures developed there.

Consider two plane-wave beams of light that are normally incident on a plane photoelectric surface, and assume for the moment that the waves are perfectly sinusoidal and that their frequencies ω_1 and ω_2 are only slightly different. If the waves are linearly polarized in the same direction, and the amplitude of their electric fields is E_0, the total electric field $E_T(t)$ is a wave of the average frequency modulated at half the difference frequency,

$$E_T(t) = E_0 \cos(\omega_1 t) + E_0 \cos(\omega_2 t) = 2E_0 \cos\left[\tfrac{1}{2}(\omega_1 + \omega_2)t\right] \cos\left[\tfrac{1}{2}(\omega_1 - \omega_2)t\right]$$

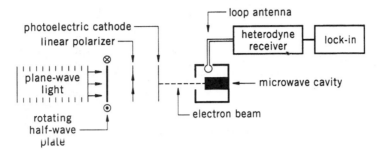

Figure 13-50. Simplified layout of the Forrester–Gudmundsen–Johnson experiment (Exercise 13-13).

The probability per unit time for emission of a photoelectron is proportional to the square of the total electric field averaged over a few cycles, so that the photocurrent has a component at the difference or *beat* frequency $\omega_1 - \omega_2$,

$$I(t) = I_0 \cos^2\left[\tfrac{1}{2}(\omega_1 - \omega_2)t\right] = \tfrac{1}{2}I_0 + \tfrac{1}{2}I_0 \cos\left[(\omega_1 - \omega_2)t\right]$$

Light waves are not perfectly sinusoidal, but they are nonetheless correlated or *coherent* over many cycles, enough so that beats in the photocurrent are observable. In this exercise you will see, in a language with which you are familiar, a stripped-down version of the FGJ experiment and the calculations required to obtain the signal-to-noise ratio.

Consider the experimental setup shown in Fig. 13-50. Plane-wave light passes through a rotating half-wave plate and a linear polarizer before striking a semitransparent photoelectric cathode. The electrons emitted from the cathode are accelerated by a constant voltage and enter a microwave cavity. The light has equal power in orthogonal linear polarizations, but in one of the polarizations—call it the *matched* polarization—the power is equally shared by two components at frequencies f_1 and f_2 such that the beat frequency $f_0 = f_2 - f_1$ matches the resonant frequency of the microwave cavity. Assume for the moment that the half-wave plate and the polarizer are set so that they transmit only the matched polarization. The beat-frequency component of the photocurrent then delivers power to the cavity, and a fraction of this power is picked off by a loop antenna and delivered to a heterodyne receiver.

The components of the electric field E in the matched polarization are shown in Fig. 13-51a. All lines have width $w \ll f_0$. In the notation of Exercise 13-11, the power spectrum of E can be expressed as

$$\mathscr{S}_E(f) = \tfrac{1}{4}\langle E^2\rangle\left[G(f+f_2)+G(f+f_1)+G(f-f_1)+G(f-f_2)\right]$$

Assuming that the electric field is a gaussian process, you can now follow the procedures used in Exercise 13-11 to obtain the power spectrum of the

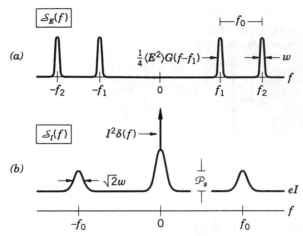

Figure 13-51. Power spectra of (a) the electric field and (b) the photocurrent in the matched polarization (Exercise 13-13).

current. Denoting the average current by I rather than $\langle I \rangle$, you should get the spectrum shown in Fig. 13-51b,

$$\mathcal{S}_I(f) = I^2\delta(f) + \tfrac{1}{2}I^2 G*G(f) + \tfrac{1}{4}I^2 G*G(f+f_0) + \tfrac{1}{4}I^2 G*G(f-f_0) + eI$$

Since the bandwidth B of the receiver is much less than w, you can consider that the in-band signal power spectrum is constant and equal to its value $\mathcal{P}_s$ at $\pm f_0$, given by

$$\mathcal{P}_s = \tau_0 I^2$$

where

$$\tau_0 = \frac{1}{4}G*G(0) = \frac{\sqrt{8\ln 2}}{4\sqrt{4\pi}\,w}$$

The in-band shot-noise power spectrum is of course also constant,

$$\mathcal{P}_n = eI$$

In the actual experiment, the light—of wavelength λ—had a spread in solid angle Ω, so that coherence extended over areas of order λ^2/Ω much smaller than the area A of the photocathode. In addition, there were factors α_1, α_2, and γ, all of order 1 and properly defined in FGJ, that affected the signal power. As a result, the actual value of the signal power spectrum near $\pm f_0$ was

$$\mathcal{P}_s = \eta\tau_0 I^2$$

where

$$\eta = \frac{\alpha_2^2 \gamma \lambda^2}{\alpha_1 A \Omega} \approx 10^{-8}$$

Assume that B is much smaller than the bandwidth f_0/Q of the cavity. The in-band photocurrent—signal plus noise—then sees the cavity as a resistor, and it is therefore transmitted without distortion to the receiver. If you also assume that the photocurrent is large enough that the cavity thermal noise is negligible, the signal-to-noise power ratio at the receiver input is the same as it is at the photocathode.

$$\frac{S^2}{N^2} \approx \frac{\mathcal{P}_s}{\mathcal{P}_n} = \eta \frac{\tau_0 I}{e}$$

Using the values $I \approx 4 \ \mu A$ and $w \approx 1$ GHz given in FGJ, you obtain $S^2/N^2 \approx 10^{-4}$. According to the discussion on square-law and linear-law detectors in Section 13-4-2, the signal-to-noise power ratio at the receiver output is of the order of the square of the input signal-to-noise power ratio, that is, about 10^{-8}. Modulation of the source and a lock-in amplifier after the receiver thus become mandatory: this explains the presence of the half-wave plate and the linear polarizer.

The linearly polarized components of the electric field rotate around the axis of the beam at twice the frequency of rotation f_p of the half-wave plate. Since the polarizer selects one direction of polarization, the total power incident on the photocathode is constant, and so therefore is the shot noise in the photocurrent, but the beat-frequency power in the matched polarization is multiplied by

$$\cos^2(2\omega_p t) = \tfrac{1}{2} + \tfrac{1}{2}\cos(4\omega_p t)$$

That is, it is modulated at a frequency $f_m = 4f_p$. [If you want to learn more about polarizers and half-wave plates, you will find excellent descriptions in the *Waves* volume of the Berkeley Physics Course (Crawford, 1965).] The beat-frequency signal at the receiver input is also modulated at a frequency f_m, and as you will see immediately, so is the signal at the receiver output. To increase the signal-to-noise ratio, the lock-in amplifier selects narrow bands of frequencies around $\pm f_m$.

Assume that the receiver is noiseless and that the i-f amplifier has a brick-wall response. Assume also that the second detector (actually the third in this experiment, the first being the photocathode) is square law. If $x(t)$ and $y(t)$ represent the unmodulated and modulated signals at the input to the square-law device, $n(t)$ the shot noise, and $z(t)$ the output of the square-law device, you have

$$y(t) = \cos^2(\tfrac{1}{2}\omega_m t)x(t) \quad \text{and} \quad z(t) = [y(t)+n(t)]^2$$

The signal-to-noise ratio is small, so you have

$$z(t) \simeq y^2(t) + n^2(t) = \cos^4\left(\tfrac{1}{2}\omega_m t\right) x^2(t) + n^2(t)$$

Observing that

$$\cos^4\left(\tfrac{1}{2}\omega_m t\right) = \tfrac{3}{8} + \tfrac{1}{2}\cos(\omega_m t) + \tfrac{1}{8}\cos(2\omega_m t)$$

and recalling that the lock-in amplifier selects frequencies around $\pm f_m$, for simplicity you can write

$$z(t) = \tfrac{1}{2}\cos(\omega_m t) x^2(t) + n^2(t)$$

(If this simplification is not obvious, it will quite probably become so in the course of the calculations that follow.) The modulation process, the signal, and the noise are independent of one another, so you have

$$S_z(f) = \tfrac{1}{4} S_m * S_{x^2}(f) + S_{n^2}(f)$$

where $S_m(f)$ is the power spectrum of $\cos(\omega_m t)$,

$$S_m(f) = \tfrac{1}{4}\left[\delta(f+f_m) + \delta(f-f_m)\right]$$

Since $x(t)$ is a bandlimited gaussian process, $S_{x^2}(f)$ is identical at low frequencies to the noise–noise spectrum $S_{nn}(f)$ in Fig. 13-15. $S_m(f)$ and the relevant piece of $S_{x^2}(f)$ are shown in Fig. 13-52; the only significant terms in their convolution are the ones involving delta functions, so that at low frequencies you have

$$S_z(f) \simeq \tfrac{1}{4}\mathcal{P}_s^2 B^2\left[\delta(f+f_m) + \delta(f-f_m)\right] + S_{n^2}(f)$$

The noise term $S_{n^2}(f)$ is also identical at low frequencies to the noise–noise spectrum in Fig. 13-15. Figure 13-52 shows the relevant piece of $S_z(f)$. If B' is the power bandwidth of the low-pass filter in the lock-in amplifier, you have

$$\frac{S^2}{N^2} = \frac{\tfrac{1}{4}\mathcal{P}_s^2 B^2}{4\mathcal{P}_n^2 BB'} = \frac{\mathcal{P}_s^2 B}{16\,\mathcal{P}_n^2 B'}$$

For an RC filter, such as was actually used, $B' = 1/4RC$, so you finally get

$$\frac{S}{N} = \frac{\mathcal{P}_s}{2\mathcal{P}_n}\sqrt{RCB}$$

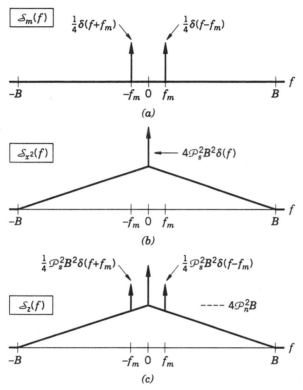

Figure 13-52. Power spectra used to calculate the signal-to-noise ratio in the Forrester–Gudmundsen–Johnson experiment (Exercise 13-13).

Using the values $B = 7$ MHz and $RC = 240$ s given in FGJ, you get $S/N \simeq 1$, which is just enough to establish the presence of beats.

It is worth noting that FGJ use an odd (but convenient) mixture of frequency and time analysis in which the unmodulated signal power spectrum is multiplied by $\frac{1}{2}\cos(\omega_m t)$. Pulling this factor out, you obtain that what they call K_B is equal to $\mathcal{P}_s/2\mathcal{P}_n$, and you can then verify that S/N is as given for a noiseless receiver in Eq. (22) of FGJ.

REFERENCES

General

Crawford, F. S., Jr. (1965). *Waves* (Volume 3 of *The Berkeley Physics Course*). New York: McGraw-Hill.

Feynman, R. P., R. B. Leighton, and M. Sands (1964). *The Feynman Lectures on Physics*. Reading, MA: Addison-Wesley.

Horowitz, P., and W. Hill (1989). *The Art of Electronics*, 2nd ed. New York: Cambridge University Press.

Purcell, E. M. (1985). *Electricity and Magnetism* (Volume 2 of *The Berkeley Physics Course*), 2nd ed. New York: McGraw-Hill.

Reif, F. (1965). *Fundamentals of Statistical and Thermal Physics*. New York: McGraw-Hill.

Reif, F. (1967). *Statistical Physics* (Volume 5 of *The Berkeley Physics Course*). New York: McGraw-Hill.

Roberts, G. E., and H. Kaufmann (1966). *Table of Laplace Transforms*. Philadelphia: W.B. Saunders.

Chapter 1

Cheng, D. K. (1963). *Analysis of Linear Systems*. Reading, MA: Addison-Wesley.

Chapter 2

Lighthill, M. J. (1958). *Introduction to Fourier Analysis and Generalized Functions*. New York: Cambridge University Press.

Roddier, F. (1971). *Distributions et Transformation de Fourier*. Paris: Ediscience.

642

Schwartz, L. (1965). *Méthodes Mathématiques pour les Sciences Physiques*. Paris: Hermann.

Zemanian, A. H. (1965). *Distribution Theory and Transform Analysis*. New York: McGraw-Hill (also Mineola, NY: Dover, 1987).

Chapter 4

Fano, R. M., L. J. Chu, and R. B. Adler (1960). *Electromagnetic Fields, Energy and Forces*. New York: Wiley.

Feynman (1964). See General References.

Griffiths, D. J., and M. A. Heald (1991). *Am. J. Phys.* 59, 111.

Heald, M. A. (1984). *Am. J. Phys.* 52, 522.

Jackson, J. D. (1996). *Am. J. Phys.* 64, 855.

Jackson, J. D. (1999). *Classical Electrodynamics*, 3rd ed. New York: Wiley.

Jefimenko, O. (1989). *Electricity and Magnetism*, 2nd ed. Star City, WV: Electret Scientific.

Marcus, A. (1941). *Am. J. Phys.* 9, 225.

Saslow, W. M., and G. Wilkinson (1971). *Am. J. Phys.* 39, 1244.

Tolman, R. C., and T. D. Stewart (1916). *Phys. Rev.* 8, 97.

Tolman, R. C., and T. D. Stewart (1917). *Phys. Rev.* 9, 164.

Chapter 5

Black, H. S. (1934). *Electr. Eng.* 53, 114.

Black, H. S. (1977). *IEEE Spectrum*, December.

Chapter 6

Conwell, E. M. (1958). *Proc. IRE* 46, 1281.

Feynman (1964). See General References.

Fonstad, C. G. (1994). *Microelectronic Devices and Circuits*. New York: McGraw-Hill.

Gray, P. R., and R. Meyer (1993). *Analysis and Design of Analog Integrated Circuits*, 3rd ed. New York: Wiley.

Hall, R. N. (1952). *Phys. Rev.* 87, 387.

Muller, R. S., and T. I. Kamins (1986). *Device Electronics for Integrated Circuits*. New York: Wiley.

Sah, C. T., R. N. Noyce, and W. Shockley (1957). *Proc. IRE* 45, 1228.

Shockley, W., and W. T. Read (1952). *Phys. Rev.* 87, 835.

Chapter 7

Brokaw, A. P. (1974). *IEEE J. Solid-State Circuits* SC-9, 388.

Gilbert, B. (1974). *IEEE J. Solid-State Circuits* SC-9, 364.

Radeka, V., et al. (1991). *IEEE Trans. Nucl. Sci.* NS-38, 83.

Russell, R. W., and D. D. Culmer (1974). *Digest of Technical Papers*, International Solid-State Circuits Conference, Philadelphia, pp. 140–141, February.

Widlar, R. J. (1971). *IEEE J. Solid-State Circuits* SC-6, 2.

Chapter 8

Hearn, W. E. (1971). *IEEE J. Solid-State Circuits* SC-6, 20.

Chapter 11

Adler, R. B., L. J. Chu, and R. M. Fano (1962). *Electromagnetic Energy Transmission and Radiation*. New York: Wiley.

Bramham, P. (1961). *Electron. Eng*. January, p. 42.

Davidson, C. W. (1989). *Transmission Lines for Communications*, 2nd ed. London: MacMillan.

Magnusson, P. C. (1970). *Transmission Lines and Wave Propagation*, 2nd ed. Needham Heights, MA: Allyn and Bacon.

Montgomery, C. G., R. H. Dicke, and E. M. Purcell (1948). *Principles of Microwave Circuits*. New York: McGraw-Hill.

Pound, R. V. (1948). *Microwave Mixers*. New York: McGraw-Hill.

Somlo, P. I. (1960). *IEEE Trans. Microwave Theory Tech.* MTT-8, 463.

Wigington, R. S., and N. S. Nahman (1957). *Proc. IRE* 45, 166.

Chapter 12

Davenport, W. B., and W. L. Root (1987). *An Introduction to the Theory of Random Signals and Noise*. New York: IEEE Press (also New York: McGraw-Hill, 1958).

Einstein, A. (1905). *Ann. Phys.* 17, 549.

Einstein, A. (1906). *Ann. Phys.* 19, 371.

Johnson, J.B. (1928). *Phys. Rev.* 32, 97.

Langevin, P. (1908). *C.R. Acad. Sci. Paris* 146, 530.

Lee, Y. W. (1960). *Statistical Theory of Communication*. New York: Wiley.

MacDonald, D. K. C. (1962). *Noise and Fluctuations: An Introduction*. New York: Wiley.

Chapter 13

Brown, R. H., and R. Q. Twiss (1956a). *Nature* 177, 27.

Brown, R. H., and R. Q. Twiss (1956b). *Nature* 178, 1447.

Buckingham, M. J. (1983). *Noise in Electronic Devices and Systems*. Chichester, West Sussex: Ellis Horwood.

Buckingham, M. J., and E. A. Faulkner (1974). *Radio Electr. Eng.* 44, 125.

Crawford (1965). See General References.

Dicke, R. H. (1946). *Rev. Sci. Instrum.* <u>17</u>, 268.

Forrester, A. T. (1956). *Am. J. Phys.* <u>24</u>, 192.

Forrester, A. T., R. A. Gudmundsen, and P. O. Johnson (1955). *Phys. Rev.* <u>99</u>, 1691.

Goulding, F. S. (1972). *Nucl. Instrum. Methods* <u>100</u>, 493.

Knoll, G. F. (1989). *Radiation Detection and Measurement*, 2nd ed. New York: Wiley.

Lawson, J. L., and G. E. Uhlenbeck (1950). *Threshold Signals*. New York: McGraw-Hill.

Mott, R. B., and J. J. Friel (1994). Improving EDS performance with digital pulse processing, in *X-Ray Spectrometry in Electron Beam Instruments*, edited by D. B. Williams, J. I. Goldstein, and D. E. Newbury. New York: Plenum.

Pound (1948). See References for Chapter 11.

Purcell, E. M. (1956). *Nature* <u>178</u>, 1449.

Radeka, V. (1984). *Nucl. Instrum. Methods* <u>226</u>, 209.

Robinson, F. N. H. (1974). *Noise and Fluctuations*. Oxford: Clarendon Press.

Shockley, W., and J. R. Pierce (1938). *Proc. IRE* <u>26</u>, 321.

Wilson, R. (1950). *Phil. Mag.* <u>41</u>, 66.

INDEX